The Future of Ocean Regime-Building

Douglas Millar Johnston (1931–2006)
Photography courtesy of Judith Johnston

The Future of Ocean Regime-Building

Essays in Tribute to Douglas M. Johnston

Edited by
Aldo Chircop, Ted L. McDorman, Susan J. Rolston

MARTINUS
NIJHOFF
PUBLISHERS

LEIDEN • BOSTON
2009

This book is printed on acid-free paper.

Library of Congress Cataloging-in-Publication Data

The future of ocean regime-building : essays in tribute to Douglas M. Johnston / edited by Aldo Chircop, Ted McDorman, Susan Rolston.
p. cm.
Includes bibliographical references and index.
ISBN 978-90-04-17267-8 (hardback : alk. paper) 1. Johnston, Douglas M. 2. Law of the sea. 3. Ships—Safety regulations. 4. Merchant mariners—Legal status, laws, etc. 5. Fishery law and legislation. 6. Marine resources conservation—Law and legislation. 7. Marine pollution—Law and legislation. 8. Waste disposal in the ocean—Law and legislation. 9. Environmental law, International. I. Chircop, Aldo E. II. McDorman, Ted L. III. Rolston, Susan.
KZA1145.F88 2009
344.04'6343—dc22

2008044834

ISBN 978 90 04 17267 8

PRINTED IN THE NETHERLANDS

Contents

DISPUTE SETTLEMENT AND REGIONAL COOPERATION

Foreword

Douglas Johnston was a modest man. As such, I believe he would have been taken wholly by surprise that so many scholars sought to pay tribute to him and his work through the medium of this publication.

Tendering honour or appreciation by way of publication is well founded in the history of the printed word. Sometimes such tribute is expressed simply by way of dedication. On other occasions the intention to honour is in the publication itself, as is the case here.

In the early days of publishing the friends and patrons of the author (or authors!) of the plays commonly attributed to William Shakespeare had printed limited numbers of what now is known as The First Folio. The First Folio was printed in 1623, seven years after Shakespeare's death. Those who commissioned the collecting of the scripts, the printing and, for the period, the use of very fine materials, are judged both to have sought to honour the author and, also, to do so in a form which spoke to the quality of the writing.

For our late friend and colleague, Professor Douglas Johnston, the need to re-publish all, or selected portions, of his work is unnecessary, at least at present. However, the desire to honour Douglas has resulted in a variation on the reasons for publishing The First Folio of Shakespeare's plays. That variation is, of course, the writing and publication of some two dozen essays by 35 authors which comprise this particular book.

These authors are a cross-section of just some of those whose lives have shared both personal and professional interactions with Douglas Johnston. The coming together of so many in this common cause to pay tribute to Douglas speaks for itself. That the authors encompass such a wide spectrum of learning in matters of the sea, its law, environment, present applicable technologies, and commerce speaks to quality of the enterprise. Thus, we trust, this act pays appropriate tribute to Douglas Johnston for the excellent man he was and for the very high quality of his legacy of researching, writing, and teaching. Douglas' legacy is particularly important at Dalhousie University, in Halifax, Nova Scotia.

Douglas Johnston and others had been working on maritime law and policy for some years at the Dalhousie Law School. But it was not until 1977 that the then Academic Vice-President, Dr. Guy MacLean, urged both the need and the opportunity to pursue significant longer term funding so as to bring about cohesion and focus through a new university programme to be established. In addition to the work being undertaken in the Law School, modest funding had been secured that year from both governmental and industry sources to support a Marine Transportation Centre. Professors Edgar Gold and Douglas Johnston from the Law School joined with me, from the Graduate Business School, to prepare the new funding application.

In 1978 Dalhousie was advised that the funding application was successful. This multi-year and significant funding provided the foundation for what became the Dalhousie Ocean Studies Programme (DOSP). In addition to the expanded teaching options provided for students in law, business and, with the arrival at Dalhousie of Professor Arthur Hanson, the environment, DOSP's research and publications were important additions to this growing field. Several associates with the Ocean Studies Programme have gone on to assume senior positions in universities, government and the private sector. Indeed, and not surprising, some of these are among the essay contributors to this publication.

I believe that Guy MacLean sensed that Dalhousie was on the cusp of having a critical mass of academic and practical capability in marine affairs. Funding for the Ocean Studies Programme was the catalyst which resulted in both programme form and substance. When I look back, I realize how much Douglas Johnston's intellect and incisiveness contributed both to the success of the funding application and, subsequently, to the Ocean Studies Programme itself. This important part of Douglas' legacy, which lives on at Dalhousie University, is alive, well and flourishing. Necessarily, the shape of today's marine-related activities at Dalhousie has changed, and will continue to change. However, let me record here that Douglas Johnston was, as I believe the Scots say, a "begetter" without whom I doubt it would have been possible to deliver on the first and critical initiatives.

A great pleasure of mine in the late 1970s and early 1980s was my regular association with Douglas Johnston. In addition to his knowledge, insights and intellect, Douglas was unfailingly warm, friendly, and courteous. Sometimes these attributes masked a well-developed sense of humour. Once, in Hawaii, when a British colleague's luggage failed to arrive, Douglas assisted in addressing this man's immediate needs by procuring a pair of emblazoned undershorts which are best not described!

When Douglas moved to the West Coast, as too often happens, Canada's size and time zones, coupled with my moving to England, resulted first in a diminished contact with Douglas, and then none. It is my deep regret, and my responsibility, that this should have been so.

Since his passing, and looking forward to this publication, I have been endeavouring to determine that which I consider to be the essence of Douglas Johnston's view of the laws and the businesses impacting the world's oceans to which so much of his work was devoted. I have concluded, importantly, that Douglas was a much greater optimist than many who labour in this area. I see Douglas Johnston as having had a solid belief that a comprehensive and agreed upon order of things in the maritime arena, an overall governance if you will, was inevitable. This thinking man, despite all the frustrations arising out of the day-to-day functioning of governments and oceans users, believed that, over time, rationality would prevail because the alternatives were unthinkable. Douglas would point to specific progress that was being made and, I remember on the occasion of a serious oil spill, his saying, as a result, that "the cause" had penetrated the consciousness of average citizens and hence now there would be no retreat from governments having to address the various root causes of ship-related oceans pollution.

I see Douglas Johnston not as a detached academic thinker who contemplated issues from a safe distance, but, rather, as a man who cared deeply about all the facets of the area in which he studied, wrote, and taught. What I judge to be his confidence in the positive future of the governance of the oceans, and the businesses which are conducted in and on them, was founded on a profound belief that the benefits achievable are so great and important to the basic lives of us all that no rational, caring person could do other than support progressive governance initiatives. There is an old saying in Britain that when a Scot goes south to England the IQ of both Scotland and England rises! When the Scottish gentleman, Douglas Johnston, came to Canada, Canada's IQ rose.

The effort of those who contributed to this publication is testimony both to the regard in which Douglas Johnston is held and the desire to pay tribute to him in a way in which he would understand and value and where, in his memory, further contributions to matters maritime are made.

Sir Graham Day
April, 2008
Hantsport, Nova Scotia

Co-Editors' Preface

Preparing "Essays in Tribute to Douglas M. Johnston" is a modest homage to an individual that profoundly influenced the careers and personal lives of the three co-editors.

Each of us came to know Douglas as students at Dalhousie University. For one of us, he was masters and doctoral theses advisor; for another, he was a professor of a law course; for another he was a professor who readily accepted a political science student into a law course. Nevertheless, for all of us, our individual relationships with Douglas followed three distinct phases. In the beginning, there was Dr. Johnston – our professor dispensing his unique insights from the front of the classroom or across his office desk, somewhat intimidating but always amiable, helpful, and generous with time. This was followed by the Dr. Doug (an in-house nickname) phase, when the relationship was more one of tutelage between a senior scholar and a junior learner. This phase existed, for example, during the Dalhousie Ocean Studies Programme (DOSP) days when Douglas was one of the programme directors. The Dr. Doug phase was characterized by chalkboards, multiple levels of analysis, crafting concepts, conversation, and gentle guidance. Finally, there came a time when each of us were working alongside Douglas in project planning and execution, teaching and graduate student supervision, and being involved in the mulling that proceeded his writing or co-writing. This was the Douglas phase. This phase still involved chalkboards and conversation, with a heavier emphasis on the latter, and no small amount of laughter.

The subject matter of this book deals with only one aspect of Douglas' impressive career – the ocean law and policy component. It is this part of Douglas' academic life which most directly touched the co-editors. As the Bibliography Note at the end of this book attests, Douglas' intellectual and writing pursuits were not confined to salt water themes, but included matters Asian, international environmental law, international law theory and treaties. His collaborative work over many years with Ronald St. J. Macdonald defies easy classification as it embraced legal theory, constitutionalism, the structure and process of international law, human welfare, and Canadian approaches to international law.

The relationship between Douglas and the contributors of the chapters to this book is varied. Almost half of the contributors are former students of Douglas and may have had similar "phases" of their relationship as did the co-editors. The next largest category of contributors is those whose relationship with Douglas was primarily as a colleague. Finally, there are several contributors who knew Douglas more by reputation than through personal contact and appreciated his work. We are grateful to all of the authors for their contributions.

We are particularly grateful to Graham Day, Brian Flemming and Ed Miles for penning personal reflections about collaborating with Douglas and his work. In this regard, note should be made of the preface to the chapter by Edgar Gold who, along with Graham, Arthur Hanson and Douglas, were the "four wise men" that established DOSP through which several of the contributors were inducted. DOSP was also the temporary home for a large number of others who wished they had been able to meet the timelines set by the co-editors for contributions.

The co-editors wish to specially thank and acknowledge the contribution to this book of Christian L. Wiktor. Christian, the former law librarian at Dalhousie Law School, was a close colleague and friend of Douglas for over three decades. For the co-editors, he provided encouragement and advice on various aspects of this book.

Judith Johnston also provided encouragement and advice on this book project to the co-editors. This was consistent with her support of Douglas' work. Judith often provided the indispensable support Douglas needed to convert the meticulous strokes of his longhand into orderly typed text. Those that admired Douglas' work may not always have appreciated that behind the great man was a great woman. Judith also witnessed all of us go through the above-noted three phases of our relationship with Douglas and played an important role for each of us along the way. For all of these things we are truly grateful to her.

For each of us, the connection with Douglas significantly shaped the path that we followed. Words do not do justice to the impact he had on us and many others. This tribute is a small memento to our friend and colleague and the intellectual heritage he left in his wake for others to appreciate.

Aldo Chircop, Halifax
Ted L. McDorman, Victoria
Susan J. Rolston, Halifax
July 2008

Part I

Douglas Millar Johnston (1931–2006) Biographical Note

Douglas M. Johnston has been recognized as a leading teacher, scholar, writer and internationalist in the field of public international law, especially in the law of the sea and international environmental law.

He was a modest man, very generous to students and colleagues, with an inspiring personality and a sharp theoretical mind. Johnston was a true educator, with excellent communication and writing skills. He profoundly influenced the future careers of many of his students.

Born in Dundee, Scotland, in 1931, where he received his early schooling, he obtained from the University of St Andrews a M.A. in 1952 and a LL.B. in 1955. Moving to Canada in 1955, Johnston completed a M.C.L. from McGill University in 1958 where he was a student of one of Canada's most distinguished international law academics, Maxwell Cohen. Moving on to Yale University, Johnston received his LL.M. in 1959 and a Doctor of the Science of Law (J.S.D.) in 1962. At Yale, he was a student of Myres McDougal and Harold Lasswell, an experience which resulted in a life-long affinity with the policy science approach.

In 1959 he married Judith Elizabeth Fuess, eventually having three children, Keith in 1961, Murray in 1964, and Caroline in 1970. In 1959 he also started his academic career as an assistant professor at the Law Faculty of University of Western Ontario in London, Canada, along side Ronald St. J. Macdonald, with whom he was to be a life-long collaborator.

In 1964, Johnston accepted a one-year position as a visiting associate professor at Louisiana State University in Baton Rouge. In 1965 he returned to Yale Law School to publish his J.S.D. thesis (*The International Law of the Fisheries: A Framework for Policy-oriented Inquiries* (Yale University Press, 1965)) and to be a special student in Chinese studies. This was followed by three years at Harvard Law School as a senior research associate where, amongst other things, he did research for the United States Arms Control and Disarmament Agency. In 1968–1969, Johnston spent a year as an associate professor in political science at the New School for Social Research in New York.

Johnston returned to Canada in 1969 taking a position as a professor of political science in the Department of Political Economy of the University of Toronto and renewed his relationship with the then-dean of the Law Faculty, Ronald Macdonald.

In 1972, the ocean beckoned and Johnston and family moved to Halifax, Nova Scotia, where he took up a position with the Law School at Dalhousie University. Over the next 15 years, Johnston was central in launching the Law School's doctoral programme and expanding the graduate programme; initiated within the Law School the Marine and Environmental Law Programme (MELP), which today is the Marine & Environmental Law Institute (MELAW); produced a torrent of articles, book chapters, books and other publications including collaborating with Macdonald on three momentous edited collections; provided advisory and consultative services to, amongst others, the Government of Canada and the International Union for the Conservation of Nature and Natural Resources (IUCN) in Switzerland; and co-founded and directed both the multi-disciplinary Dalhousie Ocean Studies Programme (DOSP) and the Southeast Asian Programme in Ocean Law, Policy and Management (SEAPOL) centred in Bangkok, Thailand. These initiatives and activities gave to Dalhousie Law School an international reputation as a centre of legal education and research in international law, especially in marine and environmental law.

In 1987, Johnston switched coasts and took up the Chair in Asia-Pacific Legal Relations at the University of Victoria, in Victoria, British Columbia. At Victoria, Johnston continued his leadership of SEAPOL; published numerous works including co-editing with Gerry Ferguson, *Asia-Pacific Legal Development* (UBC Press, 1998); taught courses on Asian and Chinese law as well as international law; and assisted in the firm establishment of the multidisciplinary Centre for Asia-Pacific Initiatives (CAPI) on the University of Victoria campus.

Johnston retired from North American teaching in 1995, but proceeded to spend the next four years as a visiting professor at the National University of Singapore in the Faculty of Law and the Policy Research Programme. His mandate in Singapore was to encourage cross-campus interdisciplinary teaching, research and publication. During these years he continued his engagement with SEAPOL in Bangkok editing a number of books on Southeast Asian ocean law and policy, which are seen as creating the main literature on these topics within the region.

In 1999 Johnston returned to Victoria where, in semi-retirement and as a University of Victoria emeritus professor of law, emeritus Chair in Asia-Pacific Legal Relations, and a senior research Fellow with CAPI, he continued to write, participated in the work of the Maritime Awards Society of Canada (MASC), and collaborated with Macdonald on what was to be their last co-edited book, *Towards World Constitutionalism: Issues in the Legal Ordering of the World Community* (Martinus Nijhoff, 2005).

His passion during his last few years was *The Historical Foundations of World Order: The Tower and the Arena* published posthumously in 2008.

Douglas Johnston passed away in May of 2006 from cancer one week after his 75th birthday.

Aldo Chircop, Halifax
Ted L. McDorman, Victoria
Susan J. Rolston, Halifax
Christian L. Wiktor, Halifax

Douglas M. Johnston: Postscript for a Polymath

Brian Flemming*

Visitors to Sir Christopher Wren's tomb in his architectural masterpiece, St. Paul's Cathedral in London, England, will see these words chiselled on it: *Lector, si monumentum requiris, circumspice* (Reader, if you seek his monument, look around you). Henceforth, anyone who wants to know what an extraordinary, integrative and erudite mind the late Douglas M. Johnston possessed need only obtain his final work – his intellectual St. Paul's Cathedral – and open it anywhere. *The Historical Foundations of World Order: The Tower and the Arena* is long – 844 pages, not including the index – but a reading of this seminal work will leave the reader limp with the sheer audacity and astounding breadth of the mind and vision of my friend, and former colleague, Douglas. The very density of Douglas's book makes it read more like a multi-volume treatise, not one of less than 1000 pages. The "Tower" in the title is, of course, the "ivory" one of the Academy; the "Arena" is the dusty, international gladiatorial pit where governmental and private paladins fight it out. Douglas inhabited both locales during his lifetime, and was always a builder of bridges between those two solitudes.

To have a history of international law written at the start of the 21st century is rare enough. To have one written by a polymath like Douglas is a gift from the gods. The stream of learnedness that pours from its every page is Amazon River-like in its relentless power. As Douglas properly notes in his Introduction, histories of international law are not thick on the ground in the 21st century, and most of the histories that are gathering dust on law library shelves around the world verge on the unreadable. Douglas's crystal clear prose – combined with his incomparable grasp of history, legal and non-legal; philosophy from all corners of the world; archaeology; music; politics: local, national and international; ethics and religion; international law and relations plus his Scottish granite-like

* International lawyer and policy advisor, Brian Flemming, CM, QC, of Halifax, Canada, was a friend and colleague of Douglas Johnston for many years.

common sense – ensure this book will still be read, and highly valued, not just by those of us lucky enough to be alive to read it today, but for grateful generations to come. That is what Douglas wanted. That is what Douglas achieved.

At one point, Douglas describes the four stages of life (*ashramas*) for each person in ancient Hindu teachings. They are: "student, married householder, retiree and sage" (p. 248). Douglas passed all those stages with flying colours and, with this culminating work, deserves the title "sage" more than anyone I have met or whose works I have read. Douglas's book was written during a time in which international law was not as highly valued as it was, say, in the late 19th century, or following the Second World War. The attitude of the rulers of the current world hegemon, the United States of America, is best encapsulated by the crude words President George W. Bush spoke to Richard Clarke, his anti-terrorism tsar, following 9/11: "I don't care what the international lawyers say. We are going to kick some ass." (Little did George W. realize that the ass he would wind up kicking would be his own, but that is another wench in another country.) After reading Douglas's wide-ranging review of world cultures and leaders throughout many millennia, one can easily imagine Alexander the Great, Genghis Khan or Adolf Hitler uttering the same words on a similar occasion. Douglas knew, deep in the marrow of his bones, that the more things appeared to change, the more they followed in the well-worn footsteps of those who have gone before.

Douglas was not pessimistic about the situation that obtains in today's world, just more clear-headed than most. As he wrote in 1973, long before the events of September 11, 2001 led to many strange outcomes, international law could be seen "as having reached a stage of 'parametric stress'" (p. 769). Parametric stress occurs when the fundamental concepts of any discipline can no longer accommodate "the dynamics of change inherent in the problems they purported to treat" (p. 769). Douglas believed the discipline of international law had reached the same stage that astronomy had reached after Copernicus; or physics after Rutherford and Einstein; or zoology after Darwin; or anthropology after Leakey.

One of the key challenges facing the future of international law is that of "managing" the Arctic region of Spaceship Earth, a favourite subject of Douglas's, and one he touches upon in Part One of his book in a section entitled "Protection: the Arctic Environment" (p. 62 *et seq.*). After describing the rise of environmental protection as a new, functionalist imperative in international law, Douglas poetically says about the Arctic Ocean: "This island-studded sea, largely ice-covered for most or all of the year, is often seen as a symbol of pristine purity, but is now becoming the focus of various, often conflicting, interests and aspirations... Perhaps no other setting illustrates so clearly the difficulties of protecting the human environment" (p. 64). Douglas praises the way Pierre Trudeau's government reacted to the "innocent passage" of the U.S. ship, *Manhattan*, through the Northwest ("international strait"?) Passage in 1969. The *Arctic Waters Pollution Prevention Act* of 1970 that asserted Canada's jurisdiction over all shipping in adjacent zones up to 100 nautical miles off Canada's Arctic coastline was the unprecedented Canadian government response, made "necessary because of the uniquely vulnerable and fragile status of the Arctic marine environment" (p. 69). For Douglas, and many influential Canadians of the day, the old black letter law of freedom of innocent passage through international straits just would not do any more in this unique situation. And the Canadian response would never have been possible if Trudeau had not had Ivan Head, his "in-house" international lawyer, at his side (p. 66 *et seq.*).

The fussy formalists of the old External Affairs Department would never have dreamed up this Arctic initiative although one of External's "new boys" of the late 1960s, Alan Beesley, did much of the important departmental intellectual lifting during this period before going on to become the head of Canada's impressive and influential delegation to the Third United Nations Conference on the Law of the Sea (UNCLOS III) (p. 66). Beesley and his department wisely appointed Douglas to an Advisory Committee on Marine and Environmental Conferences to advise and help to guide the Canadian government through the intellectual shoals of UNCLOS III. (The other members of the committee were Professor Donat Pharand of the University of Ottawa, Professor Maxwell Cohen of McGill University, and this writer.)

One of the many charming riffs Douglas inserts throughout his text is a personal, diary-like account of the 1970 annual meeting of the American Society of International Law (ASIL) where two of the then-giants of American international law – Columbia's Wolfgang Friedmann for the formalist, black letter gang and Yale's Myres S. McDougal representing the functionalist, policy-oriented New Haven school – clashed on the floor of the meeting over Canada's new Arctic initiative. I was present for this tournament of titans and paid rapt attention to the denunciation by scholar after scholar of the distasteful "illegal unilateralism" of its northern neighbour. Given the pre-emptive attack by the "coalition of the willing" in 2003 in Iraq, and the illegal bombing of Bosnia in 1999, this debate looks quite quaint and otherworldly in retrospect. Douglas was not surprised: it all depended on whose ox was being gored, not pure principle. No such exchange occurred at an ASIL annual meeting immediately following the Iraqi invasion. Q.E.D.

The new challenge for the five littoral States that are adjacent to the "rediscovered" Arctic Ocean is how to meld the old legal criteria of acquisition of Arctic territory – what Prime Minister Stephen Harper is really talking about when he suggests Canada "use it or lose it" – with the kind of imagination shown by the international community in 1959 when territorial ambitions in the Antarctic gave way to a multi-functional, trans-national regime. Douglas believed the international law community had a major role to play in ensuring that the amoral historic international law system of norms and criteria regarding the acquisition of territory are superseded by newly-created norms that will allow for the full participation of the peoples and non-governmental organizations of the North. Douglas would be dismayed by the incipient militarization of the North that the Harper government embarked upon in the summer of 2007.

Douglas's sensitivity to, and broad knowledge of, world cultures beyond the modern European experience was profound. This book is laced with references to ancient cultures and the legal accomplishments of Egypt, Greece, Rome, China, India and Africa. Douglas was always at pains to make readers of his opinions recall "... that only two or three generations have passed since the last barriers of cultural bias in the theory of international law were broken down" (p. 183). He then, characteristically, muses: "perhaps we have scarcely advanced beyond Aristotle's limited vision of the universe" (p. 183). Douglas believed strongly that "international lawyers should have no difficulty living in a world of cultural diversity" (p. 17). Indeed, the most difficult "trick" for international lawyers to "perform" may be to see issues from another perspective than their own country's national interests. Another problematic "trick" is to make "international law generally intelligible. If a history of world order is to be written at all for a non-technical readership, it seems essential, to this writer, that it be disengaged from the mainstream of legal terminology" (p. 124). And Douglas manages this brilliantly in this prodigious work as he elegantly tells how the world got from there – antiquity – to here – the post-modern world.

Douglas tries at the very beginning of his treatise to grapple with the historic bias of international law towards Europe and its once-extensive imperial reach by using the example of how the "settled principle" of diplomatic immunity played out in the Iranian occupation of the American embassy in Tehran in 1979. Douglas believed this crisis illustrated "both the strengths and deficiencies of international law" (p. 13). "On the negative side, it showed that even the most honoured of legal principles underpinning the diplomatic process can occasionally be undermined by political events, although admittedly in this case by revolutionaries running out of control rather than by the Iranian government" (p. 13). This incident led eventually to litigation before the International Court of Justice (ICJ) in the *United States Diplomatic Staff in Tehran* (*United States v. Iran*).

The competing values before the court were the principle of inviolability of embassies, i.e., the American one versus the principle of non-interference (by the United States) in the internal affairs of another State (Iran). Unfortunately, Iran never appeared before the ICJ to argue the latter side so it fell primarily to Judge Manfred Lachs to provide what Douglas believed to be a balanced view of these competing principles in his separate opinion. As he does so many times throughout his book, Douglas said "a more modern analysis [of factual situations like the Iran case] could be further enriched by 'policy-oriented' jurists who, unlike 'rules-oriented' lawyers, see international law much more broadly as a process of constantly shifting claims and counterclaims within a context such as this that reflects a tension between strictly legal principles, on the one hand, and 'world order values' on the other hand" (p. 18). That quotation encapsulates everything Douglas stood for intellectually.

Douglas is particularly adept in this book at contrasting the world-views and religious beliefs of many cultures throughout history, and analyzing why these views and beliefs are so alien to those of us living today. The many references to the issues of war and peace throughout the book illustrate his thoughtful approach to these states of human activity. His starting point is: "... all wars of early antiquity were perceived as 'holy', and therefore as 'just'" (p. 164). The United Nations Charter is our new "sacred text" – one taking the place of ancient religious texts – and declares war, other than in self-defence, to be illegal and therefore unjust. How we got to today partly rotates around a constant human quest for "perpetual peace," a subject Douglas attacks, starting at page 349. Notables, such as the early 14th century poet Dante Alighieri, put forward plans for achieving perpetual peace. Dante's involved establishing a coalition of Christian States that would then join an international federation. Other plans – by France's Pierre du Bois, Erasmus, Sir Thomas More, King Podiebrad of Bohemia, the Duc de Sully and Emeric Cruce – are summarized. (Curiously, even the great Douglas doth nod: he does not mention the elaborate plan written by philosopher Immanuel Kant.)

One of the most valuable features of Douglas's work is the appearance at the end of every significant part of the book of sections, entitled "Reflections," in which Douglas editorializes about what has gone before. In his "Reflections" following the part of the book on plans for perpetual peace, Douglas begins with these magical words: "World history reeks with the stale breath of incessant warfare" (p. 351). He then goes on to note how "ideas of *international* government" (p. 351) were to be found in peace plans, at a time when "the contours of *national* government had not yet been clearly discerned" (p. 351). Those kinds of original insights abound in every chapter.

Douglas returns to the problem of regulating the use of force in Chapter 9, "Contemporary World Order in Profile". He mourns the fact that "the impressive trends in our own time toward an inclusive, almost universal, system of world law cannot be separated from two uniquely perilous challenges to international order... [and that] nothing accomplished in the development of international law and diplomacy came close to warding off the world calamity in military history [i.e., two devastating world wars]" (pp. 689–690). In the post-20th century world, world order cannot be based on any one "model" or "mode of idealism." Neither the "scholars of the tower" nor the "technocrats of the arena" will be able to deliver any meaningful consensus on this front. Douglas applauds the world's ability in the late 20th century to have developed new forms of legal regimes, as it did with the Antarctic Treaty, the UNCLOS III agreement on the status of the international seabed and subsoil, and the demilitarization of the Moon. Those developments, together with the tentative steps taken toward an "organized world community" – to enhance, not replace, the old post-Westphalian States system – was something that Douglas thought could "break the normative ice," particularly in places like the thawing Arctic Ocean.

One of Douglas's great fears was that the current hegemonic status of the United States had wrecked the constitutionalism inherent in the United Nations Charter. The world had to find a new and creative way to renew the principle of state equality in the face of the "encroachment of political hegemony" (p. 711). But "hegemony is not a legal concept... It falls outside the domain of international law as a threatening fact of contemporary political life" (p. 711). "For those who look upon international law most darkly as a history of moral failure, hegemony is not so much an evil as a guarantee that history will continue, as always, to miss opportunities for world reform along collaborative lines" (p. 711). That cry for a "cooperative ethic" (p. 719) is part of Douglas's final jeremiad to a world he has now, sadly, left behind.

One thing is certain. In the future, regardless of whether the world lives up to Douglas's high standards or ideals, no international lawyer will be able to judge himself or herself educated unless and until they – intimately and fully – study and know Douglas's final, magnificent and original *locus classicus.* Well done, old friend.

Remembering Douglas Johnston as a Practicing Diplomat as well as a Scholar

Edward L. Miles*

Brian Flemming and the other colleagues paying tribute to Douglas as a scholar and teacher in this volume have done a marvellous job describing the immense breadth and depth of Douglas' scholarship. Douglas was not known as an organization man, but I write to pay tribute to his skills as a meticulous practical diplomat in putting to use his skills as a scholar and organizer in events that followed the Third United Nations Conference on the Law of the Sea (UNCLOS III) between 1990 and 1994. The stakes were high, since they involved working out issues related to high seas fisheries which had not been settled before the conclusion of UNCLOS III. However, the nature of the settlement, had it been injudicious, would have threatened the stability of the 1982 United Nations Convention on the Law of the Sea (LOS Convention). The issues to which I refer included the question of straddling stocks (or stocks which straddled the boundary between the outer limits of the exclusive economic zone (EEZ) and the high seas); the use of high seas pelagic driftnets, particularly by Japan, Taiwan, and Korea; the question of enforcement on the high seas on fishing vessels by States other than the flag States; and the existing and proposed organizational arrangements for the conduct of high seas fisheries. These were matters then of extraordinary sensitivity.

By way of background, I should say that my involvement in this case with Douglas was precipitated by a paper I had published with a colleague, Professor William Burke, in 1989 on the issue of actions contemplated by coastal States in the North Pacific against foreign flag States employing

---

* Virginia and Prentice Bloedel Professor of Marine and Public Affairs, School of Marine Affairs, University of Washington, Seattle, USA.

high seas pelagic driftnets.[1] Since these coastal States involved both the United States and the Soviet Union, Burke and I warned the principals about the potential examples that would be set and the consequences that could ensue which would be prejudicial to their larger interests in the stability of the LOS Convention. This paper had generated a great deal of interest, and I had been called by the foreign policy establishments of many governments to discuss over the telephone the points Burke and I had made in that publication. Not surprisingly to me, one of the governments in question was Canada, which was then facing a rather intense straddling stocks conflict with European States in the Northwest Atlantic. Out of this cauldron there came one day a call from Douglas.

I should say at this point that Douglas and I had formed a firm friendship in 1969 at the Third Annual Conference of the Law of the Sea Institute at the University of Rhode Island. I was very excited to meet him because I had read with pleasure and profit his book on the *International Law of Fisheries* which had been published by Yale University Press in 1965 and republished in 1987.[2] At that time we had a long discussion about the functional approach to international law and I have been amused to see that he published much later an article under the same title.[3] In any event, to return to the call from Douglas, he said that the Canadian government was willing to sponsor a "back channel" meeting between the principals involved in negotiating arrangements to solve the difficult high seas fisheries issues which were being discussed officially both in the North Pacific, in the United Nations, and in the United Nations Food and Agriculture Organization (FAO). The principals included Canada, the United States, Japan, Korea, the Soviet Union, Taiwan, and Poland. The Canadian government wanted such a meeting to be hosted by the Centre for Asia-Pacific Initiatives (CAPI) and Douglas and I should be

---

1 E. L. Miles and W. T. Burke, "Pressures on the United Nations Convention on the Law of the Sea of 1982 Arising from New Fisheries Conflicts: The Problem of Straddling Stocks," paper presented to the Joint Soviet Maritime Law Association/Law of the Sea Institute Symposium on the Law of the Sea, Moscow, 28 November–2 December 1988. Reprinted in *Ocean Development and International Law Journal* 20 (1989): 343–357.

2 D. M. Johnston, *The International Law of Fisheries* (New Haven: Yale University Press, 1965; 2nd ed., 1987).

3 D. M. Johnston, "Functionalism in the Theory of International Law," *Canadian Yearbook of International Law* 26 (1988): 3–59.

co-organizers and co-chairs. I replied that I would be honoured and delighted to play such a role, and we set a date for me to go to Victoria to discuss the details.

Douglas and I met at Dunsmuir Lodge in Sidney, near Victoria, 13–14 January 1990 to plan the meeting. We had decided that, as co-chairs, he would take the lead on the organizational details and I would take the lead on the substance, but in each case we would jointly decide on the outcomes. Douglas led on the organizational details and the meticulousness of his comprehensive and systematic approach astounded me. He came prepared with several pages of notes and I learned that we had secured 45 bedrooms of which 14 contained king size beds and 31 contained double beds. Five of the bedrooms were in suites, but renting 45 bedrooms entitled us to have the living rooms in the suites free of charge. The assignment of rooms and suites was a delicate matter since there were more than seven high-level government officials who were expected to attend the meeting. Decisions also had to be made on managing early arrivals and late departures as well as choosing meals for the whole week. Since seven countries were involved, we had to provide the kitchen with information on all the cultures that would be in attendance. We also, at that time, anticipated holding a press conference early in the afternoon at the end of the week. And we discussed whether we should specify whether spouses were welcome. After the whole meeting was designed, we decided against inviting spouses.

We constructed invitations to participate in an international technical workshop on the most critical problems of high seas fisheries in the North Pacific originally scheduled for 16–20 April 1990. The meeting was to be hosted jointly by CAPI and the School of Marine Affairs, University of Washington. Funding was secured from the Asia-Pacific Foundation of Canada. The meeting was to be co-chaired by Douglas and me, and we defined the critical issues to include the problem of straddling stocks, high seas pelagic driftnets, fisheries enforcement on the high seas, and existing and proposed organizational arrangements for the high seas. The meeting was to be held under rules of strict confidentiality, which, in view of the sensitivity of the issues, we interpreted to mean a) there was to be no publicity whatever, which killed the original notion of a press conference at the end; b) there would be no record; and c) the co-chairs committed themselves not to publish any articles on the substance of what had been discussed by the representatives of the States attending the meeting.

The intention here was to facilitate a free and open discussion of all the sensitive issues and we communicated these decisions to the governments in the following language:

> The co-chairs are mindful of the fact that governments will be engaged in official discussions on these sensitive and difficult issues in the months to come. It is our hope that informal, technical discussions under non-governmental auspices will facilitate development of consensus on approaches to be taken to achieve resolution of these issues. Accordingly, the co-chairs will rule out of order any non-technical (i.e., non-scientific, non-economic, non-legal, non-commercial) comments.

We invited each of the seven participating countries to send up to six representatives to the meeting, and our budget was such that we could cover the costs of three of the six members of each delegation. In addition, we let them know that we would have on hand three experts each from the United States and Canada who we proposed to use to summarize the technical history of each issue before we would call on any of the governmental representatives who wished to speak. Douglas undertook to write the letters to the representatives from Canada, Taiwan, and Korea. I was assigned the United States, Japan, the Soviet Union, and Poland. All recipients of the invitations returned positive responses and none had any objection to the procedures we intended to use. As it happened, the workshop had to be postponed to 5–9 October 1990, at which time I made a statement on behalf of the co-chairs describing what our procedures were and why we proposed to conduct business in that fashion. I shall include this statement here because it was worked out jointly by us over a period of time and does not contain any summary of comments by the delegations attending. In this way I shall remain true to the commitment made to the delegations by the two co-chairs at the same time that I reveal what, how, and why we approached the problem in the way we did. I do so to salute a friend and colleague of very long standing with whom I worked over the years on many things and with whom I shared my first successful international mediation.

Introduction to the Workshop on Critical Issues in the Management of High Seas Fisheries in the North Pacific by the Two Co-chairs

Purpose and Objectives

Why would two academics like us seek to convene such an unusual meeting like this at this time? We do so because several of the issues with which we choose to deal have given rise to considerable controversy, as a result of which governments have taken perhaps prematurely strong positions before the necessary information has become available. We do so also because the most developed fishing nations in the world are involved here along with those who are both advanced fishing nations and simultaneously major coastal states. We do so because, as a result of the first two conditions, far more is now at stake than the specific issues around which controversy revolves. What in fact is at stake is the stability of the new regime enshrined in the United Nations Convention on the Law of the Sea of 1982 and embedded in state practice. If the major fishing nations and some of the major coastal states of the world act carelessly or loosely, we can end up with new regimes which are in nobody's interests. Finally, we do so because we do not see that any coastal states acting singly or in combination can, by themselves, find adequate solutions to these problems. On the other hand, if distant-water fishing nations do not try in good faith to meet legitimate coastal state interests in the effective management of high seas fisheries, they are likely to feed the fires of further expansion of coastal state jurisdiction to a dangerous degree. That will be neither an equitable nor a wise development.

Therefore, what is needed is international cooperation. We seek to facilitate such cooperation by focusing on the technical and informational aspects underlying the critical issues in a neutral forum. This forum is the result of the combined efforts of two Universities, one Canadian and one American. For each issue, we will seek to define what we know; what we do not know; what we need to know; and how we might attempt to learn what we need.

The co-chairs had considered whether we should try to publish a report. We decided against it because it seemed to us that preparation of such a report would come to dominate our discussions and emphasize the natural propensity to preserve one's national position. Moreover, some of these issues are so controversial that it is unlikely at this time that any wording could satisfy all sensibilities. We therefore propose to avoid such a negotiation by highlighting the technical and informational aspects without the need to be constrained by worrying about phraseology. We should give our imaginations free rein to see how we might be able to facilitate the solution of the difficult problems we all face.

We realize that the format of the workshop is unique and that some of you find it a bit puzzling. It is in fact an experiment in two phases. First we focus on defining the issues using experts to begin and then soliciting

> interaction with the whole group. In this way we hope to find consensus on each item concerning what sub-issues we need to discuss in detail. Then, in phase two, we will move to substantive discussions of the issues seriatim in which all participants may join. We are not seeking formal presentations; rather we seek constructive engagement. We would note that the workshop is more unique than just in its format. So far as we are aware, no forum exists officially to convene a meeting on these issues facing these seven states and treating them in a single agenda. We have a precious opportunity, which has been afforded us by the Asia-Pacific Foundation of Canada, and we should make use of it.

It was with deep satisfaction that we watched the meeting over the course of the week develop into an extremely useful exchange of views with both sides seeking to explore where and how they might come together. We never expected that matters would be solved at this meeting, but we did manage to move the process forward by the end of the week. Douglas and I considered this fact cause for celebration. This was not the end of the matter, however. Two years later, in 1992, I received a call this time from the United States asking, on behalf of themselves and Canada, whether Douglas and I would be willing to reconvene the group under the same rules as the previous meeting with the two governments footing the bill on this occasion. I called Douglas and we immediately agreed. The second meeting went as productively as the first, even though China had replaced Taiwan, and we were immensely gratified to learn in 2004 that the Governments of the People's Republic of China, Japan, the Republic of Korea, the Republic of Poland, the Russian Federation, and the United States had signed a Convention on the Conservation and Management of Pollock Resources in the Central Bering Sea. Taiwan was not a signatory and neither was Canada. However, this Agreement solved for both the Russians and the United States the dangerous "Donut Hole" problem in the Central Bering Sea which once threatened to precipitate coastal State unilateral extension of jurisdiction beyond the EEZ. Clearly, China was exchanged for Taiwan because the United States had other significant leverage over Taiwan and China was a new entrant into the area. While Canada had an intense interest in seeing a settlement of the Central Bering Sea problem, it had no direct stakes in that particular matter. The indirect significance, however, was very large given Canada's similar problems in several locations in the Northwest and North Central Atlantic.

Douglas and I were most interested to learn from the official press release accompanying the Final Act of the Convention that the parties had been meeting officially in different capitals in February 1991, July/August 1991, November 1991, April 1992, August, 1992, January 1993, June/July, 1993, October 1993, and November/December 1993. This information allowed us to locate our meetings in this suite of negotiations and we inferred that the Canadian and U.S. governments had recourse to our unorthodox procedures on occasions on which they wanted a thorough discussion of issues unofficially in a neutral forum. That they had returned a second time we took as a vote of confidence in our assistance. It was the United States, via the National Marine Fisheries Service, Alaska Region, who had sent me a copy of the Final Act along with the official Joint Press Release. Again, Douglas and I were immensely gratified.

Part II

Introduction: Setting the Stage

Aldo Chircop,* Ted L. McDorman** and Susan J. Rolston***

"One of the most creative innovations of the international diplomatic community in the 20th century was its invention of *the international regime*," wrote Douglas M. Johnston in his last major work published posthumously.[1] Modern regime-building can be traced back to early 20th century. In the aftermath of the First World War and the establishment of the League of Nations, David Mitrany commented on the establishment, growth and functioning of early international organizations as a new form of international cooperation in support of peace maintenance in international relations.[2] He saw functionalism as horizontal international cooperation on traditionally domestic concerns, but which were in the common interest of members of the international community of States. That cooperation would be serviced by international organizations. Established in 1919, the International Labour Organization (ILO) was one such organization that offered a new approach to international cooperation, and thereby enhanced the prospects for international peace.[3] For its contributions to peace, in 1969 the ILO was the recipient of the Nobel Peace Prize. The role of international regimes with international organizations at the core was set.

---

* Professor of Law and Associate, Marine & Environmental Law Institute, Dalhousie University, Halifax, Canada.

** Professor, Faculty of Law, University of Victoria, Canada.

*** Principal, Seawinds Consulting Services, Hackett's Cove, Canada, and Associate, Marine & Environmental Law Institute, Dalhousie University, Halifax, Canada.

1 Douglas M. Johnston, *The Historical Foundations of World Order: The Tower and the Arena* (Leiden: Nijhoff, 2008), p. 730.

2 David Mitrany, *The Progress of International Government* (London: Allen & Unwin, 1933). See also Lucian M. Ashworth & David Long, eds, *New Perspectives on International Functionalism* (New York: St. Martin's Press, 1999).

3 See Wilfred Jenks, *Human Rights and International Labour Standards* (London: Stevens, 1960).

As it turned out in the inter-world war years, there was more to peace-building than the establishment and maintenance of international organizations. States jealousy guarded their territoriality and sovereignty, conceding little to international organizations, at least prior to the establishment of the United Nations Organization. Mitrany foresaw the emergence of inter-State cooperative multilateralism and his work went on to influence the establishment of the United Nations. International cooperation would be institutionalized for the most part, and international law would see a gradual transition to a system functioning on the basis of multilateralism and a softening of the hard edges of sovereignty.

By the 1960s, Wolfgang Friedmann observed that modern international relations were no longer the exclusive domain of diplomatic inter-State relations, but rather had infiltrated many strata and sub-strata within the State itself.[4] He opined that international law was in fact moving on different levels that included not only the classical international law of peaceful coexistence based on legal norms, but also an international law of cooperation that addressed both universal concerns and the interests of regional groupings through complex structures and processes. International law had grown to include new fields, new participants and subjects, new political and socio-economic principles, and new international organizations.

These changes were to become even more pronounced with the convening of the world's first mega-conference on the environment, the 1972 United Nations Conference on the Human Environment, followed by the longest mega-conference in world history, the 1973–1982 Third United Nations Conference on the Law of the Sea (UNCLOS III), and more recently the now almost continuous "Rounds" of negotiation that take place within the World Trade Organization, among other similar social phenomena.[5] Mega-conferencing and multilateralism profoundly diversified the participation, scope and depth of the subject-matter, and mandates of existing institutions and led to the establishment of many new organizations and structures in international law. Johnston noted that major regime-building initiatives such as these would spin-off further

4 Wolfgang Friedman, *The Changing Structure of International Law* (New York: Columbia University Press, 1964), pp. 60–71.

5 Douglas M. Johnston, "The Heritage of Political Thought in International Law," in R. St. John Macdonald and D. M. Johnston, *The Structure and Process of International Law* (Dordrecht/Boston/Leiden: Martinus Nijhoff, 1986), p. 203.

sub-areas of specialization in the global oceans sector including "marine pollution, fishery management, maritime boundary-making, wildlife law, and sustainable development law and policy."[6]

In the wake of the Second World War and the establishment of the United Nations system, international organization did not proceed down the path of world federalism, as some had hoped, but rather moved towards neo-functionalism and regime-building in international relations.[7] Frequently at the centre of regime-building efforts, international organizations provided the arenas for inter-State cooperation on classical issues such as the maintenance of international peace and security and for more focused sectoral and multi-sectoral problem-oriented cooperation in the areas of "specialization" referred to by Johnston. The arenas included numerous specialized agencies and programmes of the United Nations and other intergovernmental organizations, as well as regional organizations addressing equally diverse issues. Today these arenas also include respected non-State actors, for example in the environmental and humanitarian areas, who play important facilitative roles in regime-building.

Multilateralism, certainly more than historic hegemony, is central to regime-building. Modern multilateralism in diplomacy owes much to decolonization. The international community of States was transformed into a truly global and diverse community that injected new values into the international system. The new States pressed for urgent change in classical international law and the traditional law-making processes that were perceived to be based on hegemony and custom, towards a more equitable and consensus-based law. For a long time, the law of the sea was perceived as favouring the wealthy maritime and former colonizing States. The newly independent States were very diverse: large and small, coastal or landlocked or small islands, politically strong and weak, technologically have's and have not's, rich and poor, and so on. These new members pressed the international community, and international organizations as the key arenas, to re-shape the international order. Their initiatives were pursued at both global and regional levels with the intent to tackle the common problems faced by the international community. These initiatives and

[6] Johnston, *supra* note 1, p. 695.

[7] Ernst Haas, *Beyond the Nation-State: Functionalism & International Organization* (Palo Alto, CA: Stanford University Press, 1964).

processes were greatly facilitated by the United Nations system of global and regional organizations, but also by other regional intergovernmental organizations in Africa, the Americas, and Europe, among other regions. The older States also initiated change through multilateralism. In particular, European States embarked on an unprecedented process of functional cooperation, first emerging with the Coal and Steel Communities and eventually with the European Economic Community and the European Union, that incrementally created a complex system supranational governance which moved the old world towards a form of federalism.

Thus, in essence the modern international law of cooperation consists of global and regional regimes governing virtually every area of governmental competence. Regimes are no longer purely structures for cooperation necessarily and exclusively dependent on treaties. Although treaty law as normative structure is important, frequently a seminal role in the establishment, continuance and growth of regimes depends on the role of an international organization, the participation of an increasingly diverse group of both State and non-State actors, and processes guided by values and principles that are at odds with classical approaches to international law. Rather than characterization as treaties among sovereigns, regimes are seen as principles, norms, rules and procedures around which actors' expectations converge in a given issue area.[8] The actors include epistemic communities, consisting of diverse governmental and non-governmental actors possessing knowledge, skill and willingness to advance political and technical cooperation.[9]

Until the mid-1970s, ocean and environmental regime-building were primarily treaty-making exercises. There was little regime-building with environmental content, at least until the advent of the United Nations Conference on the Human Environment, in Stockholm, in 1972. Modern oceans and environmental regimes have been profoundly influenced, if not produced, by mega-conference diplomacy. This new conference process was an ideal tool for multilateralism. Starting with UNCLOS III, the world's oceans experienced a surge in global and regional regimes. The ensuing 1982 United Nations Convention on the Law of the Sea (LOS

---

[8] Robert O. Keohane, *After Hegemony: Cooperation and Discord in the World Political Economy* (Princeton, NJ: Princeton University Press, 2005).

[9] Peter M. Haas, *Saving the Mediterranean: The Politics of International Environmental Cooperation* (New York: Columbia University Press, 1990).

Convention) laid the foundations for modern regime-building for oceans because of its vision of the unity of the marine environment, the need to consider ocean uses as a whole, and the roles allocated to existing and new international organizations. Although treaty-making in the law of the sea and international environmental law continued, they assumed a more complex character and were closely tied into institutional arrangements and regime maintenance processes. Regular meetings of contracting parties, normally supported by a secretariat and frequently with an elaborate institutional arrangement, monitor the progress of a regime against set goals, coax members into action, allocate resources, and set new courses of action. The treaty is no longer a static instrument, but rather a basis for States to build a more complex institutional framework and process to transact regime business.

Regimes are very diverse. There are global and regional regimes. Regimes are conceived differently for different tasks. Some are multi-functional, while others are single-function. The LOS Convention and United Nations Framework Convention on Climate Change, and the accompanying Kyoto Protocol, established some of the most complex regimes. At the regional level, marine environmental arrangements were adopted at the same time as the LOS Convention was being negotiated, such as those in the Baltic, North Sea and the Regional Seas Programme. In particular, the latter was one of the first activities of the newly established United Nations Environment Programme (UNEP). Following the establishment of a regime for the protection of the Mediterranean marine environment, UNEP has successfully used it as a model for the establishment of mirror regimes in many other marine regions. Many of these regional sea regimes are now evolving into comprehensive ocean governance institutions. Regional fisheries regimes are different. For one, they are mostly sectoral institutions and are not well integrated into international ocean management in sharp contrast to the regional marine environmental regimes. Serviced by an ad hoc regional fisheries management organization, regional fisheries regimes are primarily concerned with resource access and allocation. Today, regional fisheries management regimes are evolving in response to declining high seas fishery resources, demands for more principled approaches to management and transparency, ecosystem and species conservation needs, and ongoing problems with compliance and enforcement. In contrast to marine environment protection and fisheries, the regime for international shipping was established as and continues to evolve as a global regime hovering around one "competent" international organization (i.e., the International Maritime Organization), which is concerned with technical regulation and standard-setting on matters of maritime safety, environmental protection from ship activity, maritime and port security, and trade facilitation. Some regimes could be described as formal, especially when they are treaty-based, such as the World Trade Organization and the UNEP Regional Seas Programme, or informal when they are established in the form of arrangements based on political agreement, such as the Canada-United States Gulf of Maine Council on the Marine Environment.

This is the stage for this book tribute. International regimes are increasingly attracting more scholarship. The growing literature is innovative, multisciplinary and interdisciplinary. It draws heavily on international relations theory, international legal theory, policy science, and, more recently, governance theory. With diverse disciplinary backgrounds and perspectives, the distinguished contributors to this tribute follow a long tradition of scholarly inquiry into the governance, creation, operation, viability and maintenance of international regimes. Their contributions on several ocean and environmental regimes attest to the depth to which modern international law and the underlying international relations have been transformed into an international law of structured cooperation.

The organizing principle of the chapters of this book is based upon ocean regimes with the first section devoted to "Global Perspectives." The pivotal role of the 1972 United Nations Conference on the Human Environment and the output of the Conference, the Stockholm Declaration, regarding the subsequent development of the structures and processes of international environmental law, and more generally environmental regime-building, are well discussed in the opening essay of the section by Jutta Brunnée. The contemporary issue of global climate change is examined by Meinhard Doelle in his chapter, "Integration Among Global Environmental Regimes: Lessons Learned from Climate Change Mitigation," through a review of the international responses to climate change from the earliest United Nations General Assembly resolutions to the adoption and implementation of the Kyoto Protocol. The focus of Doelle's inquiry is assessing the extent of the integration of international and national efforts to address global climate change and the weakness in the existing regime because of the inadequacy of the degree of current integration. The contribution by John Gamble, Ryan Watson and Lauren Piera provides a macroscopic view of multilateral treaty-making and ocean regimes summed in their comment that "it is easier to understand individual treaties if one has some idea of the forest in which those treaty trees grow." Their careful tabulation of treaties relating to marine living resources, navigation and shipping, and marine pollution indicates that, throughout the 20th century, there was a steady increase in the number of ocean-related treaties – a pattern that was not evident in other areas of treaty-making which tended to be more episodic. Moreover, they note that, unlike in other areas, ocean treaties tend to be "plurilateral (limited by interest, geography or a combination)," rather than "multilateral (open to all States)." Jay L. Batongbacal's contribution, "The Law of the Sea, Marine Technology and Global Social Justice," looks at the heart of regimes in that it identifies and questions the role that changing marine technology has played in the development of ocean regimes and the uncertain relationship between marine technology and the promotion and attainment of global social justice. A key aspect of international regimes is the extent to which States accept and adopt into practice the principal limitations and rules of the regime framework. This is the topic of Stuart Kaye's contribution "State Practice and Maritime Claims: Assessing the Normative Impact of the Law of the Sea Convention." Kaye looks at State legislation to assess whether, as regards national maritime zones and freedom of navigation, key components of the LOS Convention, States have faithfully implemented the Convention. He concludes that with respect to the 12 nautical mile territorial sea and 200 nautical mile exclusive economic zone (EEZ), States have generally adopted the wording of the LOS Convention. As regards, freedom of navigation in national waters, Kaye found that over half of the coastal States surveyed had restrictions or regulations respecting navigation that *prima facie* appear not to track the wording of the LOS Convention. The latter situation Kaye identifies as raising questions about whether the navigational provisions in the LOS Convention "represent customary international law" and the long-term efficacy of the Convention wording in this area. John Duff, in the final chapter of the first section entitled, "Trends in Ocean Zoning – Layers of Confusion and Approaches to Clarity," tackles the confusion that surrounds the frequent use and misuse of the term "zoning." By using the examples of the domestic administration of the oceans in the United States and Canada, he brings clarity to the ocean zoning management tool.

Few areas of the law of the ocean have been fraught with more debate than the legal regime respecting straits used for international navigation. Hence, it is appropriate that the second section of this book is "International Straits and Navigation Routes." The first contribution in this section is by Jon M. Van Dyke, entitled "Transit Passage Trough International Straits," which provides a valuable review and study of the provisions of the LOS Convention respecting international straits. Van Dyke points out that each international strait has its own history and politics, and he proceeds to provide sketches of these matters respecting most of the significant straits in the world. Van Dyke concludes what also is evident from the other chapters in this section, that the balance between passage rights and coastal State concerns (environmental and security) respecting international waterways remains a tension point in the modern law of the sea. Regarding one of the most critical straits in the world, the Malacca and Singapore Straits, Robert Beckman writes about the cooperative mechanism launched by Indonesia, Malaysia and Singapore that is expected to enhance coordination and cooperation respecting management among the three strait States and also with the IMO, user States and other stakeholders. What is emerging is a "regime" for the Malacca and Singapore Straits designed to balance more appropriately the interests of all users and neighbouring States as regards security, marine environmental protection and international shipping. Perhaps the most watched strait used for international navigation at present is the Torres Strait between Australia and Papua New Guinea. As explained in Sam Bateman's contribution, Australia and Papua New Guinea have introduced compulsory pilotage for all vessels utilizing these waters and this action has drawn formal public protests from both the United States and Singapore. Bateman makes the case that the introduction of "compulsory pilotage in the Torres Strait was both a clear and necessary decision for Australia" and that "sufficient doubt" exists respecting whether the action might be inconsistent with existing international law to provide comfort for Australia's actions. The final chapter in this section is the aptly titled "Northern Sea Route Navigation: The Last Frontier?" by Edgar Gold, which deals with the waterway that crosses the Arctic waters of the Russian Federation. While there are international legal issues that arise respecting the Northern Sea Route, the matters concentrated on in this contribution are more practical in nature. Vessel technology, navigational concerns such as the presence of ice, and economics were part of the multi-year projects, the International Northern Sea Route Programme (INSROP) and the Arctic Operational Platform Project (ARCOP). The conclusion is that regular, safe and environmentally protective shipping is both inevitable and achievable in the Russian (and for that matter the Canadian) Arctic.

The third section of the book is "New Standards for International Shipping." Commercial shipping is inherently international, and throughout history has been an activity to which States and commercial interests have attempted to establish a degree of uniformity of expectations. Much of that effort pre-20th century had to do with commercial expectations and the establishment of national admiralty law that reflected the international nature of shipping. In the early 20th century, safety of ships, their crew, cargo, and passengers generated a number of global multilateral treaties. The first decade of the 21st century has witnessed some significant changes in the international regime for commercial shipping. One of these changes has arisen as a result of enhanced concerns respecting vessel security and this is commented upon in the contribution from Hugh M. Kindred and Mary R. Brooks, "Consequences of Securing Merchant Shipping for Contractual Relations in the Carriage of Seaborne Trade." The Kindred-Brooks chapter examines not just the new international and national (primarily led by the United States) security-related requirements for commercial vessels, but their effect on seaborne trade. They conclude that as a result of enhanced security measures and consequent reporting and documentation requirements, commercial shipping has become more expensive and that the movement of trade by sea has not changed that much. However, the new requirements have concentrated shipping into the hands of the largest international cargo interests and liner operators since they have the financial and technical capacity to meet the new requirements. The second major global shipping event of the first decade of the 21st century is the adoption by the ILO in 2006 of the Maritime Labour Convention. This new Convention, simultaneously a consolidation of the 20th century seafarers' treaties and instruments of the ILO and new approaches to securing minimum standards of working conditions and compensation, while reflecting the competitive environment of employers, is analyzed from different perspectives in chapters by Moira L. McConnell and Peter B. Payoyo. Payoyo's concentration is on the Convention's attempt to promote and attain "social justice for seafarers," a goal, the author asserts, that has in the past proven to be "largely elusive." McConnell's concentration is on the innovative nature of the Convention fueled by the conscious effort by the ILO "to experiment with a new strategic approach to squarely tackle the problem of ways to secure universal ratification and more effective implementation of a sector closely tied to international economic and trade matters." In this respect, the Maritime Labour Convention may have regime-building lessons well beyond the area of seafarers' rights.

The subtitle to Johnston's last book is "The Tower and The Arena."[10] The arena refers to the negotiation and operational milieu of international relations, law-making and regime operation and hence populated by so-called "realists." The tower refers to academics, commentators and others largely detached from the messy situation inherent in the creation and day-to-day management and hence populated by so-called "idealists." The authors of the three chapters within the section "Fisheries Conservation and Human Rights Protection" each have extensive backgrounds in "the arena" of international fisheries. Gordon R. Munro has had a long relationship with the Food and Agriculture Organization of the United Nations (FAO); William R. Edeson spent much of his career with FAO; and Wendell Sanford, with the Canadian Department of Foreign Affairs and International Trade, has on several occasions been at "the sharp end of the stick" in Canada's engagement with the international fisheries regime. Munro's contribution shines an important light on the role of economic incentives and disincentives, often overlooked in the zeal to use law and the new-found religion of conservation to combat over-fishing, in the creation and sustainability of management regimes for internationally shared fish stocks. Sanford's contribution tells the story of Canada's dealing with unwanted and conveniently-flagged foreign fishing vessels operating outside its 200 nautical mile zone on the Atlantic Ocean prior to the 1994 arrest of the Spanish trawler *Estai*. While the status and rights of seafarers has received attention from the ILO, less attention has been paid to those involved in the fisheries sector. Edeson's contribution provides a detailed examination of "how human rights issues relate to [national] legislation for the fisheries sector."

The "Marine Environment Protection" section is composed of four chapters with one dealing directly with the effectiveness of marine environmental regimes, two concentrating on the specifics of ocean dumping, one from an international perspective and the other from a national perspective, and the last contribution dealing with international maritime safety tools that may be obtained from the IMO to achieve marine environment protection and conservation objectives. Shelley Lexmond's chapter "Improving the Effectiveness of Environmental Regimes: 'Consilience,' Science

---

[10] Johnston, *supra* note 1.

and Common Sense" is drawn from her doctoral work done under the supervision of Dr. Johnston. She writes about the melding of the domains of law of the sea and international environmental law and identifies that dealing with environmental issues is principally about risk management. The effectiveness of the latter as a management approach is determined by, amongst other things, the clarity of the long-term objective, the adequacy of science and social information, the building of knowledge, feedback and review loops, and a good portion of common sense. The chapter by David L. VanderZwaag and Anne Daniel provides a detailed analysis of the 1996 Protocol to the London Convention. They point out that the 1996 Protocol, which entered into force in 2006, alters the control of ocean dumping by embracing a precautionary stance in that those substances that can be considered for dumping at sea are listed (the "safe list") as opposed to the approach taken in the 1972 London Dumping Convention, which listed substances that could not be dumped. This is a model that the authors see as possibly extending to other chemical agreements and regimes that operate through specific listing of substances. Zou Keyuan's contribution looks at the law and practice respecting ocean dumping in China. It is pointed out that while China is a party to the 1996 Protocol, most of China's ocean dumping takes place within its internal waters and is, therefore, beyond the reach of the 1996 Protocol. However, the author concludes that there has been a steady improvement in the regulation and management of ocean dumping with China, although more needs to be done. The final paper in the section is by Aldo Chircop entitled "The Designation of Particularly Sensitive Sea Areas: A New Layer in the Regime for Marine Environmental Protection from International Shipping." There has been a recent spurt in both the number and coverage area of IMO-designated PSSAs and with these developments there has been controversy with respect to their application in some marine regions. Chircop voices moderate concern about the effectiveness of PSSAs as opposed to their symbolic/political cache; the proliferation of PSSAs and the potential for the "watering down" of the idea of special highly-sensitive ecological areas; and the process that exists within the IMO for the adoption/approval of PSSAs, which may leave contesting, but minority, interests frustrated and thus raise practical problems concerning compliance.

The final section of the collection is entitled "Dispute Settlement and Regional Cooperation." Three of the chapters deal specifically with regional regime-building and cooperation. A fourth chapter focuses on the issue of third party engagement with maritime boundary adjudication. Finally, the fifth contribution, by Ted L. McDorman, "Notes on the Historic Waters Regime and the Bay of Fundy," is somewhat like the description given to the historic waters international legal regime in that chapter, an orphan. Alex G. Oude Elferink does a survey of the consideration given in international maritime boundary adjudications to geographically-interested third party States. He concludes that in the recent cases the tribunals have broken with the past and tended to discount or ignore the regional context and the legitimate interest of third parties in bilateral maritime delimitation by relying almost exclusively on equidistance as realizing an equitable solution, but suggests that the regional context must remain, where appropriate, an important consideration in adjudications. Clive Schofield and Ian Townsend-Gault in their chapter, "Brokering Cooperation Amidst Competing Maritime Claims: Diplomacy in the Gulf of Thailand and South China Sea," set out the legal framework, national claims and the interests of the States of the region. They detail the efforts that have been made through both Track I and Track II diplomacy to enhance the possibility of cooperation, management, and resolution of ocean issues. They conclude that there is not as yet "a culture of cooperation" in Southeast Asia that allows the States to make a break-through respecting resolutive and cooperative initiatives, but that the creation of such a "culture" is enhanced by both Track I and Track II diplomacy. Mark J. Valencia's chapter focuses on East Asia and, more particularly, Northeast Asia where he sees both positive and negative developments regarding maritime regime-building. On the positive side, Valencia points out the "diverse web of bilateral understandings and exchanges" regarding fisheries and other matters that have arisen as a result of the common interests of the key players, China, Japan, and Korea, and which is indicative of the possibility of the creation of a more formal international regime. However, Valencia also sees negative developments that would appear to forestall any serious efforts at regime-building. These include increased acquisition by the relevant States of maritime power and the numerous festering island and territory disputes. Carlyle L. Mitchell's chapter, "The Legacy of the Dalhousie Oceans Studies Programme (DOSP) in the Caribbean" involves regional cooperation in a different way, as the chapter revisits the work done by DOSP in the 1980s and indicates the impact that this work had on regional developments.

The diversity of the subject matter, content, and operation of ocean regimes demonstrated in the contributions to this book should not be surprising. As earlier explained, there is no unified meaning or construct of a regime. In international relations, regimes are about processes of cooperative decision-making and the creation and fulfilment of expectations that can be reasonably relied upon by States and others in any number of areas and concerning a large number of subject areas. As this collection of essays demonstrates, while one important function of a regime often is order and certainty and the consequent reduction in disputes and misunderstandings, regimes are driven by specific concerns, such as fisheries, ocean dumping, maritime security, seafarers' rights, or enhancement of marine environmental protection. As further evidenced in a number of contributions in this book, most often by the non-lawyers, successful regimes are built on more than just legal text. They involve real commitment of States and facilitation by relevant non-State actors, and that means taking into account process issues, politics, economics, scientific and technological expertise, and social justice. The result is that in some geographical areas, for example, Northeast Asia (as pointed out by Valencia) and some subject areas, for example, marine pollution from land-based activities, it has not yet been possible to develop significant international regime structures; whereas for other geographic areas and subjects (many described in the chapters of this book), regimes, of varying operational efficacy, exist and have had success. Despite the diversity and mixed record of international regimes addressed in this book, one clear running theme in contemporary regime-building is the idea of governance involving new and innovative arrangements engaging States and non-State actors.

Ocean and Environmental Regime-Building: Global Perspectives

The Stockholm Declaration and the Structure and Processes of International Environmental Law

Jutta Brunnée*

Introduction

In reading commentary on the Stockholm Declaration on the Human Environment[1] that was offered at the time, one is struck by the overwhelming focus on the fault lines between North and South that the conference put into the spotlight.[2] To fully appreciate why the declaration is justly seen today as a historical marker,[3] it is essential to recall that the Stockholm

---

* Professor of Law, University of Toronto, Canada. I thank Asher Alkoby for his excellent research assistance in the preparation of this article. Reprinted with permission from J. Brunnée, "The Stockholm Declaration and the Structure and Processes of International Environmental Law," in M. H. Nordquist, J. N. Moore and S. Mahmoudi, eds, *The Stockholm Declaration and the Law of the Marine Environment* (The Hague: Martinus Nijhoff Publishers, 2003), pp. 67–84.

1 Reprinted in 11 *I.L.M.* 1416 (1972).

2 See, e.g., T. E. J. Campbell, "The Political Meaning of Stockholm: Third World Participation in the Environment Conference Process," *Stanford Journal of International Studies* 8 (1973): 138–153; N. J. Faramelli, "Toying with the Environment and the Poor: A Report on the Stockholm Environmental Conferences," *Environmental Affairs* 2 (1972): 469–486; D. A. Kay and E. B. Skolnikoff, "International Institutions and the Environmental Crisis: A Look Ahead," *International Organization* 26 (1972): 469–478; The Editors, "The Stockholm Conference: A Synopsis and Analysis," *Stanford Journal of International Studies* 8 (1973): 31.

3 Textbooks tend to describe the history of international environmental law as unfolding in various phases (usually divided into "pre-Stockholm," "Stockholm to Rio," and "Rio to present"), with the Stockholm Conference marking the emergence of modern international environmental law. See, e.g., U. Beyerlin, *Umweltvölkerrecht*, Part 1 (Munich: Beck, 2000); D. Hunter, J. Salzman and D. Zaelke, *International Environmental Law and Policy*, Ch. 6 (New York: Foundation Press, 1998); P. Sands, *Principles of International Environmental Law: Frameworks, Standards and Implementation*, Ch. 2 (Manchester: Manchester University Press, 1995) (suggesting four distinct phases).

Conference and its output were, indeed, shaped by the dramatically different views of developed and developing countries on matters of environment and development. By all accounts, the initial impetus for the Stockholm Conference came from growing concerns of industrialized countries over transboundary pollution and hazardous wastes.[4] The conference was to raise awareness of the international nature of environmental concerns and to promote a sense of collective responsibility of States. By contrast, developing countries tended to see pollution as a by-product of Northern industrialization and, to the extent that it was of concern in the South, as a consequence of poverty and underdevelopment. Thus, to engage developing countries in the Stockholm Conference, it had to acknowledge the environment-development interface as at the core of a collective response to environmental degradation.[5] The historical significance of the Stockholm Declaration lies in the fact that it was the first major international document that cast environmental concerns as global concerns and that highlighted development issues as integral to these concerns.[6]

The Stockholm Declaration, therefore, laid the foundation for a shift in international environmental law from its predominant focus on transboundary pollution to a conceptual framework with a much broader outlook. I deliberately say, "laid the foundation," because, in strictly legal terms, the focus of the declaration remained on transboundary pollution. As Alexandre Kiss observes, the legal core of the Stockholm Declaration is Principle 21, the rule balancing the sovereign rights of States to use their territories and their obligations to avoid transboundary harm.[7] While

---

[4] See, e.g., D. Wirth, "The Rio Declaration on Environment and Development: Two Steps Forward and One Back, or Vice Versa?" *Georgia Law Review* 29 (1995): 599–653, at p. 604 (note 18); T. Brenton, *The Greening of Machiavelli: The Evolution of International Environmental Politics* (London: Royal Institute of International Affairs, 1994), p. 39.

[5] For detailed discussions of the South-North dynamics surrounding the Stockholm Conference, see K. Mickelson, "South, North, International Environmental Law, and International Environmental Lawyers," *Yearbook of International Environmental Law* 11 (2000): 52–81; R. K. L. Panjabi, "From Stockholm to Rio: A Comparison of the Declaratory Principles of International Environmental Law," *Denver Journal of International Environmental Law & Policy* 21 (1993): 215, 236–245.

[6] See Mickelson, *ibid.* (arguing, however, that international environmental law still has not truly integrated environmental and developmental issues).

[7] A. Kiss, "The Destiny of the Principles of the Stockholm Declaration," in M. H. Nordquist, J. N. Moore and S. Mahmoudi, eds, *The Stockholm Declaration and the Law of the Marine Environment* (The Hague: Martinus Nijhoff Publishers, 2003), pp. 53–66.

the majority of the declaration's provisions actually focus on global, intertemporal, and developmental concerns that reach far beyond the scope of Principle 21, they are cast as policy statements rather than in legal language. Strictly speaking, then, the Stockholm Declaration did not expand the conceptual framework of international environmental law. Yet, it did foreshadow, to a remarkable degree, the framework within which international environmental lawyers operate today.

My goal in this contribution is to reflect on the impact of the Stockholm Declaration in light of the evolution of the structure and processes of international environmental law since 1972.[8] I will first illustrate how the "structure" of international environmental law has changed from a transboundary, bilateral framework to a global framework, which seeks to take into account considerations of both intra- and intergenerational equity. Second, I will suggest that this change in structure has been accompanied by a broadening of the "processes" of international environmental law to encompass a range of law-making methods. We are no longer looking only to customary or treaty norms to shape international environmental relations.

The Changing Structure of International Environmental Law

In 1972, the structure of international environmental law was defined by principles derived from the balancing and mutual limitation of sovereign rights. Much as domestic environmental law tended to evolve initially from concepts such as nuisance, abuse of rights, and other principles governing neighbourhood relationships, international environmental law originated in the idea that one State's right to use its territory had to find its limit in another State's right to be free from unreasonable interference.[9] Principle 21 of the Stockholm Declaration reflects this mutual limitation paradigm.

---

[8] This discussion will not focus on principles pertaining to human rights to environment or to citizen participation. These issues are addressed in the contribution by Alexandre Kiss, *ibid.*

[9] For a detailed discussion of the conceptual origins and evolution of international environmental law, see J. Brunnée, "Toward Effective International Environmental Law – Trends and Developments," in S. A. Kennett, ed. *Law and Process in Environmental Management* (1993), pp. 217–236.

Principle 21, of course, remains one of the core principles of international environmental law and it continues to furnish a useful ground-rule for international environmental relations.[10] However, it is also evident that, conceptually, the principle falls well short of what is needed to address international environmental concerns. Suffice it to point to three factors. First, Principle 21 balances competing contemporary interests and thus cannot capture the potential intertemporal impacts of States' activities. Second, given its focus on transboundary pollution, Principle 21 is largely "bilateral" in outlook. To be sure, Principle 21 alludes to a broader context by calling for the prevention of harm to "areas beyond the limits of national jurisdiction." However, it remains unclear who would be entitled to demand compliance on behalf of these areas. In other words, Principle 21 hints at the existence of an obligation owed *erga omnes*, but remains hamstrung by the unresolved question of the extent to which States other than the "injured State" are entitled to invoke State responsibility.[11] A third factor rests in the dilemmas and practical problems of applying Principle 21 in the face of serious capacity problems on the part of developing countries. These three factors suggest only some of the barriers to applying Principle 21 to global concerns such as ozone depletion, climate change, or loss of biodiversity.

Since 1972, the structure of international environmental law has been growing and shifting in significant ways. Principle 21 is now flanked by a conceptual framework that captures more fully the global, intertemporal, and equity dimensions of the international environmental challenge. In the context of this paper, it is not possible to provide a comprehensive account of the relevant conceptual developments. I will, therefore, focus

[10] Principle 21 has since been captured in a variety of other instruments, including in the *Rio Declaration on Environment and Development*, reprinted in 31 *I.L.M.* 876 (1992). For the purposes of this contribution, "Principle 21" will be used as shorthand for the rule that States have the sovereign right to exploit their resources, and the responsibility to ensure that activities within their jurisdiction do not cause significant transboundary harm.

[11] For a discussion of this issue in the context of the International Law Commission's efforts to complete the Articles on State Responsibility, see J. Crawford, J. Peel and S. Olleson, "The ILC's Articles on Responsibility of States for Internationally Wrongful Acts: Completion of the Second Reading," *European Journal of International Law* 12 (2001): 963–991, at pp. 975–976.

on five concepts that best reflect the shifts that I have traced out: common concern of humankind, common but differentiated responsibilities, sustainable development, intergenerational equity, and the precautionary principle.

Common Concern of Humankind

Unlike Principle 21, the concept of common concern does not spell out a specific rule for the conduct of States. Nonetheless, it signals that States' freedom of action may be subject to limits even where other States' sovereign rights are not affected in the transboundary sense envisaged by Principle 21. Such limits flow from the fact that the concept identifies certain issues as of concern to all members of the international community. It entitles, and perhaps even requires, all States to participate in international efforts to address the concern.[12] Examples of issues that were explicitly identified as common concerns over the last ten years or so include climate change and its adverse effects[13] and the conservation of biological diversity.[14] As these examples illustrate, the concept of common concern has the potential to mandate international cooperation to counter the degradation of "areas beyond the limits of national jurisdiction," of global resources such as the atmosphere, and even of resources physically located within the territory of individual States. While Principle 21 might suggest duties to protect areas beyond jurisdiction, it does not apply to global resources and does not restrict a State's use of its own resources except vis-à-vis other States that suffer specific transboundary injury. In sum, the concept of common concern responds to the "globalization" of environmental concerns and thus signals several fundamental shifts in the structure of international environmental law since the adoption of the Stockholm Declaration.

---

[12] For a detailed discussion of the concept of common concern, see P. Birnie and A. E. Boyle, *International Law and the Environment*, 2nd edition (Oxford: Oxford University Press, 2002), pp. 97–100; see also, J. Brunnée, "A Conceptual Framework for an International Forests Convention: Customary Law and Emerging Principles," in Canadian Council on International Law, ed., *Global Forests and International Environmental Law* (London: Graham & Trotman, 1996), 41–77, at pp. 55–62.

[13] *United Nations Framework Convention on Climate Change*, preamble, reprinted in 31 *I.L.M.* 849 (1992) [hereinafter UNFCCC].

[14] *Convention on Biological Diversity*, preamble, reprinted in 31 *I.L.M.* 822 (1992).

Common but Differentiated Responsibilities

The concept of common but differentiated responsibilities is closely connected to that of common concern and expresses the linkages between global environmental protection and the demands of equity between South and North. These linkages are encapsulated in Principle 7 of the Rio Declaration. The principle refers to the idea of common concern in a somewhat roundabout manner, calling on States to "...cooperate in a spirit of global partnership to conserve, protect and restore the health and integrity of the Earth's ecosystems." It then continues as follows:

> In view of the different contributions to global environmental degradation, States have common but differentiated responsibilities. The developed countries acknowledge the responsibility that they bear in the international pursuit of sustainable development in view of the pressures their societies place on the global environment and of the technologies and financial resources they command.[15]

The relationship of common concerns and differentiated responsibilities is also expressed in the preamble to the Climate Change Convention, which acknowledges

> ...that the global nature of climate change calls for the widest possible cooperation by all countries and their participation in an effective and appropriate international response, in accordance with their common but differentiated responsibilities and respective capabilities and their social and economic conditions.[16]

Whereas the Climate Convention explicitly mentions common but differentiated responsibilities,[17] other global agreements, such as the Montreal Protocol on Substances That Deplete the Ozone Layer[18] and the Biodiversity Convention,[19] reflect the underlying idea through differentiation in

[15] Rio Declaration, *supra* note 10.

[16] See UNFCCC, *supra* note 13, sixth preambular recital.

[17] See *ibid.*, and Articles 3.1, 4.1.

[18] *Montreal Protocol on Substances that Deplete the Ozone Layer*, 26 *I.L.M.* 1550 (1987); as adjusted and amended June 29, 1990, 30 *I.L.M.* 539 (1990); further adjusted and amended November 25, 1992, 32 *I.L.M.* 875 (1992); further adjusted and amended September 17, 1997; further adjusted December 3, 1999; consolidated text available at <www.unep.org/ozone/montreal.shtml>, 13 May 2002).

[19] *Supra* note 14.

the obligations of developed and developing countries.[20] Taken together, existing instruments and practice suggest that States are paying attention to considerations of equity in international environmental relations and recognize that approaches to global concerns must reflect both the States' contributions to a given problem, and their abilities to address it. The practical manifestations of the concept range from differentiation of commitments in terms of timelines or degree[21] to provision of technical and financial assistance to developing countries,[22] and even to contingency of developing country compliance upon provision of such assistance.[23] Common but differentiated responsibilities thus represent another fundamental set of shifts in the conceptual structure of international environmental law. At least some of its norms, it would seem, no longer necessarily have equal application to all States. Further, international environmental law now seems to focus not only on current pollution but seems to take past pollution into account in determining the extent to which polluters must "pay."

Sustainable Development and Intergenerational Equity

Like the notion of common but differentiated responsibilities, the concept of sustainable development is focused upon the interface between environmental and developmental concerns. In addition, sustainable development places the intertemporal dimensions of environmental degradation and protection, respectively, at the centre of international

20 For a detailed discussion, see Birnie and Boyle, *supra* note 12, at pp. 100–104.

21 See, e.g., Article 5 of the Montreal Protocol, *supra* note 18, which provides different baseline years and different timelines for developing country commitments. See also, Article 4.2 of the UNFCCC, *supra* note 12, which envisages emission limitation commitments only for developed countries and countries with economies in transition.

22 See, e.g., Article 10 of the Montreal Protocol, *supra* note 18, which establishes a financial mechanism for the purposes of financial and technical assistance to developing countries to enable their compliance with protocol commitments.

23 See, e.g., Article 5.5 of the Montreal Protocol, *supra* note 18, which provides that the implementation by developing country parties of their commitments "will depend upon the effective implementation of" financial cooperation and technology transfer. Similarly, Article 7 of the UNFCCC, *supra* note 12, stipulates that the "extent to which developing country Parties will effectively implement their commitments... will depend upon the effective implementation by developed country Parties of their commitments... related to financial resources and transfer of technology."

environmental law and policy. In this regard, the concept is closely linked to the idea of intergenerational equity,[24] a linkage that finds expression in the Brundtland Commission's definition of sustainable development as "development that meets the needs of the present without compromising the ability of future generations to meet their own needs."[25] It is also expressed in the somewhat elliptical Principle 3 of the 1992 Rio Declaration, which states that "[t]he right to development must be fulfilled so as to equitably meet developmental and environmental needs of present and future generations."[26] The Rio Declaration was "animated" by the concept of sustainable development[27] and uses the term in twelve of its twenty-seven principles.[28] Yet, the precise content and contours of this basic prescription remain notoriously elusive. Rather than attempt to offer a definition of the concept, the declaration opted to outline various elements of sustainable development. Birnie and Boyle suggest that the declaration sets out the concept's substantive elements (such as sustainable utilization of natural resources, the integration of environment and development, a right to development, intra- and intergenerational equity, and the polluter pays principle), as well as its procedural elements (such as public participation in environmental decision-making and environmental impact assessment).[29]

In essence, then, the concept of sustainable development acknowledges the need for development but cautions that development must remain within the carrying capacity of the environment and, therefore, that environmental protection must be part of the development process. As a result, sustainable development imposes several potential limitations upon domestic development that reinforce the already noted expansion and shifts in the conceptual structure of international environmental law. Like the concept of common concern, it makes certain aspects of domestic resource

---

24 See E. Brown Weiss, "The Emerging Structure of International Environmental Law," in N. J. Vig and R. S. Axelrod, eds, *The Global Environment: Institutions, Law and Policy* (Washington, DC: Congressional Quarterly, 1999), 98–115, at pp. 106–107.

25 World Commission on Environment and Development, *Our Common Future* (Oxford: Oxford University Press, 1987), p. 43.

26 Rio Declaration, *supra* note 10. On the provision's wording, see Wirth, *supra* note 4, at pp. 627–628.

27 Wirth, *supra* note 4, at p. 625.

28 Rio Declaration, *supra* note 9, Principles 1, 4, 5–9, 12, 20–22, 24, and 27.

29 See Birnie and Boyle, *supra* note 12, at pp. 86–95.

management a matter of international concern.[30] Further, appropriate resource use and development is not measured solely against the extent of transboundary impact but is defined according to what the environment can sustain in the longer term. Finally, in normative terms, the pragmatic intertemporal outlook of sustainable development is reinforced by the responsibilities flowing from the requirements of intergenerational equity. In short, like the concept of common but differentiated responsibilities, sustainable development is concerned with equitable balancing of contemporary environmental and developmental interests. Both concepts also have an intertemporal dimension. However, while the former concept looks, *inter alia*, to past activities to determine current responsibilities (and, in that sense, also aims at a form of intergenerational equity), sustainable development calls upon us to factor in future needs.[31]

The Precautionary Principle

The precautionary principle complements the intertemporal concepts of sustainable development and intergenerational equity. Indeed, it is difficult to see how either of the latter goals could be achieved without the approach to "scientific uncertainty" that the precautionary principle promotes. Principle 15 of the Rio Declaration posits that

> [i]n order to protect the environment, the precautionary approach shall be widely applied by States according to their capabilities. Where there are threats of serious or irreversible damage, lack of full scientific certainty shall not be used as a reason for postponing cost-effective measures to prevent environmental degradation.[32]

Principle 15 is riddled with qualifiers and there is ongoing debate as to how great the risk of future harm and how serious the potential damage must be to trigger a precautionary obligation. Although Principle 15 of the Rio Declaration has become the most commonly cited version of the precautionary principle, some States maintain that there is no single authoritative definition and that as a result, international environmental

---

30 *Ibid.*, at p. 85.

31 For further discussion of the concept of sustainable development and its linkages to the notions of intergenerational equity and common but differentiated responsibilities, see Brunnée, *supra* note 12, at pp. 62–69.

32 Rio Declaration, *supra* note 10.

law may promote a precautionary approach but does not contain an overarching precautionary principle.[33] Be that as it may, the idea of precaution permeates the practice of States in a broad range of issue areas and at virtually all levels. It has been employed in the context of the conservation of marine resources, trade and environment matters, and an array of regional and global environmental concerns. Further, it is reflected in decisions of international tribunals and various international instruments, as well as in judicial decisions and environmental law and policy at the domestic level.[34] Thus, the precautionary principle also contributes to the shifting structure of international environmental law. The principle is crucial to the implementation of the emerging intertemporal framework. It also serves to expand the reach of the transboundary paradigm expressed in Principle 21.[35] To the extent that principle encompasses a duty to prevent transboundary harm,[36] the trigger for that duty has been a matter of debate. Following the *Trail Smelter* decision's requirement of "clear and convincing evidence,"[37] States have maintained that preventive measures must be taken only upon proof of (pending) harm. In this context, the existence of a precautionary principle, rather than general support for a precautionary approach, could have significant implications for the rights and obligations of States.

The Contributions of the Stockholm Declaration

As I suggested earlier, the extent to which the abovementioned concepts are foreshadowed in the Stockholm Declaration is remarkable and it is worth highlighting the relevant portions of the declaration. The thrust of the concept of common concern, for example, is neatly captured in one of the "proclamations" that precede the Stockholm Declaration. It reads in part:

---

[33] See, e.g., Government of Canada, *A Canadian Perspective on the Precautionary Approach/ Principle – Proposed Guidelines* (September 2001), at 5, <www.dfait-maeci.gc.ca/tna-nac/ prec-booklet-e.pdf> ("Rules of customary international law are developed by the common consent of States. Due to an absence of clear evidence of uniform State practice and *opinio juris*, Canada does not yet consider the precautionary principle to be a rule of customary international law").

[34] See, generally, Birnie and Boyle, *supra* note 12, at pp. 115–121.

[35] For further discussion, *see* Brunnée, *supra* note 12, at pp. 71–74.

[36] See *Corfu Channel* Case *(United Kingdom v. Albania)*, [1949] I.C.J. Rep. 4.

[37] *Trail Smelter* Case *(United States v. Canada)*, (1941) 3 R.I.A.A. 1905.

> A growing class of environmental problems, because they are regional or global in extent or because they affect the common international realm, will require extensive co-operation among nations and action by international organizations in the common interest...[38]

The ideas underlying the concept of common but differentiated responsibility, in turn, are reflected in several provisions of the Stockholm Declaration. Suffice it for present purposes to cite Principle 12, which suggests that

> [r]esources should be made available to preserve and improve the environment, taking into account the circumstances and particular requirements of developing countries....

and Principle 23, which stresses the need to consider

> ...the extent of the applicability of standards which are valid for the most advanced countries but which may be inappropriate and of unwarranted social cost for the developing countries.[39]

Although the Stockholm Declaration does not name the concept as such, it also captures various facets of the concept of sustainable development. The declaration speaks to the idea that natural resources must be used sustainably and to the need for integration of environmental protection and development. With respect to the first aspect, Principle 3 provides that "[t]he capacity of the earth to produce vital renewable resources must be maintained and, wherever practicable, restored or improved," whereas Principle 5 notes that "[t]he non-renewable resources of the earth must be employed in such a way as to guard against the danger of their future exhaustion."[40] With regard to the second aspect, Principle 13 stipulates:

> In order to achieve a more rational management of resources and thus to improve the environment, States should adopt an integrated and coordinated approach to their development planning so as to ensure that development is compatible with the need to protect and improve the environment...

---

[38] Stockholm Declaration, *supra* note 1, proclamation 7.

[39] *Ibid.* See also proclamation 4 (noting that "in the developing countries most of the environmental problems are caused by under-development"); Principle 9 (referring to the need for "the transfer of substantial quantities of financial and technological assistance"); and Principle 11 (calling for appropriate steps "with a view to reaching agreement on meeting the possible national and international economic consequences resulting from the application of environmental measures" in developing countries).

[40] Stockholm Declaration, *supra* note 1.

Principle 14 adds:

> Rational planning constitutes an essential tool for reconciling any conflict between the needs of development and the need to protect and improve the environment.[41]

The concept of intergenerational equity also finds prominent expression in the Stockholm Declaration. According to proclamation 6, "[t]o defend and improve the human environment for present and future generations has become an imperative goal for mankind." Further, Principle 1 underlines the "solemn responsibility to protect and improve the environment for present and future generations," while Principle 2 stresses that "[t]he natural resources of the earth ... must be safeguarded for the benefit of present and future generations."[42]

Finally, the ideas underlying the precautionary principle are reflected in the Stockholm Declaration as well. Proclamation 6 observes in remarkably strong language that

> ... we must shape our actions throughout the world with a more prudent care for their environmental consequences. Through ignorance or indifference we can do massive and irreversible harm to the earthly environment on which our life and well-being depend.

In the same vein, Principle 6, in the context of threats posed by toxic substances and other pollutants, refers to the need to end their release in

> ... quantities or concentrations [that] exceed the capacity of the environment to render them harmless ... in order to ensure that serious or irreversible damage is not inflicted upon ecosystems.[43]

In a number of programmatic or policy statements, then, the Stockholm Declaration captured the changing scope of the international environmental challenge and the attendant shifts in the parameters for policy responses. The declaration could not have cast these statements in legal terms at the time, but it is of note that it correctly identified the main contours of the developments in international environmental law and policy that have occurred since 1972. As I hope to have illustrated, there has been conceptual "accretion" around all of the issue markers contained

41 *Ibid.*
42 *Ibid.*
43 *Ibid.*

in the Stockholm Declaration that I highlighted above. The legal nature and effects of this accretion remain to be explored.

The Processes of International Environmental Law

Up to this point, I have focused only on the shifts in the conceptual structure of international environmental law and I have left aside the thorny questions surrounding the legal status of the concepts that I canvassed. I will offer a brief survey of the state of debate on the legal status of each concept. I then want to reflect on how international environmental law operates and how its norms influence the conduct of States and shape international environmental relations. I will suggest that the legal status of structural concepts such as the ones that I discussed may not be the primary indicator of their impact.

Customary Law: Five New Concepts – Thirty Years after Stockholm

An assessment of the progress of customary law since the Stockholm Conference has the potential to be a sobering experience. In his contribution, Professor Kiss concludes that two of the Stockholm principles, Principle 1 ("right to environment") and Principle 21, have developed into customary law. Professor Kiss' conclusion with respect to Principle 21 finds confirmation, *inter alia*, in the International Court of Justice's (ICJ) *Advisory Opinion on the Legality of the Threat or Use of Nuclear Weapons*. Given the extensive reliance on the rule in Principle 21 over the last thirty years, the conclusion of the Court that it constituted a "general obligation" of States should hardly be surprising.[44] What is perhaps more noteworthy is the fact that the ICJ, both in the *Advisory Opinion* and in the subsequent *Gabcikovo-Nagymaros* decision,[45] did not attribute customary law status to any of the other concepts discussed in this essay, although it does allude to them. For example, in the *Advisory Opinion*, the Court raises the intertemporal and intergenerational dimensions of international environmental law when it notes that "the environment

[44] ICJ, *Advisory Opinion on the Legality of the Threat or Use of Nuclear Weapons*, reprinted in 35 *I.L.M.* 809 (1996), at para. 29 [hereinafter *Advisory Opinion*].

[45] ICJ, *Case Concerning the Gabcikovo-Nagymaros Project*, reprinted in 37 *I.L.M.* 162 (1998) [hereinafter *Gabcikovo*].

is not an abstraction but represents the living space, the quality of life and the very health of human beings, including generations unborn,"[46] and refers to the "serious danger to future generations" posed by nuclear weapons.[47] In *Gabcikovo*, the ICJ provides the most oblique of references to the *erga omnes* quality of international environmental law when it prefaces a reference to its *Advisory Opinion* statement on the general obligation regarding transboundary harm by stressing "the significance it attaches to respect for the environment, not only for States but also for the whole of mankind."[48] Also in *Gabcikovo*, the Court gives a nod to the idea of precaution but does not endorse the precautionary principle when it declares itself to be "mindful that, in the field of environmental protection, vigilance and prevention are required on account of the often irreversible character of damage to the environment and of the limitations inherent in the very mechanism of reparation of this type of damage."[49] Finally, with respect to sustainable development and the evolution of international environmental law in general, the ICJ notes that

> ... new norms and standards have been developed, set forth in a great number of instruments during the last two decades. Such new norms have to be taken into consideration, and such new standards given proper weight, not only when States contemplate new activities but also when continuing with activities begun in the past. This need to reconcile economic development with the protection of the environment is aptly expressed in the concept of sustainable development.[50]

Thus, some thirty years after Stockholm, the ICJ found itself in a position to confirm the customary nature of Principle 21 and to acknowledge the emergence of "new norms," but not to identify any of the principles discussed in this essay as customary law. As disappointing as this state of affairs may seem at first glance, the ICJ is hardly to blame. Measured against the prevailing criteria for the existence of customary law,[51] it is,

---

46 *Advisory Opinion, supra* note 44.

47 *Ibid.*, at para. 35.

48 *Gabcikovo, supra* note 45, at para. 53.

49 *Ibid.*, at para. 140.

50 *Ibid.*

51 For a recent discussion, see A. E. Roberts, "Traditional and Modern Approaches to Customary International Law: A Reconciliation," *American Journal of International Law* 95(4) (2001): 757–791.

indeed, impossible or at least difficult to conclude that the five concepts highlighted here are customary law.

With respect to the concept of common concern, a first difficulty rests in the fact that its normative content is hard to pinpoint. As discussed, the concept is in the first instance a participation rule, giving States standing regarding the concern, perhaps even requiring their cooperation. In the second instance, the concept has the potential to limit States' rights to use their resources beyond the constraints imposed by Principle 21 if that use creates or aggravates a common concern. Important legal implications of the common concern concept might thus flow from the identification of cases – the common concerns – in which rights and obligations exist *erga omnes*.[52] In this respect, the concept of common concern could connect international environmental law to the law of State responsibility, and the possibility of invocation of responsibility for breaches of obligations owed to a group of States, or to the international community as a whole.[53] However, while it may be possible to posit potential legal consequences of the existence of a common concern, it remains unclear according to precisely what criteria such concerns are to be identified.[54]

The inquiry into the legal status of the concept of common but differentiated responsibilities leads to similarly inconclusive results. As noted earlier, it is possible to identify a number of practical manifestations of the concept and to extrapolate its potential legal implications from these. Overall, the concept is less a rule of conduct than a principle that affects the structuring or consequences of other obligations, notably in the treaty context.[55] Although the concept is widely employed, it is not clear

---

52 See also Birnie and Boyle, *supra* note 12, at pp. 98–100.

53 See International Law Commission, *State responsibility – Titles and texts of the draft articles on Responsibility of States for internationally wrongful acts adopted by the Drafting Committee on second reading*, U.N. Doc. A/CN.4/L.602 and Corr.1 and L.602/Rev.1, 26 July 2001, Article 48. Available at <www.un.org/law/ilc/sessions/53/53docs.htm>, 14 April 2002; J. Peel, "New State Responsibility Rules and Compliance with Multilateral Environmental Obligations: Some Case Studies of How the New Rules Might Apply in the International Environmental Context," *Review of European Community and International Environmental Law* 10 (2001): 82–97.

54 It has been suggested that the concept of common concern has no "normative effect" whatsoever. See Beyerlin, *supra* note 3, at p. 60.

55 See, generally, D. French, "Developing States and International Environmental Law: The Importance of Differentiated Obligations," *International & Comparative Law Quarterly* 49(1) (2000): 35–60.

that it has acquired customary law status.[56] Indeed, recently, some States have shown reluctance to include the principle as such in environmental agreements. It has been argued, for example, that the principle is "not necessary" because its goals can be achieved simply through appropriate structuring of an agreement, and "not helpful" because its meaning and legal implications are unclear.[57] Of course, this effort to resist the principle may be a rear-guard action that actually suggests less the principle's weakness than its potential strength. After all, the principle may point to developed country obligations to assume the main burdens of global environmental protection.

The perhaps most lively debate has taken place with respect to the legal status of the concept of sustainable development. Assessments range from the conclusion that the concept is not even normative in nature,[58] to the view that it is an umbrella concept that is fleshed out by other principles,[59] all the way to the assertion that it is a normative concept with *erga omnes* implications and part of "modern international law."[60] A nuanced examination of the legal impact of sustainable development was recently offered by Vaughn Lowe. He suggested that the concept has normative force as an element of legal reasoning. He aptly described sustainable development as

> ... a legal concept exercising a kind of interstitial normativity, pushing and pulling the boundaries of true primary norms when they threaten to overlap or conflict with each other.[61]

---

[56] Although some suggest that it is well on the way to becoming customary law. See Beyerlin, *supra*, note 3, at p. 60.

[57] Remarks of Susan Biniaz, Assistant Legal Adviser, U.S. Department of State, 96th Annual Meeting of the American Society of International Law, Washington, D.C., 16 March 2002 (notes on file with author).

[58] Beyerlin, *supra* note 3, at p. 59.

[59] Sands, *supra* note 3, at pp. 198–208. See also, Commission for Sustainable Development, *Report of the Expert Group Meeting on Identification of Principles of International Law for Sustainable Development*, U.N. Doc. E/CN.17/1996/17/Add.1 (1 March 1996).

[60] *Gabcikovo*, *supra* note 45, at pp. 213, 217 (Separate Opinion of Vice-President Weeramantry).

[61] V. Lowe, "Sustainable Development and Unsustainable Arguments," in A. Boyle and D. Freestone, eds, *International Law and Sustainable Development: Past Achievements and Future Challenges* (Oxford: Oxford University Press, 1999): 19–38, at p. 31.

Although the concept of intergenerational equity is more explicitly normative in outlook, its legal status is also unclear at this stage. It has been argued that the obligation of one generation to pass on to the next generation the natural and cultural resources of the earth in no worse conditions than they were received requires the "conservation of options," "the conservation of quality," and the "conservation of access."[62] However, it is not clear that the concept has evolved in a manner that spells out these requirements. As suggested earlier, while the concept finds expression in most contemporary international environmental instruments, it appears to have become melded into the concept of sustainable development. In terms of State practice and *opinio juris*, the distinctive legal content of intergenerational equity may, therefore, have become obscured rather than clarified.[63]

Finally, although the precautionary principle may be the most likely candidate for customary law status, doubts linger in this case as well. To be sure, an increasing number of observers conclude that the principle is binding custom,[64] including a recent and very carefully executed study of international practice and academic literature.[65] Yet, there are also sceptical voices and,[66] as previously mentioned, a number of States continue to assert that there exists no legally binding precautionary principle. Rather, they argue, international environmental law merely reflects various manifestations of a precautionary approach. I will not enter into the fray of this debate, except to say that, given the manifest impact of the precautionary principle, the debate has assumed a slightly surreal flavour.[67]

---

[62] E. Brown Weiss, *In Fairness to Future Generations: International Law, Common Patrimony and Intergenerational Equity* (1989), pp. 37–45.

[63] For skeptical assessments, see Birnie and Boyle, *supra* note 12, at pp. 90–91; Beyerlin, *supra* note 3, at p. 60.

[64] See, e.g., J. Cameron, "The Precautionary Principle in International Law," in T. O'Riordan, J. Cameron and A. Jordan, eds, *Reinterpreting the Precautionary Principle*, 113 (London: Cameron May, 2001): 113–142.

[65] A. Trouwborst, *Evolution and Status of the Precautionary Principle in International Law* (The Hague: Kluwer Law, 2002), at p. 284.

[66] See, e.g., C. D. Stone, "Is There a Precautionary Principle?" *Environmental Law Reporter* 31 (2001): 10790–10799.

[67] See also, D. Freestone, "Caution or Precaution: 'A Rose By Any Other Name...'?" *Yearbook of International Environmental Law* 10 (1999): 25–32, at p. 26. (noting that the debates about the legal status of the precautionary principle "have a distinctly 1990s feel about them"); and Birnie and Boyle, *supra* note 12, at p. 120 (concluding

Indeed, the story of the precautionary principle is one illustration of the argument that I now wish to develop: that the influence of the principles that I have discussed may not, in fact, depend on whether or not they are formally legally binding.

Beyond Custom

So far, this essay may appear to suffer from a disjuncture. Having spent considerable time on outlining elements of the new structure of international environmental law, I proceeded to conclude that none of the structural concepts I discussed have clearly established themselves as customary law – with the possible exception of the precautionary principle. One might ask whether the structural shifts I have sketched, then, are more apparent than real. If there has been little change in the rules of customary international environmental law, can its structure really have shifted? While we may have understood – and have understood since the Stockholm Conference – that certain issues required our attention and efforts, have we developed international environmental law so as to enable it to meet the attendant challenges, to remain relevant?

If we were to make formal bindingness the measure of legal success, we would have to answer these questions largely in the negative. Indeed, some have mourned the end of a "golden era of international law,"[68] and international environmental law's "regression" to growing reliance on

> ...supposedly agreed-upon "rules" of so-called soft law, which are so deprived of any mandatory or imperative language and of any reciprocal rights and obligations that they could hardly be recognized as rules of international law at all.[69]

In my view, this type of assessment does not do justice to the diverse processes through which international environmental law actually operates. And to the extent that there has been a shift away from formally

---

that "the propositions that it is, or that it is not, customary international law are too simplistic").

[68] A. Székely, "Compliance with Regulatory Treaties: The Empirical Evidence. A Commentary on the Softening of International Environmental Law," in American Society of International Law, *Proceedings of the 91st Annual Meeting: Implementation, Compliance and Effectiveness* (Washington, DC: American Society of International Law, 1997), at p. 236.

[69] *Ibid.*, at p. 235.

binding processes and outcomes, this shift may, in fact, not weaken but expand international environmental law's ability to influence. To be sure, I do not mean to deny that formally binding rules may be important in certain contexts or that enforceability may have potential benefits in certain settings. My point is, rather, that international environmental norms are rarely used in a litigation or enforcement-oriented context. In light of this fact, Daniel Bodansky, in an essay entitled "Customary (and not so Customary) International Environmental Law," questioned the preoccupation with whether particular rules are legally binding or not.[70] Whatever one's views in this regard, for the most part, international environmental law actually influences through horizontal rather than hierarchical processes. It can exert "compliance pull" and influence States in the absence of enforcement mechanisms, notably where norms are seen as legitimate.[71] Or it shapes international discourse, such as negotiation or diplomacy, helping to justify positions and to persuade other actors.[72] In my view, international environmental law furnishes an array of examples that suggest that its ability to operate in this fashion does not necessarily depend on its formal bindingness.[73]

In the treaty context, we have witnessed how broad frameworks serve to shape the context within which detailed rules are generated and interpreted.[74] Indeed, the adoption of a multilateral environmental agreement typically is not the endpoint of the international legal process but the beginning. What is more, many of the rules, principles, and guidelines generated within multilateral environmental agreements are adopted through decisions of Conferences of the Parties or similar bodies and are thus not formally binding. Yet, the relevant standards appear to be treated

---

70 D. Bodansky, "Customary (and not so Customary) International Environmental Law," 3 *Indiana Journal of Global Legal Studies* 3 (1995): 105–120.

71 See T. M. Franck, "Legitimacy in the International System," *American Journal of International Law* 82 (1988): 705–759.

72 See J. Brunnée and S. J. Toope, "International Law and Constructivism: Elements of an Interactional Theory of International Law," Colorado Journal of Transnational Law 39 (2000): 19–74, at pp. 64–74.

73 I will sidestep the more complex question of whether binding legal status gives norms a distinctive ability to influence States.

74 For an overview, see Birnie and Boyle, *supra* note 12, at 209–215.

as a sort of quasi-law, and it is not clear that they are any less influential than "hard" law would be.[75]

In the context of the principles of international environmental law, there is ample evidence of their ability to exert influence even while their legal status remains contested. The precautionary principle may serve as an example. At the international level, the principle (or approach) has found expression in an array of environmental agreements, as well as in trade agreements.[76] As a result, the principle has come to influence the evolution, interpretation, and implementation of these agreements.[77] The concept of precaution has also been taken into consideration in international dispute settlement processes.[78] The principle's impact at the domestic level is perhaps best illustrated through the practice of one of the States that continues to deny its customary law status – Canada.[79] Notwithstanding this official position, the precautionary principle is shaping Canadian law and policy in decisive ways. For example, the preamble to the *Canadian Environmental Protection Act* expresses a commitment to "implementing *the* precautionary principle" (emphasis added) and makes reference to Canada's international obligations.[80] Similarly, the Supreme Court of Canada invoked the precautionary principle in the context of a decision regarding the jurisdiction of a municipality to regulate lawn chemicals. Relying on various scholarly authors, the Court noted that there might be "currently sufficient State practice to allow a good argument that the precautionary principle is a principle of customary international law."[81] The Court also confirmed that, in any case, the values reflected in international law "inform the contextual approach to statutory interpretation

[75] See J. Brunnée, "COPing with Consent: Lawmaking under Multilateral Environmental Agreements," *Leiden Journal of International Law* 15(2002): 1–52.

[76] See Birnie and Boyle, *supra* note 12, at 115–121.

[77] See, generally, A. E. Boyle, "Some Reflections on the Relationship of Treaties and Soft Law," *International and Comparative Law Quarterly* 48(4) (1999): 901–913.

[78] See WTO Appellate Body, *Measures Concerning Meat and Meat Products*, WTO Doc. WT/DS26/AB/R (16 January 1998); *Southern Bluefin Tuna Cases (Provisional Measures)*, (1999) ITLOS Nos. 3 & 4.

[79] Government of Canada, *supra* note 33.

[80] *Canadian Environmental Protection Act*, S.C. 1999, c. 33.

[81] *114957 Canada Ltée (Spraytech, Société d'arrosage)* v. *Hudson (Town)*, [2001] S.C.C. 40, at para. 32.

and judicial review."[82] Finally, at the level of government policy, suffice it to point to a recent discussion paper. While the paper reaffirms Canada's position on the legal status of the precautionary principle,[83] it notes that Canada supports the statement in Principle 15 of the Rio Declaration.[84] The very purpose of the paper is to outline guiding principles to "support consistent, credible and predictable policy and regulatory decision-making when applying the precautionary approach/ principle."[85]

The precautionary principle, due to its applicability to all levels of decision-making and planning, has shown itself capable of exerting influence both nationally and internationally. Other principles, like common concern or common but differentiated responsibilities, are by definition focused on international relations. These differences notwithstanding, my sense is that it would be possible to illustrate concretely how all of the principles that I have identified in this essay have reshaped the structure of international environmental law in the manner I suggested earlier, and do influence the conduct of States. Through diverse processes, they shape approaches to international treaty-making, guide and constrain legal arguments, and influence judicial decision-making. In all of these respects, their precise legal status appears to be of secondary importance. Whether they are formally binding or not, it seems unlikely that arguments or decisions in blatant disregard of any of the five principles would be persuasive and could be sustained in the long run.

Conclusion

The Stockholm Declaration did mark an important moment in the evolution of international environmental law. In my view, its significance should not be measured against the degree to which its principles have, or have not, become binding as customary international law. Rather, the declaration should be appreciated as the beginning of a normative process. Of necessity, such a process begins with the identification of issues

---

82 *Ibid.*, at para. 30. See also, *Baker v. Canada (Minister of Citizenship and Immigration)*, [1999] 2 S.C.R. 817.

83 See Government of Canada, *supra* note 33.

84 *Ibid.*, at p. 1.

85 *Ibid.*, Foreword.

of common concern. The perhaps most significant contribution of the Stockholm Declaration, then, was that it looked beyond the transboundary concerns of the day and highlighted the global nature of environmental degradation and, most importantly, the unavoidable interface between environment and development issues. In so doing, the declaration also provided a reference point, a marker of the issues that the structure of international environmental law was not equipped to address. Thirty years after the Stockholm Declaration, we may not have formally binding rules to fill all of the gaps identified at Stockholm. But we can say that shared normative understandings have begun to emerge that have had a significant impact on both the structure and processes of international environmental law.[86]

[86] See also, L. B. Sohn, "The Stockholm Declaration on the Human Environment," *Harvard Journal of International Law* 14 (1973): 423–515, at pp. 513–515 (predicting that the declaration, its non-binding and policy-oriented character notwithstanding, would have this type of impact).

Integration among Global Environmental Regimes: Lessons Learned from Climate Change Mitigation

Meinhard Doelle*

Introduction

The topic of integration was one of so many topics Douglas Johnston thought and wrote about long before the terms became common place in international law circles.[1] As far back as 1985 Douglas was contemplating the need for a systemic response to a range of global environmental challenges. He saw then what has in the interim become abundantly clear, that treating individual symptoms of our troubled relationship with nature is inadequate, and that the global community has to find ways to develop a more integrated approach that uses individual symptoms of the problem to collectively address the root causes.[2]

Few symptoms make the case for integration as effectively as climate change, an issue Douglas described in 1985 as one that "may be beyond human capability to deal effectively with."[3] This contribution will look back at the evolution of the international response to climate change from

* Associate Director, Marine & Environmental Law Institute; Director, Marine & Environmental Law Programme; and Professor of Law, Dalhousie University, Halifax, Canada.

[1] See, for example, D. Johnston, "Systemic Environmental Damage: The Challenge to International Law and Organizations," *Syracuse Journal of International Law and Commerce* 12 (1985–1986): 255.

[2] *Ibid.*, p. 257. The issue has more recently been the subject of deliberations at the International Law Commission. See "Conclusions of the work of the Study Group on the Fragmentation of International Law: Difficulties arising from the Diversification and Expansion of International Law," adopted by the International Law Commission at its Fifty-eighth session, in 2006, reprinted in *Yearbook of the International Law Commission, 2006*, Vol. II, Part Two.

[3] *Ibid.*, p. 258.

the early resolutions of the U.N. General Assembly to the implementation of the Kyoto Protocol to consider how successful the international community has been in developing an integrated approach to mitigating human interference with the climate system. Specifically, the linkages between climate change and other issues are considered to determine to what extent the climate change regime has evolved in isolation and to what extent it has been successful in integrating selected other issues into the chosen approach to climate change mitigation.[4]

Integration is used in many different ways in various contexts. Perhaps the most well known use of the concept of integration in the environmental context is that contained in Principle 4 of the Rio Declaration. It states that "in order to achieve sustainable development, environmental protection shall constitute an integral part of the development process and cannot be considered in isolation from it."[5] Integration is used in many other contexts. It is used for integration among levels of government, encouraging federal, provincial and local levels of government to work together. It is used similarly to describe inter-departmental cooperation and coordination. An interdisciplinary approach to problem solving is yet another form of integration that has received considerable attention in literature and practice alike. Regional integration has been the focus of much attention with respect to trade. In short, there is a long list of uses of the term integration that are considered in the literature. In this essay, integration is primarily used in one sense, integration of international efforts to address a range of global environmental challenges in an overall context of sustainability.[6]

---

[4] This paper builds on an assessment of integration conducted using Canada's Kyoto implementation plan as a case study. See M. Doelle, "Linking the Kyoto Protocol and other Multilateral Environmental Agreements; From Fragmentation to Integration?" *Journal of Environmental Law and Practice* 14 (2004): 75.

[5] United Nations General Assembly, *Report of the United Nations Conference on Environment and Development, 3–14 June 1992, Annex 1, Rio Declaration on Environment and Development*, A/CONF 151/5/Rev.1, 31 *I.L.M.* 874, <www.un.org/documents/ga/conf151/aconf15126-1annex1.htm> [hereinafter Rio Declaration].

[6] For a review of this and other forms of integration, see D. VanderZwaag et al., "Canadian Aquaculture and the Principles of Sustainable Development: Gauging the Law and Policy Tides and Charting a Course," *Queen's Law Journal* 28 (2002–2003): 279, at 286. See also, J. L. Dunoff, "Levels of Environmental Governance," in D. Bodansky, J. Brunnée and E. Hey, eds, *The Oxford Handbook of International Environmental Law* (Oxford: Oxford University Press, 2007), pp. 98–106.

The focus of this contribution is on the substantive outcomes of the climate change regime. The following section will briefly consider linkages between human activities responsible for greenhouse gas emissions, the atmosphere, the climate system, natural systems, and human systems. This will be followed by a review of the evolution of the climate change regime to determine to what extent it has identified and integrated appropriate linkages to other environmental issues. Using the assessment of the regime and missed opportunities for integration as a starting point, the essay then discusses how international integration of global environmental challenges may be improved.[7]

Integration in the Climate Change Context

The starting point proposed for understanding the relationship between the human influence on the climate system and environmental, social and economic sustainability is to identify human activities that result in the release of greenhouse gas emissions. There are two categories of human activities that contribute to greenhouse gas emissions, energy use and non-energy activities. On the energy side, the main culprit is the burning of fossil fuels. For non-energy sources, key activities include deforestation, forest management, various agricultural practices, and disposal of organic material in landfills. Key drivers of these activities are population and economic growth. Other factors include a range of issues dealing with the use of energy, such as the sources of energy available, technologies, transportation choices, energy efficiency and conservation, and land use choices.

These are some of the key factors that influence the release of greenhouse gases from human activities. These activities each contribute significantly

[7] For a detailed discussion of the problem of fragmentation generally in international law, see International Law Commission, "Fragmentation of International Law: Difficulties Arising from the Diversification and Expansion of International Law," A/CN.4/L.682, 13 April 2006. The focus of the study is on legal conflicts among the multitude of sources of international law. Normative conflicts are also considered, but are not the primary focus of the study. See also, A. Boyle, "Relationship between International Environmental Law and Other Branches of International Law," in Bodansky, et al., *ibid.*, pp. 128–136, 138–140, and T. Gehring, "Treaty Making and Treaty Evolution," in Bodansky, et al., *ibid.*, pp. 475–495.

to the concentration of greenhouse gases in the atmosphere, but they also have a range of other impacts on natural and human systems. A full discussion of these links is not possible here, but an exploration in the context of the use of fossil fuels as a source of energy illustrates the process.

Consider the impact of fossil fuels on natural systems. The exact nature and significance of the impacts will, of course, vary with the fuel, the manner of extraction, and the care with which it is refined, transported and used. Nevertheless, the extraction of fossil fuels clearly has some negative impacts on natural systems in terms of habitat loss, pollution and other impacts. Similarly, the refinement process has land-use implications and involves pollution in the form of air emissions and effluent. The end use again involves air emissions that contribute to air pollution and will in other ways impose stresses on natural systems.[8] In addition to these direct impacts on natural and human systems, the emission of greenhouse gases from the burning of fossil fuels affects the climate system through changes in temperature, precipitation, wind patterns, ocean currents and sea level. These changes to the climate system in turn affect natural and human systems through floods and droughts, and by affecting species health, ecosystems and habitats.[9]

An integrated approach to climate change mitigation would consider the range of impacts, the various interactions between human activities, the atmosphere, the climate system, natural and human systems. It would similarly consider all these impacts and interactions for any mitigation strategy to ensure the solutions proposed not only are effective in addressing climate change, but also enhance human and natural systems. An integrated approach would minimize any adverse consequences of mitigating climate change and maximize enhancement of natural and human systems.

---

[8] These processes also impact on humans and human systems. The extraction, refinement, transportation and use of fossil fuels contribute to the economies of many States. Transportation and energy choices have social and health implications.

[9] For a good discussion of how climate change is undermining efforts in the United States to protect biodiversity, water resources and clean air, see L. Hansen and C. R. Pyke, "Climate Change and Federal Environmental Law," *Sustainable Development Law & Policy* 7 (2007): 26.

For purposes of the review of the climate change regime, it is important to be clear about the evaluation criteria to be applied. Taking the outcomes of the Rio Conference as a guide, sustainable development is offered as a reasonable test for whether climate change mitigation measures promoted through the international regime offer an integrated solution to climate change.[10] For purposes of this study, the following criteria are offered as key indicators of sustainability:

- Does it increase or decrease biodiversity?[11]
- Does it increase or decrease the amount of pollution in natural systems?
- Does it generate or reduce waste?
- Does it increase or decrease the carrying capacity of renewable resources?
- Does it decrease the availability of non-renewable resources?
- Does it increase or decrease the equitable distribution of access to resources?

The basic premise of this review of the climate change regime is that how climate change is mitigated is at least as important as when or how quickly it is addressed. Therefore the central question posed here is whether climate change mitigation commitments under the climate change regime offer integrated solutions, particularly with respect to the six environmental criteria offered as a test for sustainability. Does it encourage, enable or discourage integration? Does the climate change regime address climate

---

[10] Given that sustainable development has been the underlying principle for international environmental law since the early days of the climate change regime, it would appear reasonable to expect integration efforts to be consistent with the concept of sustainable development. For a discussion of the relationship between integration and sustainable development, see D. Barstow Magraw and L. D. Hawke, "Sustainable Development," in Bodansky et al., *supra* note 6, at p. 628. The debate over what it means and over its utility is, of course, ongoing. See, for example, B. Stark, "Sustainable Development and Postmodern International Law: Greener Globalization?" *William and Mary Environmental Law and Policy Review* 27 (2002): 137.

[11] For an assessment of links between climate change and biodiversity, see Intergovernmental Panel on Climate Change, *Climate Change and Biodiversity* (Technical Paper, IPCC Working Group II, Technical Support Unit) (Geneva: IPCC, April 2002). For a discussion of the central role of biodiversity and the health of natural systems for human health and sustainability, see J. D. Brown, "The Integration of Man and the Biosphere," *Georgetown International Environmental Law Review* 14 (2002): 741.

change in isolation without considering these other environmental issues or does it embrace opportunities to address these other issues at the same time? Does it give adequate weight to the risk of creating new problems in the course of solving climate change? In answering these questions, the following mitigation options for climate change are considered:

- energy conservation and improved efficiency[12]
- solar energy[13]
- wind energy[14]
- energy from biomass, including wood and ethanol[15]
- hydropower[16]
- nuclear power[17]
- use of forests as sinks[18]

---

[12] This is also referred to as demand side management. It includes any measures that reduce the consumption of energy. For a discussion of the feasibility of a mitigation strategy focussed on demand side management, see R. Torrie, R. Parfett, and P. Steenhof, *Kyoto and Beyond: The Low-emission Path to Innovation and Efficiency*, Report prepared for the David Suzuki Foundation and Climate Action Network Canada (Vancouver: David Suzuki Foundation, October 2002), <www.davidsuzuki.org/files/Kyoto_Beyond_LR.pdf>, 31 March 2008. The same principle can be applied to non-energy sources of greenhouse gas emissions by focussing on reducing or eliminating human activities that lead to emissions. A good example would be the diversion of organic waste from landfills as a way to reduce or eliminate methane emissions resulting from the anaerobic decomposition of organic material in landfills.

[13] Solar energy options include photovoltaic solar energy, or electricity from solar, thermal solar energy (or the collection of heat from solar energy), and passive solar (or the direct use of the sun's energy for heating).

[14] Convention on Biological Diversity, *Biological Diversity and Climate Change*, Report of the Ad Hoc Technical Expert Group on Biodiversity and Climate Change, Ninth meeting, Subsidiary Body on Scientific, Technical and Technological Advice, Montreal, 10–14 November 2003, UNEP/CBD/SBSTTA/9/INF/12 (30 September 2003), Annex III, pp. 68–69 [hereinafter Ad Hoc Expert Report].

[15] *Ibid.*, pp. 65–66.

[16] *Ibid.*, p. 66.

[17] Eligibility of nuclear power is limited for the Clean Development Mechanism, but not for domestic emission reductions. Thus Canada is free to use nuclear power in Canada to meet its Kyoto obligations, but has agreed not to use emission reductions achieved by supporting nuclear power in developing countries.

[18] Ad Hoc Expert Report, *supra* note 14, at pp. 50–60.

- use of soils, such as cropland and grazing land, as sinks[19]
- deep sea ocean storage of carbon[20]
- carbon injection in oil wells and other geological formations for long-term storage[21]

As a starting point, it is important to note the limitation of this kind of general assessment of these mitigation options. Much will depend on their implementation in a particular geographic, ecological, social, and economic context. The focus here is on the potential for desired and undesired consequences beyond climate change mitigation, leaving aside the opportunity to mitigate negative and enhance positive consequences in the manner a particular mitigation option is implemented in a given location.[22]

Of the measures listed, conservation and efficiency clearly offer effective ways to address climate change without risk of contributing to the loss of biodiversity, the generation of waste and pollution, the depletion of resources and the inequitable distribution of resources. The focus is on directly reducing the human activity that leads to the emissions rather than finding a replacement process. This means that as long as the conservation and efficiency gains are achieved responsibly, these mitigation options make a net contribution to sustainability. To the extent that the energy sources eliminated or reduced as a result of conservation and efficiency have biodiversity, waste, resource and distributive impacts in addition to their climate change impacts, these measures can make a significant contribution to sustainability. This is particularly true if decisions about where to reduce energy supply to reflect reduced demand are made based on what sources of energy are most problematic from a combined climate change and sustainability perspective.

The fundamental benefit of the use of energy efficiency and conservation as a climate change mitigation strategy is that it reduces the level of human

[19] *Ibid.*, pp. 61–63.

[20] This offset option is generally recognized not to be eligible under Kyoto rules, see *ibid.*, p. 63.

[21] *Ibid.*, p. 64.

[22] For an assessment of climate change mitigation measures from a biodiversity perspective, see K. Choudhury, C. Dziedzioch, A. Häusler and C. Ploetz, *Integration of Biodiversity Concerns in Climate Change Mitigation Activities* (Berlin: Federal Environmental Agency, 2004).

interference with natural systems overall. As a result, it has the potential to contribute to the mitigation of numerous environmental challenges, including climate change, resource depletion, air and water pollution, and loss of biodiversity.[23] Reducing greenhouse gas (GHG) emissions with the use of other sources of energy instead of reducing the demand for energy can lead to similar results. The main difference here is that we have to consider the impact of producing the energy from these alternate sources. In other words, the overall benefit can only compete with energy conservation if the alternative source of energy has no negative impact on other sustainability criteria. Otherwise, the net impact is the difference between the impact of traditional sources and these alternative sources.[24]

In case of solar energy, the two main areas of concern would be the range of impacts of production of the solar panels and the space required to operate the panels. In case of wind power, there are similar considerations. In both cases, the net impact clearly points to these options being a net contribution to solving both the climate change and biodiversity problems, but because of the impact of production and siting, reducing energy consumption where possible is still a preferable option.

The next group of mitigation measures includes the use of biomass, hydropower, and nuclear energy. In each case, the switch from the use of fossil fuels would reduce GHG emissions, but the reduction would come at a cost. In case of the use of biomass, there are two issues to consider. One is the potentially competing land use between fuel production from biomass and protection of biodiversity. The other is the pollution from the burning of biomass.[25]

In case of hydropower, there are again significant reductions in GHG emissions possible compared to the use of fossil fuels for power generation. At the same time, hydropower is a "competing land use," and therefore a threat to biodiversity. In addition, hydropower leads to mercury and other water contamination.[26] Nuclear power also results in much lower GHG emissions than power production using fossil fuels. It carries with it a significant threat to human health and biodiversity from accidents and from the waste generated by nuclear power plants.

---

[23] See Torrie, et al., *supra* note 12.

[24] For a discussion of the biodiversity impacts of wind and solar energy, see IPCC, *supra* note 11, at p. 40.

[25] *Ibid.*, p. 38.

[26] For an overview of impacts from hydro-power, see *ibid.*, p. 39.

The last category is a group of climate change mitigation measures to offset emissions by taking greenhouse gases back out of the atmosphere. They include the use of forests, soils[27] and oceans[28] to take CO_2 out of the atmosphere and store it in the form of carbon. They also include measures to capture the CO_2 during combustion of fossil fuels and storing it in oil wells and other geological formations. This group of mitigation measures has the potential to contribute to solving the climate change problem, but is subject to the issue of permanence. To the extent that the storage is temporary, these measures delay the problem rather than solve it.

With respect to biodiversity and other sustainability criteria, the answer is mixed. On the one hand, mitigating climate change through storage of carbon in living things provides an incentive to protect life. Done properly, this can clearly contribute to biodiversity. On the other hand, if the storage of carbon is pursued in isolation from the need to protect biodiversity, climate change mitigation through carbon storage in forests, soils and oceans can be a significant threat to biodiversity. Much again turns on whether countries choose to consider climate change in the form of carbon storage and biodiversity as two separate issues, or whether they become two motivations toward the objective of ensure the health of forests, soils and oceans. In other words, do we look at each service nature provides separately and try to manage that service, or do we protect natural systems overall in the general recognition that overall ecosystem health provides the best hope for nature to continue to perform these services, including biodiversity protection and climate change mitigation.

Applying the above assessment of the range of mitigation options would lead to a conclusion that as much of the mitigation effort as possible should focus on reducing energy consumption through conservation and efficiency. The next obvious choice appears to be fuel switching to solar, wind, and geothermal energy. Beyond these measures, we must at best acknowledge that other mitigation measures carry with them a significant risk of creating other environmental problems. At worst, we may find that they make a

27 For an overview of the environmental costs and ancillary benefits of storing carbon in forests and soils as a way to offset greenhouse gas emissions, see B. Metz, O. Davidson, R. Swart and J. Pan, eds, *Climate Change 2001: Mitigation: Contribution of Working Group III to the Third Assessment Report of the Intergovernmental Panel on Climate Change (IPCC)* (Cambridge: Cambridge University Press, 2002), p. 326. See also, *ibid.*, pp. 35–37.

28 *Ibid.*, p. 41.

net negative overall contribution to efforts to become sustainable in that they may result in substantial undesirable consequences.

This would lead to a conclusion that climate change mitigation efforts should focus on reducing energy demand and the promotion of solar, wind and geothermal energy as alternatives to fossil fuels. A consideration of other mitigation measures would have to precede a further analysis of whether and how these measures fit into an overall sustainability strategy. A detailed consideration of criteria that might be applied to such an analysis is not possible here, but considerable thought has been given to the issue of sustainability criteria elsewhere.[29]

Having briefly considered how human influence on the climate system and efforts to mitigate this effect relate to key sustainability criteria, such as biodiversity, waste pollution, resources and access to resources, the next section considers the key steps in the climate change regime in light of these links. Specifically, the focus of the next section will be on acknowledgements of these links, and any sign of preference for climate change mitigation options that offer integrated solutions to sustainable development.

Integration and the Evolution of the Climate Change Regime

Breaking the evolution of the climate change regime into stages is inevitably arbitrary. The steps selected here are driven mainly by the documentation of developments in the regime. The stages that are relatively well documented are the initial U.N. General Assembly resolutions, the negotiating mandate, the United Nations Framework Convention on Climate Change (UNFCCC), the Berlin mandate, the Kyoto Protocol, and the Marrakech Accords on the implementation of the Kyoto Protocol.

The first phase of the climate change regime considered is the pre-convention phase, consisting of resolutions of the General Assembly of the United Nations from 1988 to 1990 that collectively provided the context

[29] See C. George, "Testing for Sustainable Development through Environmental Assessment," *Environmental Impact Assessment Review* 19(2) (1999): 175. This article lists 18 criteria to test the sustainability of proposed development, including equity in various contexts, social impacts, public participation, precaution, biodiversity, climate change, and overall local and global impacts.

for the negotiations that lead to the framework convention in 1992. The first resolution in 1988 recognized that climate change posed economic and social consequences resulting from environmental impacts such as sea level rise. The link between climate change and ozone layer depletion was identified in the 1988 resolution, and the identification of links to other existing international instruments was encouraged.[30]

Follow-up resolutions in 1989 quickly identified the need for a new international regime to deal with climate change, to be developed in parallel with preparations for the Rio Conference on the Environment and Development in 1992.[31] The principle of common but differentiated responsibilities was introduced, and the resolutions otherwise encourage States and existing institutions to cooperate in seeking solutions to climate change without much direction from the General Assembly. Similarly, resolutions dealing with preparations for the Rio Conference in 1992 mention climate change, but do not directly address linkages between climate change mitigation and other environmental issues. Interestingly, the following are listed as environmental issues of major concern:

- climate change
- quality and supply of freshwater resources
- protection of the oceans
- land protection against deforestation, desertification and drought
- conservation of biological diversity
- biotechnology
- sound management of waste
- living and working environments of the poor
- human health and quality of life[32]

---

[30] United Nations, "Protection of global climate for present and future generations of mankind," General Assembly Resolution A/RES/43/53, 70th plenary meeting (6 December 1988).

[31] United Nations, "Possible adverse effects of sea-level rise on islands and coastal areas, particularly low-lying coastal areas," General Assembly Resolution A/RES/44/206, and "Protection of global climate for present and future generations of mankind," General Assembly Resolution A/RES/44/207, 85th plenary meeting (22 December 1989).

[32] United Nations, "United Nations Conference on Environment and Development," General Assembly Resolution A/RES/44/228, 85th plenary meeting (22 December 1989).

In 1990, the General Assembly then passed a resolution to initiate negotiations for a framework convention on climate change to be ready for signature at the Rio Conference.[33] The negotiating mandate incorporates previous resolutions, and generally recognizes links between climate change and the social and economic well-being of current and future generations of humans. There is little indication in any of these early resolutions of any direction on how to identify appropriate mitigation measures to address climate change. Other than a reference to the need for developed States to take the lead as a result of their capacity and their contribution to the problem to date, the choice of mitigation measures is not discussed. In short, there is nothing in the initial negotiating mandate for the framework convention that would restrict or prevent an integrated solution to climate change mitigation, but there is little to encourage it either. By default, it is left to negotiators to determine whether and how to pursue integrated solutions to climate change.

The end product of these negotiations initiated in 1990 was the UNFCCC.[34] The convention carries forward the themes identified in the negotiating mandate, notably the concept of common and differentiated responsibility, and the basic notion that unmitigated climate change will have adverse consequences for human and natural systems. The preamble does recognize the importance of ensuring that climate change is addressed in a manner that is environmentally, socially and economically most effective. The preamble also specifically recognizes that some climate change mitigation measures have economic and other environmental benefits and can therefore be justified regardless of whether they are needed for climate change mitigation. Finally, the preamble to the convention recognizes that climate change mitigation may have adverse social and economic consequences, particularly in developing States.

What is missing is the recognition that climate change mitigation can cause other environmental challenges either in the preamble or the text

---

33 United Nations, "Protection of global climate for present and future generations of mankind," General Assembly Resolution 45/212, 71st plenary meeting (21 December 1990).

34 United Nations, General Assembly, *Annex 1: United Nations Framework Convention on Climate Change*, Report of the Intergovernmental Negotiating Committee for a Framework Convention on Climate Change, 5th session, second part, New York, 30 April–9 May 1992, UN Doc. A/AC.237/18 (Part II)/Add.1 (1992), 31 *I.L.M.* 849, <unfccc.int/resource/docs/a/18p2a01.pdf> [UNFCCC or Framework Convention].

of the convention. Even Article 3.3, dealing with the application of the precautionary principle, makes reference to the potential socio-economic consequences of climate change mitigation, but not possible environmental consequences. Article 4 includes a commitment by developed States to return to 1990 levels of greenhouse gas emissions by 2000, but does not direct them to select or give priority to mitigation measures based on their environmental, social or economic consequences. In short, the UNFCCC does not prevent an integrated approach to climate change mitigation, but it does not require it, and only identifies it as an issue with respect to the social and economic consequences of climate change mitigation in developing States. Otherwise, it is left to parties to decide how to implement the voluntary commitment to return to 1990 levels of greenhouse gas emissions.

In fairness, the focus up to this point in the development of the climate change regime was not on mitigation measures but on building the basic infrastructure and on objectives and principles. Serious negotiations on mitigation measures got under way after the UNFCCC came into force in 1994. While identification of this form of integration in the UNFCCC could have sent a clear signal to the negotiators, there was certainly nothing in the framework convention to prevent negotiators from giving priority to integrated solutions to climate change.

The next phase of the climate change regime was initiated with the Berlin Mandate in 1994 and concluded with the Kyoto Protocol in 1997. There is little new in the Berlin Mandate to direct parties to the negotiations on the social, environmental and economic consequences of climate change mitigation.[35] There are, however, clear signs in some of the Articles of the Kyoto Protocol that parties considered some of the links. Others are surprisingly silent on the issue. [36] The treatment of climate change mitigation in the Kyoto Protocol can only be understood in the context of the overall obligation it imposes on developed nations to reduce greenhouse gas emissions based on country specific targets that are relative to 1990

[35] United Nations, Framework Convention on Climate Change, *Report of the Conference of the Parties on Its First Session, held at Berlin, from 28 March to 7 April 1995*, FCCC/CP/1995/7/Add.1/1/CP.1, p. 4 [Berlin Mandate].

[36] *Kyoto Protocol to the United Nations Framework Convention on Climate Change*, 10 December 1997, U.N. Doc. FCCC/CP/1997/L.7/Add.1, 37 *I.L.M.* 22 (1998) [Kyoto Protocol]. See, for example, Arts. 2, 3, 6, 12.

levels of emissions.[37] The key issue for purposes of this contribution is therefore what restrictions are imposed on developed nations in terms of mitigation measures to meet their individual targets.

Article 2 of the Kyoto Protocol makes a general link between climate change mitigation and sustainable development, but it offers no guidance on how this link is to be implemented.[38] It lists a range of mitigation options without either prioritizing them or directing any further work to identify which mitigation option should be given priority under what circumstances to maximize desired and minimize undesired environmental, social and economic consequences of climate change mitigation. The options listed include energy efficiency, identified above as one of the most likely measures to offer desirable consequences with minimal risk of undesirable ones. Also included are some that clearly have a significant risk of undesired consequences, such as greenhouse gas sinks and reservoirs and carbon dioxide capture and storage.[39] Article 2.3 then specifically assigns responsibility to strive to minimize adverse social, environmental and economic impacts on other parties, particularly developing States.[40]

Articles 3.3. and 3.4 are also of particular interest.[41] These provisions collectively provide for the use of sinks as a climate change mitigation option for the first commitment period under the Kyoto Protocol. Article 3.3 is relatively uncontroversial in that it essentially discourages land use changes that deplete natural carbon stocks in the form of forests. There

---

37 For a detailed review of the Kyoto Protocol, see M. Doelle, *From Hot Air to Action? Climate Change, Compliance and the Future of International Environmental Law* (Toronto: Carswell, 2005), p. 17.

38 The reference to sustainable development was in fact a later addition to the text by the Group of 77 and China. See J. Depledge, *Tracing the Origins of the Kyoto Protocol: An Article-by-Article Textual History* (Bonn: UNFCCC, 2000) (FCCC/TP/2000/2), pp. 18–29. There was also considerable pressure from the European Union for certain policies and measures to be mandatory, which would have been an opportunity to require States to give priority to policies and measures more likely to offer integrated solutions. The idea was rejected by other developed States.

39 Kyoto Protocol, *supra* note 36, at Art. 2.1(a)(ii) and (iv). For similar language, see also Art. 10(d).

40 The issue was therefore clearly in the minds of negotiating parties, however, there was no collective will to require parties in any meaningful way to focus mitigation efforts on solutions that have desired consequences beyond climate change mitigation.

41 For the negotiating history of these provisions, see Depledge, *supra* note 38, at p. 48.

is an opportunity for aforestation in a manner that may be a threat to biological diversity, but otherwise there are no obvious undesired consequences. Article 3.4 is different. It provides for credits resulting from an increase carbon uptake in forests and soils as a result of changes in their management. While there may be compatibilities between forest management for carbon storage and forest management for biodiversity and sustainable yields, Article 3.4 clearly introduces a potentially competing use for forests, creating the real risk of undesired consequences.

Articles 6 and 12 are the other two key provisions in the Kyoto Protocol dealing with the range of mitigation measures available for parties to meet their emission reduction obligations.[42] These provisions deal with Joint Implementation (JI) and the Clean Development Mechanism (CDM) respectively.[43] Given their similar design, only the CDM is considered here. The basic concept behind the CDM is to allow developed States to fund emission reduction projects in developing nations to offset emissions within their own territory. A stated objective of the CDM is to help developing States to develop sustainably while assisting developed States to comply with their emission reduction obligations. Other than this general statement indicating a desire to encourage sustainable projects in developing States, there is nothing in Article 12 to restrict projects to those with desirable environmental, social and economic consequences.

The bottom line on Kyoto itself appears to be that it did not require, did little to enable, but also did not prevent, an integrated approach to mitigation.[44] There is considerable uncertainty in the Kyoto Protocol itself on how the emission reduction obligations would be implemented as much of the detail on mechanisms such as sinks and the clean development mechanism was yet to be negotiated. To more fully appreciate whether the Kyoto approach offers a path toward integrated mitigation solutions, it is necessary to consider the detailed rules for its implementation, most

---

[42] For an overview of the negotiating history of JI, see Depledge, *supra* note 38, at p. 61. For the history of the CDM negotiations, see Depledge, *ibid.*, p. 75.

[43] For a more detailed discussion of JI and CDM, see Doelle, *supra* note 37, at p. 30.

[44] For a general discussion of linkages between the Kyoto Protocol and MEAs, see The United Nations University Institute for Advanced Studies, *Global Climate Governance: Inter-linkages between the Kyoto Protocol and other Multilateral Regimes, Final Report* (Tokyo: United Nations University, 1999), p. 62.

of which are contained in the Marrakech Accords adopted in 2001 and formally implemented in Montreal in 2005.[45]

The Marrakech Accords continue the trend established in the UNFCCC and the Kyoto Protocol, of placing the burden mainly on individual parties to ensure that mitigation measures chosen offer the best possible combination of minimum undesirable and maximum desirable consequences. Little is done through the Marrakech Accords, for example to limit the use of Article 3.4 to management practices for soils and forests that also protect biodiversity. Ensuring CDM projects make a net positive contribution to sustainable development in the host country is left to developing countries often desperate for any development, no matter how short term, and no matter what the environmental cost. In short, in spite of clear recognition of the issue since the lead up to the Kyoto Protocol, the international regime has done little to ensure that climate mitigation offers integrated solutions taking into account a range of social, environmental and economic consequences. The international regime does not enable, nor does it prevent, an integrated approach to climate change mitigation.[46]

Taking Stock: Opportunities for Integration

Assuming the evolution of the climate change regime is representative of other environmental regimes, it is safe to conclude that international

---

[45] United Nations, Framework Convention on Climate Change, *Report of the Conference of the Parties on its Seventh Session, Conference of the Parties, United Nations Framework Convention on Climate Change, 29 October–10 November 2001*, FCCC/CP/2001/13/Add.1 (Decisions 1/CP.7–14/CP.7), FCCC/CP/2001/13/Add.2 (Decisions 15/CP.7–19/CP.7), FCCC/CP/2001/13/Add.3 (Decisions 20/CP.7–24/CP.7), FCCC/CP/2001/13/Add.4 (Decisions 25/CP.7–39/CP.7 and Resolution 1/CP.7–2/CP.7) [Marrakesh Accords].

[46] Efforts to encourage integration among MEAs is ongoing, but with few concrete results. At the 9th Session of the Conference of the Parties to the UNFCCC, for example, the IUCN, the Deputy Secretary General to the Ramsar Convention, and the Executive Secretary of the Convention on Biological Diversity (CBD) each submitted letters encouraging cooperation and exploration of linkages with the climate change regime. Copies of these letters were distributed at the 9th Conference of the Parties and are on file with the author. More recently, efforts have focussed on integrating climate change into U.N. activities outside the climate change regime. See United Nations, "Overview of United Nations Activities in Relation to Climate Change," General Assembly Report A/62/644 (10 January 2008).

environmental law is still mainly developing on an issue-by-issue basis. Significant efforts to integrate have been made, but essentially in the form of soft law, through instruments such as the Rio Declaration as well as non-binding principles in treaties designed to encourage parties to integrate at a domestic level.[47] Clearly, there is a case to be made for integration at the domestic level, and there are limits to what can be done at a global level to maximize the benefits of integration.[48] A more integrated approach to climate change at the international level would have required more work up front to identify integrated solutions. The result, if successful, would have been a more focussed approach both to the international response and national investments in climate change solutions. Assuming therefore, that better integration internationally is desirable, the questions posed in this section are what is preventing this from taking place, and how international integration might be improved.

The review of integration at key stages of the development of the climate change regime suggests a number of possible reasons for the lack of integration internationally. It is certainly possible that some nations simply took the view that integration was a domestic implementation issue. Whatever the underlying reasons, the lack of attention to the issue in the negotiating mandate from the General Assembly in 1990 set the stage for the issue never making it to the top of the agenda in the evolution of the regime. Certainly, the UNFCCC itself was an opportunity lost. Its purpose was to put into place the infrastructure and the context for the future development of the regime. If a principle, such as the need to ensure integrated mitigation strategies, was to guide the more substantive negotiations in the form of Kyoto, the UNFCCC was perhaps the opportunity to ensure integration would be taken seriously internationally. Having missed the opportunity to make integration a central principle of the regime, it is not surprising that the issue was not a focus on the negotiations leading up to Kyoto or the follow-up negotiations of how to implement the Kyoto Protocol.

A number of factors work against effective integration internationally. First, it can only work in combination with domestic implementation

47 See Doelle, *supra* note 4, at p. 98, on the effectiveness of integration at the domestic level. See also, M. Nilsson and K. Eckerberg, eds, *Environmental Policy Integration in Practice* (London: Earthscan, 2007).

48 *Ibid.*

efforts. Secondly, the absence of agreed upon criteria and principles for integration make it difficult to agree on what can and should be achieved internationally. Institutions with a motivation to encourage integration, such as the secretariats of existing international environmental regimes, generally do not have the necessary means to improve integration. Integration among environmental regimes is complicated by the difference in membership. While the difference in membership is often small, even a few countries can veto the implementation of effective integration based on the fact that they are not party to all regimes involved.

As was apparent with the Kyoto Protocol, once substantive obligations are negotiated without taking an integrated approach, it may be difficult for parties to meet their obligations in an integrated manner. Once a party has a GHG emission reduction target that assumes that it will use all mitigation measures available to the fullest, it may not have the luxury to only choose mitigation measures that make sense from an integrated perspective. Time may also be an issue here; the luxury of choosing mitigation measures which are associated with more desired than undesired consequences may have existed in the 1990s. As time has passed, the threat of climate change may be forcing the international community to choose more and more mitigation measures that are creating other environmental, social and economic problems. In other words, effective integration requires early action.

There are clearly opportunities to enhance integration by placing the issue on the negotiating agenda from the start. The basic idea is that links between issues should be identified on an ongoing basis, starting with the negotiating mandate. This would mean the initial mandate, such as the General Assembly resolution in 1990 to initiate negotiation for a climate change convention, identifies the expectation that parties work toward integrated solutions. The question is how? Can it be left to parties to bring this forward within the U.N. process?[49] Can it be left to secretariats of existing regimes to identify and bring forward links? Is it the responsibility of the General Assembly, or the United Nations Environment Programme (UNEP)? Can the Office of the U.N. Secretary-General play a constructive

---

[49] See, for example, J. Drexhage, D. Murphy, O. Brown, A. Cosbey, P. Dickey, J.-E. Parry, J. Van Ham, R. Tarasofsky, and B. Darkin, *Climate Change and Foreign Policy: An Exploration of Options for Greater Integration* (Winnipeg: International Institute for Sustainable Development, 2007).

role in improving integration? Is a new World Environment Organization needed to ensure effective integration?

Assuming integration is on the negotiating agenda from the outset, the next step is to ensure integration is given sufficient weight by parties in the heat of the negotiations on a specific issue. The need for integrated solutions was clearly recognized in the Kyoto negotiations, but the issue tended to be drowned out by more immediately pressing issues, such as the cost of mitigation and efforts to minimize the cost of meeting emission reduction targets. Perhaps secretariats from existing regimes can play a more significant role in the negotiations? It may also be possible to develop general principles that will encourage or require these issues to be taken seriously in the negotiation process. An important question is whether existing institutions such as UNEP should play this role in the negotiations, or whether new institutions are needed.

Once the obligations are negotiated, international efforts to ensure integration are likely to be limited to education, awareness raising and other forms of motivation. Integration is now in the hands of the parties. Where nations are parties to multiple regimes that provide opportunities for integrated implementation, secretariats can facilitate this through targeted education efforts. Otherwise, opportunities for international efforts to encourage integration at this stage appear limited. There have been some efforts to establish links among various marine environmental agreements in recent years. However, it is not clear that they are able to accomplish much other than to educate on the value of integrated approaches and to encourage States to take an integrated approach to implementation domestically.[50]

This leaves integration at the domestic level. Clearly, this is where much of the work on integration has been and will continue to be done. Up to now, integration has almost exclusively been left to individual States.[51] These efforts will continue to be essential, both in light of limited success internationally to date and the fact that many opportunities for integration are specific to the local social, environmental and economic conditions. If all nations took this responsibility seriously, there would be less need

---

[50] See, for example, UNEP/CBD/COP8/31, p. 270, where parties to the CBD encourage States to integrate biodiversity into national policies on issues such as climate change.

[51] Doelle, *supra* note 4, and Nilsson, et al., *supra* note 47.

for integration at the international level, and individual States would be free to integrate in a manner most appropriate to local conditions. At the same time, it is clear that there is insufficient motivation for individual States to seek integrated solutions, given that there is often an additional cost associated with such solutions and the benefits are often distributed geographically and inter-generationally.

Assuming therefore that the status quo is inadequate, and that more effective integration requires increased focus on the negotiation phase, integration can still be championed by parties, by the various existing U.N. institutions from UNEP to the General Assembly, or by a new World Environment Organization. Individual nations can make a difference by giving priority to integration in their negotiating positions for any future agreements. Examples, such as attempts by the European Union in the negotiations leading up to the Marrakech Accords to link UNFCCC and the Convention on Biological Diversity (CBD), teach us that efforts by individual parties can make a difference, but diplomacy is critical. The United States and other non-CBD parties reacted strongly against efforts by the European Union to formally link the CDB and the UNFCCC, and the effort failed.[52] A more subtle approach of incorporating objectives into negotiating positions may have been more effective.

A step could formally be added to the process to invite secretariats of all existing regimes to provide input into how the negotiations on the new problem should be framed to ensure it is addressed in an integrated manner. Beyond the negotiating mandate, secretariats of relevant or existing regimes could have some formal involvement in negotiations under other regimes. For example, the biodiversity regime could have been given a formal voice in the climate change negotiations. Similar connections would be possible during the implementation stage, but their effectiveness would depend on the appropriate context having been set through the negotiation phase.[53]

---

[52] *Ibid.*

[53] See, for example, *Options for Enhanced Cooperation among the three Rio Conventions*, FCCC/SBSTA/2004/INF.19 (2 November 2004). See also efforts under the United Nations Convention to Combat Desertification (UNCCD) to explore relationships with other MEAs. See, for example, UNCCD, Committee for the Review of the Implementation of the Convention, *Review of Activities for the Promotion and Strengthening of Relationships with Other Relevant Conventions and Relevant International Organizations*, ICCD/CRIC(1)/9 (15 October 2002), and the follow-up report to the 6th Conference

If secretariats of existing regimes are not in a position to sufficiently influence the negotiation mandate, the process or the outcome, other institutions to affect this influence could include UNEP or a new global environmental governance structure. In either case, to be effective, the institution would have to have some power to ensure that the issue of integration was taken seriously throughout the negotiations by all parties involved. Whether UNEP or a new institution should be equipped with such powers is beyond the scope of this discussion, and has been the subject of much debate and consideration elsewhere. What appears clear is that UNEP's current influence is inadequate to ensure effective integration.[54]

One of the critical barriers remains the differences in membership among treaties. This is a fundamental problem for integration at an international level because only parties to a treaty are generally bound to comply with its terms and obligations. There is, however, one exception. When important principles within treaties reach the status of customary international law, those principles are binding on all States, not just those who have ratified the given treaty. A formal mechanism that clearly identifies when treaties have been followed by a significant number of nations for a sufficient amount of time for its substance to have attained the status of customary law would at least allow for better integration with those well

---

of the Parties, *Review of Activities for the Promotion and Strengthening of Relationships with Other Relevant Conventions and Relevant International Organizations, Institutions and Agencies, in accordance with Article 8 and Article 22, paragraph 2(i) of the Convention*, ICCD/COP(6)/4 (27 June 2003).

[54] A full analysis of the implications of establishing a new global environmental governance structure goes well beyond the issue of integration and the scope of this essay. For a more detailed analysis of the potential for global governance on the environment, see J. Hierlmeier, "UNEP: Retrospect and Prospect – Options for Reforming the Global Environmental Governance Regime," *Georgetown International Environmental Law Review* 14(4) (2002): 767; D. Marshall, "An Organization for the World Environment: Three Models and Analysis," *Georgetown International Environmental Law Review* 15(1) (2002): 79; and N. A. Robinson, "Befogged Vision: International Environmental Governance a Decade After Rio," *William and Mary Environmental Law and Policy Review* 27 (2002): 2. See also, T. Marauhn, "Changing Role of the State," in Bodansky et al., *supra* note 6, at pp. 738–748, and E. Hey, "International Institutions," in Bodansky et al., *supra* note 6, at pp. 749–769.

established treaties. An example might be the United Nations Convention on the Law of the Sea.[55]

Conclusion

The need for better integration seems difficult to dispute. If there is one conclusion to be drawn, it may be that all participants in the process of negotiating obligations to address a particular environmental issue must rethink their role in that process to identify how they can ensure more effective integration.[56] The participating States clearly have a crucial role to play, as have those involved in existing international regimes responsible for addressing issues that are in some way linked to the issue under negotiation. The roles of UNEP and other possible institutional structures of the United Nations system such as the General Assembly, and the Office of the Secretary General also may have to be rethought if integration is to improve internationally. There is a critical need for the establishment of principles and criteria for integration to guide the negotiation and implementation of international environmental agreements. The potential roles of the full range of international institutions need to be explored to ensure the development and effective application of such principles and guidelines.

One alternative or complementary way to move at least some of these issues forward would be through a separate treaty on integration that would address issues on how parties should behave internationally and domestically to improve integration. Such a treaty could provide the impetus to resolve a range of unanswered questions on how to encourage the global community to move forward on integration. A treaty could formalize obligations to implement treaties in an integrated manner to the extent possible. It could establish a general duty on all States to take

[55] See M. Doelle, "Climate Change and the Use of the Dispute Settlement Regime of the Law of the Sea Convention," *Ocean Development and International Law* 37 (2006): 319–337.

[56] Not surprisingly, Douglas Johnston wrote about these issues a decade ago. See D. M. Johnston, *Consent and Commitment in the World Community* (New York: Transnational Publishers Inc, 1997).

an integrated approach to treaty obligations.[57] It could establish principles for integration; it could clarify the relationship among interrelated regimes during negotiation and implementation stages. Such a treaty could consider the powers and responsibilities to be allocated to secretariats of existing regimes to enable them to play a more effective role. Finally, a separate treaty on integration could clarify the roles and responsibilities of existing institutions such as UNEP and the General Assembly of the United Nations.

57 The Aarhus Convention (*Convention on Access to Information, Public Participation in Decision-Making and Access to Justice in Environmental Matters*, Aarhus, 25 June 1998) has demonstrated that global acceptance is not always essential for the effective global penetration of the substance of a treaty. If enough States ratify a convention on integration and incorporate its principles into their negotiating positions, the treaty would achieve its objective of becoming an international norm on integration, much like the Aarhus Convention is becoming an international norm on public access to information and justice.

Ocean Regimes as Reflected in 500 Years of Multilateral Treaty-Making

John Gamble,* Ryan Watson** and Lauren Piera***

Introduction

The idea of "regime" has been prevalent in the political science and, to a lesser extent, international legal literatures for decades. However, it is necessary to be certain we are on the same page, as it were, when discussing regimes. One of the broadest definitions of regimes was provided by Charlotte Ku and Paul Diehl: "the rules agreed to by states to regulate their exchanges."[1] This definition is parsimonious and can accommodate a wide range of situations. Further, it can be applied easily to international law, e.g., if the "rules" are treaties. Others have added requirements to their definitions of regime. The most widely used text in international relations defines a regime as "the rules agreed to by two or more states, aimed at working collectively on shared problems, even at the short-term risk of suffering relative losses, which are offset by the expectation that all parties will benefit and realize absolute gains."[2] An even more elaborate formulation was provided by Edward Miles who defined regimes in a similar way – "the norms and rules which govern expectations and behaviour in particular issue areas and stipulate decision making processes

* Distinguished Professor of Political Science and International Law, The Behrend College, The Pennsylvania State University, Erie, USA.

** Schreyer Scholar, The Behrend College, The Pennsylvania State University, Erie, USA.

*** Department of Government, Georgetown University, Washington, D.C., USA.

1 C. Ku and P. Diehl, "International Law as Operating and Normative Systems: An Overview," in C. Ku and P. Diehl, eds, *International Law: Classic and Contemporary Readings* (Boulder: Lynne Rienner, 2nd ed., 2003), p. 5.

2 C. Kegley, *World Politics: Trends and Transformation* (Belmont, CA: Thomson, 10th ed., 2006), p. 157.

and procedures,"[3] but subsequently discussed what must occur after rules, norms, or laws are agreed upon.

Once agreement is reached at the constitutive level, these norms and rules, processes and procedures must be implemented by States parties at their respective national levels. Hence there are very strong connections between regime implementation, compliance and effectiveness.[4]

If implementation, understood here to include enforcement, is completely ineffective, does a regime exist? There are areas of ocean policy, e.g., whaling and fisheries, where some have argued that this is the case.[5]

Miles' notion of "particular issue areas" is pertinent here. How does one delimit "ocean regimes" and distinguish between them and other kinds of regimes? A major goal of this essay is to ascertain how ocean regimes differ from those existing in other areas. One of the earliest examples of an ocean regime, although the word regime was not used, can be seen in claims for freedom of the seas championed most prominently by Grotius. "Grotius wrote that the seas must be free for navigation and fishing because natural law forbids the ownership of things that seem to have been created by nature for common use."[6] A regime of freedom of the seas may still exist today, but it certainly is asserted for less of the world's oceans and has been qualified and limited in ways that would astound Grotius.

The problem of a workable definition of regimes cannot be dismissed. Is there one macroscopic, global ocean regime? If so, are there also sub-regimes for fisheries, pollution, navigation, etc? Are the norms advanced globally and regionally compatible with one another? William Burke provides a good example of a possible clash between global and regional norms, specifically between the 1995 U.N. Fish Stocks Agreement and

---

[3] E. Miles, "Implementation of International Regimes: A Typology," in D. Vidas and W. Østreng, eds, *Order for the Oceans at the Turn of the Century* (The Hague: Kluwer Law International, 1999), p. 327.

[4] *Ibid.*

[5] H. Scheiber, "Historical Memory, Cultural Claims, and Environmental Ethics: The Jurisprudence of Whaling Regulation," in H. Scheiber, ed., *Law of the Sea: The Common Heritage and Emerging Challenges* (The Hague: Kluwer Law International, 2000), p. 127.

[6] J. Van Dyke, "Sharing Ocean Resources – In a Time of Scarcity and Selfishness," in H. Scheiber, ed., *Law of the Sea: The Common Heritage and Emerging Challenges* (The Hague: Kluwer Law International, 2000), p. 3.

the fisheries provisions of the 1982 United Nations Convention on the Law of the Sea (LOS Convention).[7] Burke writes:

> The inclusion of the precautionary approach in the new Convention on Straddling and Highly Migratory Fish Stocks is an extremely important step toward improvement in conservation and management since it can facilitate coastal and high seas fishing states in promulgating effective measures over an entire stock... While the precautionary approach is an important component of the SSA [U.N. Fish Stocks Agreement], it is not entirely clear that it is consistent with UNCLOS [LOS Convention] under which the coastal state's sovereign rights entitle it alone to determine the substantive policies to be sought in regulating stocks within the EEZ. Under UNCLOS a coastal state was not bound by any need to exercise caution in accord with specific reference points which would automatically trigger specific actions regulating exploitation. Under UNCLOS the coastal state has discretion in adopting conservation measures but is obliged to refrain for endangering stocks by overexploitation.[8]

The most accurate description of ocean regimes may be that of many, some overlapping, regimes, operating simultaneously, often prescribing unclear, even contradictory, norms. It also is important to specify if one's approach to regimes is descriptive or normative. Both approaches are valid, in fact complementary, but the goals and assumptions undergirding them can be quite different. Our macroscopic approach using multilateral treaties as an indicator of ocean regimes falls squarely in the descriptive category.

The Comprehensive Statistical Database of Multilateral Treaties (CSDMT), developed at Pennsylvania State University Erie over almost a decade, is based on the premise that treaties are important and affect State behaviour and sets out to assess the aggregate effect of multilateral treaties. Bilateral treaties provide insurmountable problems if one's goal is complete coverage, i.e., it is impossible to locate all bilateral treaties or to make reasonable assumptions about the representativeness of the

---

[7] *Agreement for the Implementation of the Provisions of the United Nations Convention on the Law of the Sea of December 10, 1982 Relating to the Conservation and Management of Straddling Fish Stocks and Highly Migratory Fish Stocks*, New York, 4 August 1995, 34 *I.L.M.* 1542 (1995), entered into force 11 December 2001; *United Nations Convention on the Law of the Sea*, Montego Bay, 10 December 1982, 1833 *U.N.T.S.* 396, entered into force 16 November 1994.

[8] W. Burke, "Compatibility and Precaution in the 1995 Straddling Stock Agreement," in H. Scheiber, ed., *Law of the Sea: The Common Heritage and Emerging Challenges* (The Hague: Kluwer Law International, 2000), p. 119.

sample identified. In short, we make the *a priori* judgment that treaties are significant.

When Professor Peter Rohn began his career-long effort to quantify treaties he wrote:

> Our ignorance of basic facts is impressive. For example, nobody knows how many treaties there are in the world today or how many there were 10 or 20 or 50 years ago. Nor is it known how the world's treaties subdivide by signatories, topics, title, duration or reliance on international institutions, dispute settlement procedures or by any other criterion. Most of these matters are considered relevant to a legal analysis of any one treaty. And yet they are ignored in the global context of all treaties.[9]

This lack of a broad, macroscopic perspective on treaties is due in part to the fact that treaty scholarship has been dominated by lawyers and law professors, most of whom do not engage in quantitative research and often do not acknowledge the value of such research. The response is that understanding a 300-page treaty is enhanced by macroscopic analyses; it is easier to understand individual treaties if one has some idea of the forest in which those treaty trees grow. Prudence is the order of the day. Harvard professor and Judge on the Permanent Court of International Justice Manley O. Hudson, who, among myriad other accomplishments, tabulated treaties, said, "Count, by all means count, but count things that count."[10]

The CSDMT originated in 1998 with a positive review one of us (Gamble) prepared of Christian Wiktor's *Multilateral Treaty Calendar, 1648–1995* (1998) for the *American Journal of International Law*.[11] Wiktor's work is a comprehensive record of all multilateral treaties and covered 350 years and more than 6,000 instruments. One of the best honours students ever to attend Penn State Erie was examining the *Calendar* and asked a simple question – how many multilateral treaties have been signed (sic) over the last 400 years – a question that could not be answered. It was fortuitous that, in the fall of 1999, Behrend College of Penn State had a group of honours students willing to begin the project,

[9] P. Rohn, "The U.N. Treaty Series Project As Computerized Jurisprudence," *Texas International Law Forum* 2 (1966): 168.

[10] Cited in P. Rohn, *Treaty Profiles* (Santa Barbara: Clio Press, 1976), p. 4.

[11] J. Gamble, "Multilateral Treaty Calendar, 1648–1995," *American Journal of International Law* 93(2) (April 1999): 565–566.

a project to produce a database starting with the Wiktor book that, over the years, has grown in breadth and depth. We developed a reasonable approach to what constitutes a multilateral treaty, e.g., omitting declarations and final acts and checked all major treaty series and indices to be sure we have included virtually all multilateral treaties.

Development of the CSDMT had to be realistic. We did not have external funding and could not compete with research centres that specialize in particular subfields of international law, e.g., human rights law at the University of Minnesota.[12] Since this was a cottage industry financed only with internal Penn State funds, we could only afford to spend on average 30 minutes examining the text of each treaty. We did not collect the two kinds of information most likely to come to mind when one hears the word treaty: full text and a complete list of parties. We have collected basic information (variables) for 6,000–7,000 multilateral treaties signed during the 500-year period, 1500–1999. Variables, selected by balancing importance, significance from an international law perspective, and ease of collection, include:

- headnote
- name of instrument (e.g., convention)
- nature of instrument (e.g., amendment, protocol, and questionably binding)
- regional focus (13 options, e.g. Asia/Pacific)
- treaty series and location
- laterality (plurilateral/general)
- signature date
- force date
- relation to intergovernmental organizations (none, create, action by/on)
- relation to International Labour Organization, i.e., is it an "ILO treaty"
- topic category #1 (United Nations' about 300, e.g., whaling)
- topic category #2 (about 30, e.g., maritime/oceans)

---

[12] The University of Minnesota Human Rights Library has one of the largest collections of human rights materials, including 25,000 documents, several hundred treaties and other international human rights instruments. The website provides access to other search engines on human rights, and has translations of the materials on their website in six languages (English, French, Spanish, Russian, Japanese and Arabic) and is working on adding Swedish as well, <www1.umn.edu/humanrts>, 28 February 2008.

- topic category #3 (about 10, e.g., economic)
- dispute settlement provisions (e.g., none, vague, binding)
- reservations provisions (e.g., none, "object and purpose," prohibited)
- duration clause for the treaty (e.g., automatic lapse)
- duration clause for parties (e.g., withdraw with notice)
- length of text (number of articles and number of words)
- official languages (36 values, e.g., French and English)
- number of languages
- languages (10 variables tracking these languages individually: Latin, French, English, Spanish, Russian, Chinese, Arabic, German, Dutch and Italian)
- total number of parties
- code number (a rational numbering system for all multilateral treaties)

The Impossible Dream – Mutual Exclusivity

We believe the CSDMT is an important research tool for understanding broad patterns in international law. That said, treaties, constructed as they are by people representing States operating in an emotionally-charged, complex political milieu are anything but simple. For the purposes of this essay, this complex problem can be simply stated: What is an ocean treaty? In some cases, the answer appears straight forward. However, even the 1982 LOS Convention, certainly the quintessential ocean treaty of the 20th century, presents some interesting issues. If one had a dozen substantive categories in which to classify treaties, and one of the categories were oceans, certainly the LOS Convention would be placed there. However, even this easy case might present problems. These sections of the LOS Convention present especially strong cases for placement in a category other than oceans:

XII. Protection & Preservation of the Marine Environment
XIII. Marine Scientific Research
XIV. Development and Transfer of Marine Technology
XV. Settlement of Disputes
Annex IV. Statute of the Enterprise
Annex V. Conciliation
Annex VI. Statute of the International Tribunal for the Law of the Sea

Annex VII. Arbitration
Annex VIII. Special Arbitration

For example, dispute settlement is a principal concern of international law. The innovative dispute settlement provisions in the LOS Convention might cause scholars, especially if they are not law of the sea experts, to view this treaty more in a dispute settlement light.[13] A comparable point might be made about intergovernmental organizations. The LOS Convention could be the focus of a text on international organizations. "The global international organizations concerned with the implementation of the LOS Convention are not limited to the specially identified 'competent international bodies' but rather include the whole range of the United Nations system."[14]

Another illustration of the difficulty of single category can be seen in the various treaties in the developmental sequence resulting in the 27-member European Union. One can see a huge change in emphasis, i.e., provisions that go far beyond economics, as one views the evolution of the European Economic Community to the European Union. The Treaty of Rome's emphasis is mostly economic, e.g., "The Community shall have as its task, by establishing a common market and an economic and monetary union and by implementing common policies or activities referred to in Articles 3 and 4..."[15] Moving ahead 35 years to the Treaty on European Union, 2 July 1992, one finds much more attention to issues that transcend economics, for example "attachment to the principles of liberty,

---

13 "The Law of the Sea Convention can be seen as a turning point as regards acceptance by States of the obligation to submit disputes to the binding decision of a court or tribunal at the request of any other party to that dispute." T. Treves, "New Trends in the Settlement of Disputes and the Law of the Sea Convention," in H. Scheiber, ed., *Law of the Sea: The Common Heritage and Emerging Challenges* (The Hague: Kluwer Law International, 2000), p. 61.

14 G. Kullenberg, "Global International Organizations and the Implementation of the Law of the Sea Convention," in D. Vidas and W. Østreng, eds, *Order for the Oceans at the Turn of the Century* (The Hague, Boston: Kluwer Law International, 1999), p. 343.

15 *Treaty Establishing the European Economic Community*, Rome, 25 March 1957, 294 *U.N.T.S.* 17, entered into force 1 January 1958. Article 2 lists the goals of the treaty, which are mostly economic, as well as the environment and social equality.

democracy and respect for human rights and fundamental freedoms and of the rule of law."[16]

If one were limited to a small number of topic categories, one would think that the European Economic Community should be placed in an economic treaty category. That probably is accurate, but as one moves from the Treaty Rome to the Treaty of Nice, it seems that the treaties are becoming more political (or administrative) and less economic. The point is that many treaties are a complex mixture with diplomatic, economic and political elements. Our analyses represent a compromise and some oversimplification since we endeavour to place each treaty in a single category.

It should be apparent that it is not easy to define ocean treaty. However, we adopted what we believe is a workable solution. As one can see from the above discussion, there may be ocean attributes in many multilateral treaties. Our task here is to isolate and try to understand those treaties whose principal focus is the oceans. We have taken a broad definition of ocean-related but exclude inland uses such as canals and rivers. These treaties deal with a bewildering array of subjects among them: certain claims, fisheries, hydrography, certain International Labour Organization (ILO) conventions, landlocked States, marine pollution, maritime matters, ocean mineral resources, naval warfare, navigation, seabeds, seals, straits, and whaling. Our catch-all, ocean-related group, contains about 450 treaties, roughly eight per cent of all multilateral treaties, but a higher percentage of those signed since World War II.

In the next section, we compare this ocean-related group to overall treaty trends. However, it is desirable to try to establish some subcategories of ocean-related treaties to get a better understanding of this diverse group. These subcategories must be intrinsically meaningful, consistent, and must have enough treaties to discern trends and patterns. For example, we could not use as a subcategory major ocean-related constitutive instruments, quintessentially, the 1982 LOS Convention, as important as these few treaties are, since macroscopic analysis would be impossible with such small numbers. We believe that the three subcategories set out in Table 1, with examples, are relatively clear and meet the sizeable number of treaties criterion.

---

[16] *Treaty on European Union*, Maastricht, 7 February 1992, 31 *I.L.M.* 253 (1992), entered into force 1 November 1993.

Table 1. Sub-categories of ocean-related treaties with examples

Treaty Title	Date Signed	Date Entered into Force	Citation
Living Resources – 159 treaties			
Geneva Convention on Fishing and Conservation of the Living Resources of the High Seas	29 April 1958	20 March 1966	559 *U.N.T.S.* 285
Agreement Concerning Protection of the Salmon Stock in the Baltic Sea	20 December 1962	1 March 1966	955 *U.N.T.S.* 259
Agreement between the Government of the United Kingdom of Great Britain and Northern Ireland, the Kingdom of Norway and the Government of the Union of Soviet Socialist Republics on the Regulation of the Fishing of Northeast Arctic (Arcto-Norwegian) Cod	15 March 1974	15 March 1974	925 *U.N.T.S.* 3
Convention respecting Measures for the Preservation and Protection of Fur Seals in the North Pacific Ocean	7 July 1911	15 December 1911	214 Parry 80
Navigation including SOLAS – 177 treaties			
Convention on the International Regulations for Preventing Collisions at Sea	19 November 1987	19 November 1989	1558 *U.N.T.S.* 436
International Convention for the Safety of Life at Sea	20 January 1914	1 July 1915	219 Parry 177
Convention Relative to the Establishment and Maintenance of a Lighthouse on Cape Spartel	31 May 1865	31 May 1865	131 Parry 203

Table 1 (*cont.*)

Treaty Title	Date Signed	Date Entered into Force	Citation
Marine Pollution – 69 treaties			
International Convention for the Prevention of Pollution of the Sea by Oil	12 May 1954	26 July 1958	327 *U.N.T.S.* 3
International Convention Relating to Intervention on High Seas in Cases of Oil Pollution Casualties	29 November 1969	6 May 1975	970 *U.N.T.S.* 211
Convention for the Prevention of Marine Pollution by Dumping from Ships and Aircraft	15 February 1972	7 April 1974	932 *U.N.T.S.* 3

Ocean Regimes Over 500 Years

Figure 1, "500-Year Trend in Ocean-Related Treaties," provides the broadest overview. Treaties are placed on a timeline according to signature date, the most accurate, least ambiguous way of displaying trends in treaties.[17] Since multilateral treaties were made at a far faster rate after 1900, we have lumped all the treaties before that date into a single group (the leftmost bar(s) in all graphs). There were only 475 treaties signed over the 400-year period, 1500–1899. Treaties from 1900–1999 are displayed in equal time periods of 20 years, making comparison and analysis easier. Of course, Figure 1 does not try to account for the importance of treaties, it merely tabulates new treaties according to signature date. Several things stand out from Figure 1. First, the ocean-related group, while certainly important, is a relatively small percentage of multilateral treaties for each of the six time intervals, but the *percentage* of ocean treaties has increased steadily:

[17] The date when a treaty is signed is the most accurate representation since it shows when initial agreement on the instrument occurred. The date of entry into force can be confusing since it can occur at the same time as signature or decades later. Further, it can be unclear if one means in force for the treaty itself or for a particular party.

Time Interval	*Percentage*
1500–1899	2
1900–1919	3
1920–1939	5
1940–1959	5.5
1960–1979	8
1980–1999	12.5

This provides unequivocal proof that the proportion – and one would assume aggregate importance – of ocean-related treaties has increased steadily over 500 years. For the earliest time period, only about one treaty in fifty related to the oceans. For the period 1980–1999, the figure is one in eight.

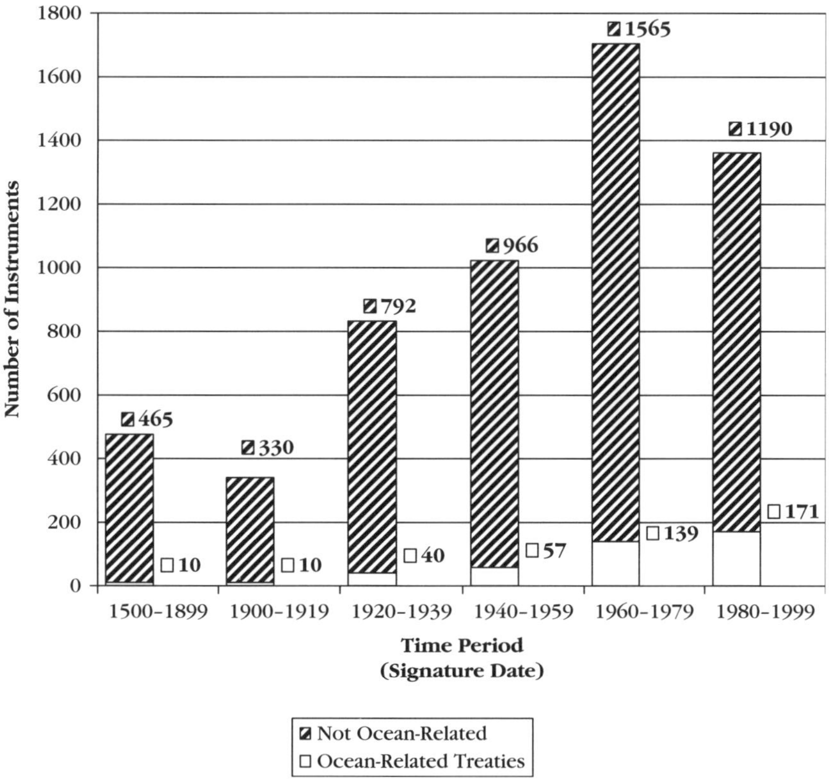

Figure 1. 500-year trend in ocean-related treaties

Ocean-related treaties did much more than keep up with a rising tide of treaty-making especially pronounced since World War I. Ocean-related treaties have steadily increased as a percentage of the total. Further, ocean-related treaties display a pattern counter-cyclical to that of multilateral treaty-making in general. Many scholars assert – often based on no more than a hunch – that the glory days of treaty-making have passed.[18] A much more precise description is possible. The greatest level of activity in terms of new treaties occurred around 1970 with the rate declining somewhat thereafter but remaining at a high level compared to the rest of the 20th century. Levels of activity in ocean-related treaties have continued to increase throughout the 20th century.

Those who specialize in ocean law are accurate when they state there has been a marked increase in the number of ocean treaties. The increase is there, in relative and absolute terms, and it has been sustained over 500 years. Similar claims made about most other subfields of international law are not supported by the empirical treaty data.

Figure 2, "Trends in Ocean Treaty Subsets," examines the three principal subgroupings of ocean-related treaties discussed in the previous section: living resources, navigation, including SOLAS, and marine pollution. There was very little activity before 1900 and none in the marine pollution category before World War II. Perhaps the most remarkable thing about the figure is that, from 1945 onwards, one sees marked increases in numbers of treaties in all three areas. All three lines have about the same slope indicating comparatively high rates of increase. Thus we can add to our conclusion from Figure 1 and say that the marked increase in ocean treaty activity occurs across all three subcategories, although marine pollution treaties got off to a late start, as it were.

Table 2, "The Plurilateral/General Dimension," illustrates an important, but often overlooked, aspect of treaties. Laterality in the CSDMT is represented by a variable that categorizes each treaty as general multilateral (open to all States) or plurilateral (participation restricted by purpose, geography or both). General treaties aim for the widest possible participa-

[18] Ambassador Allan Gottlieb and Prof. Thomas Franck, Statements to Panel, "Engaging Stakeholders in Treaty-Making and Implementation," 72nd Biennial Conference of the International Law Association, Toronto, 4–8 June, 2006, <www.ila2006.org>.

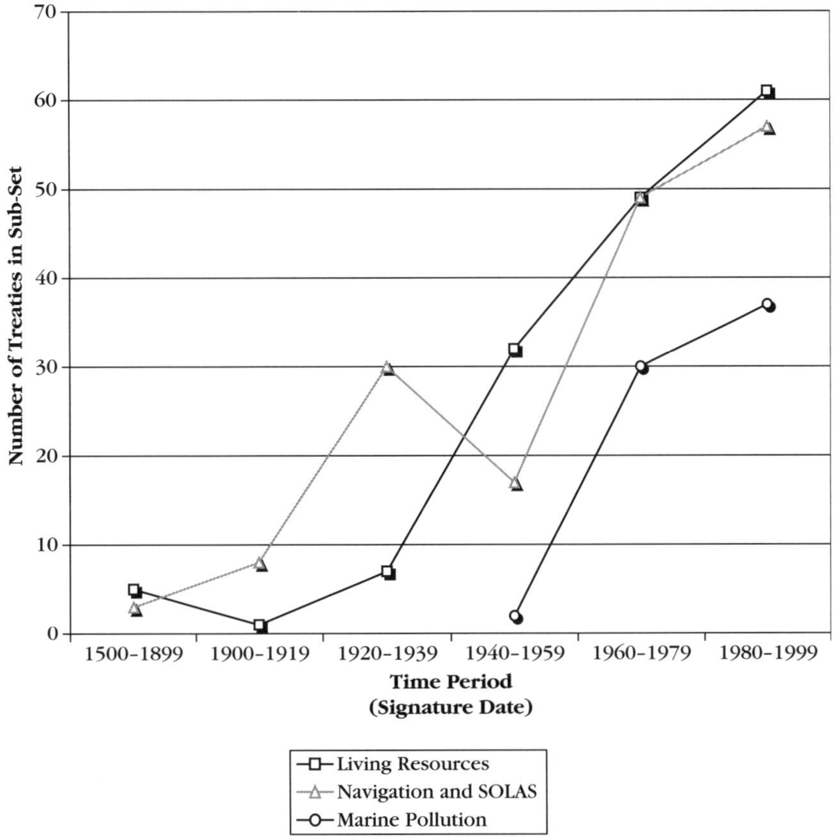

Figure 2. Trends in ocean treaty sub-sets

tion globally.[19] The archetypical examples are the Charter of the United Nations (general)[20] and the 1957 Treaty of Rome (plurilateral).[21]

We know from CSDMT data that plurilateral treaties are far more numerous than general multilateral treaties, on average by about a 5/1 ratio. Thus, the pertinent question for ocean treaties is the size of the

[19] K. Holloway, *Modern Trends in Treaty Law: Constitutional Law, Reservations and the Three Modes of Legislation* (London: Stevens and Sons, 1967), p. 582.

[20] *Charter of the United Nations*, San Francisco, 26 June 1954, 9 *B.F.S.P.* 327, entered into force 24 October 1945.

[21] *Treaty Establishing the European Community*, *supra* note 15.

plurlateral/general divide that we know will tilt heavily in favour of plurilaterlal treaties. The findings from Table 2 are striking. There are virtually no general treaties prior to the 20th century. The living resources category is overwhelmingly (about 97 per cent) plurilateral. Perhaps this is not so surprising when one considers that most living resources are concentrated in a geographic region and managed accordingly. The navigation category was dominated by plurilateral treaties during the 1920–1939 period, but subsequently shows an almost equal distribution between the plurilateral and general modes.

The marine pollution category has significant numbers of treaties only for the 1960–1979 and 1980–1999 periods. One finds an almost equal distribution for 1960–1979 but for the 1980–1999 interval almost 90 per cent of the treaties fall into the plurilateral grouping. These appear to be different norm-creating tracks; if one track is unable to function, the other may compensate. It is reasonable to hypothesize that certain pollution problems could be dealt with globally; many of these were resolved in the decades of the 1960s and 1970s. Since 1980, States have been much more likely to use the plurilateral mode.

Figure 3, "Trends in Binding Dispute Settlement," is derived from a CSDMT categorical variable that has four possible values: 1) no dispute settlement provisions (4,531 treaties/79 per cent); 2) very vague provisions (115 treaties/2 per cent); 3) specific, non-binding (230 treaties/4 per cent); and 4) specific, binding (803 treaties/14 per cent). Provisions that require binding dispute settlement certainly represent an increased level of commitment, an important step towards effective implementation. Looking at all multilateral treaties, States are chary of binding dispute settlement, using it in only about 14 per cent of treaties. The contrast between plurilateral and general treaties is stark. States are much more willing, from 2 1/4 (1940–1959) to 6 1/2 (1960–1979) times, to negotiate binding dispute settlement provisions in their plurilateral treaties. Reliable dispute settlement is a cornerstone of effective treaty-based ocean regimes with credible commitments. States are far more likely to accept this restriction on their behaviour when they are acting in a plurilateral mode. While this result probably is not surprising, its magnitude may be.

Figure 3 shows the per cent of treaties with binding dispute provisions for each of our three ocean subsets. The actual numbers of treaties are relatively small since we have 388 treaties, only about ten per cent of which have binding dispute settlement provisions. The tyranny of small numbers

Table 2. The plurilateral/general dimension of ocean-related treaties by number (No.) and percent (%) of instruments according to signature date

	Time Period (Signature Date)											
	1500–1899		1900–1919		1920–1939		1940–1959		1960–1979		1980–1999	
	No.	%	No.	%	No.	%	No.	%	No.	%	No.	%
Living Resources	5		1		7		32		49		61	
General	0	0	0	0	1	14.29	1	3.13	0	0	3	4.92
Plurilateral	5	100	1	100	6	85.71	31	96.88	49	100	58	95.08
Navigation and SOLAS	3		8		30		17		49		57	
General	0	0	5	62.50	8	26.67	9	52.94	22	44.90	29	50.88
Plurilateral	3	100	3	37.50	22	73.33	8	47.06	27	55.10	28	49.12
Marine Pollution	0		0		0		2		30		37	
General	0	–	0	–	0	–	1	50.00	13	43.33	4	10.81
Plurilateral	0	–	0	–	0	–	1	50.00	17	56.67	33	89.19

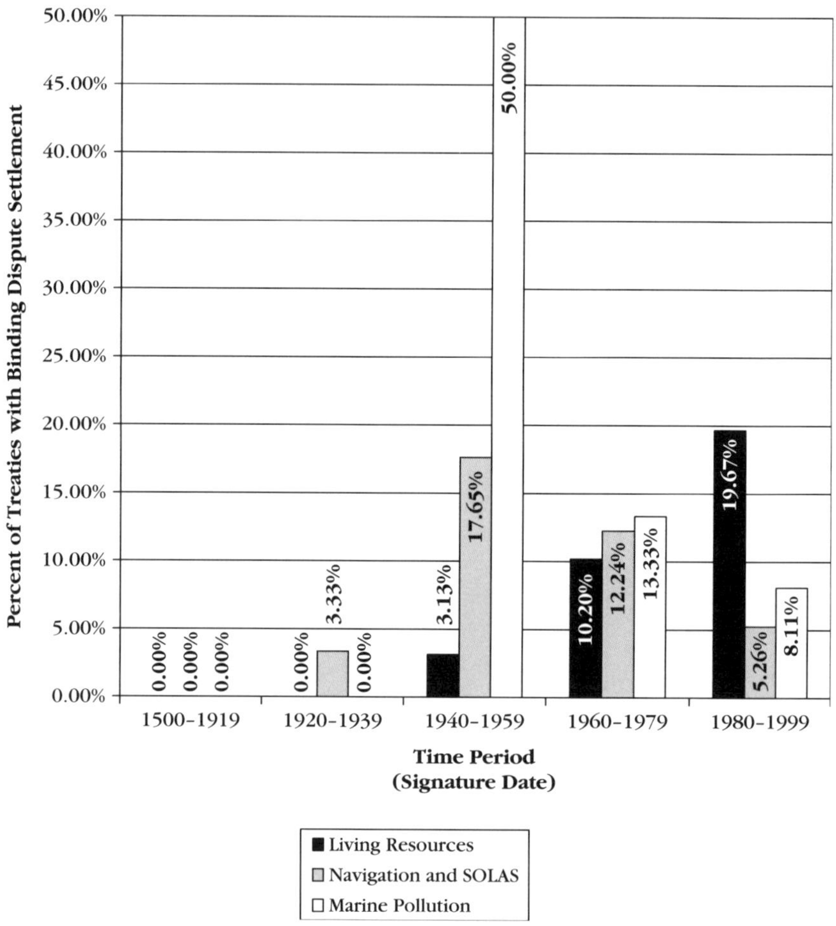

Figure 3. Trends in binding dispute settlement

is especially dangerous for the marine pollution subset for 1940–1959. Fifty per cent of those treaties have binding dispute settlement. But that 50 per cent is one of two treaties! The two most recent time intervals have adequate numbers of treaties from which certain conclusions can be drawn. Overall, these ocean-related treaties are somewhat less likely to have binding dispute settlement provisions than are multilateral treaties writ large. The exception is in the living resources category for the most recent time interval, where almost 20 per cent of treaties stipulate binding dispute settlement. As we have pointed out previously, the living resource group is skewed towards plurilateral treaties. Thus, most of the explanation probably lies with the fact that, especially in recent years, States have managed living resources through regional treaties where they are more comfortable with meaningful dispute settlement provisions.

Summary and Conclusion

An ocean regime, or regimes, is discernable in multilateral treaties. Of course, relying solely on the macroscopic approach adopted here is no substitute for a careful examination of individual treaties. Borrowing from a technique used by Pierre-Marie Dupuy in a very clever hypothetical conversation with the late Wolfgang Friedmann,[22] we might ask how Douglas Johnston might react to this approach. Based on the experiences of one of us (Gamble), a close friend of Douglas Johnston for 30 years, we can offer the following. Douglas would be fascinated by the broad patterns and trends of multilateral treaties especially when they provide a way to test conventional wisdom and unproven assumptions. However, Douglas probably would feel that more attention should be paid to system-altering treaties such as the 1982 LOS Convention that have enormously more influence, far more than one multilateral treaty among 500.

The idea of a macro-ocean regime along with subregimes could be operationalized. However, there are elements of ocean law in hundreds, perhaps thousands, of other multilateral treaties. The ideal solution would be a careful analysis of the content of all treaties in order to assess the amount of ocean-related content. This might be feasible for smaller groups of treaties, but the strength of the CSDMT lies in its coverage of virtually all multilateral treaties. We sacrificed depth to achieve breadth and complete coverage.

Probably the clearest and most significant finding from this research is a steady, consistent increase in the portion of multilateral treaties dealing with the oceans. This increase has been sustained for several hundred years. Other areas of international law might make comparable claims, most of which would not stand up to empirical scrutiny. There is some

---

[22] Pierre-Marie Dupuy, "International Law: Torn Between Coexistence, Cooperation and Regionalism. General Conclusions," *European Journal of International Law* 9 (1998): 282.

evidence that the peak time for the creation of new multilateral treaties was between 1960 and 1980 with the rate of new treaty creation declining somewhat thereafter. This is not the case for ocean treaties. The absolute and relative numbers of ocean treaties increased for each 20-year period during the 20th century.

We analyzed three important subsets of ocean treaties: living resources, navigation including SOLAS, and marine pollution. While certain of these subgroups were rare before 1945, all three accelerated rapidly and comparably after World War II. We examined how a very important distinction manifests itself in ocean treaties. Multilateral treaties divide quite clearly into general, open to all States, and plurilateral, limited by interest, geography or a combination. Since we know that, for all multilateral treaties, there are about five plurilateral treaties for each general treaty, the issue is whether ocean treaties deviate from this norm. In several important respects, they do. The living resources category is overwhelmingly plurilateral – 97 per cent – one of the highest for any subset of treaties. In the case of navigation and SOLAS, we find almost an equal distribution from 1940–1959. Marine pollution had no treaties before 1940. From 1940–1979, there was almost an even distribution between plurilateral and general treaties. Since 1980, marine pollution treaties have been almost 90 per cent plurilateral. It is clear that this plurilateral/general dichotomy is a fundamental characteristic of ocean treaties especially when we examine the three subcategories.

The CSDMT provides a broad context within which to place ocean treaties. The analogy of understanding the entire woods rather than concentrating on just one tree is apt. Douglas Johnston would have improved on this analogy. Perhaps this broader perspective is more important for ocean law because of the shadow cast by one treaty, the 1982 LOS Convention. Of course, these analyses could be sharpened if we could accommodate the importance of a few super treaties. But that raises two major research challenges: weighting according to the importance of treaties, and accounting for the fact that some treaties affect the international system for years or decades beyond the time when they were signed. These are major methodological challenges that will take years to meet. In the mean time, we feel we have made a good start by counting something that is important – multilateral treaties – and counting almost all of them.

The Law of the Sea, Marine Technology, and Global Social Justice

Jay L. Batongbacal*

Introduction

Technological progress inevitably prompts legal developments, but how does it influence social justice? At the international level, the first relationship has long been recognized by authorities, particularly with respect to the law of the sea, while the latter question remains an unfocused inquiry. But this is something that needs to be considered carefully, especially in this age of rapidly advancing knowledge and the ubiquity of new technologies in everyday life. This essay reflects on the relationship between the law of the sea, marine technology, and its possible connections with global social justice. It is argued that the law's uncritical acceptance of and preoccupation with marine technology may be providing an unseen basis for the perpetuation and reinforcement of inequalities between coastal States despite the avowed purpose of the United Nations Convention on the Law of the Sea (LOS Convention) and its related agreements to distribute the benefits of the ocean and its resources equitably among all nations. By tracing connections between marine technology and social justice, and highlighting three examples of how the law of the sea and related agreements may be leading to the continued marginalization of developing countries from full and equal access to the assumed benefits of the legal regimes, it is hoped that closer attention and deeper reflection may be called upon the role of marine technology in future developments in international maritime law.

* A lawyer from the Philippines, Jay L. Batongbacal holds a Master of Marine Management degree from Dalhousie University, Canada, and has begun doctoral studies at Dalhousie Law School through a scholarship awarded by the Pierre Elliot Trudeau Foundation.

The Double Edge of Technology

In a piece simply entitled *Law, Technology and the Sea* written forty years ago,[1] Douglas Johnston saw the nascent international law of the sea as developing along a distinctly technological path. The sea represented a frontier of complex physical resources and unoccupied space open for the expansion of human activities. Fully equitable and rational solutions to the problems of allocation and management of these resources and unoccupied space, he reasoned, "must accommodate themselves to the overriding value of efficiency inherent in modern technology." He expected these problems to become more and more technical in nature, and demand solutions devised with what he called "hypertechnical logic." This referred to "reasoning that places maximum emphasis on the value of efficiency, skill and economy." He perceived this logic in turn to be in conflict with the traditional structure of international law. He implied that ultimately a "technological order of the sea" based on those three values could eventually form the basis of restructured international legal regimes. For Douglas, whose body of work manifested a distinct sense of internationalism and revolved around global cooperation, there was clearly a hope that the development of these new international legal regimes represented a transition to an objectively more equitable and prosperous world community. But, he also noted that this was not going to be an easy or rapid transition. Fully aware of the great disparities in the distribution of knowledge and resources in marine science and technology among the countries of the world, he conceded that the strict optimality demanded by hypertechnical logic would have to give way to "tolerably rational suboptimal practices" in ocean resource allocation and management that would be more consistent with a subjective sense of equity and fairness.

Looking back on Douglas' optimistic piece, written at the height of the Cold War in a period of high-profile technological achievements,[2] we realize that the faith he placed on technology understandably reflected

[1] D. M. Johnston, "Law, Technology and the Sea," *California Law Review* 55 (1967): 449–472.

[2] The United States and Soviet Union were racing to the moon, commercial air travel had become common and affordable, atomic power was being tapped for electrification and transportation, the possibilities of genetic engineering had begun to be explored, more and more powerful computers were being built or planned, and applications of automation were increasingly diversified and spread throughout agriculture and industry.

what a conservative Canadian political pundit, George Grant, called "the central fact of the North American dream – progress through technological advance."[3] According to Grant, technology as the key to material prosperity has always been one of the central assumptions of the market economy practised in the West. He acknowledged that technological achievements had indeed expanded human freedom and equality by overcoming the impact of random "chance" and of nature in the development of Western civilization. But Grant believed this to actually be a universal assumption shared by both Right and Left, the West and East, that "the control of nature through modern techniques" held the key to their respective visions of the good society. And unlike Douglas Johnston, Grant believed the assumption to be fraught with danger, because ultimately, a society devoted to the mastery and perfection of technique had no values or standards by which to judge them beyond their own effectiveness in comparison with other techniques.[4] Or in other words, never beyond standards of "efficiency, skill, and economy" – "hypertechnical logic."

Johnston's and Grant's contrasting opinions on the ultimate impact of technology, revolving around the pre-eminent value of efficiency to which technology is directed, represent a snapshot of the then-nascent debate on the role of technology in national and global society. To be sure, that role is pivotal, as recent history has shown. One of the most profound effects of the end of the Cold War is the globalization of the market economy[5] borne upon the technological bridges established by electronic communications and seaborne transportation. Communications technology combined with computerization and automation are used to coordinate all kinds of economic activities, while rapid and efficient marine transportation[6] comprise worldwide networks that convey goods

[3] G. Grant, "In Defence of North America," *Technology and Empire: Perspectives on North America* (Toronto: House of Anansi Press, 1969), p. 30.

[4] *Ibid.*, p. 15.

[5] A market economy, as defined by Karl Polanyi, is "an economic system controlled, regulated, and directed by market prices; order in the production and distribution of goods is entrusted to this self-regulating mechanism." See K. Polanyi, *The Great Transformation: The Political and Economic Origins of Our Time* (Boston: Beacon Press, 1944; reprint, 2001), p. 71.

[6] This is not to ignore the significant roles of land and air transportation. However, neither land or air transportation currently can establish a truly global flow of goods beyond continental borders at the same low prices and large volumes afforded by shipping.

across product chains that span continents. Although the basic foundations of these product chains were laid centuries ago in the colonial era, the final fall of ideological barriers allowed the rapid spread and convergence of technological advances, and permitted the consolidation and expansion of global electronic and commodity highways that now serve as the backbone of the world economy. This may be seen as an eminent example of how technology plays a pivotal role in the establishment of a particular global order: without either information and communications technology or modern marine transportation, the world economy as it exists now would simply not be possible. One could say that it is a perfect demonstration of hypertechnical logic at work in support of a sea-change in international society.

However, whether technology actually works toward the common good is a much more subtle question than it seems. The fundamental assumption of the ideas of technology-for-progress and technology-transfer is that technology is inherently neutral and passive. Even when technology is the cause of a problem or used malevolently, the common perception is that the ill effects are due to the motivations and intentions of the user rather than the tool, characterizing technology as largely as an extension of human decision and action. Most people are beguiled by the convenient benefits of ever-advancing technology and accept the neutrality of technology without a second thought, but this notion has also been challenged as technology becomes more and more intrusive and seemingly inseparable from daily life. Academic work to this end written by historians appeared as early as the 1950s, though it was not until the 1980s when a wider range of disciplinary lenses and a broader scope of studies were brought to bear on the relationship between technology and human society, particularly as it related to the environment.[7]

At the core of counter-perspectives on technology are the critiques of industrial society, with its penchant for the mass production of goods and single-minded concern for growth, purchased at the cost of the transfor-

---

The International Maritime Organization (IMO) estimates that more than 90 per cent of total world trade in goods is borne by sea. See IMO Library Services, ed., *International Shipping and World Trade: Facts and Figures* (London: IMO, 2006), <www.imo.org/InfoResource/mainframe.asp?topic_id=1075&doc_id=5417>, 31 January 2008.

[7] J. K. Stine and J. A. Tarr, "At the Intersection of Histories: Technology and the Environment," *Technology and Culture* 39(4) (1998): 601–640.

mation of nature and the standardization of thought and culture. In *Tools for Conviviality*,[8] Ivan Illich lamented about industrial society's response to contemporary problems:

> It has become fashionable to say that where science and technology have created problems, it is only more scientific understanding and better technology that can carry us past them. The cure for bad management is more management. The cure for specialized research is more costly interdisciplinary research, just as the cure for polluting rivers is more costly non-polluting detergents. The pooling of stores of information, the building up of a knowledge stock, the attempt to overwhelm present problems by the production of more science is the ultimate attempt to solve a crisis by escalation.[9]

Illich's critique might be perceived by some as being primitivist and anti-technological. But closer reading shows that it is not the existence of technology that he questions, but rather the way in which society is adjusting and conforming to technology, rather than the other way around. Beyond generating dependence and addiction due to the convenience it offers, technology can also subtly and invisibly hinder society from exploring options. It may limit social choices in the search for solutions to those which rely on technology. And this proposition is often acceptable because usually it is offered as a bargain: maximum results with minimum effort, one of the characteristic aims of efficiency and economy.

However, there is more to this reorientation than just promoting laziness. In *The Real World of Technology*, Ursula Franklin describes technology as a synergistic system that "involves organization, procedures, symbols, new words, equations, and most of all, a mindset."[10] She argues that this has resulted in the reordering and restructuring of all social relations between individuals, social groups, nations, and even the environment.[11] Franklin regards technology as referring to the ways of doing things, not merely the tools or knowledge used.[12] By encompassing the systems of practice that accompanied or were necessitated by devices, she is able to

---

[8] I. Illich, *Tools for Conviviality* (1990 ed., London: Marion Boyars Publishers, 1973).

[9] *Ibid.*, p. 9.

[10] U. M. Franklin, *The Real World of Technology* (Toronto: House of Anansi Press, 1999), p. 12.

[11] *Ibid.*, p. 24.

[12] U. M. Franklin, "New Issues of Access to Justice Raised By Modern Technology," *The Ursula Franklin Reader: Pacifism As A Map* (Toronto: Between The Lines, 2006), p. 183.

differentiate between two kinds of technology: work-related technologies that make methods easier, and control-related technologies that are not mainly concerned with method, but whose purpose is to increase control over the work process. This then leads her to distinguish between two broad types of technology. The first are holistic technologies which organize work in a way that leaves an individual with control over what s/he is creating or doing from start to finish. These technologies allow the users to make decisions on their own as they work, i.e., to retain their autonomy which impacts not only on what they do or produce but also in directing how they are done or produced and more importantly, what the activity or production is for. The second type are prescriptive technologies, that divide a process into clear and simple steps or tasks which confine the user's skill and autonomy to only that task which s/he is assigned. They tend to limit, if not eliminate, the ability for independent judgement and principled decision-making, since the goal of the technology is already incorporated into the design of the tasks or system.[13]

The implications of this distinction are extraordinarily enlightening. Whereas holistic technologies, which have characterized most pre-industrial societies, require users to maintain an extensive and integrated knowledge base to enable them to independently control the entire technological process, prescriptive technologies require detailed divisions of labour coordinated through discipline, planning, or organization, and cannot work without the existence of some form of hierarchy. Franklin contends,

> Today's real world of technology is characterized by the dominance of prescriptive technologies. Prescriptive technologies are not restricted to materials production. They are used in administrative and economic activities and in many aspects of governance, and on them rests the real world of technology in which we live. While we should not forget that these prescriptive technologies are often exceedingly effective and efficient, they come with an enormous social mortgage. The mortgage means that we live in a culture of compliance, that we are ever more conditioned to accept orthodoxy as normal, and to accept that there is only one way of doing 'it'.[14]

By expanding technology to mean *practices*, or the way things are done, Franklin establishes a close relationship between technology and culture, since technology can then be considered as a behaviour, and a web of

---

[13] Franklin, *supra* note 10, at pp. 12–13.

[14] Franklin, *supra* note 10, at p. 17.

behaviours form the foundations of culture. Culture then also defines the social structure by establishing the relationships of social actors. From the structure and distribution of power within, Franklin believes that it can then be determined whether a technology enables a society to come closer or farther from establishing a social order characterized by justice.

Both Illich and Franklin point toward the dehumanizing impacts of prescriptive technologies, and how subordination into these types of machine-processes serves to create and maintain an invisible hierarchy in society. Taken in this light and considered under the rubric of globalization, it then comes as no surprise why some see technology as the instrument of modern-day imperialism. As Otto Ulrich succinctly states,

> For the cultures of other countries, the requisite psycho-social preparation of people and the cultural transformation looks much more traumatic because it confronts them with an essentially alien culture. Through technological 'development aid' more euphemistically called technical assistance, from industrialized countries, they receive 'trojan machines' (to use Robert Jungks's phrase), which conquer their culture and society from within. They are forced to gradually absorb an alien industrial work ethic, to subordinate themselves completely to unaccustomed time rhythms, to value objective relations higher than human relations, to experience increasing stress and to regard it as normal, and to accept jobs without regard for motivation or meaning. Wage labour and commodity fetishism expand, and they define the competitive struggle of all against all as the social synthesis. It becomes self-evident that everyone is to be a mechanical cog in a great production apparatus dominated by the world market.[15]

Technology and Social Justice

If technology does influence social patterns as suggested, then there can indeed be a relationship between the adoption of technological systems and the promotion or limitation of social justice. Social justice, as Miller puts it "very crudely," is about

> ... how the good and bad things in life should be distributed among the members of a human society. When, more concretely, we attack some policy or some state of affairs as socially unjust, we are claiming that a person,

[15] O. Ullrich, "Technology," in W. Sachs, ed., *The Development Dictionary: A Guide to Knowledge as Power* (London and New York: Zed Books, 2001), p. 275.

> or more usually a category of persons, enjoys fewer advantages than that person or group of persons ought to enjoy (or bears more of the burdens than they ought to bear), given how other members of the society in question are faring.[16]

As indicated above, social justice is often regarded as synonymous with distributive justice.[17] Young has counselled against taking this "distributive paradigm" too rigidly and exclusively, pointing out that a preoccupation with the distribution of goods and bads brings with it the dangers of mistakenly overextending the concept of distribution to immaterial and intangible social goods such as rights, overlooking the influence of the processes that lead to the distribution, and most importantly, conceal the patterns of behaviour and systems of power that keep reproducing injustices over time.[18] For her, apart from distribution, social justice should be concerned with "the degree to which society contains and supports the institutional conditions necessary for the realization of... values comprised in the good life."[19] These values she further identified as "(1) the development of one's capacities and expressing one's experience, and (2) participation in one's action and the conditions of one's actions."[20]

Even though most theories of justice implicitly consider only singular societies,[21] and do not squarely contend with social justice between States,[22]

---

16 D. Miller, *Principles of Social Justice* (Cambridge, Mass. and London: Harvard University Press, 1999), p. 1.

17 *Ibid.*, p. 2.

18 I. M. Young, *Justice and the Politics of Difference* (Princeton: Princeton University Press, 1990), pp. 15–38.

19 *Ibid.*, p. 37.

20 *Ibid.*

21 Miller, *supra* note 16, at pp. 4–5.

22 Brian Barry argues that the dominant liberal theories of justice, which are either utilitarian or contractual by nature, are insufficient frameworks to provide a satisfactory basis for States to establish principles of justice upon which international social justice is to be based. The problem is that ultimately, despite official statements and declarations to the contrary, States are not equals in international society, and to conceive of justice as the result of either utilitarian motivation (where it is maximum aggregate utility that justifies the outcome) or contractual agreement (where it is mainly the fact of agreement justifies the outcome) in such a situation allows the relative strengths and weaknesses of States to determine the end result, rather than being based purely on mutually-accepted principles of justice. B. Barry. *Theories of Justice Volume I: A Treatise on Social Justice* (Berkeley and Los Angeles: University of California Press, 1989), pp. 183–189.

recent years have witnessed the establishment of a truly global economic system and the expansion of international legal regimes throughout many aspects of international life. These provide a rudimentary framework for a global society from which it is possible to draw some intriguing parallels. The reality of continuing social, economic, and political disparities between nations despite the establishment of the United Nations, the acceptance of the principle of sovereign equality, and the promotion of international cooperation, however, all provide sufficient basis for continued inquiries into questions of social justice between the members of the international community.[23]

Many international endeavours have generally been justified, promoted and embraced in the name of mutual advantage and benefit. However, the continued failure of these expected benefits to materialize in many countries, calls into question how well the agreements and institutions negotiated by the world community have effected or are effecting a just distribution of the supposed benefits from their cooperation. This is the very same issue of social justice posed by Miller. Meanwhile, Young's values of the good life remarkably parallel the ultimate objectives of self-determination and participation in the international community. Her call for attention to the processes and patterns that perpetuate inequity is very relevant to questioning the negotiation and operation of international agreements. Since agreements establish expectations and patterns of behaviour among the parties by providing for their respective duties and obligations, it is therefore reasonable to inquire whether such agreements constitute these inequity-inducing processes. And when such agreements involve the adoption and transfer of technology, it is even more important to discover whether these are actually two-pronged systems of perpetuating inequity: a formal one through the enforcement of an overt system of rules and obligations, and an informal one through the unseen modification of systems of thought and behaviour.

---

[23] For some explorations into the concept of global social justice, see J. Rawls, *The Law of Peoples: With the Idea of Public Reason Revisited* (Boston: Harvard University Press, 2001); T. Pogge, *Global Justice* (Oxford: Blackwell Publishers, 2001); and C. Jones, *Global Justice: Defending Cosmopolitanism* (Oxford and New York: Oxford University Press, 1999).

Marine Technology and the Law of the Sea

The key to avoiding and defending against the technological encroachment into the social and cultural sphere, as Franklin notes, is a healthy scepticism, to question technologies and technological practices on a principled basis. This includes a critical assessment of the patterns by which technology is transmitted and disseminated, and their possible impact on the individual social and cultural systems that receive or adopt them. These almost invisible impacts of technology are often irreversible, which makes it all the more important to establish an awareness of these unnoticed potentials. Franklin contends that we need to ask not only what any given technology allows us to do, but also what it prevents us from doing.[24]

The need for this critical inquiry is especially relevant to the law of the sea,[25] as well as to the significant number of multilateral agreements that were negotiated in its wake. Increasingly, these agreements have implicated marine technologies in their implementation. For the purposes of this essay, "marine technology" refers generally to any and all of the technologies available for the use of or interaction with the marine environment and its resources. This includes shipping and fishing technologies, as well as non-sea based technologies applied to them (e.g., remote sensing technologies on satellites and aircraft).

Developments in marine technology have always been the main drivers in the evolution of international maritime law. As Miles notes, there has always been a distinct correlation between technological advancement and the development of international law[26] and authors often highlight this

[24] Franklin, *supra* note 10, at p. 51.

[25] *United Nations Convention on the Law of the Sea*, Montego Bay, 10 December 1982, UN/Doc. A/CONF.62/122, 7 October 1982 [hereinafter LOS Convention].

[26] E. L. Miles, "Preparations for UNCLOS IV?" in L. M. Alexander, S. Allen and L. C. Hanson, eds, *New Developments in Marine Science and Technology: Economic, Legal and Political Aspects of Change, Proceedings of the 22nd Annual Conference of the Law of the Sea Institute, Narragansett, Rhode Island, 12–16 June 1988* (Honolulu: Law of the Sea Institute, 1988), 491–501, at p. 492.

connection.[27] The background of the original limits of the territorial sea[28] and the continental shelf[29] are particularly good illustrations of this correlation. They also exemplify the most prominent relationship of marine technology to the law, which is to establish the foundation for claims to, or counter-claims against, some form of control, jurisdiction, or sovereignty over the marine environment and its resources. The mere expectation of future technological developments may itself provide a great impetus, as the interest in seabed mining in the 1960s so vividly demonstrated in convincing the international community to begin the Third United Nations Conference on the Law of the Sea (UNCLOS III).[30]

[27] See, for example, R. P. Anand, *Origin and Development of the Law of the Sea* (Hague/Boston/London: Martinus Nijhoff Publishers, 1983), pp. 162–163; D. M. Johnston, *The Theory and History of Ocean Boundary-Making* (Kingston and Montreal: McGill-Queen's University Press, 1988), pp. 61–74, and 252; and E. Gold, *Maritime Transport: The Evolution of International Shipping Policy and Maritime Law* (Lexington, Mass.: Lexington Books, 1989).

[28] The statement of the cannon-shot rule upon which it was first based, "*Imperium terrain finiri ubi finitur armorum potestas*," devised by Cornelius van Bynkershoek in 1737, may be read immediately as a boundary based on technology. Thus, when the first-assumed absolute limit of one league or three geographic miles was surpassed decades later by improved gunnery, the limits of State sovereignty into the adjacent sea again became a contested concept, and would not be settled for centuries until the LOS Convention. See P. C. Jessup, *The Law of Territorial Waters and Maritime Jurisdiction* (New York: G. A. Jennings Co., 1927), pp. 5–7; R. R. Churchill and A. V. Lowe, *The Law of the Sea*, 2d ed. (Manchester: Manchester University Press, 1988), pp. 59–84.

[29] See U.S. President Harry S. Truman's Proclamation No. 2667, 28 September 1945, entitled "Policy of the United States with respect to the Natural Resources of the Subsoil and Sea Bed of the Continental Shelf," issued after the successful development of offshore oil drilling technology. See also, Churchill and Lowe, *ibid.*, pp. 120–131.

[30] Malta's 1967 proposal for a new treaty on the seabed was accompanied by a memorandum pointing out that the seabed had been beyond national jurisdiction because of the lack of technology to access the resources of the seabed. But by the late 1960s it was envisaged that that technology barrier was about to be breached, and the technologically advanced countries would soon be progressively and competitively appropriating the seabed for their own use. This, Malta feared, "...would result in the militarization of the accessible areas of the ocean floor, through the establishment of fixed military installations and in exploitation and depletion of resources of immense potential benefit to the world, for the national advantage of technologically developed countries." See R. P. Anand, *Legal Regime of the Sea-Bed and the Developing Countries* (New Delhi: Thomson Press (India) Ltd., 1976), described and quoted at p. 181.

In the aftermath of UNCLOS, we may further discern that particular marine technologies are implicitly and increasingly becoming the prescribed means by which the marine environment and its resources are to be accessed, used, regulated, and managed. This second relationship is most apparent and pervasive in the case of shipping. The activities of the International Maritime Organization (IMO) have increasingly revolved around the prescription of technological standards for a wide range of purposes, such as safety, pollution prevention, facilitating trade, and ensuring national security. This trend is also reflected in other maritime sectors such as fishing and offshore petroleum exploration and development.

The third relationship of marine technology with the law of the sea concerns its perceived role as an "equalizer" among the maritime nations of the world. It is upon this assumption that technology-transfer provisions are justified and agreed upon because technology transfer is seen as the means to compensate for inequalities in capacity and capability to implement international agreements. The law of the sea clearly manifests this relationship, and the framers of the LOS Convention certainly had great hopes for the spread of marine science, research, and technology through the inclusion of Part XIV on the development and transfer of marine technology. The exciting prospects of seabed mining that had galvanized the international community for most of the UNCLOS III negotiations were accompanied by the immediate recognition that its benefits could only be assured by a certain level of scientific and technological advancement. Promoting fair access to such knowledge and technology therefore became the centrepiece of provisions to promote cooperation in marine scientific research[31] and marine technology transfer and capacity-building.[32]

These three functions of establishing foundations, prescribing terms of use, and levelling the field among States, emphasize that marine technology per se plays a pivotal role in determining whether a coastal State would be able to fully and effectively benefit from its recognized rights and entitlements, both exclusively within its maritime territories and jurisdictions or shared with other States in areas beyond national jurisdiction. There is also a distinct correlation between access, use and systems of technology and a coastal State's participation in the oceans' bounty. Thus, it is all the more timely and important that a critical perspective be taken to reviewing how

[31] LOS Convention, *supra* note 25, at Arts., 123, 143, 239, 242 and 243.

[32] *Ibid.*, Arts. 266–278.

and why the adoption and use of marine technology might also lead to the opposite result. Marine technology may be as likely to contribute to perpetuating the inequalities originally intended to be addressed by the international legal regime for the oceans. Such a reflection might point toward the necessary safety nets and solutions that bring the technology back under the control of its users and ensure that it fulfils the promises of prosperity by which the technology was promoted.

It must be noted, however, that at least three factors make this task a rather complex undertaking. First, as a framework instrument, the LOS Convention naturally does not contain all the law that relates to marine technology. Subsequent implementation of the provisions of the Convention may be contained in the myriad other international agreements that have been negotiated since. We must also consider that the sea may have been subjected to the application of other conventions independently of the law of the sea itself, such as the Convention on Biological Diversity, or the IMO conventions. Second, given the expanded definition of technology, technology and technology transfer may often be embedded in activities and agreements that are not identified as such. "There is a technology of prayer as well as of ploughing, there are technologies to control fear as well as to control flood," as Franklin quotes Kenneth Boulding.[33] Thus, technical assistance, development aid, cooperative activities and joint ventures, education and training of any kind involve the transfer of technologies, and with them, the transfer of prescribed social and international relations. Often these relations are not overtly identified as such, but are assumed as requisite conditions in order for the system to be properly implemented.

Third, the continued and unabated advocacy for technology transfer and capacity-building since the 1960s has led to an accumulation of related duties and obligations scattered across international law, and their convergence upon the overlapping subjects of agreements needs to be considered. The United Nations Conference on Trade and Development (UNCTAD) reports, "over 80 international instruments and numerous sub-regional and bilateral agreements contain measures related to transfer of technology and capacity-building."[34] Their interaction with the global

[33] Franklin, *supra* note 10, at p. 6.

[34] United Nations Conference on Trade and Development, *Compendium of International Arrangements on Transfer of Technology: Selected Instruments* (New York: United Nations, 2001).

intellectual property regime now in place, which is comprised of a network of separate multilateral agreements,[35] also needs to be reckoned with. The problematic issues may come from unexpected quarters, which can not be seen until after the interactions of all the related agreements have actually emerged in practice. These factors establish a very broad scope of possible technology impacts that pose difficulties for any attempt to comprehensively foresee and assess the implications of continued adoption and transfer of marine technology.

Notwithstanding these difficulties, however, perhaps one way to begin is by considering three cases in which the use or adoption of marine technology results in broader questions of social justice between maritime nations: technological limits in delineating the extended continental shelf, high seas fisheries management, and seafaring. These vignettes provide concrete examples of how marine technology is an important factor in the allocation of the expected common benefits of the oceans.

Technological Limits in Delineating the Extended Continental Shelf

As previously noted, the development of the legal regime of the continental shelf is almost inseparable from the development of marine technology, particularly offshore petroleum exploration and development. From its classical formulation in treaty law, a technology-based criteria, depth-of-exploitability, supplemented the arbitrary depth of 200 metres as a maritime boundary limit.[36] The renegotiation of the continental shelf regime in UNCLOS served to clarify these limits into an admittedly still non-scientific standard breadth of 200 nautical miles, but also

[35] The 1883 *Paris Convention for the Protection of Industrial Property*, the 1886 *Berne Convention on the Protection of Intellectual and Literary Works*, and the 1994 *Agreement on Trade-Related Aspects of Intellectual Property Rights* (TRIPS) annexed to the *Marrakesh Agreement Establishing the World Trade Organization*, are just three of the most prominent and well-known of these intellectual property conventions. The World Intellectual Property Organization lists at least 17 major international conventions on intellectual property. See World Intellectual Property Organization, *WIPO Intellectual Property Handbook: Law, Policy and Use*, WIPO Publication No. 489(E) (Geneva: WIPO, 2004), pp. 240–363. Online at <www.wipo.int/about-ip/en/iprm/>, 31 January 2008.

[36] *Convention on the Continental Shelf*, Geneva, 29 April 1958, 499 *U.N.T.S.* 312, Art. 1.

established a distinctly technological basis for the "extended continental shelf" beyond that limit.

Under paragraph 4 of Article 76 of the LOS Convention, a coastal State may delineate the outer limits of its continental shelf beyond 200 nautical miles from its baseline through two formula that refer to either the thickness of sediments or a fixed distance in relation to the properties of the "foot of the slope," which in turn is determined either by a "maximum change in gradient" or, more importantly, "evidence to the contrary." Both sediment thickness and the evidentiary basis for determining the foot of the continental slope rely on a scientific interpretation of the properties of the seabed itself, which can be undertaken only if the coastal State has access to the requisite geo-scientific and physical oceanographic technology. The "evidence to the contrary" acceptable to the Commission on the Limits of the Continental Shelf includes "modern" geodetic, bathymetric, seismic, magnetic, and gravimetric data acquired through specialized scientific surveys, instrumentation and techniques.[37] A perusal of the data requirements for submission also indicates that the quality and reliability of the evidence, and impliedly its acceptability, appear to be directly related to the submitting coastal State's technological sophistication. For example, a coastal State which includes as evidence multi-beam seismic and bathymetric data, complete computer models of the seabed, gravity and magnetic anomaly maps, supplemented by core sampling and remote sensing data, may be more likely to persuade a body of scientific experts of its entitlement to an extended continental shelf area than a coastal State similarly situated but submitting only bathymetric data. The technological advantage is even more crucial where various underwater features such as trenches, troughs and depressions could easily bring into question a coastal State's claim to a "natural prolongation" beyond 200 nautical miles, or where the physical properties of the seabed itself are crucial.

This seems to imply that the extended continental shelf is ultimately a privilege of technological capability and expertise: despite the reference to the use of scientific criteria and evidence, the very act of acquiring and interpreting such data is entirely dependent on the availability of relevant technical and human resources to the coastal State. The legal characterization

---

[37] See Sections 6.3.6 to 6.3.10, and 9.2 to 9.3.17, Commission on the Limits of the Continental Shelf, *Scientific and Technical Guidelines of the Commission on the Limits of the Continental Shelf*, CLCS/11, 13 May 1999.

of the coastal State's right to the continental shelf *ipso facto* and *ab initio* masks the practical difficulty that the extent of the entitlement is now either limited or facilitated by the technology that was adopted by the LOS Convention for its measurement.

Article 76 has already been seen as posing serious questions of just allocation for a developed country like Canada. Scholars have framed very interesting questions on the intertwining of legal and technical ambiguities resulting in policy questions that have yet to be resolved in the practice of coastal States and the CLCS.[38] Chief among their concerns was how these uncertainties and their resolution could be a factor in defining what would be required for Canada to marshal its continental shelf claim. However, these concerns would be even more serious for countries not as endowed with fiscal and human resources as Canada.

Less developed coastal States are faced with particularly thorny domestic policy questions. The first issue that faces their governments is whether they should allocate significant financial resources to acquiring the technology and data on their own for a relatively limited purpose, compared to other urgent social needs. If they do not, and instead resort to engaging foreign expertise on a commercial basis, they must then contend with questions of how to determine if they are being supplied with correct advice and competent services, or if their payments are commensurate to the services offered and undertaken. Certainly questions will be raised as to whether the investment in time, effort, energy, finances and people in preparing a submission is actually worth pursuing, especially if the coastal State does not even have adequate information on existing, let alone prospective, resources in its maritime spaces. Countries fortunate enough to have had a reasonably busy history of offshore exploration and corresponding data archives would at least have some idea of their potential resources, but those that have not had active offshore petroleum sectors and less strong economies would likely not. Given that the LOS Convention establishes a deadline for submission of extended continental shelf claims, it would

[38] See S. Cockburn, S. Nichols, D. Monahan, and T. McDorman, "Intertwined Uncertainties: Policy and Technology on the Juridical Continental Shelf," Paper presented at the at the Accuracies and Uncertainties in Maritime Boundaries and Outer Limits, Advisory Board on the Law of the Sea to the International Hydrographic Organization (ABLOS) Conference, Monaco, 18–19 October 2001, <www.gmat.unsw.edu.au/ablos/ABLOS01Folder/ablos01_papers.htm>, 31 January 2008.

seem that less technologically developed countries could be vulnerable to situations of imprudent expenses, unnecessary claims, or premature abandonment.[39]

High Seas Fisheries Management Technology

A second example to consider is the rise to prominence of regional fishery management organizations (RFMOs) in the wake of the UN Fish Stocks Agreement. Since 1995, RFMOs have emerged as the technology of choice, figuratively and literally, for the management and conservation of straddling and highly-migratory fish stocks. It has been noted that currently, RFMO jurisdictions have covered practically all of the waters of the world.[40] Since RFMOs have a number of major issues and problems in common, the techniques and technologies that have been used or proposed for application in their respective regulatory areas are fairly similar. One example of a popular group of technologies is the use of a combination of satellite transponders, aerial or sea-based electronic surveillance, and shore-based tracking stations to form a vessel monitoring system (VMS) that keeps tabs on fishing activities taking place within the convention area. This type of technological system was contemplated in the establishment of the RFMO for the Western and Central Pacific Ocean,[41] a major tuna fishing ground encompassing a considerable area of the Pacific Ocean, including the waters of the developing States in Southeast Asia and the South Pacific island States. In fact, the use of the technology is expressly required as a flag State duty in Article 24(8) of the WCPFC.

---

39 In this respect, it is commendable that the United Nations has established a trust fund to provide training, advice, and personnel, and assistance in preparation of submissions, to developing and least developed countries to assist them in meeting the obligations under Article 76 of the LOS Convention (United Nations, *General Assembly Resolution No. 55/7*, UN Doc. No. A/RES/55/7, 27 February 2001). While this supports the technical tasks of preparing the claim, it does not appear to extend very far into the more demanding work of acquiring the actual evidence and data that the technical personnel must work with.

40 E. Meltzer, "Global Overview of Straddling and Highly Migratory Fish Stocks: Maps and Charts Detailing RFMO Coverage and Implementation," *International Journal of Marine and Coastal Law* 20, No. 3–4 (2005): 571–604.

41 *Convention on the Conservation and Management of Highly Migratory Fish Stocks in the Western and Central Pacific Ocean*, Honolulu, 5 September 2000 [hereinafter WCPFC], <www.wcpfc.int>, 31 January 2008.

The negotiations on the WCPFC represented a problem for the Philippines. Some Philippine fishing companies had been operating within the convention area since at least the post-war period, and the local tuna fishing industry had experienced a massive boom in the 1990s. Artisanal tuna fishers and fishing cooperatives using thousands of hand-liners and wooden pump-boats formed the backbone of the tuna canning and export industry based in southern Mindanao, contributing significantly to that region's recovery from the ravages of two decades of secessionist rebellion. Though relatively small and primitive compared to modern fishing vessels,[42] they were known to venture well beyond Philippine waters and into the deep waters of and around neighbouring Indonesia, Palau and Papua New Guinea. Unlike the established Philippine fishing companies using purse seiners and carriers, these cooperatives could not bear the costs of outfitting their pump-boat fleets with the satellite transponders required by the WCPFC. Neither could the Philippine government, which at the time was faced with more pressing economic problems.

As the negotiations went forward, it was clear to the tuna industry that the adoption by the WCPFC of VMS as a standard would effectively convert a major fisheries sub-sector into an illegal industry simply because of its inability to conform to the technological requirement. It is for this reason that Philippine positions in the negotiations tended to be protective of what the Convention called the "traditional" fisheries.[43] Due to the lack of adequate and detailed information about Philippine tuna hand-

[42] Although Philippine tuna fishers use a variety of boats, the most popular for the small- to medium-sized operators are wooden outrigger canoes ranging from 12 to 14 metres in length and powered by 160 horsepower diesel engines, and larger motorized pump-boats called *palakaya* of about 27 to 30 metres and powered by 225 horsepower diesel engines. Even though they may weigh less than three gross tons dead weight, and thus are classified under Philippine law as artisanal (or municipal) fishing, they are known to be used for long-distance fishing in the open ocean. See V. Aprieto, *Philippine Tuna Fisheries: Yellowfin & Skipjack* (Quezon City: University of the Philippines Press, 1995).

[43] Toward the end of the negotiations for the WCPFC, this author participated in a number of meetings of the Philippine National Tuna Industry Council, a multi-sectoral council established by the Department of Agriculture as the main forum for discussions and recommendations about policies and programmes affecting the tuna industry, including bilateral and multilateral fishing and trade negotiations. Up to the last meeting he attended in 2002, there were continuing concerns about VMS technology and its impact on the smaller fishing co-operatives.

liners and pump-boat operations, whether these should be considered a concern of the WCPFC, or a problem in exclusive economic zone (EEZ) management has been a subject of vigorous debate. Since the technical description of the WCPFC regulatory area includes the EEZs of the littoral countries, and there are attendant obligations to ensure consistency and compatibility of conservation measures applicable in the EEZs and high seas of the Convention area, Philippine fisheries authorities believe that the problem of a possible "technological de-legitimization" of this sector will have to be faced again in the future.

At the root of the Philippines' problem with the RFMO model and the "traditional" tuna fishing sector is the fact that RFMOs and the management techniques that have been developed for them have historically been based on the assumption that the fishing effort is conducted by a highly industrialized fishing sector. Pursuing the critique of the RFMO model as an entire technological system then leads to a number of questions about the ultimate purposes of RFMOs. For instance, if RFMOs as a whole are geared to support an industrialized fishery supplying distant voracious markets, who are the real beneficiaries and what are their actual effects in distributing the benefits of the seas? It is generally agreed that most fish stocks worldwide are fully- or over-exploited,[44] and in response some RFMOs have advised prospective new members that they cannot expect fully equal participation in the catch quota allocations and shall be limited to smaller segments of the fishery in their respective convention areas.[45]

---

[44] See, for example, Food and Agriculture Organization, *State of World Fisheries and Aquaculture 2007* (Rome: Food and Agriculture Organization, 2007); R. Hilborn, T. A. Branch, B. Ernst, A. Magnusson, C. V. Minte-Vera, M. D. Scheuerell, and J. L. Valero, "State of the World's Fisheries," *Annual Review of Environment and Resources* 28 (2003): 359–399; Food and Agriculture Organization, "The Rome Consensus on World Fisheries," Adopted by the FAO Ministerial Conference on Fisheries, 14–15 March 1995, FAO Doc. AC441/E.

[45] See North Atlantic Fisheries Organization, *Resolution No. 1/99 adopted by the General Council of NAFO to guide expectations of future new members with regard to fishing opportunities in the NAFO Regulatory Area*, NAFO/GC Doc. 99/8, 17 September 1999, <www.nafo.int/about/frames/about.html>, 31 January 2008. See also, North East Atlantic Fisheries Commission, "Guidelines for the expectation of future new Contracting Parties with regard to fishing opportunities in the NEAFC Regulatory Area," 22nd Annual Meeting, November 2003, <www.neafc.org/about/becoming-cp.htm>, 31 January 2008.

In this light, it could be argued that RFMOs and their associated management techniques have become a means for appropriating a global commons for the benefit of a few advanced countries in the name of conservation and management. This implies that developing or least developed countries that seek economic opportunities by developing high seas fisheries sectors, are then practically cut off from all future access to global fisheries, or worse, de-legitimized at inception, primarily because their fishing technologies are not as sophisticated as those of others, or simply due to their late entry into the high seas fishery.

Technology Transfer and Seafaring

Due to the declining popularity of the seafaring life in the industrialized countries, the major shipping nations, chiefly the Organisation for Economic Co-Operation and Development (OECD) countries, now face an extreme shortage of senior officers.[46] As early as 2000, studies had shown that that by 2010 shipowners would experience a serious shortage of competent officers at both the deck and engineering levels, numbering in the tens of thousands, to command their fleets.[47] As there was no improvement in the recruitment rates for seafaring careers in the West, it was realized that given the maritime human resource trends, there would be little choice but to accept that the officer corps would soon be drawn from seafarers of mostly developing countries of the Far East. This was one of the major incentives for the urgent revision of the International Convention on Standards of Training, Certification and Watchkeeping for Seafarers (STCW) (1979), to produce the rigorous education and training standards and requirements under STCW-95. The IMO defined the minimum standards and acceptable processes for the education and training of maritime human resources acceptable to the shipowning coun-

[46] See "Shortage of Seafarers to Worsen in the Future: Drewry Consultants," *The Economic Times*, 26 May 2008, <economictimes.indiatimes.com/News/News_By_Industry/Transportation/ Shipping_Transport/Shortage_of_seafarers_to_worsen_in_future_Drewry_Consultants/rssarticleshow/3071572.cms>, 10 July 2008; "Crew Shortage Fueling P & I Claims," *Fairplay Daily News*, 28 March 2008; and Warwick Institute for Employment Research, *The Worldwide Demand for and Supply of Seafarers BIMCO/ISF 2000 Manpower Update* (London: BIMCO/ISF, 2000 and Warwick Institute for Employment Research, 2000).

[47] See *BIMCO/ISF 2000 Manpower Update, ibid.*, and *The Worldwide Demand for and Supply of Seafarers: BIMCO/ISF Manpower 2005 Update* (London: BIMCO/ISF, 2005).

tries. All labour-supplying countries hoping to tap the lucrative market would have to implement these stringent international standards in their respective national maritime education and training systems.

The Philippines, a developing country in Southeast Asia, is currently the world's largest supplier of seafarers with over 250,000 deployed in the international fleet as of 2006.[48] The country was among the first labour-supplying nations to benefit from the now-common practice of outsourcing foreign crews for out-flagged vessels. In addition to ensuring implementation of and compliance with STCW-95 standards through a range of regulatory measures, the Philippines also accepted overseas development assistance projects and/or joint venture agreements with traditional shipping nations such as Norway, Denmark, the Netherlands, and Japan, or with their respective shipowners' organizations, to establish or upgrade maritime schools in the Philippines, or to implement special curricula tailored to the partner States' requirements. Coupled with agreements for various modes of sponsorship and placement, this would assure the human resource supply for the partner nations' fleets in the years to come.[49] These cooperative arrangements also involve the transfer of finances, materials, equipment, and expertise. The successful implementation of the initial programmes encouraged the government and the manning sector to promote this as a model of cooperation that produced a win-win situation for both sides: it generated lucrative employment for Filipinos, and in turn provided the shipping nations with much needed officers.

The whole practice, as well as the individual projects and arrangements, may be viewed as a complete technological system, encompassing the entire range of tools, equipment, techniques, methodologies, skills, knowledge systems, and organizations, being transferred from ship-owning to labour-supplying countries. To be sure, the scale of this technology transfer goes well beyond the minimums implied in the LOS Convention, as well as relevant IMO conventions. It is indeed an excellent example of very close State-to-State and private-sector-to-private-sector cooperation.

---

48 V. Uy, "Filipino Seafarers Deployed in 2006 Breach 250,000 Mark," *Philippine Daily Inquirer*, 16 January 2006.

49 C. C. Salinas, President, Filipino Shipowners Association, "Challenges and Opportunities for the Future: The Role of the Philippines and Its Impact on the European Shipping Industry," Paper presented at the 10th European Manning and Training Conference, Croatia, March 2007.

But even as the mutual benefits of these arrangements are acknowledged, critical questions need to be asked about the long-term impact of these arrangements on the overall situation of the parties in terms of national economic opportunities in the world market of shipping. The upgrading of maritime education and training institutions in response to STCW-95 requirements has led to the improvement of course content. This allows Filipino seafarers to be qualified operators of prescribed shipping machinery and equipment, but does not allow much room for innovation. It must also be considered that current trends in ship design are leading to smaller crews relative to the size and weight of new vessels. Aside from increasing the workload of the crew per capita, this also tends to require increasing degrees of technical proficiency and specialization in tasks and equipment. Increasingly, seafarers are becoming technicians within prescribed competencies.

While formal and technical programmes are emphasized, other substantive maritime courses such as naval architecture and ship-building, related subjects such as shipping investment and financing, and ancillary maritime courses that would directly help to strengthen and expand the indigenous shipping sector are neglected. Ship-building, ship-owning, and ship management are not emphasized in technology transfer efforts with the Philippines because these are the traditional domains of the maritime industrial powers in which there are currently no shortages.

While Filipino seafarers are sent abroad in ever-increasing numbers to man foreign fleets, the industrialized sector of the Philippine domestic shipping fleet has difficulty attracting and maintaining its own human resource requirements. The management and regulation of the informal, small-scale, non-industrialized shipping sector (made up of tens of thousands of wooden outrigger boats carrying smaller volumes of inter-island and cross-border trade) is largely ignored.

It could thus be argued that the maritime education and training system has been oriented to primarily service the needs and ensure the viability of the industrialized shipping industries of other countries, rather than the needs of the Philippine's own shipping sector. This serves largely to maintain the dominant position of industrialized countries as ship-owners and employers, while labour-supplying countries such as the Philippines and countries similarly situated remain relatively less developed. It preserves the present distribution of capital and labour in the global shipping industry, notwithstanding the increasing lack of seafarers in the ship-owning countries. Several labour-supplying countries with over a million seafarers combined are left to compete with each other for the few thousand opportunities that are opened by the attrition and retirement of the older generation of Western officers. Senior officers are assimilated from the developing countries to support the dominant position of the few industrialized countries.

Whatever opportunity the ship-owning sectors of developing countries might have had to expand and compete with the established maritime nations by filling in the gaps caused by the lack of human resources in the latter is effectively nipped in the bud. Any possible benefits that the local shipping industry might derive from having an experienced corps of merchant marine officers is likewise obviated because they are absorbed by foreign employment. Thus, the shift in demographics of seafarers from the OECD countries to the countries of Far East is certainly not going to be accompanied by an appreciable and similar shift in the latter's control and ownership, or even market share, in proportion to the world fleet. And it is likely to be a one-way shift for years to come. This is one example of how technology transfer, well-meaning though it may be, can be both an obvious boon but also a hidden bane on a maritime country's national development and place in the world economy.

Marginalization and Marine Technology

The previous three examples are good illustrations of how marine technology may have a marginalizing impact on those less fortunate than their developers or promoters. This is not a new phenomenon; at the start of the Industrial Revolution, workers displaced by machinery bore the brunt of exclusionary impacts by the thousands. However, today, this kind of technological displacement can happen at much larger scales, in more subtle ways, and span the entire global community. It is no longer simply a matter of certain jobs and professions becoming obsolete. Today, the options and opportunities of entire societies and countries are being placed on a lower and subordinate priority in a global social and economic order that is constantly being structured by technology.

It is thus worth engaging in similar reflections about any number of international agreements and activities that rely heavily on the application and transfer of technology. This questioning is especially urgent for the marine environment: any activity in the oceans requires the application of technology because humanity is not innately suited for life in the sea. Technology serves as the medium and the tool through which society interacts with the marine environment. It is technology which defines a society's relationship with the seas and allows it to benefit from the ocean's bounty. Technology also invisibly influences how those benefits are allocated and distributed. While the technology versus environment debate became more prominent in 1960s and 1970s,[50] it has since advanced and been brought into sharper focus, particularly in the light of the technological innovations of the past 15 years. As far as the LOS Convention is concerned, those early stages of the polemic have been frozen in time through its marine technology transfer provisions. However, these provisions have a longer lasting normative influence on the growth and development of ocean management that has probably not been as critically examined.

Today, the convergence of cutting-edge technologies in machinery, materials, software, automation, and engineering are pushing the boundaries of our knowledge and use of marine spaces and resources. Deep ocean exploration reveals completely new forms of life in extreme environmental conditions that are thought to contain exotic genetic raw materials that could be used to bio-engineer new compounds and promise yet unknown advances in medicine, life and chemical sciences. Information technologies are allowing more and more data to be accessible to indiscriminately larger numbers of people. This information ranges from that whose discovery provides the missing piece in a long unsolved scientific puzzle, to information whose isolation or concealment was the last remaining defence against over-exploitation and destruction of the resources and culture of a marginalized people. At the fringes of the seas, aquaculture and mariculture technologies relying on breeding techniques, genetics, and engineering are being employed, improved, invented, tried, and tested in myriad conditions in the hopes of providing reliable alternative food sources. Multi-platform and multi-level surveillance and monitoring devices combining aerospace and remote sensing technologies are employed to scrutinize and document everything from natural sea conditions to movements of maritime commercial and military transportation, whose data can be employed for anything from academic research to military action. Seabed-based energy

[50] Stine and Tarr, *supra* note 7, at p. 603.

resources such as petroleum are being plumbed from ever greater depths, while surface-accessible wind and wave power potentials have begun to be tapped at increasing distances from the shore. Coastal frontage is being artificially expanded, or planned to be so, through reclamation and the creation of artificial islands to provide additional land spaces for economic activities, or perhaps in an attempt to strengthen claims to territory. Living resources of the sea are being systematically and efficiently traced, chased, captured and processed to supply voracious markets half a world away from their points of origin using an extraordinarily coordinated system of fishing, trading, and transportation technologies. Greater and greater volumes of goods and materials are moving across the oceans due to continued advances in logistics, supply chain management, and port or shipping technologies, providing the lifeblood of an expanding global economy.

In all of these activities, the common questions about technology rarely go beyond how much they can extend sensory perceptions or analytical capabilities, or how much faster or easier they can perform or permit difficult operations, or what kinds of positive and negative bio-physical or economic effects they are expected to have. Few, if any, would challenge the benefits of new technology for any given task, and it will only be after the technology has already been adopted and set in place that its deeper impact may be slowly realized. But by then it may already be too late to undo what technology has already changed. International marine law has always been the follower, not the leader, to technological developments in human use of the oceans; marine law has always been a technological regime as much as a legal one. History has already demonstrated that it was the traditional systems of practice developed with ancient shipping technologies that provided the foundation for modern international maritime law. In the same way, the practices that arise out of the application of new technologies to the last remaining frontiers of the oceans will shape the marine legal regimes of the future.

Many of these developments, and more, are taking place under the rubric of international cooperation in accordance with the general duties that were initially enshrined in the LOS Convention. Multinational efforts have been organized at many different levels in various areas on the basis of firm conviction that joint and cooperative scientific and technological endeavours will benefit all of humankind. Marine science and technology, after all, can be such a huge endeavour in an enormous arena that it is almost axiomatic among all people and professions concerned with the marine environment that cooperation is a must, and technology is indispensable. While these cannot be denied, it does not mean that they cannot also be questioned. Again, Franklin notes:

> There is a lot of talk about global crises and "our common future." However, there is far too little discussion of the structuring of the future which global applications of modern technologies carry in their wake. What ought to be of central concern in considering our common future are the aspects of technological structuring that will inhibit or prevent future changes in social and political relations.[51]

Conclusion

Just as Douglas Johnston had predicted back in 1967, technology is continually re-ordering humanity's relationship with the seas. Indeed, there is a hypertechnical logic at work in identifying issues and offering solutions to an enormous number of interesting technical and technological hurdles. Forty years after his initial assessment, many of the great strides he expected have come to pass, and the values placed on efficiency, skill and economy have risen to prominence, especially since they conform so well with the values of the global market economy. There can be little doubt that a technological order has emerged and continues to be built over the sea.

But we must also take a step back and ask ourselves: how is technology re-ordering our humanity and our societies, and what other purpose does that hypertechnical logic wittingly or unwittingly serve? These are important questions not only for international marine policy, but also for international environmental discourse generally since every environmental issue that is now on the table involves some form of technological mediation or intervention. What the critical view of technology provides is a way to assess technology not only on the bio-physical impacts of its application, whether expected or not, beneficial or detrimental, but also on the question of what the technology itself means for the very nature of society that adopts and uses it. It allows us to reflect more deeply on whether the tools, techniques, and practices we use to solve problems or

[51] Franklin, *supra* note 10, at p. 42.

to establish order may either prevent or allow the pursuit and promotion of global social justice.

This is not to say that efficiency and technology, or their advocacy, do not have their place. The task of ocean management is a huge and complex undertaking, one that does indeed require the application of different philosophies and perspectives. As Douglas Johnston pointed out, much later in his scholarly and incredibly productive life,

> It may be that the ocean dilemma lies not so much in the challenges to idealism as in the coexistence of differing, and sometimes competing, brands of idealism, which in turn can be seen to project different visions of the future.[52]

He was always optimistic and open to the constant clashing and melding of contrasting and contesting viewpoints, including the revisiting and testing of old ideas and comparing them with new ones. He had always known that the best solutions are likely to be found somewhere in the middle ground, rather than from a singular and close-minded idealism that historically had proven to always be inadequate. Hopefully, through a continuing engagement with his life's worth of thoughtful contributions by the present and future generations of scholars, his ultimate vision of the oceans shared in common by all the peoples of the world will be realized.

---

[52] D. M. Johnston, "Ocean Governance: Converging Modes of Idealism," in J. M. Van Dyke, D. Zaelke and G. Hewison, eds, *Freedom for the Seas in the 21st Century: Ocean Governance and Environmental Harmony* (Washington D.C. and Covelo, CA: Island Press, 1993), p. 471.

State Practice and Maritime Claims: Assessing the Normative Impact of the Law of the Sea Convention

Stuart Kaye*

Introduction

The United Nations Convention on the Law of the Sea (LOS Convention)[1] entered into force on 16 November 1994, and was the culmination of many decades of negotiation between States. The Convention remains one of the greatest attempts at codification of customary international law, as well as the creation of new law and institutions, ever undertaken. It was intended to be a "constitution for the oceans" and circumscribe and regularise State maritime practice into the 21st century.[2]

The purpose of this essay is to examine whether the objective of creating a "constitution for the oceans" has been realized, as demonstrated through the prism of State practice. States have had over ten years since the entry into force of the LOS Convention, and over 25 years since it was opened for signature, some 155 States have become parties to it.[3] This would seem a reasonable period in which to regularize their practices, and ensure they are compliant with the LOS Convention, if indeed the Convention is exerting a normative effect on State behaviour, in the fashion one might expect of a "constitution."

* Professor of Law, University of Melbourne, Australia.

1 *United Nations Convention on the Law of the Sea*, Montego Bay, 10 December 1982, 1833 U.N.T.S. 396 [hereinafter LOS Convention].

2 This phrase was used by the President of the Third United Nations Conference on the Law of the Sea (UNCLOS III) in his statements at the final session of the Conference on 6 and 11 December 1982: T. T. B. Koh, "A Constitution for the Oceans," <www.un.org/Depts/los/convention_agreements/texts/koh_english.pdf>, 6 March 2008.

3 As at 31 May 2007.

This essay will consider two key regimes of the LOS Convention and examine State practice, with a slight emphasis on the Asia-Pacific region, to determine whether State compliance with the Convention reflects its important status, and to demonstrate any normative effect the Convention has exerted. While these regimes represent only a small fraction of the Convention, a wider study of the text as a whole would be a vast undertaking, given the over 300 articles and nine annexes contained within it. Consequently, in the context of this book, length permits only these specific regimes within the LOS Convention to be explored.

The regimes selected are the maritime zones prescribed under the Convention and the guarantees of freedom of navigation. These have be chosen because each represents provisions which are in high usage and are applied by all coastal States. As well, their application is of interest not merely to coastal States, but also a wide range of users of the world's oceans. The analysis of State behaviour will consider the practice of coastal States only, as it will be based on legislative responses to the LOS Convention.[4] Landlocked States rarely have relevant legislation indicating their attitude to aspects of the Convention, and certainly not in respect of maritime zones and freedom of navigation. At best, such States exert a benign influence on these aspects of the law of the sea, and so they are not considered here.

Maritime Zones

Prior to the Third United Nations Conference on the Law of the Sea (UNCLOS III), the width of the territorial sea was an issue of great contention in the law of the sea. Disagreement over the issue of the width of the territorial sea was principally responsible for the failure of the 1930 Hague Conference[5] and of the Second United Nations Conference on

[4] Tables have been compiled from data on the United Nations DOALOS website: <www.un.org/Depts/los/LEGISLATIONANDTREATIES/index.htm>, 2 October 2007.

[5] See M. O. Hudson, "The First Conference on the Codification of International Law," *American Journal of International Law* 24 (1930): 447–466; D. P. O'Connell (I. A. Shearer, ed.), *The International Law of the Sea*, Vol. 1 (Oxford: Oxford University Press, 1982), pp. 158–159; J. Colombos, *The International Law of the Sea* (London: Longman, 1967), pp. 103–106.

the Law of the Sea,[6] where it was the principal issue for the Conference to resolve. By UNCLOS III, the width of the territorial sea was less of an issue, with the emergence of the exclusive economic zone (EEZ) as an accepted concept. Consequently, there was little contention in the adoption of 12 nautical miles as the maximum breadth of the territorial sea in Article 3 of the LOS Convention.[7]

It is apparent from the selection of State practice that the Convention reflects a consensus on the distance of 12 nautical miles. Table 1 shows that 12 nautical miles was the favoured distance for the width of the territorial sea by 1982, with 68 per cent of States using that distance, and a further 14 per cent using distances of less than 12 nautical miles, which would still be consistent with Article 3. This means that only 18 per cent of littoral States were using distances for their territorial sea that were inconsistent with the Convention. By 2007, 89 per cent of littoral States were using 12 nautical miles, and a further 4 per cent using lesser distances. This is a significant improvement and demonstrates some normative effect around Article 3.

This conclusion is also supported by the State practice of some of those States whose territorial sea claims are inconsistent with the Convention. For example, Benin, Congo, Somalia, Liberia, and Togo assert a territorial sea of 200 nautical miles, but do so from legislation that predates the Convention.[8] This suggests these States have been lax in updating their legislative regimes rather than deliberately flouting the Convention.

---

[6] A. H. Dean, "The Second Geneva Conference on the Law of the Sea: The Fight for Freedom of the Seas," *American Journal of International Law* 54 (1960): 751; D. Bowett, "The Second United Nations Conference on the Law of the Sea," *International and Comparative Law Quarterly* 9 (1960): 414.

[7] Article 3 of the LOS Convention provides:

> Every State has the right to establish the breadth of its territorial sea up to a limit not exceeding 12 nautical miles, measured from baselines determined in accordance with this Convention.

[8] Benin: Decree No. 76–92 *Extending the Territorial Waters of the People's Republic of Benin to 200 Nautical Miles*, 1976, <www.un.org/Depts/los/LEGISLATIONANDTREATIES/PDFFILES/BEN_1976_Decree.pdf>, 6 March 2006; Congo: *Ordinance No. 049/77 of 20 December amending Article 2 of Ordinance 26/71 of 18 October 1971*, <www.un.org/Depts/los/LEGISLATIONANDTREATIES/PDFFILES/COG_1971_Ordinance.pdf>, 6 March 2008; Liberia: *Act to Approve Executive Order of 24 Dec 76*, <www.dtic.mil/whs/directives/corres/20051m_062305/Liberia.doc>, 6 March 2008; Somalia: *Law No. 37 on the Territorial Sea and Ports, of 10 September 1972*, <www.un.org/Depts/los/

Table 1. State practice – Width of the territorial sea

1982		2007	
Distance (nm)	No. of States	Distance(nm)	No. of States
3	20	3	6
12	94	12	136
13–199	8	13–199	1
200	15	200	6
Other	2	Other	1

Source: Derived principally from tables in Germany, *Commander's Handbook: Legal Bases for the Operations of Naval Forces*; additional material compiled by the author from data publicly available at <www.un.org/Depts/los/index.htm>.

Table 2. State practice – Width of the territorial sea (Asian and Pacific States only)

1982		2007	
Distance (nm)	No. of States	Distance (nm)	No. of States
3	6	3	1
12	38	12	46
200	5	200	2
Other	1	Other	1

Source: Derived principally from tables in Germany, *Commander's Handbook: Legal Bases for the Operations of Naval Forces*; additional material compiled by the author from data publicly available at <www.un.org/Depts/los/index.htm>.

Similar results can be seen in a selection of Asia Pacific States using the same criteria, as is evident in Table 2. If anything, the dominance of 12 nautical miles in a contemporary context is more pronounced in the Asia Pacific. Ninety-four per cent of States in the region were in compliance with the LOS Convention in 2007, up from 88 per cent in 1982.

LEGISLATIONANDTREATIES/PDFFILES/SOM_1972_Law.pdf>, 6 March 2008; Togo: *Ordinance No. 24 Delimiting the Territorial Waters and Creating a Protected Economic Maritime Zone of 16 August 1977*, <www.un.org/Depts/los/LEGISLATION-ANDTREATIES/PDFFILES/TGO_1977_Ordinance.pdf>, 6 March 2008.

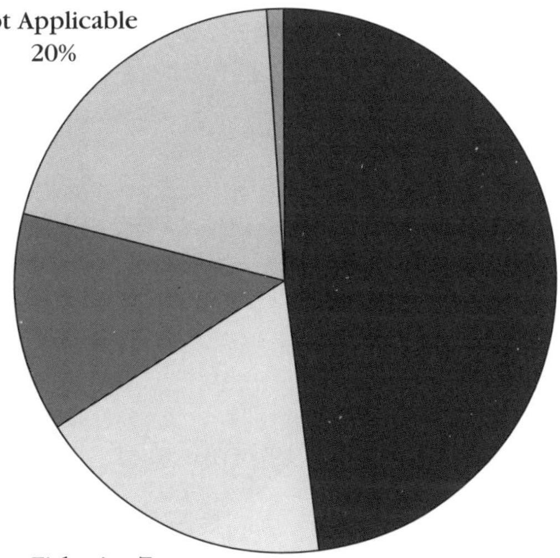

Figure 1. EEZ practice 1985

A similar tale is reflected in respect of the implementation of the EEZ regime under the Convention, with growing dominance of the EEZ. Article 55 of the LOS Convention permits States to assert jurisdiction over an EEZ for a variety of purposes as outlined in Article 56, including fisheries, exploration and exploitation of the seabed, marine scientific research and environmental protection. The dimensions of the EEZ are described in Article 57 as being those waters beyond the territorial sea to a distance of 200 nautical miles from the baselines of the territorial sea.

As can be seen in Figure 1, in 1985 the adoption of an EEZ by a coastal State was the most common response by States, although significant numbers were still maintaining more limited fisheries zones, or had yet to implement any offshore regime of this nature beyond the territorial sea. As is evident from a comparison of Figures 1 and 2, this situation has changed. The implementation of a full EEZ, as opposed to a limited fisheries zone, has become dominant in international practice, with a substantial rise from 48 per cent of States to 80 per cent of coastal States adopting an EEZ. Relatively few States maintain purely fisheries zones, and most of the coastal States that are yet to pass legislation dealing with the

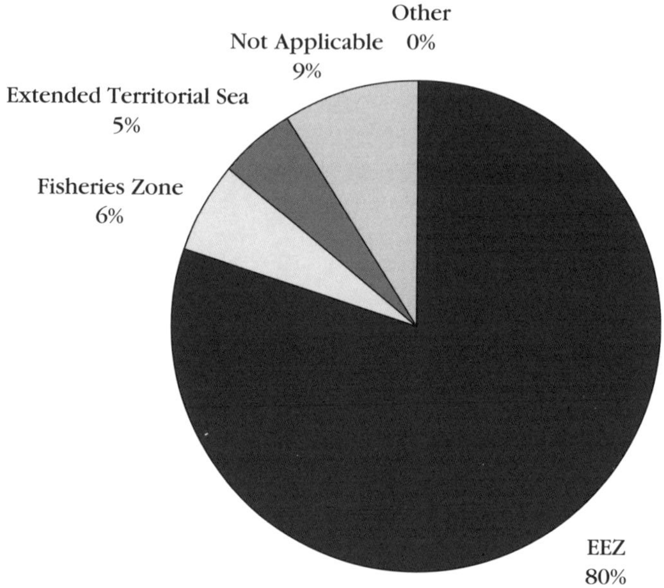

Figure 2. EEZ practice 2007

offshore beyond the territorial sea have relatively small areas of potential jurisdiction that they could proclaim in any case.[9]

That this is the case is not surprising. In spite of perceived opposition to an extended area of national jurisdiction beyond the territorial sea through the 1960s,[10] as is in part evidenced by the litigation before the International Court of Justice in the *Fisheries Jurisdiction Case*,[11] the EEZ was adopted at UNCLOS III relatively quickly, with the strong support of almost all States.[12] The EEZ also gives a coastal State a high level of

[9] For example, such States include The Gambia, Malta and Palau.

[10] See Colombos, *supra* note 5, at pp. 108–109; O'Connell, *supra* note 5, Vol. 1, at pp. 535–536.

[11] *Fisheries Jurisdiction (United Kingdom v. Iceland)*, [1974] *I.C.J. Reports* 3.

[12] Support for the EEZ was indicated by Costa Rica, Barbados, South Vietnam, the USSR, Trinidad and Tobago, Iran, Argentina, Egypt, Sweden, East Germany, Honduras, Kenya, Samoa, Norway, Australia, South Korea, Mongolia, India, Canada, Jamaica, Albania, Bangladesh, Colombia, Bolivia, Mauritius, Nicaragua, the United Kingdom, Fiji, Czechoslovakia, Yemen, Poland, Uruguay, Iceland, Democratic Yemen, Romania, Greece, Finland, Zambia, The Bahamas, Libya, Lebanon, Denmark, Liberia,

jurisdictional control over much of what it would perceive as its most important interest in its immediate offshore, that is, jurisdiction over economic activity.[13] Some of the States maintaining fisheries zones have had the relevant legislation in place for such zones since the 1970s,[14] when the scope of the EEZ regime was still fluid, and they chose a conservative assertion of jurisdiction in the zone adopted. Presumably these States have not felt the need to update their legislation to date.

Based on the results for the territorial sea and EEZ, it might be expected that the regime of the continental shelf, as found in Part VI of the LOS Convention, might also have exerted a normative influence over State practice. However, the situation is more complicated than that, affecting the width of the territorial sea or the EEZ. The concept of the continental shelf predates the LOS Convention, although the definition of the continental shelf adopted by the international community in the 1958 Convention on the Continental Shelf[15] underwent substantial revision at UNCLOS III.[16]

Article 1 of the Continental Shelf Convention provides the previous definition of the continental shelf. It is a two-prong definition, reflecting the lack of consensus in an appropriate approach to the definition of the concept. Article 1 provides:

West Germany, Nigeria, United Arab Emirates, Hungary, Switzerland, Saudi Arabia, Equatorial Guinea, Pakistan, Thailand, Iraq, Bulgaria, Israel, Oman, Tunisia, France, Kuwait, The Gambia, Ireland, Paraguay, Khmer Republic, Senegal, Belgium, Uganda, Bahrain, Cyprus, Sierra Leone, Morocco, Italy, Dominican Republic, Japan, Upper Volta, Venezuela, Mexico, Brazil, Guinea, Chile, Ecuador, Panama, Peru, Malta, and Somalia: United Nations, *Third United Nations Conference on the Law of the Sea: Official Records* (New York: United Nations, 1975), Vol. 1, pp. 59–187.

13 LOS Convention, *supra* note 1, at Art. 56.

14 For example, see The Gambia: Fisheries Act 1977, <www.dtic.mil/whs/directives/corres/20051m_062305/Gambia.doc>, 6 March 2008; Malta: *Act No. XXIV of 1978*, <www.dtic.mil/whs/directives/corres/20051m_062305/Malta.doc>, 6 March 2008; Palau: *Palau National Code*, para. 144: <www.un.org/Depts/los/LEGISLATIONANDTREATIES/PDFFILES/PLW_1990_PNC.pdf>, 6 March 2008; Papua New Guinea: *Offshore Seas Proclamation*, <www.un.org/Depts/los/LEGISLATIONANDTREATIES/PDFFILES/PNG_1978_Proclamation.pdf>, 6 March 2008.

15 *Convention on the Continental Shelf*, Geneva, 29 April 1958, 499 U.N.T.S. 311.

16 See M. H. Nordquist, ed., *United Nations Convention on the Law of the Sea, 1982: A Commentary* (Dordrecht: Martinus Nijhoff, 1985), Vol. 2, pp. 841–890.

> For the purpose of these articles, the term "continental shelf" is used as referring (a) to the seabed and subsoil of the submarine areas adjacent to the coast but outside the area of the territorial sea, to a depth of 200 metres or, beyond that limit, to where the depth of the superjacent waters admits of the exploitation of the natural resources of the said areas; (b) to the seabed and subsoil of similar submarine areas adjacent to the coasts of islands.

The above definition reflected a desire of States to have both certainty in the extent of the continental shelf, with the adoption of a depth-based criterion, while at the same time leaving open their ability to exploit more of the sea floor in time, once technology permitted.

The emergence of technology to exploit the seabed at greater and greater depths ultimately rendered the 200 metre portion of the definition in the Continental Shelf Convention increasingly irrelevant, and by UNCLOS III, there was a substantial mood for change.[17] That mood did not easily resolve itself into a new consensus, with States supporting several different positions. Some States preferred the abandonment of the continental shelf regime altogether, and to subsume the regime into that of the EEZ, with the only criteria for the exploitation of the seabed being based around distance, specifically 200 nautical miles.[18] Other States, those possessing wide continental margins, wanted to retain the benefits that the exploitability criterion in the Continental Shelf Convention provided. These States, dubbed the "margineers'" at UNCLOS III preferred a definition based on criteria that would permit extension of the continental shelf beyond 200 nautical miles based on criteria objectively ascertainable from the configuration of the seabed and its subsoil.[19]

The result of these positions was a complex compromise embodied in the definition of the continental shelf now contained in Article 76 of the LOS Convention. The definition contains a number of distinct elements:

[17] The Nepalese delegate noted there were "more than 90 States." United Nations, *supra* note 12, Vol. 7, at p. 38.

[18] See the statements of the delegates of Austria, Libya, Paraguay, Zaire, Finland, Japan, Singapore, Uganda, Pakistan, Egypt, Romania, Switzerland, Iraq, The Gambia, Kenya, Denmark, Tunisia, Lesotho, Ghana, Mali, Jamaica, Panama, and Malta. United Nations, *supra* note 12, Vol. 2, at pp. 142–169.

[19] See Nordquist, *supra* note 16, Vol. 1, at pp. 72–73.

1. A basic criterion of distance, where a coastal State will have a continental shelf that always extends 200 nautical miles from the territorial sea baselines.
2. Two criteria based on the calculation of the continental margin, based on the configuration of the seabed. These are determined through the location of a point described as the foot of the slope, and either calculating the edge of the legal continental shelf through distance alone or distance based upon the geology of the seabed.
3. Limitations on the width of the continental shelf based upon distance or distance and depth.
4. Coastal States seeking to assert a continental shelf beyond 200 nautical miles are obliged to lodge data in support of their assertion with the Commission on the Limits of the Continental Shelf within a 10 year period of becoming a party to the Los Convention.[20]

States have had access to the definition in Article 76 for about 25 years. If the Convention is exerting a normative impact upon State practice, it might be expected that a significant number of States had moved to implement the definition therein. Certainly in 1985, relatively few States had moved to implement the new definition, and this is borne out in Figure 3.

Several observations can be made concerning Figure 3. First, a relatively large percentage of States had not moved to adopt any specific measures for their continental shelf. This is not surprising, as a large number of States would not be in a position to claim any extended continental shelf, and would be restricted to jurisdiction within 200 nautical miles. Such States, if they had asserted an EEZ jurisdiction, would not materially benefit in specifically establishing a continental shelf over the same area as their EEZ, and therefore would not have taken the trouble to adopt legislation.

Second, just under a third of coastal States in 1985 were still maintaining the definition based on the two limbs in Article 1 of the 1958 Continental Shelf Convention, with a further 8 per cent using either the 200 metre isobath or the exploitability criteria. Such a significant percentage could reflect a "lag time" in State implementation of the new definition. If the LOS Convention is exerting a normative impact upon

[20] LOS Convention, *supra* note 1, at Art. 76.

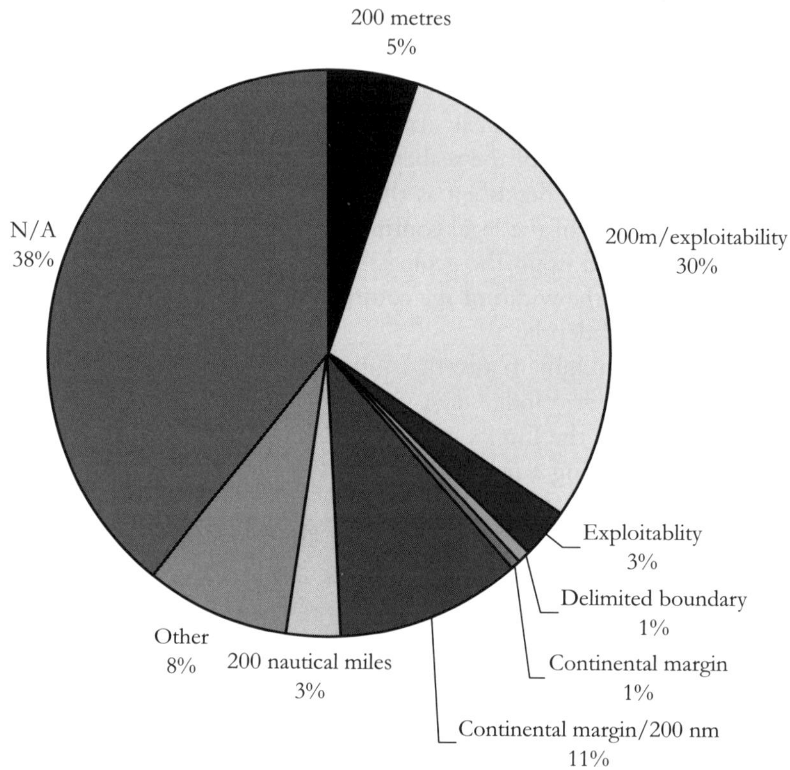

Figure 3. Continental shelf practice 1985

State behaviour, it would be reasonable to assume that this figure would reduce substantially by 2007.

Third, unlike the EEZ and territorial sea regimes, there was relatively limited take up of the new Article 76 definition of the continental shelf by 1985, with only 11 per cent adopting the full definition, and a further 4 per cent adopting elements of it. Again, one might expect this figure to have risen with time as more States moved to implement its provisions and replace the 1958 definition.

The difference between the EEZ and territorial sea on one hand, and the continental shelf on the other, can be explained by events at UNCLOS III. It was apparent relatively early in UNCLOS III that there was a high level of international support for a 12 nautical mile territorial sea, combined with a 200 nautical mile EEZ. This can be contrasted with the emergence of a consensus on the continental shelf, where Article 76 was one of the last texts to be finalized, and around which there had been a great deal of contention.[21] In such circumstances, within three years of the opening of the LOS Convention for signature, it would be reasonable to expect that relatively few States would have moved to implement the new text.

State practice on the continental shelf regime in 2007 is of great interest when compared to the breakdown from 1985 (Figure 4). Again a number of observations can be made. First, the percentage of coastal States in the "not applicable" category has essentially remained the same. Such a result bears out the conclusion that a large number of States have no real interest in the definition of the extended continental shelf, and have all the jurisdiction they need from the EEZ. Since 80 per cent of the same group of coastal States have EEZs, and a further 11 per cent claim some other form of extended jurisdiction out to 200 nautical miles, most of the 40 per cent of States who have taken no steps to assert a continental shelf are asserting some form of jurisdiction out to 200 nautical miles.

Second, there has been a noticeable decline in the proportion of States using the criteria from the 1958 Continental Shelf Convention based on the 200 metre isobath or exploitability. While there has been a large decline, it is interesting that this figure is still a statistically significant 14 per cent. That an essentially defunct convention is still exerting an influence in the definition of the continental shelf in almost one sixth of coastal States 25 years after a new definition was arrived at is surprising. This finding might again be explained by a lack of urgency engendered by the EEZ regime dealing with all the significant jurisdictional issues these States might face.

Third, and perhaps most surprising, is that the Article 76 definition of the continental shelf has less penetration among coastal States 25 years after the LOS Convention was opened for signature in 1982 than the definition of the continental shelf in Article 1 of the Continental Shelf Convention had just over 25 years after its being opened for signature in 1958.

Finally, a growing number of States have defined their continental shelves in terms of delimited points. That this category would have risen substantially since 1985 is also to be expected, as the number of delimited maritime boundaries agreed throughout the world has continued to rise. Interestingly, few of these agreements appear to deal with the extended

[21] See Nordquist, *supra* note 16, Vol. 2, at pp. 841–890.

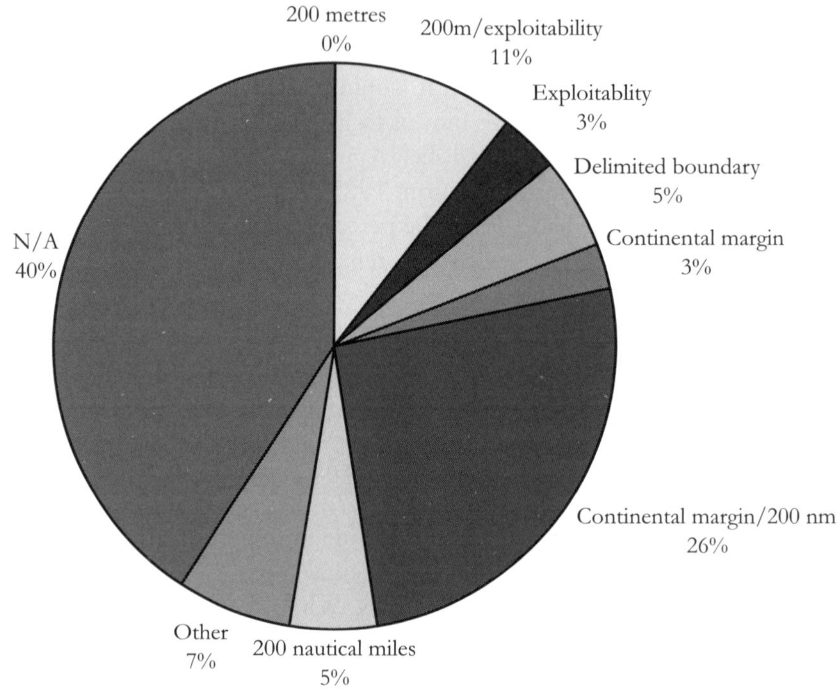

Figure 4. Continental shelf practice 2007

continental shelf beyond 200 nautical miles. Perhaps this reflects some diffidence by States to commit to such boundaries until they have been the subject of some consideration by the Commission on the Limits of Continental Shelf.[22]

Freedom of Navigation – Security

In the law of the sea, freedom of navigation was perceived as a critical issue and this was reflected in the now superseded Geneva conventions on the law of the sea. A limited right of freedom of navigation in the

[22] For an example of a boundary treaty dealing with a continental shelf boundary beyond 200 nautical miles see *Treaty between the Government of Australia and the Government of New Zealand Establishing Certain Exclusive Economic Zone Boundaries and Continental Shelf Boundaries*, Adelaide, 25 July 2004, [2006] A.T.S. 4.

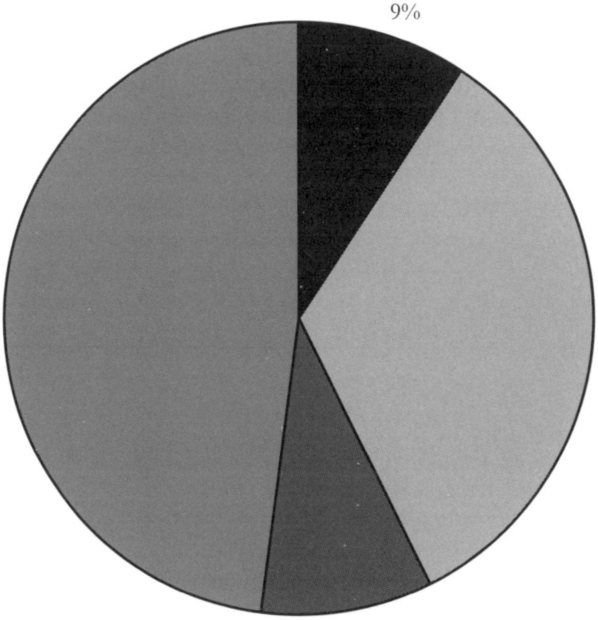

Figure 5. Freedom of navigation and position on restriction

territorial sea, in the context of a right of innocent passage was guaranteed in Section III of the Convention on the Territorial Sea and Contiguous Zone.[23] Freedom of navigation on the high seas was also guaranteed in Article 2 of the Convention on the High Seas.[24] Article 3 of the Continental Shelf Convention ensured that the status of waters above a State's continental shelf remained as high seas and therefore enjoyed freedom of navigation.[25] Figure 5 summarizes the number of States restricting freedom of navigation in territorial waters or beyond for either security or environmental reasons.

---

[23] *Convention on the Territorial Sea and Contiguous Zone*, Geneva, 29 April 1958, 516 U.N.T.S. 205.

[24] *Convention on the High Seas*, Geneva, 29 April 1958, 450 U.N.T.S. 82.

[25] Continental Shelf Convention, *supra* note 15.

Freedom of navigation was a significant issue at UNCLOS III. The incorporation of various guarantees of freedom of navigation was part of an overall package of measures to make the extension of maritime jurisdiction acceptable to States. This "package deal" approach was integral to the acceptance of the LOS Convention, so it is significant to see whether States have acted in accordance with both the spirit and the letter of the provisions.

Article 17 of the LOS Convention provides a right of innocent passage.[26] The nature of what is considered innocent is explored in Article 19, which indicates the activities of a vessel that are considered inconsistent with a right of innocent passage:

1. Passage is innocent so long as it is not prejudicial to the peace, good order or security of the coastal State. Such passage shall take place in conformity with this Convention and with other rules of international law.
2. Passage of a foreign ship shall be considered to be prejudicial to the peace, good order or security of the coastal State if in the territorial sea it engages in any of the following activities:
   (a) any threat or use of force against the sovereignty, territorial integrity or political independence of the coastal State, or in any other manner in violation of the principles of international law embodied in the Charter of the United Nations;
   (b) any exercise or practice with weapons of any kind;
   (c) any act aimed at collecting information to the prejudice of the defence or security of the coastal State;
   (d) any act of propaganda aimed at affecting the defence or security of the coastal State;
   (e) the launching, landing or taking on board of any aircraft;
   (f) the launching, landing or taking on board of any military device;
   (g) the loading or unloading of any commodity, currency or person contrary to the customs, fiscal, immigration or sanitary laws and regulations of the coastal State;

[26] "Subject to this Convention, ships of all States, whether coastal or land-locked, enjoy the right of innocent passage through the territorial sea." LOS Convention, *supra* note 1, at Art. 17.

(h) any act of wilful and serious pollution contrary to this Convention;
(i) any fishing activities;
(j) the carrying out of research or survey activities;
(k) any act aimed at interfering with any systems of communication or any other facilities or installations of the coastal State;
(l) any other activity not having a direct bearing on passage.

Beyond the restrictions on innocent passage indicated in Article 19, there are few restrictions on a foreign vessel from passing through the territorial sea of a coastal State. The Convention gives some small concession to security concerns, allowing a coastal State to temporarily suspend innocent passage for the purposes of essential security protection.[27] This provision has not often been used by States, and in recent times only Mexico has provided published notifications.[28]

It is notable that navigation does not per se render passage by warships through the territorial sea non-innocent. This is supported by the view taken by the International Court of Justice in the *Corfu Channel Case* of the passage of the four British warships along the Channel that led to the damage to HM Ships *Saumarez* and *Volage*.[29] As such, the LOS Convention does not provide a basis for coastal States to unilaterally prevent the passage of warships through its territorial sea other than in circumstances where the passage cannot be characterized as innocent.

Beyond the territorial sea, the LOS Convention also confirms there is freedom of navigation for all vessels. Article 87 provides:

---

[27] Article 25(3) of the LOS Convention provides:

> The coastal State may, without discrimination in form or in fact among foreign ships, suspend temporarily in specified areas of its territorial sea the innocent passage of foreign ships if such suspension is essential for the protection of its security, including weapons exercises. Such suspension shall take effect only after having been duly published.

A provision with similar effect extends the same right to an archipelagic State in respect of archipelagic waters where innocent passage may be exercised. See LOS Convention, *supra* note 1, at Art. 52(2).

[28] See <www.un.org/Depts/los/convention_agreements/innocent_passages_suspension.htm>, 8 March 2008.

[29] *Corfu Channel Case (United Kingdom v. Albania)*, [1949] *I.C.J. Reports* 3.

1. The high seas are open to all States, whether coastal or land-locked. Freedom of the high seas is exercised under the conditions laid down by this Convention and by other rules of international law. It comprises, *inter alia*, both for coastal and land-locked States:
   (a) freedom of navigation;
   (b) freedom of overflight;
   (c) freedom to lay submarine cables and pipelines, subject to Part VI;
   (d) freedom to construct artificial islands and other installations permitted under international law, subject to Part VI;
   (e) freedom of fishing, subject to the conditions laid down in section 2;
   (f) freedom of scientific research, subject to Parts VI and XIII.
2. These freedoms shall be exercised by all States with due regard for the interests of other States in their exercise of the freedom of the high seas, and also with due regard for the rights under this Convention with respect to activities in the Area.

While the LOS Convention makes it clear there is freedom of navigation on the high seas, the same freedom is extended to the EEZ by Article 58(1).[30]

The right of freedom of navigation on the high seas and in the EEZ is limited so a vessel must have "due regard" for the rights of others. As such, there is no explicit limitation based upon security to the benefit of the coastal State beyond that associated with the rights of others. For example, the only restriction on the undertaking of military exercises on the high seas and EEZ of another State would be subject to the non-interference with the rights of other users.[31]

Within the territorial sea, and in some circumstances beyond it, several States assert what may best be described as a "security jurisdiction." That is to say, they purport to regulate, restrict or exclude third State

---

[30] "In the exclusive economic zone all States, whether coastal or land-locked, enjoy, subject to the relevant provisions of this Convention, the freedoms referred to in article 87 of navigation and overflight and of the laying of submarine cables and pipelines, and other internationally lawful uses of the sea related to these freedoms, such as those associated with the operation of ships, aircraft and submarine cables and pipelines, and compatible with the other provisions of this Convention." LOS Convention, *supra* note 1, at Art. 58(1).

[31] See B. Oxman, "The Regime of Warships under the United Nations Convention on the Law of the Sea," *Virginia Journal of International Law* 24 (1984): 809.

vessels from their adjacent waters, even though the only support for such a jurisdiction in the LOS Convention is implicit and derives from the Convention's definition of innocent passage not being prejudicial to the peace and security of the coastal State.

A coastal State has a right to regulate certain matters with respect to a vessel exercising a right of innocent passage, although these do not refer to security interests.[32] The rights of the coastal State are essentially directed at ensuring the territorial sea has safe navigation, criminal activity affecting the coastal State, including immigration and customs, is prohibited, and unauthorized fishing and pollution do not occur.

Nor does the regime of the contiguous zone give rise to a right to restrict or regulate passage. There is no justification within the text of the LOS Convention that permits a jurisdiction based around security concerns to be included within the regime of the contiguous zone, especially as it is part of the EEZ which explicitly has guarantees of freedom of navigation. The contiguous zone is dealt with in a single article, and does not refer to security directly, or even by implication.[33]

---

[32] Article 21 of the LOS Convention, *supra* note 1, provides:

> The coastal State may adopt laws and regulations, in conformity with the provisions of this Convention and other rules of international law, relating to innocent passage through the territorial sea, in respect of all or any of the following:
> (a) the safety of navigation and the regulation of maritime traffic;
> (b) the protection of navigational aids and facilities and other facilities or installations;
> (c) the protection of cables and pipelines;
> (d) the conservation of the living resources of the sea;
> (e) the prevention of infringement of the fisheries laws and regulations of the coastal State;
> (f) the preservation of the environment of the coastal State and the prevention, reduction and control of pollution thereof;
> (g) marine scientific research and hydrographic surveys;
> (h) the prevention of infringement of the customs, fiscal, immigration or sanitary laws and regulations of the coastal State.

[33] Article 33 of the LOS Convention, *supra* note 1, provides:

> In a zone contiguous to its territorial sea, described as the contiguous zone, the coastal State may exercise the control necessary to:
> (a) prevent infringement of its customs, fiscal, immigration or sanitary laws and regulations within its territory or territorial sea;
> (b) punish infringement of the above laws and regulations committed within its territory or territorial sea.

A large number of coastal States assert security zones in their territorial sea and beyond, into the EEZ. The range of measures varies considerably, and does not easily lend itself to the type of statistical representation undertaken above. Overall, in excess of 60 coastal States have asserted some form of restriction or notification. Table 3 summarizes the nature of various coastal State measures.

The restrictions noted in Table 3 fall into two broad categories. The first places restrictions upon freedom of navigation in the territorial sea, requiring warships, or even third State merchant vessels, to seek permission before entering the territorial sea or contiguous zone. An example of this is the requirement asserted by China. [34] As considered above, there is no clear justification in the LOS Convention, and it is inconsistent with established international law predating the Convention, in particular the *Corfu Channel Case*.[35]

The second category relates to certain activities purported to have been prohibited by the coastal State, usually weapons testing or military exercises. While the innocent passage regime makes it explicitly clear that weapons firing and other forms of military exercise cannot be undertaken in the coastal State's territorial sea and be consistent without that State's permission,[36] establishing a similar restriction for the waters beyond is more difficult.

Military exercises, including the live firing of weapons outside of the territorial sea, and a broader range of other military activities are not dealt with in the LOS Convention.[37] There is also no authorization with respect to such exercises, as military exercises and related activities are not

---

[34] Article 13 of the *Law of the Territorial Sea and the Contiguous Zone of 25 February 1992* provides that the Chinese exercise jurisdiction over security within the contiguous zone and seeks prior notification. This is rejected by the United States as inconsistent with Article 33 of the LOS Convention, which has no reference to a security jurisdiction in respect of the contiguous zone. The 1992 U.S. protest is cited in Naval War College, *Annotated Supplement to the Commander's Handbook on the Law of Naval Operations* (Newport: Naval War College, 1997), para. 1–90.

[35] *Corfu Channel Case, supra* note 29.

[36] See *infra* note 26 and following text.

[37] For example, see G. V. Galdoresi and A. G. Kaufman, "Military Activities in the Exclusive Economic Zone: Preventing Uncertainty and Defusing Conflict," *California Western Journal of International Law* 32 (2002): 253.

Table 3. State practice – Freedom of navigation and security

State	Type of Rights Asserted
Albania	Warships require prior special authorization
Algeria	Authorization must be obtained for warships 15 days prior to their passage; exception: force majeure
Antigua and Barbuda	Warships require prior authorization
Bangladesh	Warships require prior authorization; coastal zone 18 nm – security interests
Barbados	Warships require prior authorization
Brazil	Prohibition of the boarding, searching and capturing of vessels in the EEZ; military exercises and manoeuvres may be conducted in the EEZ only with the consent of Brazil
Bulgaria	24 nm – "control rights"
Cambodia	Control of all foreign activities on the continental shelf, irrespective of their purpose; coastal zone 24 nm – security interests
Cape Verde	Warships require prior authorization; prohibition of "non-innocent use" of the EEZ, including weapons exercises
China	Requires prior notice for transports of waste in the territorial sea and EEZ; warships require prior authorization; coastal zone 24 nm – security interests
Congo	All ships require prior authorization
Costa Rica	Fishing vessels must announce their passage through the EEZ beforehand
Croatia	Warships must announce their passage; the number of warships is limited
Denmark	Warships and government ships are required to notify the Danish authorities prior to their passage through territorial waters if that involves passage through the Great Belt, the Samsø Belt or the Øre Sound; prior authorization is required for more than three warships passing through at the same time
Djibouti	Prior notice required of any passage of nuclear-powered ships and ships carrying nuclear or other radioactive material
Ecuador	"Special area to be avoided"
Egypt	Warships have to announce their passage in advance; ships carrying nuclear material or other hazardous substances require prior authorization; coastal zone 24 nm – security interests
El Salvador	Expressed concern at UNCLOS III in respect of military activities in the EEZ

Table 3. (*cont.*)

State	Type of Rights Asserted
Estonia	Warships and research vessels must announce their passage 48 hours in advance; authorization must be applied for nuclear-powered ships 30 days prior to their passage
Finland	Warships and government ships have to announce their passage in advance
Gambia	Asserts the right to prohibit navigation on certain areas of its continental shelf
Greece	Claims only a 6-nm territorial sea but 10 nm of airspace for air traffic control purposes
Grenada	Warships require prior authorization
Guinea	Taking photographs and transporting toxic or hazardous material are considered a criminal offence
Guyana	Warships have to announce their passage in advance
Haiti	Passage prohibited to ships carrying waste or materials with an inherent health or environmental hazard; prohibition of the passage of all vessels carrying waste or materials that are environmentally harmful or detrimental to health; claims the right to exercise the control required in the EEZ in order to ensure navigational safety and to prevent violations of financial, customs, health and environmental protection regulations; coastal zone 24 nm – security interests
India	Warships have to announce their passage in advance; prior consent to military exercises and manoeuvres in the EEZ and on the continental shelf; coastal zone 24 nm – security interests
Indonesia	Warships and all vessels other than merchant ships must announce their passage in advance; within 100 nm ships are not allowed to stop, anchor or cruise "without legitimate cause"
Iran	Warships, submarines, nuclear-powered ships, as well as ships carrying nuclear or other hazardous materials, require authorization; prohibition of "military activities and practices" in the EEZ and on the continental shelf; coastal zone 24 nm – security interests
Latvia	Reserves the right to regulate the passage of warships
Libya	Innocent passage to be announced in advance and allowed during daylight hours only; four exclusion zones
Lithuania	Warships require prior authorization if this is required by the flag State
Malaysia	Prior consent to military exercises and manoeuvres in the EEZ and on the continental shelf

Table 3. (*cont.*)

State	Type of Rights Asserted
Maldives	Warships require prior authorization; with regard to the EEZ, acknowledges only the right of innocent passage; makes entry of fishing and research vessels into the EEZ conditional upon prior consent
Malta	Asserts the claim for warships to obtain prior authorization
Mauritania	Reserves the right to restrict navigation and aviation in or above the EEZ for reasons of national security
Mauritius	Warships must announce their passage; apparently makes the passage of warships and submarines through the EEZ conditional upon prior approval
Myanmar	Warships require prior authorization; claims the right to restrict the freedom of navigation and overflight in its EEZ; coastal zone 24 nm – security interests
Namibia	Claims sovereign rights with regard to financial, customs, immigration and health regulations in the EEZ
Nicaragua	25 nm – security interests; 15 days advance notification for warships and military aircraft, seven days for civilian traffic
North Korea	62 nm military zone 50 nm seaward of the territorial sea; all ships and aircraft require prior approval
Oman	Warships, nuclear-powered ships, submarines and ships carrying hazardous loads require prior authorization
Pakistan	Warships require prior authorization; supertankers, nuclear-powered ships and ships carrying nuclear materials are required to announce their passage in advance; claims authority to regulate transit through parts of the EEZ and to enact and enforce all regulations required for controlling activities in the EEZ; coastal zone 24 nm – security interests
Peru	Prior consent to military exercises and manoeuvres in the EEZ and on the continental shelf
Philippines	Expressed concern at UNCLOS III in respect of military activities in the EEZ
Poland	Reserves the right to regulate the passage of warships
Portugal	In the EEZ, acknowledges only the right of innocent passage
Romania	Reserves the right to regulate the passage of warships
São Tomé and Príncipe	Reserves the right to regulate the passage of warships
Saudi Arabia	Reserves the right to regulate the passage of nuclear-powered ships; coastal zone 18 nm – security interests and navigation
Senegal	Expressed concern at UNCLOS III in respect of military activities in the EEZ
Seychelles	Warships are required to announce their passage in advance
Slovenia	Reserves the right to regulate the passage of warships

Table 3. (*cont.*)

State	Type of Rights Asserted
Somalia	Warships require prior authorization
South Korea	Warships and government ships have to announce their passage three days in advance
Sri Lanka	Warships require prior authorization; coastal zone 24 nm – security interests
St. Vincent and Grenadines	Warships require prior authorization
Sudan	Warships require prior authorization; the right of innocent passage may be suspended for security reasons; coastal zone 18 nm – security interests
Syria	Warships require prior authorization; 41 nm – security interests
United Arab Emirates	Warships require prior authorization; nuclear-powered ships and ships with nuclear or hazardous loads must announce their passage in advance; coastal zone 24 nm – security interests
Uruguay	Asserts the right to prohibit military exercises in the EEZ
Venezuela	15 nm – national and security interests
Vietnam	Warships require authorization to be applied for at least 30 days prior to passage; passage restricted to three warships at a time; coastal zone 24 nm – security interests; submarines are required to navigate on the surface and to show their flag; aircraft are not allowed to land on board ships or be launched from them; on-board weapons have to be set in "non-operational" mode prior to the entry into the zone
Yemen	Warships require prior authorization; nuclear-powered ships or ships carrying nuclear materials must announce their passage in advance; coastal zone 24 nm – security interests
Yugoslavia	Warships must announce their passage 24 hours in advance

Source: Derived from S. B. Kaye, "Freedom of Navigation in a Post 9/11 World: Security and Creeping Jurisdiction," in R. Barnes, D. Freestone and D. Ong, eds, *The Law of the Sea: Prospects and Problems* (Oxford: Oxford University Press, 2006), p. 347.

included in the list of freedoms. This of itself should not be taken as negating such rights, as the rights listed in Article 87(1) are not exhaustive, as demonstrated by the use of the phrase "*inter alia*." Further, the freedoms described are subject to the conditions set down in the Convention and "other rules of international law." This language makes it clear that the LOS Convention is not intended to be the only source of law in relation to the use of the high seas or EEZ.

Certainly if a third State is able to exercise and fire weapons in another State's EEZ, it is subject to qualification. The critical limitation would seem to come from Article 87, which notes that the existing freedoms of the high seas are to be exercised "with due regard." It would seem strange if military exercises on the high seas were not subject to the same qualification, as they have the potential of being far more prejudicial to the rights and uses of others than other high seas freedoms.[38]

Freedom of Navigation – Environmental Protection

Restrictions with respect to security are not the only attempt by some States to restrict freedom of navigation through their territorial sea or EEZ. Several States also purport to deny vessels carrying ultra-hazardous cargoes, most notably nuclear materials, passage through their territorial sea as well as their EEZ (Figure 5). These States are motivated by a concern that were an accident to occur, and the ultra-hazardous cargo be exposed to the marine environment, it may cause irreparable harm. Typically, these States are not the point of origin or destination of the cargo, but rather their waters are along the route of these vessels.

The basis of a coastal State purporting to exclude or regulate vessels carrying ultra-hazardous cargo through its territorial sea or EEZ is based upon their jurisdiction over environmental matters in these waters. Both the territorial sea and EEZ regimes explicitly give a coastal State the right to regulate environmental matters,[39] and international law has long recognized the liability of a State for environmental harm it causes to another State.[40] At the very least, the coastal State would contend it has a right to be notified of a voyage carrying ultra-hazardous cargo, if only to be prepared to respond appropriately to an accident or other disaster.

A preventative approach to environmental protection is not without some appeal, but the LOS Convention provides little direct assistance for States who wish to deny passage to vessels carrying ultra-hazardous cargoes. Certainly the scope of environmental regulation does not explicitly authorize a coastal State to direct vessels in the EEZ on the basis of their cargo. This is borne out by the content of Article 23 of the Convention,

---

38 See F. Francioni, "Peacetime Use of Force, Military Activities, and the New Law of the Sea," *Cornell International Law Journal* 18 (1985): 203 at 215; see also, Nordquist, *supra* note 16, Vol. 2, at pp. 553–565.

39 LOS Convention, *supra* note 1, at Arts. 21(f) and 56.

40 *Trail Smelter Arbitration (United States v. Canada)*, 3 RIAA 1907 (1941).

which deals with vessels carrying ultra-hazardous cargo in the territorial sea of a coastal State:

> Foreign nuclear-powered ships and ships carrying nuclear or other inherently dangerous or noxious substances shall, when exercising the right of innocent passage through the territorial sea, carry documents and observe special precautionary measures established for such ships by international agreements.

Rather than indicate there are duties owed to the coastal State by a vessel carrying ultra-hazardous cargo, to obey its instructions with respect to routeing, and the option of coastal State to deny passage, Article 23 indicates that special precautionary measures drawn from other instruments ought to be complied with. In the absence of such other instruments, or compliance with them, it would seem the coastal State has no more authority over such vessels in its territorial sea than any other vessel. Further, if this is the case for the territorial sea, it is difficult to see how a right of regulation over the EEZ can be asserted as no equivalent provision to Article 23 exists for the EEZ.

The legal position has not prevented State practice from developing in this area, and this covers a variety of situations. Examples include the assertions of Spain and France that they would inspect single-hulled oil tankers at sea, and if they found them to be unseaworthy, would exclude them from their EEZs.[41] Further examples can be drawn from the international responses to shipments of radioactive materials around the world in the last two decades. The voyages of the *Pacific Pintail, Pacific Teal, Pacific Swan* and the *Atatsuki Maru* carrying radioactive cargoes attracted protests from a significant number of States. Several States, including Argentina, Chile, Antigua and Barbuda, Colombia, the Dominican Republic, New Zealand, South Africa and Mauritius, purported to exclude vessels carrying radioactive ultra-hazardous cargo from their EEZs. These voyages were also condemned by CARICOM, representing the Caribbean States, and

[41] This was as a result of the break up of the oil tanker *Prestige* in November 2002. See J. M. Van Dyke, "Military Ships and Planes Operating in the Exclusive Economic Zone of Another Country," *Marine Policy* 28 (2004): 29 at 37.

the South Pacific States.[42] The following States[43] have asserted that they will not permit nuclear cargo vessels in the territorial sea or EEZ:

Antigua and Barbuda	Malaysia
Argentina	Malta
Brazil	Nauru
Chile	New Zealand
Colombia	Oman
Dominican Republic	Papua New Guinea
Egypt	Peru
Fiji	Philippines
Guinea	Saudi Arabia
Haiti	Singapore
Indonesia	South Africa
Iran	Venezuela
Kiribati	Yemen

Conclusion

In relation to maritime zones, it is apparent that the territorial sea and EEZ regimes dealing with the extent of these zones in the United Nations Convention on the Law of the Sea have exerted a normative impact upon State behaviour. States are now far more regular in their practice, asserting 12 nautical mile territorial seas and 200 nautical mile EEZs to an extent that dominates other practice.

The same cannot be said of the continental shelf regime, where there has been limited implementation of the provisions within Article 76, and a surprising level of retention of the old definition from the Continental Shelf Convention. The diminishing of the normative impact of the LOS

---

[42] See generally, M. Roscini, "The Navigational Rights of Nuclear Ships," *Leiden Journal of International Law* 15 (2002): 251; J. M. Van Dyke, "Balancing Navigational Freedom with Environmental and Security Concerns," *Colorado Journal of International Environmental Law and Policy* 15 (2003): 19; J. M. Van Dyke, "The Legal Regime Governing Sea Transport of Ultrahazardous Radioactive Materials," *Ocean Development and International Law* 33 (2002): 77.

[43] Derived from S. B. Kaye, "Freedom of Navigation in a Post 9/11 World: Security and Creeping Jurisdiction," in R. Barnes, D. Freestone and D. Ong, eds, *The Law of the Sea: Prospects and Problems* (Oxford, Oxford University Press, 2006), p. 347.

Convention in this regard is not surprising given the lesser importance many States might give the continental shelf regime within 200 nautical miles.

The LOS Convention does not appear to have exerted a similar normative impact in the context of freedom of navigation. A large number of States purport to restrict or regulate navigation in their territorial seas or beyond without reference to the Convention, and in apparent contravention of it. Much of this activity is concentrated in the area of security, but there are significant numbers of States also purporting to limit or regulate the navigation rights of vessels carrying ultra-hazardous cargos (Figure 5). Surprisingly, when considered in the context of coastal States only, those States seeking restriction or regulation make up 52 per cent of the international community. This raises the question as to whether the LOS Convention in the context of freedom of navigation represents customary international law, and whether such behaviour might serve in the long term to undermine the efficacy of the Convention in this or other areas. It would certainly be a matter of concern that so large a proportion of the international community have taken the view that restriction of innocent passage in certain circumstances is permissible.

In conclusion, while it is apparent that certain aspects of the LOS Convention do appear to have exerted a normative influence on State behaviour, the results are patchy. States appear uninterested, in the case of the continental shelf, or non-compliant, in the case of freedom of navigation. What impact these results have on the efficacy and acceptance of the Convention in the long term remains to be seen.

Trends in Ocean Zoning – Layers of Confusion and Approaches to Clarity

John Duff*

Introduction

In Douglas Johnston's book *The Theory and History of Ocean Boundary Making*, he included a chapter entitled "Trends in Ocean Zoning."[1] In the two decades since the publication, the phrase ocean zoning has continued buzzing through academic literature and public policy discussions, with loose definitions that have varied in their meaning and level of detail. In light of the increasingly widespread use and often conflicting employment of the term over the years, Johnston can bee seen as having been prescient in detecting the trend regarding not only what he considered appropriately named "zoning," but a corollary trend related to the problematic use of the phrase. The latter trend, Johnston observed, was born of a "more or less continuous confusion between two rival modes of thought about the distribution and administration of authority in the ocean."[2]

The purpose of this contribution is to integrate Johnston's observations into the current discussion on ocean zoning as part of an effort to illustrate the merit of employing a set of appropriate terms and constructs designed to dispel a second layer of confusion regarding the appropriate use of zoning (the term and the tactic) within a governing authority's administration of ocean space. To do so, a brief explanation of two layers of confusion are set forth followed by a series of observations related to

---

* Assistant Professor, Department of Environmental, Earth and Ocean Sciences, University of Massachusetts Boston, USA. This contribution is an expansion upon a few paragraphs taken from a briefer piece, "Is it Time to Zone the Ocean?," *The Coastal Society Bulletin* 26(3) (2004): 1.

1 D. Johnston, *The Theory and History of Ocean Boundary-Making* (Kingston: McGill-Queen's University Press, 1988), pp. 47–60.

2 *Ibid.*, p. 58.

how the dissipation of confusion at the primary level has been followed by confusion at a secondary level.

The Primary Confusion: The Regime-Zone Distinction

The continuous confusion to which Johnston referred stemmed from two distinct but complementary aspects of ocean governance. Those rival aspects, he suggested, are regime-oriented and zone-oriented. At the risk of oversimplifying, ocean regimes can be thought of as those institutions, principles and rules that are constitutive in nature and which find their origins in public international law as it relates to ocean jurisdiction. In contrast, the zone concept is, in Johnston's words, "essentially administrative" in nature.[3] To boil it down even further, ocean regime construction determines who (i.e., among States) may govern ocean spaces while ocean zone construction deals with how a State, having established its jurisdictional authority, will manage allocations and activities within its ocean spaces.[4] Johnston pointed out that those concepts are, respectively and primarily, international and intra-national in nature.

With the entry into force and widespread acceptance of the United Nations Convention on the Law of the Sea,[5] international ocean regime-related rules and institutions have taken firm root. The exclusive economic zone (EEZ) regime as a foundation upon which States may rely to engage in zone-oriented administrative ocean planning and management (and domestic law development) gives sufficient substance to each regime/zone aspect so as to reduce the confusion by way of moving the abstractions to more concrete form. The trends Johnston observed related to increasing use and development of ocean resources and the national administrative efforts that those uses have prompted. The observations also imply that the emphasis of ocean governance would shift from a regime dominated phase played out to a large degree on the international stage to a more

[3] *Ibid.*, p. 42.

[4] References to upper case term "States" in this article refer to nation-States while the lower-case term "states" refers to certain sub-national units of government (i.e., U.S. states).

[5] *United Nations Convention on the Law of the Sea*, Montego Bay, 10 December 1982, UN/Doc. A/CONF.62/122, 7 October 1982, 21 I.L.M. 1261 (1982) [hereinafter LOS Convention].

practical administrative phase played out at the national stages of coastal States. And that shift demonstrates the increasing understanding of the zone-regime distinction such that it is fair to say that resulting intra-national ocean administrative activities of States have effectively served to chipped away at some of the confusion that Johnston observed prior to the LOS Convention's entering into force.

States understand that they may claim extended ocean space via the EEZ regime but few States manage their nominal EEZs per se. Rather most coastal States embark upon efforts to allocate and administer particularly described areas or activities within that EEZ-legitimized space. Confusing as the terms and concepts may have been twenty years ago, State practice has arguably reduced some of that confusion. Nonetheless, Johnston's observation still merits recitation particularly to students and researchers who would benefit from being reminded that, to paraphrase Johnston, the EEZ is a regime. But while Johnston's rule-of-thumb observation that "the law developed specifically for a zone is likely to be composed largely of national elements, whereas a regime is an expression of international law,"[6] employing a national-international filter for zone-regime discrimination does not bring critical analysis of all things called ocean zoning to an end. Rather it allows that assessment to proceed at a more localized level. For while ocean zoning is a means by which a coastal State might manage the ocean space within its jurisdiction, not all such administration is properly characterized as zoning.

Secondary Confusion: The Zoning-Management Confusion

In the United States two high profile ocean commissions have issued reports with dozens of references to zones and zoning.[7] The term has

---

[6] Johnston, *supra* note 1, at p. 43.

[7] Pew Oceans Commission, *America's Living Oceans: Charting a Course for Sea Change* (2003) [hereinafter Pew Report], <www.pewtrusts.org/pdf/env_pew_oceans_final_report.pdf>, 5 October 2006; U.S. Commission on Ocean Policy, *An Ocean Blueprint for the 21st Century: Final Report of the U.S. Commission on Ocean Policy* (Washington, D.C.: U.S. Commission on Ocean Policy, 2004) [hereinafter USCOP Report], <oceancommission.gov/documents/full_color_rpt/000_ocean_full_report.pdf>, 5 October 2006.

also been employed in a host of policy documents and scholarly articles.[8] Ocean zoning as a domestic law tool has been the focus of at least one international workshop and rears its head as the title or sub-title of a growing number of panel sessions and speaker presentations at conferences and meetings around the world.[9] A search of the academic literature for the notion of ocean zoning results in hundreds of "hits" in dozens of articles. In many cases, use of the term ocean zoning with regard to an aspect of ocean management is used in a well-defined contextualized discussion that suggests something similar, if not identical, to land based zoning. In too many cases though, the phrase seems to be employed as a bromide suggesting that the mere act of delineating spaces is sufficient to achieve management objectives.

One effort to shine a light on the confusion elicited by the term "ocean zoning" can be found in a simple question: Is it time to zone the ocean? If possible alternative answers are considered, it becomes apparent that the absolutes of 'yes' and 'no' are both appropriate depending upon the meaning that is applied to the phrase. Further, the question cannot be asked entirely prospectively because coastal States have already engaged in a variety of ocean-zoning type efforts suggesting an equally valid response of 'too late'. As a result, an all of the above response is most appropriate. Consider those various responses.

Yes, We Need to Zone the Ocean

Even those who do not see eye-to-eye with the reports and recommendations touting the use of ocean zoning are likely to agree with some of their basic findings, in particular that some ocean spaces are increasingly subject to a host of users and activities giving rise to conflicts over what the best use for a given area may be. Simply stated, all uses in all places at all times cannot be accommodated. The preference is for the use of some areas for certain activities/development. Some areas should be maintained for low-impact/no impact use. There is an opportunity to increase economic productivity and efficiency. And, it is desirable to sustain many traditional uses. Proponents of ocean zoning contend that it will allow

---

8 See, e.g., D. Russ and G. Zeller, "From *Mare Liberum to Mare Reservum*," *Marine Policy* 27(1) (2003): 75–78.

9 P. Doherty and M. Butler, "Ocean Zoning in the Northwest Atlantic," *Marine Policy* 30(4) (2006): 389–391.

for the reduction of conflicts, appropriate allocation of access and use of offshore space and resources, and ensure adequate protection of the marine environment. As a result, in those ocean areas that exhibit user conflict, overexploitation, and/or environmental vulnerability signals, management efforts that include detailed land-based zoning-type prescriptions may be appropriate. Johnston suggested as much when he opined that an activity, such as aquaculture, that was "likely to become a major use of coastal waters [would support the application of] ocean zoning."[10]

No, We Do Not Need to Zone the Ocean

The frequent employment of the word zoning in discussions of national ocean governance and management proves problematic in two ways. First, the term zoning, which has a very technical legal meaning particularly in the United States and Canada, is often employed by people unfamiliar with the legal constructs and context that give it practical application. Often that group of users employs the term when what they actually mean is domestic ocean governance in general, or if more particular, ocean management and planning. In either of those cases, general terms would be more appropriate when the larger general idea is being discussed.

Even if the more general terms planning and/or management are substituted for the arguably misused zoning, an argument can be made that there is no need to thus plan and manage entire ocean spaces under the jurisdiction of a coastal State. The fact is vast expanses of the ocean are not understood. If continuous and consistent use of ocean space is deemed the equivalent of settlement, then most of the ocean space under most coastal States' jurisdiction remains frontier – unsettled areas, unknown in any great detail. The suggestion that what is needed is to actively manage vast expanses of unmapped, unsettled, infrequently visited (i.e., frontier) national ocean space, whether or not that space might be exhibiting symptoms of something amiss, likely will not withstand scrutiny when the costs of doing so are considered. The ocean areas under the jurisdiction of the United States (including those expanses claimed pursuant to the EEZ regime) constitute over 3.4 million square nautical miles and exceeds the land mass of the country.[11] Detailed planning and

[10] Johnston, *supra* note 1, at p. 49.

[11] USCOP Report, *supra* note 7, at pp. 30–31.

management of the country's entire ocean area could be a waste of time, energy and money.

Too Late

A too late response to the question suggests that the query posed is a trick question, since the situation is not a blank slate when it comes to ocean zoning. Many coastal States have been employing domestic law to delineate national ocean space, claimed under international law of the sea principles, in ways to, as Johnston observed, assign intra-national administrative jurisdiction and authority. And the U.S. Ocean Commission report attempts to illustrate that point when it notes that,

> Although invisible to the naked eye, governments have carved the world's oceans into many zones, based on both international and domestic laws. These zones are often complex, with overlapping legal authorities and agency responsibilities.[12]

The too-lateness response to the question is illustrated by the fact that in the waters of the United States and many other maritime nations, a random walk through national, sub-national, and municipal laws highlights the fact that ocean and coastal waters are already buried under multiple monikers. In some cases, boundaries are set to distinguish relevant political-legal jurisdictions. In the United States, there are lines separating federal waters from state waters, lines separating adjacent states' waters, lines in internal waters (e.g., tidally influenced rivers and estuaries) that serve as boundaries between the states themselves, and municipal harbour lines.

In addition to all of that political administrative zone-boundary marking are the myriad designations that proscribe or allow particular activities in particular ocean areas. An examination of nautical charts illustrates this fact in the form of shipping channels, marine sanctuaries, recreational diving spots, oil and gas platforms, and security zones. Finally, the U.S. federal and state governments impress vast ocean areas with considerable additional "dos and don'ts" in the form of environmental, navigation, immigration, and public health and safety laws.

As a result, the most appropriate response to the question – Is it time to zone the ocean? – is 'all of the above', since it reflects the practical reality

[12] *Ibid.*, p. 70.

as well as varying perspectives. Unfortunately, such a response does little to untangle the confusion that exists at the national level.

Untangling the Zoning – Management Confusion at the National Level

A pair of concerns arises as the buzz of ocean zoning turns into an ongoing drone. First, the term zoning is too frequently used in a generic form (i.e., to delineate based on features) in legal and public policy discussions. As students of zoning law understand, the zoning tool resides within the larger realm of the land use management and policy process. That larger process supports the manner by which national and sub-national units of government determine how land may be used via two distinct capacities: sovereign as protector of the public's interests (characterized as an element of the police power in the United States and attributed as derivative of constitutionally-based property and civil rights authority in Canada), and sovereign as proprietor of public lands (i.e., the State/Crown as owner). In the former capacity the government's policies constrain private activities on private property. Examples include building dimension restrictions, use limitations, and infrastructure requirements. Private owners are free to use their property so long as they do so within the zoning constraints. And the counter-balancing laws that protect private property interests limit the amount of zoning that can be applied to reflect the kind and degree of public interests sought to be protected. In the latter capacity, governments as owners play a substantially different role in that they are the actually users and managers of the land. While some observers might suggest that the oceans, as public space, fall exclusively into the category by which governments need only exercise their authority as owners, the increasing expanses of ocean space that have acquired private property characteristics, particularly in the United States and Canada, suggests that both governmental capacities must be engaged for comprehensive ocean management.

For those intent on seeing ocean zoning ultimately applied to some offshore areas, two suggestions merit consideration. First, conduct the necessary preliminary steps that facilitate the determination of whether zoning principles or some higher level governance principles are appropriate. Second, take some lessons (good and bad) from the history of public land management.

If zoning is the wrong mechanism and therefore the wrong term to be used in general ocean governance, what are the appropriate phrases and constructs which represent more appropriate management scales? Some of the apt appellations being employed include sea use planning, marine spatial planning, integrated sea use management, and variations on those themes.[13]

Further, many of those familiar with zoning's legal meaning who are keen on pushing the concept out to sea, seem to be under the impression that zoning is a panacea for any ocean governance concern, ignoring the larger public policy-making process within which zoning might ultimately be employed. That group might be likened to those who suggest a scalpel as a remedy for a stomach ache. To progress from symptom to surgical instrument requires the intervening steps of diagnosis, alternative remedy evaluation, informed decision-making, and consideration of operational protocols prior to any actual use of a particular instrument.

Given that there may be sound reasons to zone some ocean areas while refraining from zoning others and in light of the fact that ocean exploration, boundary delineation and resource management have been going on in ad hoc fashion for centuries, it might be useful to look to U.S. land settlement history as demands for ocean use evolve. Four significant stages merit consideration: 1) exploration and pioneer settlement of the frontier; 2) gross-scale disposition, withdrawal and reservation of the public domain; 3) wide-scale settlement and use in accordance with the withdrawal and reservation objectives of stage 2; and 4) zoning applied to that settlement and use to protect health, welfare and proximate private property interests. Most of the confusion regarding ocean zoning today stems from a lack of discrimination between stages two and four. Fortunately those stages, which are stated above in general terms, may be illuminated by specific examples which highlight the important chronological and scalar distinction between them. Offshore energy development and marine conservation constitute such examples.

Offshore energy development in the United States and Canada provides a helpful model for the increasingly detailed levels of governance

[13] See Johnston, *supra* note 1, at p. 57; F. Douvere and C. Ehler, "Conference Report – International Workshop on Marine Spatial Planning, UNESCO, Paris, 8–10 November 2006," *Marine Policy* 31(4) (2007): 582–583; H. D. Smith and A. Vallega, eds., *The Development of Integrated Sea Use Management* (New York: Routledge, 1991).

that has moved offshore in recent decades. In both countries, the federal governments (in conjunction, as appropriate, with their state/provincial governments) have decades-long histories of planning offshore energy development, then implementing management systems (including leasing systems whereby private firms may engage in the development) to direct such activities; assessing the impacts of early stage development and use; and finally, when circumstances dictate, zoning (in the closest sense of the term as it is used with respect to private property holders on land) the manner in which the development activities take place. And just as the amount of zoning that can be applied to a community of users is influenced by the scope of activities sought to be constrained, so too it seems, the level of ocean governance suitable for ocean areas ought to be rationally related to the scope of actual or contemplated use taking place therein.

Marine parks and sanctuaries have similarly moved through the governance process from general exploration to site suitability determination and selection via the reservation process to initial use and management to the more finely detailed zoning within reserved space. The chronology and progression through finer scale governance methods of marine reserves such as the Florida Keys National Marine Sanctuary in the United States and the Great Barrier Reef Marine Park in Australia are illustrative.[14] Under the auspices of the *Marine Protection, Research and Sanctuaries Act*, the United States adopted a process for identifying and designating important ocean areas.[15] Today, the United States manages thirteen national marine sanctuaries. While the sanctuaries are designated under a set of general principles set forth in the statute, the objectives of the different sanctuaries vary greatly. As a result, each sanctuary is governed by a set of regulations tailored to the area's resources and goals (a parallel to classification). Zoning is considered but one of a variety of sanctuary management tools.[16] In addition to clarity that can be gained from examining appropriately characterized ocean zoning in specific contexts such

---

[14] See B. Causey, "Zoning the Florida Keys National Marine Sanctuary," in P. Doherty, ed., *Ocean Zoning: Can it Work in the Northwest Atlantic-Workshop Proceedings. Marine Issues Committee Special Publication No. 14* (Halifax: Ecology Action Center, 2005), pp. 102–111; J. Day, "Zoning – Lessons from the Great Barrier Reef Marine Park, Ocean & Coastal Management 45 (2002): pp. 139–156.

[15] *Marine Protection, Research and Sanctuaries Act*, 16 U.S.C. §§1431–1445.

[16] Causey, *supra* note 14, at p. 104.

as offshore energy or marine reserves, public lands history also shines a light on chronological and scalar distinctions that help reduce the zoning-management confusion. In the United States a well documented public lands history proves helpful.

Exploring the U.S. Ocean Frontiers – Lessons from the Land

Considering the historical phases that preceded land-use oriented zoning in the United States, it becomes apparent that for much of the U.S. ocean writ large, i.e., the expanses of U.S.-claimed continental shelf areas and EEZ afforded waters, Lewis and Clark mode still prevails. When President Thomas Jefferson commissioned that pair of explorers to go out into the frontier to survey and examine a swath of the continent's terrestrial unknown, he was not determined to zone the U.S. claimed portions of it. Jefferson wanted to explore the frontier in the hopes of understanding what lay out there and perhaps one day opening it to beneficial use and settlement. Lewis and Clark's effort are recorded as paramount in that opening act.[17] Explorations of what would become the U.S. lower forty-eight states were not influenced by anything akin to zoning laws. Rather exploration and pioneer settlement played a significant role in shaping the political and legal landscape of the expanding country.[18] This phase occupied much of the space and time of 19th century America. Indeed the frontier space gave way to more and more settlement until, as the 1890 U.S. Census report indicated, the frontier had effectively disappeared.[19]

Given that very little of the oceans governed by U.S. law have been explored or mapped in any great detail, it is fair to characterize those little known but far reaching expanses as frontier spaces. As such, for

[17] For those unfamiliar with early U.S. history, President Jefferson dispatched a team of explorers led by Meriwether Lewis and William Clark in 1804 shortly after the United States acquired the Louisiana territory (roughly the middle vertical portion of the current contiguous 48 states). A scholarly overview of the journey, including many passages from the journals of the explorers can be found in the form of M. Lewis (B. DeVoto, ed.), *The Journals of Lewis and Clark* (Boston: Houghton Mifflin, 1953). A recent and more popular characterization of Lewis and Clark's exploratory effort can be found in S. Ambrose, *Undaunted Courage: Meriwether Lewis, Thomas Jefferson, and the Opening of the American West* (New York: Simon & Schuster, 1996).

[18] F. J. Turner, "The Significance of the Frontier in American History," *Proceedings of the State Historical Society of Wisconsin*, December 14, 1893, reprinted in F. J. Turner, *The Frontier in American History* (New York: H. Holt and Company, 1920).

[19] *Ibid.*, p. 1.

those areas, it makes little sense to suggest that they ought to be zoned at any fine level. Taking a page from Thomas Jefferson's book, it might be wiser to enhance U.S. efforts to advance our understanding of the unknown before embarking on any particular effort to physically occupy or economically allocate those spaces.

Gross-scale Disposition, Withdrawal and Reservation of the Public Domain
Even as settlers transformed large expanses of North America from frontier lands into communities and U.S. territories by way of public domain lands being made available, the concept or application of modern zoning was non-existent. Rather, only as areas of the West were explored and mapped did the U.S. federal government begin to employ gross-scale disposition, withdrawals and reservations of the public domain.[20] With basic information at hand in the form of maps, natural resource descriptions, climatic conditions and possible transportation corridors, the U.S. government enacted a series of land and resource grant making laws in an effort to populate the areas. Homesteading legislation and other public domain disposition laws motivated people to risk their lives and families in exchange for the opportunity to acquire land, build homes, raise crops, graze livestock and mine for gold and silver in some of the most inhospitable areas and circumstances. Under these disposition laws, acceptable (and sometimes unacceptable but unknown) uses of public lands allowed users to claim them and transform them into private property. As settlements took hold the federal government determined that some expanses of public land, while still available to those who could put them to productive use, ought to be "withdrawn" from certain activities deemed unsuitable for the area or held for future examination to determine their ultimate preferred use. Areas that had been open for a wide variety of uses might continue to be open for some while withdrawn from others. As the U.S. federal government's objective to settle the West gave way to the multiple objectives of securing settlements and maintaining and managing remaining public domain lands, the United States began to reserve large expanses of land for particular purposes (e.g., national parks, forests, wildlife conservation areas). As those reserved lands were placed into the jurisdiction of particular federal authorities, they might even be

---

[20] G. Coggins, C. Wilkinson and J. Leshy, *Federal Public Land and Resources Law* (5th ed.) (New York: Foundation Press, 2005), Ch. 2, pp. 34–137.

classified at a finer level as more suitable or less suitable for the particular activity normally allowed (e.g., even some tree-covered areas of U.S. Forest Service land may not be cut).

In U.S. ocean waters, the general rule has been to refrain from disposing of near-shore ocean space in light of public trust doctrine principles that generally prohibit such conveyances, while history and the murkiness of ocean property rights further offshore forestall fee-simple conveyances of ocean space beyond territorial sea areas.[21] As a result, there is no reasonably foreseeable likelihood that the United States will posit the possibility of wide-scale ocean disposition. Nonetheless, a wealth of valuable offshore resources (e.g., fish, oil and gas), have been made available to those capable of capturing them and introducing them into commerce with little or limited compensation sought by the state or federal government in whose jurisdiction they reside. And just as those "availabilities" in the form of usufruct extractions might be considered roughly equivalent to the claim of widely available public domain lands of the 19th century, shifts in U.S. ocean policies akin to the withdrawals and reservations, in the forms of fishery closure areas and marine protected area designations respectively, can be observed.

When ocean areas within a State's jurisdiction have been examined to determine their environmental condition and economic value, environmental management policies have been employed to protect areas deemed to be particularly sensitive to the type of development contemplated.[22] In areas that may contain valuable extractable resources, mechanisms rather than zoning have been employed when government policy suggests a wait-and-see approach. Canada and the United States have both employed moratoria on oil and gas exploration or extraction on their respective portions of Georges Bank in light of environmental concerns. Johnston's

---

21 Even attempts by a coastal State to allow in fee simple conveyances of ocean space within the State's territorial sea might attract the scrutiny and perhaps objection of other States concerned with the impact of such attempted conveyances on their rights of innocent passage.

22 Canadian Centre for Energy Information, "Canada's Evolving Offshore Oil and Gas Industry," (2004), <www.em.gov.bc.ca/DL/offshore/Reports/OffshoreFAMar2004_low.pdf >. Noting that "[t]he Royal Society of Canada has recommended the creation of coastal exclusion zones to protect sensitive marine environments such as sponge reefs off the West Coast," *ibid.*, p. 20.

thoughts seem particularly germane along this line of ocean management in that he "suggested that the best treatment of environmental security concerns in the ocean may be by evaluating the zoning techniques of ocean management that might be applied to specific threats to identifiably vulnerable coastal and marine areas."[23]

To the degree that the United States has embarked on an offshore effort similar to the public domain dispositions of the 19th century, it might be reflected in the wide-scale leasing of outer continental shelf lands in the U.S. portion of the Gulf of Mexico. The similarity is qualified in that the "disposition" of those lands is in the form of leases that may be used to explore for and extract oil and gas. The leases are limited in duration, competitively bid for, and if productive result in the payment of royalties for oil and gas extracted. Nonetheless, the U.S. outer continental shelf leasing system is the closest thing to the wide-scale public domain land conveyances of the 19th century.

Continental shelf claims by coastal States have been widely recognized and respected in international law for over half a century. Those claims afford States sufficient governance interests such that the coastal State may lease the property interests they hold in those offshore lands to private entities. While legal scholars may debate whether such leases amount to property rights or contractual rights, there is little doubt that the transactions result in the transfer of legal interests that carry ownership attributes.

In 2007, the U.S. agency in charge of managing offshore lands for oil and gas activity, the Minerals Management Service, reported that almost 37 million acres of outer continental shelf lands were being leased out in the Gulf of Mexico.[24] Having finally approached some offshore interests capable of being owned and transferred, i.e., spatially explicit tracts of land and a set of rights to develop and use such land, it might finally be appropriate to discuss the employment of zoning tools in the form of conditions and restrictions that are placed on private property interests

---

[23] D. Johnston, "Vulnerable Coastal and Marine Areas: A Framework for the Planning of Environmental Security Zones in the Ocean," *Ocean Development and International Law* 24 (1993): 63–79, at p. 73.

[24] U.S. Minerals Management Service, *MMS Gulf of Mexico Region Planning Areas and Active Leases October 1, 2007*, <www.gomr.mms.gov/homepg/lsesale/mau_gom_pa_20071001.pdf>, 15 March 2008.

as part of an effort to protect the general health, welfare and environment of the public. In a similar vein, Canadian authorities seem to have waited until offshore areas are examined and tested for potential uses prior to applying any thing akin to ocean zoning principles. In those Canadian waters where use restrictions and conditions are illustrated in maps and described in accompanying text, maps and narrative being the chief means of articulating a zoning scheme, the system has followed and accommodated rather than preceded and dictated the private property interests that have emerged.

Moving from Exploration to Reservation to Use to Zoning

In the United States, Congress enacted a set of principles that have been applied to the management of federal public lands. The *Federal Land Policy and Management Act* (FLPMA) merits consideration as the United States faces the growing demand for ocean development. The FLPMA principles can be summarily described as:

- Do not sell public lands, do inventory them;
- Plan for present and future use;
- Establish rules and regulations after considering views of general public;
- Establish goals and objectives as guidelines for public land use planning;
- Manage on basis of multiple use and sustained yield;
- Manage to protect scientific/scenic/historical/ecological/environmental/ air and atmospheric/water resource/archeological values;
- Preserve and protect certain public lands in their natural condition; and,
- Obtain fair market value of the use of the public lands and resources.[25]

Federal public lands have been studied and assessed with an eye toward determining what they ought to be relieved from (withdrawal), what they ought to be employed for (reservation), and how they might be characterized and managed on an individual basis (classified or zoned).

[25] *Federal Land Policy and Management Act*, 43 USC §1701; see also J. Duff, "Offshore Management Considerations: Law and Policy Questions Related to Fish, Oil, and Wind," *B.C. Environmental Affairs Law Review* 31 (2004) 385, at p. 402 (summarizing FLPMA principles).

The application of these practices and the FLPMA principles serve as a template for an ocean space withdrawal-reservation-designation-zoning process that will better serve the beneficiaries of the public ocean resources. And while there is no parallel Federal Ocean Policy and Management Act, some of the FLPMA principles can be seen in the management activities of departments and agencies charged with administering U.S. ocean space and resources.

Similarly, Canada has public land parallels that merit consideration. Parks Canada employs relatively fine scale zoning of lands under its jurisdiction, but only after potential park sites have been identified and evaluated, parks designated, and general management principles applied. Those necessary and preliminary steps provide a sufficient foundation and appropriate context then for zoning.[26]

Conclusion

Douglas Johnston's thoughts on ocean zoning, including his efforts to clarify some of what is characterized here as primary level confusion arising from the phrase along with his observations on zoning trends in ocean management, provide an excellent foundation upon which continuing examination of ocean administration can be built. This contribution

---

[26] See Parks Canada Agency, "Parks Canada Guiding Principles and Operational Policies, Part II – Activity Policies: National Parks Policy," <www.parcscanada.pch.gc.ca/docs/pc/poli/princip/sec2/part2a4_e.asp>. Section 2, Management Planning, 2.2. Zoning, states: "The national parks zoning system is an integrated approach by which land and water areas are classified according to ecosystem and cultural resource protection requirements, and their capability and suitability to provide opportunities for visitor experiences. It is one part of an array of management strategies used by Parks Canada to assist in maintaining ecological integrity through providing a framework for the area-specific application of policy directions, such as for resource management, appropriate activities, and research. As such, zoning provides direction for the activities of park managers and park visitors alike. The application of zoning requires a sound information base related to both ecosystem structure, function and sensitivity, as well as the opportunities and impacts of existing and potential visitor experiences." See also, Parks Canada, "Fundy National Park of Canada – Management Plan – November 2005," <parcscanada.pch.gc.ca/docs/v-g/nb/fundy/pd-mp/plan-2005_6_E.asp>. Section 6, Park Zoning Plan and Wilderness Area Declaration, states the "zoning system is a resource-based policy tool used to allocate land use within national parks."

suggests that as ocean policy develops and as ocean zoning discussions evolve, new confusions arise that merit the type of scrutiny that Johnston suggested, i.e., do not let ambiguities of language or abstractions of concept cloud the ocean governance horizon. While Johnston may not have anticipated the secondary level of ocean zoning confusion that would arise as the primary level of confusion subsided, his call for matching zoning techniques to particular threats provides practical advice and suggests an observational approach to ocean zoning as it continues to evolve.

International Straits and Navigation Routes

Transit Passage Through International Straits

Jon M. Van Dyke*

The Third United Nations Conference on the Law of the Sea (1974–1982)[1]

The Third United Nations Conference on the Law of the Sea (UNCLOS) began in 1974 in Caracas, Venezuela, amid great fanfare and high expectations. The delegations gathered to negotiate a comprehensive treaty that would clarify and bring certainty to the many ocean issues that had divided nations over the years. Eight years later, after long negotiating sessions that alternated between New York and Geneva, the United Nations Convention on the Law of the Sea (LOS Convention) was completed, and on 10 December 1982, 119 nations signed the document in Montego Bay, Jamaica. The Convention came into force in July 1994 after a sufficient number of countries had formally ratified the treaty.[2]

One of the central disputes among the countries negotiating this treaty concerned the width of the territorial sea, coastal State control of its adjacent offshore resources, and the navigational rights of commercial

* Professor of Law, William S. Richardson School of Law, University of Hawai'i at Manoa, USA.

[1] A few sections of this paper are adapted and updated from Jon M. Van Dyke, "Legal and Practical Problems Governing International Straits," in E. Mann Borgese, N. Ginsburg and J. R. Morgan, eds, *Ocean Yearbook 12* (Chicago: University of Chicago Press, 1996), p. 109, also published in H. Ahmad ed., *The Straits of Malacca: International Co-Operation in Trade, Funding & Navigational Safety* (Kuala Lumpur: Maritime Institute of Malaysia, Pelanduk Publications, 1997), p. 305 [hereinafter *Straits of Malacca*]; and from J. M. Van Dyke, "The Legal Regime Governing the Gibraltar Straits," in B. Ozturk and R. Ozkan eds, *The Proceedings of the Symposium on the Straits Used for International Navigation* (Istanbul: Turkish Marine Research Foundation, 16–17 November 2002), p. 72 [hereinafter *Istanbul Straits Symposium*].

[2] *United Nations Convention on the Law of the Sea*, Montego Bay, 10 December 1982, U.N. Doc A/Conf.62/122 (1982), 21 *I.L.M.* 1261 (1982) [hereinafter LOS Convention].

and military vessels to pass through straits and island archipelagos.[3] The United States and other maritime powers initially resisted efforts to allow coastal countries to claim extended fisheries zones because they were concerned that such a zone could limit navigational freedoms. The United States was particularly concerned about its continuing ability to navigate its warships, including submerged submarines, through key international straits such as the Strait of Gibraltar (into the Mediterranean Sea), the Strait of Hormuz (into the Persian/Arabian Gulf), the Strait of Bab al-Mandeb (into the Red Sea), the Strait of Malacca (connecting the Indian Ocean with the Pacific), the Dover Strait (through the English Channel), the Bering Strait (in the Arctic), and the Strait of Lombok (through the Indonesian archipelago). The United States was worried that if countries were allowed to extend their territorial seas from three to twelve nautical miles, no high seas corridors would remain in these narrow straits and control over passage might arguably fall under the control of the countries bordering on these key waterways. The United States maintained that free movement through these straits was essential to its national security and protested claims of expanding territorial seas.[4] The U.S. position on navigational freedoms was supported during this period by the Soviet Union, which was also a major maritime power.

Opposing the position of the maritime countries was a group of strait States that included Malaysia, Indonesia, Spain, the Philippines, Cyprus, Egypt, Morocco, and Yemen. Spain and Malaysia argued that oil tankers presented serious pollution dangers to coastal countries.[5] Malaysia argued that the passage of oil tankers should be viewed as "non-innocent" and that coastal countries should be allowed to regulate their passage.[6] Draft

---

[3] See generally, S. N. Nandan and D. H. Anderson, "Straits Used for International Navigation: A Commentary on Part III of the United Nations Convention on the Law of the Sea 1982," *1989 British Year Book of International Law* (1990): 159–204.

[4] G. Galdorisi, "The United States and the Law of the Sea: Decade of Decision," in G. Galdorisi, D. Bandow, and M. C. Jarman, eds, *The United States and the 1982 Law of the Sea Convention: The Cases Pro and Con* (Law of the Sea Institute Occasional Paper No. 38, 1994), pp. 7, 16–17.

[5] United Nations Office of Legal Affairs, Division for Ocean Affairs and the Law of the Sea, *The Law of the Sea: Straits Used for International Navigation: Legislative History of Part III of the United Nations Convention on the Law of the Sea* (New York: United Nations, 1992), p. 30 [hereinafter *Straits Legislative History*].

[6] *Ibid.*, p. 36 (statement by Lal Vohrah, 12 August 1971, U.N. Doc. A/AC.138/SC.II/SR.11).

articles submitted by Malaysia, Morocco, Oman, and Yemen proposed a regime of innocent passage for travel through straits that would have required warships to seek authorization from coastal States prior to exercising innocent passage through territorial seas in straits.[7] These proposals were rejected by the maritime powers and failed to receive the support of many other coastal States.[8]

The compromise that emerged during the protracted negotiations consisted of (a) allowing coastal States to extend their territorial seas to 12 nautical miles; (b) recognizing the right to "transit passage through international straits;" and (c) allowing countries to establish an exclusive economic zone (EEZ) out to a distance of 200 nautical miles from their coasts, governed by Part V of the Convention, Articles 55–75. The right of "transit passage through international straits," as defined in the Convention, is nonsuspendable and applies to all vessels – military and commercial – and also to airplanes (Article 38(1)). Pursuant to the language in Article 39(1)(c), submarines are allowed to remain submerged when they exercise the right of transit passage.[9]

The position of the maritime countries that all ships should have the right to unimpeded passage through international straits was thus largely adopted in Part III (Articles 34–45) of the 1982 LOS Convention. Each strait, however, presents unique geographical and practical considerations, and some straits have historically been governed by unique legal regimes, which remain in force pursuant to Article 35(c) of the Convention. Professor Lewis M. Alexander, who served as the Geographer for the U.S. State Department during the 1980s, has identified 265 important straits around the globe,[10] but the number would be much higher if every narrow

[7] E. Franckx and A. Razavi, "The Strait of Hormuz," in *Istanbul Straits Symposium*, *supra* note 1, at pp. 53, 58 (citing UNCLOS III, *Official Records*, Vol. III, p. 129, U.N. Doc. A/CONF.62/C.2/L.16).

[8] *Ibid.*

[9] S. N. Nandan, "Legal Regime for Straits Used for International Navigation," in *International Straits Symposium*, *supra* note 1, at p. 7. "The common practice of submarines transiting some international straits while submerged is recognised in the provision that passing vessels refrain from any activities other than those 'incident to their normal mode of continuous and expeditious transit' (art. 39(1)(c))."

[10] L. M. Alexander, *Navigational Restrictions within the New LOS Context: Geographical Implications for the United States* (Peace Dale, RI: Offshore Consultants, 1986), p. 99 and pp. 188–198, Table 12-A.

waterway between bodies of land were counted. The specific rules that apply to individual straits are discussed in more detail below.

The Corfu Channel Case

The right to pass freely through international straits was not firmly established in international law until the *Corfu Channel Case*[11] in 1949, when the International Court of Justice said that ships have the right of nonsuspendable innocent passage through such straits. In 1946, the United Kingdom sent four warships through the Corfu Channel, which separates the Greek island of Corfu and the Albanian coast. Several of the vessels were seriously damaged by mines in the channel and a number of British sailors were killed. Albania argued that the channel was not a necessary route between two parts of the high seas and, therefore, that no right of passage existed. Albania explained that this waterway was "only of secondary importance and not even a necessary route between two parts of the high seas, and that it is used almost exclusively for local traffic to and from the ports of Corfu and Saranda."[12] The International Court agreed that it was not a necessary route, but said that "[i]t has nevertheless been a useful route for international maritime traffic."[13] The decisive criteria is simply "its geographical situation as connecting two parts of the high seas and the fact of its being used for international navigation" and hence that "the North Corfu Channel should be considered as belonging to the class of international highways through which passage cannot be prohibited by a coastal State in time of peace."[14] As long as the passage

[11] *Corfu Channel Case (United Kingdom v. Albania)*, [1949] I.C.J. Reports 4.

[12] *Ibid.*, at 28.

[13] *Ibid.*

[14] *Ibid.* The International Court also noted that British warships could pass through with crew and guns at the ready when Albanian shore batteries had previously fired at British ships, since "the measure of precaution" was not "unreasonable" under the circumstances. The Court ruled, however, that subsequent sweeping of the Corfu Channel of mines without Albanian consent was an impermissible use of force or enforcement jurisdiction in Albanian territorial waters. Especially unpersuasive was a British claim that it was acting to gather evidence (before it disappeared) concerning the illegal placement of mines in the channel.

through the waterway is innocent, passage of warships is permissible and the coastal State cannot require prior authorization.

> States in time of peace have a right to send their warships through straits used for international navigation between two parts of the high seas without the previous authorization of a coastal State, provided the passage is innocent. Unless otherwise prescribed in an international convention, there is no right for a coastal State to prohibit such passage through straits in time of peace.[15]

This decision clarified a number of rules governing passage through straits, but disagreements continued regarding what activities qualified as "innocent," and commentators differed on whether warships were entitled to pass under this regime.[16] The text of Part III of the LOS Convention was designed to clarify these uncertainties.

What Controls Can Coastal States Exercise over Vessels Engaged in Transit Passage through International Straits?

The rules recognized in the LOS Convention do not allow suspension of transit passage (Article 44) and do not require innocence,[17] but they do impose, *inter alia*, the following restrictions on transit passage: (1) transit passage must be solely for the purpose of continuous and expeditious transit (Article 38(2)); (2) transiting ships must comply with generally accepted international regulations, procedures, and practices for safety at sea (Article 39(2)(a)) and for the prevention, reduction, and control of

---

[15] *Ibid.*, p. 28.

[16] S. C. Truver, *The Strait of Gibraltar and the Mediterranean* (Dordrecht: Martinus Nijhoff, 1980), p. 153 (citing "Commentary on Article 24 of the Draft Articles Concerning the Law of the Sea," *Yearbook of the International Law Commission* 2 (1956), pp. 276–277).

[17] The regime of nonsuspendable innocent passage continues to apply, according to Articles 38(1) and 45, to passage through as strait separating an island and mainland of a State (e.g., the Pemba and Zanzibar Channels) and between a part of the high seas or an EEZ and the territorial sea of a foreign State (e.g., the Strait of Tiran, and the entrance to Passamaquoddy Bay separating the United States and Canada). Article 36 says that the right of normal innocent passage is applicable to the territorial seas in straits wider than 24 nautical miles (e.g., the waters between the United States and Cuba).

pollution from ships (Article 39(2)(b)); and (3) ships exercising the right of transit passage must proceed without delay through the strait, must not engage in any research or fishing activities, and must refrain from any threat or use of force (Articles 39(1), 40, and 42(1)(c)).

Article 38(3) of the LOS Convention states explicitly that: "[a]ny activity which is not an exercise of the right of transit passage through a strait remains subject to the other applicable provisions of the Convention." Any such "nontransit" activity, if undertaken in the territorial waters of a coastal State, would have to comply with the innocent passage provisions of Articles 17–26 of the Convention and the activity could be prevented if "noninnocent."

The LOS Convention allows countries bordering on straits to establish certain types of regulations. Traffic separation schemes and other safety measures can be established under Articles 41 and 42(1)(a) of the LOS Convention, but Article 41(4) indicates that the International Maritime Organization (IMO) must approve a traffic separation scheme before it can be put into force.[18] These must be developed in coordination with other adjacent or opposite States, must conform to generally accepted international regulations, must be submitted to the component international organization (the IMO) for adoption, and must be widely publicized. Traffic separation schemes have been adopted for many of the important straits, including the Baltic, Dover, Gibraltar, Kerch, Bab al-Mandeb, Hormuz, Malacca-Singapore, and Kurile Straits.[19]

Pollution control regulations can be adopted under Article 42(1)(b), which allows States bordering international straits to adopt laws and regulations with respect to "the prevention, reduction and control of pollution, by giving effect to applicable international regulations regarding the discharge of oil, oily wastes and other noxious substances in the strait."[20] However, such laws and regulations shall not be discriminatory,

---

[18] See Nandan, *supra* note 9, at p. 6. Nandan states that traffic separation schemes "must first have been adopted by the competent international organization (art. 41), which would normally be the International Maritime Organization."

[19] Alexander, *supra* note 10, at p. 129.

[20] During the final negotiating session in 1982. Spain objected to the word "applicable" in this provision because it meant that the regime that could be imposed on a ship would change with the flag of the ship, and urged instead that the phrase "generally accepted" be used in order to ensure a uniform standard. A vote was taken on Spain's

nor "in their application have the practical effect of denying, hampering or impairing the right of transit passage" (Article 42(2)), and must have been duly publicized (Article 42(3)).

Fishing regulations can be adopted under Article 42(c) to prevent fishing, including the requirement to stow all fishing gear. Further, regulations can be adopted to control the loading, unloading, or transfer of any goods, any currency, or any person in contravention of the "customs, fiscal, immigration or sanitary laws and regulations" of the coastal State, under Article 42(d).

The question whether countries adjacent to straits can act to control pollution of their coasts was discussed during the Convention negotiations, and it has continued to be a difficult one. During the debates that created the transit passage regime, Norway, supported by Turkey, suggested establishing a mandatory insurance pool covering all shippers to guarantee that coastal States would be compensated when other rules of liability were inadequate.[21]

The regulations issued by strait States cannot discriminate against foreign ships nor can they have the effect of "hampering or impairing the right of transit passage" (Article 42(2)), and due publicity must be given to these regulations. Nonetheless, they can be promulgated, and foreign States whose flag vessels do not comply are responsible for "any loss or damage which results to States bordering straits" (Article 42(5)). Strait States also have a duty to "give appropriate publicity to any danger

---

proposal, with 60 countries voting in favour, 29 against, and 51 abstentions. Because the proposal did not receive the affirmative votes of two-thirds of those voting, it was deemed to have been defeated. *Straits Legislative History, supra* note 5, at pp. 136, 141–142. After Spain was defeated on this vote, it issued an "understanding" to the effect that "it considers that the provisions of [Article 42(1)(b)] do not prevent it from issuing, in accordance with international law, laws and regulations giving effect to generally accepted international regulations." *Ibid.*, p. 156. Later the phrasing of Spain's "understanding" became: "The regime established in Part III of the Convention is compatible with the right of the bordering State to enact and enforce in straits used for international navigation its own regulations, provided that such regulations do not interfere with the right of transit passage." *United Nations Division of Ocean Affairs and the Law of the Sea, Convention on the Law of the Sea of 10 December 1982 – Overview*, <www.un.org/Depts/los/convention_agreements/convention_overview_convention.htm>, 23 July 2006) [hereinafter *Convention Overview*].

21 *Straits Legislative History, supra* note 5, at pp. 3, 13.

of navigation or overflight within or over the strait of which they have knowledge" (Article 44).[22]

With regard to passage through the territorial sea, the LOS Convention recognizes in Articles 22 and 23 that nuclear cargoes present unusual risks and allows them to be regulated accordingly. Article 23 states that restrictions should be arranged pursuant to "international agreements," but no such agreements have yet been developed. Several countries (including Malaysia and Saudi Arabia) have concluded, therefore, that they are authorized to regulate such transports directly, until international agreements are completed.[23]

Straits States Have Limited Abilities to Enforce Their Regulations

During the final negotiating sessions in 1982, Indonesia, Malaysia and Singapore issued a joint statement that: "States bordering the Straits may take appropriate enforcement measures in accordance with article 233, against vessels violating the laws and regulations referred to in article 42, paragraph 1(a) and (b) causing or threatening major damage to the marine environment of the Straits."[24] This assertion challenged the principle of unimpeded transit passage, but was deemed necessary by the straits States to protect and preserve their marine environment.

Under the language of the LOS Convention, straits States are limited in their abilities to enforce their regulations, because under Articles 38(1), 42(2) and 44 they cannot impede, impair, hinder, deny, or suspend the right of transit passage. But they can, under Article 233, "take appropriate enforcement measures" in the event transiting vessels violate the regulations in a manner "causing or threatening major damage to the marine

[22] This proposal was presented by the United Kingdom, with reference to the *Corfu Channel* decision, discussed *supra* text at notes 11–16. Nandan and Anderson, *supra* note 3, at p. 194.

[23] See declarations filed by Saudi Arabia and Malaysia, *Convention Overview, supra* note 20.

[24] *Straits Legislative History, supra* note 5, at p. 144; see also Malaysia's declaration issued 14 October 1996 in *Convention Overview, supra* note 20, where Malaysia refers to the statement made on 28 April 1982 regarding Article 233.

environment of the straits."[25] Enforcement actions under Article 233 are limited to "exceptional" cases,[26] but the Malacca Straits States have interpreted Article 233 to allow them to take appropriate enforcement measures against ships passing through the Straits that fail to meet the 3.5 metre under-keel clearance requirement which they have established.[27] Oman issued a statement when it signed the Convention in 1983 that it understood "that the application of the provisions of the Articles 19, 25, 34, 38 and 45 of the Convention does not preclude a coastal State from taking such appropriate measures as are necessary to protect its interests of peace and security."[28] Professor Tullio Scovazzi of the University of Milan has suggested that the term "appropriate enforcement measures" should be interpreted "as allowing bordering States to forbid the passage of such ships."[29] He has criticized the immunity given to warships as a "questionable exception to the principle that, in the field of protection of the environment, prevention is preferable to compensation."[30] Satya Nandan, one of the major architects of the LOS Convention and now the Secretary-General of the International Seabed Authority, has said that, except in an Article 233 major damage situation, "the only means of enforcing international standards or laws and regulations against passing ships is through the flag State unless the ship voluntarily enters the

[25] Article 233 "emerged from consultations in Kuala Lumpur between the United States and Malaysia, represented at the ministerial level" and was central to winning the support of Malaysia, Indonesia, and Singapore for the transit passage regime. B. Oxman, "Observations on the Interpretation and Application of Article 43 of the United Nations Convention on the Law of the Sea with Particular Reference to the Straits of Malacca and Singapore 7–8," paper presented at Conference on Navigational Safety and Control of Pollution in the Straits of Malacca and Singapore: Modalities of International Cooperation, sponsored by the IMO and the Institute of Policy Studies, Singapore, 2–3 September 1996, Singapore [hereinafter Oxman, "Observations on Article 43"].

[26] Nandan and Anderson, *supra* note 3, at p. 192.

[27] M. George, "The Regulation of Maritime Traffic in Straits Used for International Navigation," in A. G. Oude Elferink and D. R. Rothwell, eds., *Oceans Management in the 21st Century: Institutional Frameworks and Responses* (Leiden: Martinus Nijhoff, 2004), pp. 19 and 34 (citing 1982 Interpretive Statement on Straits of Malacca and Singapore, U.N. Doc A/CONF./62/L/145).

[28] *Convention Overview, supra* note 20.

[29] T. Scovazzi, "Management Regimes and Responsibility for International Straits," in *Straits of Malacca, supra* note 1, at p. 335.

[30] *Ibid.*

port of the strait State."[31] The opposition in 1992 of Malaysia, Singapore and Indonesia to the proposed passage of the Japanese plutonium ship through the Malacca Straits[32] presents an example where the straits States opposed passage because of the threat of major damage to the marine environment of the straits.

Is the Regime of Transit Passage through International Straits Now Binding Customary International Law?

The United States has not yet ratified the Law of the Sea Convention, but has argued vigorously that the regime of transit passage through international straits is now part of binding customary international law. On 17 August 1987, the United States said:

> The United States particularly rejects the assertions that the right of transit through straits used for international navigation, as articulated in the LOS Convention, are contractual rights and not codification of existing customs or established usage. The regimes of transit passage, as reflected in the Convention, are clearly based on customary practice of long standing and reflects the balance of rights and interests among all States, regardless of whether they have signed or ratified the Convention.[33]

Other countries have, however, viewed transit passage as emanating directly from the LOS Convention and thus not invocable by countries that are not contracting parties.[34] Some commentators have suggested that the transit passage regime in the LOS Convention may not yet have been confirmed as customary international law because of "the attitude taken

---

[31] Nandan, *supra* note 9, at p. 7.

[32] M. Eager and I. Steward, "Freighter Heads for Nuclear Shipment," *South China Morning Post Wire Service*, 21 September 1992; "Malaysia May Cite Security Laws to Block Japanese Plutonium Ship," *UPI Business & Financial Wire*, 24 September 1992; see generally, J. M. Van Dyke, "Sea Shipment of Japanese Plutonium Under International Law," *Ocean Development and International Law* 24 (1993): 399, 411.

[33] N. Űnlű, *The Legal Regime of the Turkish Straits* (The Hague: Martinus Nijhoff Publishers, 2002), p. 74.

[34] The Turkish scholar Nihan Űnlű lists the countries that "consider the regime of transit passage as an exclusive part of the UNCLOS" as Chile, Denmark, Egypt, Greece, Iran, Indonesia, Italy, Japan, South Korea, Malaysia, the Netherlands, Oman, and Spain. *Ibid.*, p. 75.

by a significant number of States which appear reluctant, either explicitly or implicitly, to accept the transit passage regime as a whole or some of its implications."[35] Some Greek scholars have argued, for instance, that Turkey would not be entitled to invoke the right of transit passage through the Aegean Sea (if Greece were to claim 12 nautical mile territorial seas around its Aegean islands) because Turkey is not a contracting party to the LOS Convention.[36]

Professor Scovazzi has explained that the Convention does not adequately protect the "vital concern" of States bordering straits regarding the protection of their marine environment.[37] In particular, the Convention provides only limited authority to the bordering States to enforce their environmental regulations. It does not create an adequate liability regime, nor does it require the prior notification of transit of ultrahazardous cargoes that would allow coastal States to protect their coastal populations and resources.[38] These inadequacies have led a number of straits-bordering States to promulgate regulations that appear to go beyond what is permitted by the Convention.[39] Professor Scovazzi concludes that "[i]t is therefore possible to argue that the LOS Convention transit passage regime is still far from fully corresponding to present customary international law."[40]

---

[35] T. Scovazzi, *The Evolution of International Law of the Sea: New Issues, New Challenges* (The Hague: Martinus Nijhoff, 2001) (reprinted from Recueil des Cours 286 (2000)), p. 174. Űnlű agrees that "a general right of transit passage has not yet been established in customary law." Űnlű, *supra* note 33, at p. 75.

[36] See, e.g., G. P. Politakis, "The Aegean Dispute in the 1990s: Naval Aspects of the New Law of the Sea Convention," in T. C. Kariotis, ed., *Greece and the Law of the Sea* (The Hague: Martinus Nijhoff, 1997), p. 291, 303 (similar to G. P. Politakis, "The Aegean Agenda: Greek National Interests and the New Law of the Sea Convention," *International Journal of Marine and Coastal Law* 10 (1995): 497), summarizing scholarly discussion that indicates that all aspects of the transit passage regime have not yet crystallized into customary international law, and A. Strati, "Greece and the Law of the Sea: A Greek Perspective," in A. Chircop, A. Gerolymatos and J. Iatrides, eds, *The Aegean Sea After the Cold War* (Houndsmill: Macmillan, 2000), pp. 89, 94, "it is highly questionable whether the LOS Convention provisions on transit passage in all their detail reflect customary law, thereby entitling Turkey to benefit from them."

[37] Scovazzi, *supra* note 35, at pp. 174–175.

[38] *Ibid.*, pp. 175–177.

[39] *Ibid.*, pp. 177–187, providing examples from the Malacca Strait, the Canadian Arctic Straits, the Russian Arctic Straits, and the Turkish Straits.

[40] Scovazzi, *supra* note 29, at p. 344. See also, Hamzah bin Ahmad, "Global Funding for Navigational Safety and Environmental Protection," in *Straits of Malacca, supra*

Security Concerns

Article 39(1)(b) requires vessels and planes that are exercising their right of transit passage to "refrain from any threat or use of force against the sovereignty, territorial integrity or political independence of States bordering the strait, or in any other manner in violation of the principles of international law embodied in the Charter of the United Nations." Warships can, however, according to Professor Bernard Oxman, travel through straits in "squadron formation," and they may "communicate by radio, use radar or sonar, and, where circumstances permit, travel in defensive formation and use defensive maneuvers."[41] Professor Oxman, of the University of Miami, who was a member of the U.S. delegation to the Law of the Sea negotiations, has also contended that the launching and recovery of aircraft from an aircraft carrier is a permitted activity during transit passage, so long as it does not constitute a threat or use of force against a straits State.[42]

If a vessel does engage in an activity other than those permitted under the transit passage regime, Article 38(3) states that it remains subject to the "other applicable principles" of the Convention. But it is unclear exactly what actions the coastal State can take in the light of such a threat. What should happen if an airplane that is purporting to exercise transit passage over a strait strays over land and then drops anti-government leaflets to the residents below? Professor Oxman has acknowledged that straits States can take action under Article 233 against nongovernmental ships "causing or threatening major damage to the marine environment of the straits," but he contends that no action can be taken against military ships or planes no matter how nefarious their behaviour might be.[43] Nandan has commented that the only appropriate action is to require the ship or plane to continue expeditiously through the strait.

note 1, at pp. 125, 131, "the concept of transit passage is relatively new and cannot be said to have acquired the status of customary international law."

[41] B. H. Oxman, "Transit of Straits and Archipelagic Waters by Military Aircraft," Paper delivered at Third International Law Seminar, Singapore, August 28, 1999, pp. 21–22, reprinted in *Singapore Journal of International and Comparative Law* 4 (2001): 377–434.

[42] *Ibid.*, p. 22.

[43] *Ibid.*, p. 27.

> It is implicit, therefore, that any activity threatening a coastal State would bring the ship or aircraft under the general regime of innocent passage and enable that State to require the vessel to leave the area expeditiously (art. 30), presumably in the direction it was travelling, since transit passage cannot be suspended for any reason (art. 44) – including threats to security.[44]

The recent security concerns of countries adjacent to the Malacca Straits and those using the Straits have been focused on piracy and terrorist bombings. In early July 2006, for instance, pirates boarded and robbed two ships in the Malacca Strait chartered by the United Nations carrying construction materials for tsunami-torn Aceh and attempted to board a Japanese cargo ship that was able to defend itself with floodlights and fire hoses.[45] The major military powers and main users of the Straits have frequently expressed an interest in helping to patrol the Straits, but Malaysia and Indonesia have strenuously resisted this idea, saying that the only assistance they would be willing to accept would be information, intelligence, and technical equipment. Singapore and Indonesia have established a joint radar surveillance system in the Singapore Strait, called Project Supric, to allow their navies to observe activities in the Strait.[46] These countries want to undertake the active security operations themselves. Activities such as providing armed escorting vessels to commercial vessels are expensive, however, and the question inevitably arises as to whether it would be lawful to introduce a compensation scheme so that the users of the straits would pay for the necessary security measures. As the materials below explain, payments for passage through straits have been utilized historically, in the Baltic Straits in particular[47] and also in the Malacca Straits when they were under Portuguese control,[48] payments are now required for certain services related to passage through the Turkish Straits,[49] and payments are always required for passage through canals.[50]

---

[44] Nandan, *supra* note 9, at p. 6.

[45] "Japanese Cargo Ship Foils Attack," *Honolulu Advertiser*, 5 July 2006, at A14, col. 3.

[46] M. Hand, "Joint Radar System for Singapore Straits," *Lloyd's List*, 6 February 2005.

[47] See *infra* text at notes 87–118.

[48] See *infra* text at notes 241–243.

[49] See *infra* text at notes 126–127.

[50] See *infra* text at notes 273–285.

Can Ships Be Charged for Passing through Straits?

Article 37(1) of the LOS Convention says that "the right of transit passage...shall not be impeded," but Article 42 allows countries bordering on straits to promulgate a variety of laws regulating passage. Article 43 says that "[u]ser States and States bordering a strait should by agreement cooperate" to establish navigational aids and "for the prevention, reduction and control of pollution from ships." No such formal agreements have, however, been established yet under Article 43.

Article 26 is entitled "Charges Which May Be Levied Upon Foreign Ships," and its paragraph 2 indicates that a coastal State can charge ships passing through its territorial sea "for specific services rendered to the ship." Does this provision apply to a ship in the territorial sea of a country bordering on a strait while the ship is exercising its right to transit passage through the strait? No provision in Part III on "Straits Used for International Navigation" says explicitly that it does not apply and application of Article 26(2) does not directly conflict with the purposes of Part III.

If a ship is passing through a strait narrower than 24 nautical miles wide, it must of necessity be in the territorial sea of one of the strait States, but is a ship in the territorial sea of a strait State as a juridical matter when it is exercising its right of transit passage through the strait? One author has suggested that "[t]he transit passage regime implies, as regards navigation, that the strait is no longer to be considered as part of the territorial sea of a strait State and that coastal State powers in the strait are different from those it can generally exercise in the territorial sea."[51] The strait States would generally disagree with this characterization and argue that "sovereignty and jurisdiction over the waters, air space, seabed, and subsoil of the straits are still subject to other rules of international law."[52] Nandan has sided with the straits States on this issue and has said that "[t]here is nothing in the Convention which prohibits charges for similar

---

[51] George, *supra* note 27, at p. 22 (citing T. Treves, "Navigation," in R. J. Dupuy and D. Vignes, eds, *A Handbook on the New Law of the Sea*, Vol. 2 (Dordrecht: Martinus Nijhoff, 1991), p. 950.

[52] *Ibid.*, see, e.g., W. Awang bin Wan Yaacob, "Regional Co-Operation and the Straits of Malacca," in *Straits of Malacca, supra* note 1, at pp. 15, 21, citing Art. 26(2) of the LOS Convention for the proposition that "[t]herefore international law does indeed provide for riparian states situated along straits to impose charges on foreign ships, but for specific services rendered only."

services [similar to the "specific services rendered to the ship allowed under Article 26(2)] in straits which are part of the territorial sea."[53]

The subject of compensation came up during the UNCLOS negotiations, with the strait States arguing in 1971 that they should be compensated for expenditures designed to promote navigational safety[54] and the United States opposing any requirement of compensation.[55] Professor Oxman has contended that the absence of specific language in Part III similar to that of Article 26(2) was not unintentional, that Article 43 was included to address the issue of fees in the context of straits, and hence that straits States cannot charge fees for services absent an Article 43 agreement.[56] Article 43 is cast in "conditional, non-mandatory terms,"[57] according to Nandan and David Anderson, and efforts to make its language obligatory were not accepted by the negotiating conference.[58] As Nandan has explained, the problem of compensating strait States for their services was "addressed in a perfunctory manner in article 43 of the Convention, which exhorts user States to cooperate through agreements to assist strait States, [but] the fact is that apart from the exceptional case of Japan's cooperation in respect of the Malacca and Singapore Straits, such cooperation has not materialized."[59]

Nandan has noted that the issue of compensation to strait States "has been festering for some time" and has explained that

> Straits States are legitimately concerned with the financial burdens they have to bear for establishing and managing traffic separation schemes, for installing and maintaining navigational aids, and by the pollution they must

---

[53] Nandan, *supra* note 9, at p. 8.

[54] *Straits Legislative History, supra* note 5, at p. 75, draft Art. 11(3). In 1973, Ecuador, Panama and Peru submitted a paper that suggested that strait States should be allowed to establish an "equitable charge" that would be administered by the "international ocean space institutions" to finance dredging, to promote navigational safety, and to compensate strait States for damage caused by passing vessels. *Ibid.*, p. 102, Art. 40(2)–(3).

[55] *Ibid.*, p. 85.

[56] Oxman, "Observations on Article 43," *supra* note 25, at p. 6.

[57] Nandan and Anderson, *supra* note 3, at p. 193.

[58] *Ibid.*, pp. 193–194, citing Amendment C2/22 offered by Morocco in 1978 which would have substituted "shall" for "should."

[59] Nandan, *supra* note 9, at pp. 7–8.

> endure, without receiving any corresponding benefits, since many ships transit straits en route to ports in other States.[60]

Nandan has noted that this "matter remains unresolved" and that "a meaningful global solution would be difficult to achieve." He has suggested that the special circumstances of each strait need to be examined separately and that "[a]ccount also has to be taken of the sensitivity of the straits States to any diminution in the exercise of sovereignty over the strait."[61]

If Article 26(2) does apply to straits, for what "specific services" could strait States charge? Among those that might fall into the "specific services" category would be pilotage, towing, and escorting services, because they would be applied to specific vessels to ensure the success of their voyages. If a ship requests such services from the coastal State, then the charging of a fee would certainly seem appropriate. But can the strait State require a passing ship to accept such services if it does not want them?

One strait where a compulsory pilotage scheme seems to be emerging is the Torres Strait, between Australia and Papua New Guinea. In May 1996, the IMO approved a reporting regime for the Torres Strait region between Australia and Papua New Guinea and the inner route of Australia's Great Barrier Reef.[62] In 2003, Australia and Papua New Guinea jointly applied to have the Torres Strait area declared to be a particularly sensitive sea area (PSSA),[63] which was subsequently approved. The IMO's Maritime Safety Committee (MSC) and its Marine Environment Protection Committee (MEPC) have both endorsed Australia's initiative to extend the mandatory pilotage provisions governing passage through the Great Barrier Reef (which has previously been designated as a PSSA) to the Torres Strait area.[64] Although the United States has noted that this IMO action is only "recommendatory,"[65] it will become mandatory, at least under Australian law, when Australia acts to extend its compulsory pilotage programme to the Torres Strait region. If the user States accept the duty to utilize an Australian pilot during passage through the Torres Strait pursuant to the

---

[60] *Ibid.*, p. 7.

[61] *Ibid.*, p. 8.

[62] IMO Resolution MSC 52(66), 30 May 1996.

[63] IMO Doc. MEPC 49/8 (submitted by Australia and Papua New Guinea).

[64] MSC, 79/23.

[65] MEPC, 49/22.

IMO resolutions, then their agreement would conform to the responsibility under Article 43 to reach agreements for navigational safety and "for the prevention, reduction and control of pollution from ships."

The Malacca Strait might be a logical candidate to be designated by the IMO as a particularly sensitive sea area because of the human and economic dependency on this Strait. Its economic importance as a transport channel is unquestioned,[66] and the closure of the Strait because of an accident or act of terrorism would be disastrous to the region and the world, and would cause severe harm to other economic activities to the region including offshore fishing, tourism, and mangrove harvesting. The vulnerability of the Strait to an accident, act of terrorism, or act of piracy is also clear. If the Strait were to be designated as a PSSA, the countries bordering on the Strait would have strong arguments that they could require escorts and charge users for activities related to patrolling and protecting the Strait. Under Article 211(6)(a), they could petition the IMO to approve "laws and regulations for the prevention, reduction and control of pollution from vessels."

In September 1997, the United Kingdom submitted an information paper to the Legal Committee of the IMO entitled "Developing Principles for Charging Users the Cost of Maritime Infrastructure," which suggested that the Organization should develop a set of fair and equitable principles governing the charges that strait States could levy on users for navigational aids and other services to passing vessels.[67] The paper argued that such aids were important to the shipping industry and that it was unfair to impose the expenses solely on the coastal States.[68] The paper suggested that charges should be exacted on a nondiscriminatory basis and that they "would be linked to the recovery of costs, including capital investment and improvements, but there would be no element of profit since that

[66] See, e.g., G. Naidu, "The Straits of Malacca in the Malaysian Economy," in *Straits of Malacca*, *supra* note 1, at pp. 33–60.

[67] See Nandan, *supra* note 9, at p. 8.

[68] A distinguished Malaysian author has asserted this equitable perspective in stronger terms, noting that "maritime nations are concerned about collisions, pollution liability, higher navigational safety standards, low-level threats to navigation, such as piracy and thefts" and they "have demanded that navigational aids, charts and other services be secured for them to ensure the continued freedom of navigation," but "they are not prepared to contribute to the cost of providing all these public good in the straits. They insist on a free ride." Hamzah bin Ahmad, *supra* note 40, at pp. 132–133.

would amount to tax on traffic."[69] The IMO Legal Committee did not act on the proposal because of the complexity of the legal issues.

One further reason why this problem is so vexing is that

> [t]he issue cannot be left just for straits States and flag States to resolve, for many flag States, as it is well known, are mere flags of convenience with little or no capacity to assist. The term 'user States' in art. 43 must therefore include States other than flag States, whose nationals benefit from safe passage.[70]

The category of "user States" must therefore also include "exporting states, receiving States, and States of ship-owners, insurers of ships and cargoes and major oil corporations whose global trade is facilitated."[71]

The Messina Exception – Article 38(1)

Which straits are subject to the regime of transit passage through international straits? Article 38(1) says that the right of transit passage does not apply "if the strait is formed by an island of a State bordering the strait and its mainland" and "if there exists seaward of the island a route through the high seas or through an exclusive economic zone of similar convenience with respect to navigational and hydrographical characteristics." This provision is generally referred to as the Messina Clause "because it emerged from proposals put forward by Italy with implicit reference to the strait which separates the island of Sicily from the mainland."[72] But to what other straits does it apply? Two examples mentioned frequently are the Pemba and Zanzibar Channels separating the mainland of Tan-

[69] Nandan, *supra* note 9, at p. 8.

[70] *Ibid.*

[71] *Ibid.*, pp. 8–9. Professor Oxman agrees that the term "user state" includes "those whose trade is carried through the straits" and includes both the purchasers and sellers of the goods. Oxman, "Observations on Article 43," *supra* note 25.

[72] Scovazzi, *supra* note 29, at p. 338. Professor Scovazzi has said that Article 38(1) will apply to the Strait of Messina "should it be demonstrated that there exists seaward of Sicily a route of similar convenience with respect to navigational and hydrographical characteristics." *Ibid.*, p. 350. The United States appears to have accepted that the "regime of non-suspendable innocent passage" applies to the Strait of Messina. *Ibid.*, p. 351 (quoting from U.S. Dept. of State, *Limits in the Seas*, No. 112, p. 68).

zania from the offshore islands of Pemba and Zanzibar.[73] Many relatively unimportant passageways, such as the waters between Cape Cod and the islands of Nantucket and Martha's Vineyard on the East Coast of the United States, certainly also qualify.[74] Professor Alexander has used the Messina Strait as providing a yardstick for when an alternative route is of "similar convenience" and has explained that a ship going from Marseilles to Trieste would have to travel 60 miles longer, requiring an additional five extra hours of steaming time, by going around Sicily rather than through the Strait of Messina.[75] Using this guideline, he listed 22 straits that fit the Article 38(1) exception, six of which are in Canada, and three of which are in Japan.[76] The Corfu Channel would not appear to qualify because "part of the strait lies between Corfu and Albania."[77]

The People's Republic of China has taken the position that the Qiongzhou Strait between Leizhou Peninsula and Hainan Island is an Article 38(1) strait, and China's straight baselines claim in its 1958 territorial sea declaration classified the waters in this strait as internal waters.[78] The Qiongzhou Strait is about 50 miles long, ranges from 9.8 to 19 miles wide, and is an important shipping route connecting the South China Sea to the Gulf of Tonkin.[79] In 1964, China issued regulations excluding foreign warships and regulating the passage of foreign commercial vessels in this waterway.[80]

---

[73] Nandan and Anderson, *supra* note 3, at p. 181; Alexander, *supra* note 10, at p. 207, Table 14-E.

[74] Alexander, *supra* note 10, at p. 157.

[75] *Ibid.*, p. 159. When Germany acceded to the LOS Convention on 14 October 1994, it included in its declaration a statement that "Article 38 limits the right of transit passage only in cases where a route of similar convenience exists in respect of navigational and hydrographical characteristics, which include the economic aspect of shipping." *Convention Overview*, *supra* note 20.

[76] Alexander, *supra* note 10, at pp. 206–207, Table 14-E.

[77] *Ibid.* See also, Scovazzi, *supra* note 29, at p. 353, agreeing that the transit passage regime applies in the Corfu Channel. But see Nandan, *supra* note 9, at p. 5, listing "the Corfu Channel and the Pemba Strait" as examples of straits that are exempt from the transit passage regime under Article 38(1).

[78] H. Kim, "Legal Status of the Cheju Strait," in *Istanbul Straits Symposium*, *supra* note 1, at pp. 31, 33.

[79] *Ibid.*

[80] *Ibid.* Professor Alexander has accepted China's claim as legitimate. See his Table 14-E in Alexander, *supra* note 10, at p. 207.

Some Korean scholars have argued that the Cheju Strait – which separates the Korean mainland from Cheju Island to the south – is an Article 38(1) strait, but Japan argues that the route around Cheju Island is not "of similar convenience" and thus challenges this claim.[81] One Korean commentator, Hyun-Soo Kim, has observed that Cheju is "inseparable" from the Korean mainland, that reducing traffic "is essential not only for conserving living resources and ensuring the fishermen's livelihood, but also to protect the security of Korea," and that "Cheju Strait has not been used ordinarily for international navigation."[82] He explained that going around Cheju Island would add only 30–35 nautical miles to any voyage and that the route south of the island "is a safe navigational route...well-marked by lighthouses and other navigational aids."[83] Kim concluded:

> When vessels have a choice in navigational routes where there exists a route of similar convenience, it is hoped that the special interest of a coastal State should be given a little more weight over the interests of maritime states. Accordingly, it can be easily concluded that the Cheju Strait meets the requirements of Article 38(1).[84]

Kim recognized, however, that if Korea were to declare formally the Cheju Strait to be internal waters, it "would raise considerable legal conflicts with the maritime states." Thus, he has suggested establishing a traffic separation scheme as an alternative way of protecting the fragile environmental resources of this region.[85]

Straits "Regulated in Whole or in Part by Long-standing International Conventions"

Article 35(c) of the LOS Convention says that the regime of transit passage through international straits does not apply to "straits in which passage is regulated in whole or in part by long-standing international conventions in force specifically relating to such straits." A definitive list of such straits does not exist, but most commentators agree that the Baltic or Danish

---

[81] H. Kim, *supra* note 78, at p. 33.
[82] *Ibid.*
[83] *Ibid.*, p. 38.
[84] *Ibid.*, pp. 37–38.
[85] *Ibid.*, p. 40.

Straits, the Turkish Straits, and the Strait of Magellan are included in this category, and some add the Aaland Strait to this list.[86] These straits are governed by their applicable treaties, some of which are quite venerable, and thus have their own unique regimes that govern passage. The materials that follow discuss these straits in detail, and others that are governed by the transit passage regime, and then compares straits and canals.

The Baltic/Danish Straits

The right of passage through straits has had a tumultuous and contentious history, and unique legal regimes have developed for the many important straits that exist around the globe. Maritime States have always insisted on free passage for all ships, but countries fronting onto straits have likewise always tried to regulate such passage, and have frequently distinguished among types of ships and made distinctions between times of war and times of peace.[87] The Baltic or Danish Straits[88] provide an intriguing case study to illustrate the historical tensions regarding passage rights through

---

[86] W. L. Schachte, Jr. and J. P. A. Bernhardt, "International Straits and Navigational Freedoms," *Virginia Journal of International Law* 33 (1993): 527, 546, reporting that the straits negotiating group at the Law of the Sea negotiations submitted that the Danish Straits, the Aaland Strait, the Turkish Straits (Bosporus and Dardanelles), and the Strait of Magellan were the straits covered by Article 35(c). See also, R. Lagoni, "Straits Used for International Navigation: Environmental Protection and Maritime Safety in the Danish Straits," in *Istanbul Straits Symposium*, *supra* note 1, at pp. 159, 170, agreeing that these four straits are the Article 35(c) straits. Professor Alexander listed only the Turkish, Danish, and Magellan Straits as covered by Article 35(c). Alexander, *supra* note 10, at p. 205, Table 14-B. When Finland signed and ratified the LOS Convention on 10 December 1982 and 21 June 1996, it issued a declaration saying:

> that the exception from the transit passage regime in straits provided for in article 35(c) of the Convention is applicable to the strait between Finland (the Aland Islands) and Sweden. Since in that strait the passage is regulated in part by a long-standing international convention in force, the present legal regime in that strait will remain unchanged after the entry into force of the Convention.

Convention Overview, *supra* note 20.

[87] See, e.g., E. Bruel, *International Straits* (two volumes) (Copenhagen: Nyt Nordisk Forlag, 1947).

[88] Sometimes the term "Baltic Straits" is used to refer to all the straits in the Baltic, including the Aaland Strait, with the term "Danish Straits" being used for the straits through and around the Danish peninsula; sometimes these straits through and adjacent to Denmark are called the "Baltic Straits." In this essay, the term "Baltic Straits" refers to the straits leading into the Baltic through and around Denmark.

straits. These Straits link the Baltic Sea to the Kattegat, which in turn leads into the Skagerrak and out to the North Atlantic Ocean. They lie predominantly within Danish and Swedish territory, and include the Little Belt, the Great Belt, and the Sound. Some 150 ships pass through these Straits each day.[89]

The Little Belt, between Denmark's Jutland-Als and Fyn-Aero, is divided by islands into channels. The channels most used for navigation are Aaro Sund and Baago Sund. Because of the Little Belt Bridge, passage through the strait is limited to ships with a mast height of no more than 33 metres. The current in the Little Belt is strong and unpredictable. At least one author has characterized the waters in the Little Belt as "internal waters."[90]

The Great Belt lies between Fyn-Langeland and Sjaelland-Lolland. This passageway, along with the Samso Belt, the Fehmarn Belt, and the Kadet Channel, form one seaway for large vessels entering or leaving the Baltic. The Great Belt varies in width from 18.5 to 28.2 kilometres. Depths vary from 20 to 25 metres in the northern part of the Belt to 66 metres in the southern area, allowing the largest vessels to pass through.

The Sound is located between Sjaelland and Skåne in Sweden. It is divided into an eastern and western channel by the island of Ven. Traditionally, the Sound was the shortest and busiest route between the Baltic Sea and the Kattegat, but the Great Belt has replaced it as the route most often used by larger vessels because of the Sound's insufficient depth south of Copenhagen and Malmő – it is only 26 feet deep in its southern sector. The maritime boundary between Sweden and Denmark was delimited in 1932.[91]

For more than four centuries (1429–1857), Denmark collected a transit duty on ships passing through these straits. At their peak, these fees contributed about two-thirds of Denmark's budget.[92] Foreign governments and merchants protested these fees over the years, and the British challenged them directly in the first half of the nineteenth century, shelling Copen-

[89] A. Couper, ed., *The Times Atlas of the Oceans* (New York: Van Nostrand Reinhold Co., 1983), p. 151.

[90] W. Graf Vitzthum, "The Baltic Straits," in C. Park, ed., *The Law of the Sea in the 1980s* (Honolulu: Law of the Sea Institute, 1983), pp. 537, 542.

[91] 1932 Declaration of Stockholm, cited in Vitzthum, *ibid.*

[92] G. Alexandersson, *The Baltic Straits* (Dordrecht: Martinus Nijhoff, 1982), p. 70.

hagen in 1801 and capturing the Danish fleet in 1807. The Copenhagen merchants also saw these dues as limiting trade into and out of their markets, and a canal was built across southern Sweden to circumvent the Danish fees. Although the United States agreed to pay the traditional dues in a treaty with Denmark in 1826,[93] the United States announced in 1845 that it would not pay these fees as a matter of principle, citing the "public law of nations,"[94] and denounced the 1826 Treaty in 1856. These dues were discontinued in 1857 with the signing of the Copenhagen Convention on the Sound and the Belts by the European shipping nations.[95] Article I of the 1857 Convention contains the key language that: "No vessels shall henceforth, under any pretext, be subject, in its passage of the Sound or Belts to any detention or hindrance." That same year, a special strait convention between the United States and Denmark was also signed in Washington, D.C. In exchange for $393 million, Denmark granted U.S. vessels free passage "in perpetuity."[96] Since then, no other multilateral treaties or conventions have dealt with the Baltic Straits except the Treaty of Versailles, which reiterated the right of "free passage into the Baltic to all nations."[97]

It can be argued, based on the extraordinarily high sums paid by the United States and other nations for the right of "free passage" through the Danish Straits, that the maritime nations purchased this right in perpetuity with the massive lump sum they paid in 1857 and, therefore, that the right of passage cannot be characterized as "free." Even today, disputes continue whether warships are entitled to free passage through the Baltic Straits or whether prior notification and authorization can be required. Sweden allows foreign naval ships to pass through the Swedish

---

93 E. Somers, "The Legal Regime of the Danish Straits," in *Istanbul Straits Symposium*, *supra* note 1, at pp. 12, 16, citing Treaty of 26 April 1826 in Martens, NRG, 1st Series, VI, 919.

94 Alexandersson, *supra* note 92, at p. 72.

95 *Treaty between Great Britain, Austria, Belgium, France, Hanover, Mecklenburg-Schwerin, Oldenburg, the Netherlands, Prussia, Russia, Sweden, and Norway and the Hanse Towns, on the One Part, and Denmark on the Other Part, for the Redemption of the Sound Dues*, Copenhagen, 14 March 1857, 116 Consol. T.S. 357. To soften the financial blow to Denmark, the contracting parties paid an indemnity "corresponding to an annual income capitalized to the current value." Alexandersson, *supra* note 92, at p. 73.

96 *Convention between the United States of America and Denmark for the Discontinuance of the Sound Dues*, Washington, D.C., 11 April 1857, 11 Stat. 719.

97 Alexandersson, *supra* note 92, at p. 73, citing Article 195 of the Treaty of Versailles.

part of the Sound according to the rules of innocent passage – they cannot stop or anchor and submarines must operate on the surface.[98] Denmark also allows innocent passage through the Straits as long as the passage avoids claimed internal Danish waters.[99] Passage of naval vessels through the Straits is subject to advance notification through diplomatic channels. Denmark requires authorization if more than three naval vessels flying the same flag are passing though the same part of the strait together and requires submarines to pass on the surface.[100] According to Alexandersson, "the Swedish and Danish regulations on the use of the Baltic Straits are in agreement with international law, the Geneva Convention of 1958 on the Territorial Sea and the Contiguous Zone as well as customary law on the use of foreign territorial waters by navy ships."[101]

The two 1857 treaties were written with only surface navigation in mind, but U.S. commentators have argued that they should now be viewed as authorizing free transit by submerged vessels and airplanes because the regime established in 1857 "was ostensibly the broadest regime possible to grant" and should be interpreted now in light of "subsequent developments of customary international law."[102] In the U.S. view, the regime governing these straits "would preclude the Danes from applying their domestic laws to foreign flags transiting the straits, except as recognized under the LOS Convention, and from applying their internal 1976 Ordinance to foreign warships."[103]

This view is probably not widely shared by commentators in the Baltic region. Writing two decades ago, the German Professor Wolfgang Graf Vitzthum reported that the Danish and Soviet governments maintained that only merchant vessels were covered by the 1857 Conventions and that the Danes viewed their restrictions on the passage of warships described above as consistent with the 1857 Conventions and with the LOS Convention because these straits are exempt from the transit pas-

---

[98] *Ibid.*, p. 82, citing a Swedish law of 3 June 1966.

[99] *Ibid.*, citing a Danish law of 27 February 1976.

[100] *Ibid.*

[101] *Ibid.*, p. 83.

[102] Schachte and Bernhardt, *supra* note 86, at p. 546.

[103] *Ibid.*, pp. 546–547. See also, J. N. Moore, "The Regime of Straits and the Third United Nations Conference on the Law of the Sea," *American Journal of International Law* 74 (1980): 77, 111, concluding that the Baltic Straits are covered by Article 35(c), but also that the 1857 Treaties "provide for freedom of navigation."

sage regime under Article 35(c).[104] The Danish diplomat, responding to Professor Vitzthum's presentation, said that the Baltic Straits are covered by Article 35(c)[105] and that

> it is clear from the preamble and the whole purpose of the Convention that it does not cover warships. The passage of warships is regulated by the general international rule of innocent passage through international straits.[106]

Professor Rainer Lagoni has said that Denmark believes a regime of nonsuspendable innocent passage exists through the Danish Straits.[107] In 1996, the IMO approved a mandatory ship reporting system in Denmark's Great Belt Traffic Area.[108]

Although some scholars have expressed uncertainty whether a special regime established by "long-standing international conventions" and recognized under Article 35(c) of the LOS Convention exists for the Baltic straits,[109] the Finnish, Swedish, and Danish delegates stated explicitly during the final 1981 session of the UNCLOS negotiations that the Baltic Straits were covered by Article 35(c) and that their legal status should remain unchanged.[110] This understanding that the Baltic Straits are an Article 35(c) strait appears now to be generally accepted,[111] although U.S. commentators have contended that "it is somewhat academic whether

---

[104] Vitzthum, *supra* note 90, at p. 552. He cited the Danish expert Erik Bruel as having taken the position that "warships fall outside the scope of [the 1857 Treaties'] provisions," *ibid.*, citing Bruel, *supra* note 87, at pp. 41, 45. Vitzthum's own analysis of the text and context of the 1857 treaties led him to conclude that the treaties "do not pertain to the rights of passage of warships." *Ibid.*, p. 555. But he also concluded that the 1857 Treaties should be considered as dealing with commercial, fiscal and customs matters, *ibid.*, p. 575, rather than as establishing a special straits regime for the Baltic Straits, and therefore that these straits should not be considered Article 35(c) straits, *ibid.*, pp. 555–558, and that (at least as of the time he was writing (in 1980)) the regime of nonsuspendable innocent passage should apply in the Baltic Straits. *Ibid.*, p. 565.

[105] I. R. Andreasen, "Commentary," in *The Law of the Sea in the 1980s, supra* note 90, at pp. 597, 602.

[106] *Ibid.*, p. 600.

[107] Lagoni, *supra* note 86, at pp. 161–164.

[108] IMO Resolution MSC.63(67), 3 December 1996.

[109] Alexandersson, *supra* note 92, at p. 73, citing the dispute between Erik Bruel, Wolfgang Graf Vitzthum, and Ib R. Andreasen discussed *supra* in the text and notes.

[110] *Straits Legislative History, supra* note 5, at pp. 132, 149, 154, 156.

[111] See *supra* note 86.

or not the Belts are considered 35(c) straits" because the 1857 treaties "ensure free navigation."[112] Sweden and Finland ratified the LOS Convention in 1996, Germany ratified it in 1994, and Denmark ratified in 2004.[113] Upon its ratification, Denmark issued a declaration stating that the Danish Straits are an Article 35(c) strait, governed by the Copenhagen Treaty of 1857.[114]

In 1991, Finland asked the International Court of Justice to resolve a problem raised by a Danish proposal to build a bridge across the East Channel, which would have blocked ships and oil rigs requiring a clearance greater than 65 metres.[115] The issue was resolved by negotiations before the Court had time to rule. Denmark agreed to pay Finland $16 million and the two countries agreed to explore ways to use an alternative, but shallower, strait through the Sound.[116]

Environmental issues now loom large in the Baltic Straits region. The IMO has approved a traffic separation scheme and has recommended the use of a pilot when traversing the Sound and Great Belt.[117] In 1991, a mandatory vessel traffic system (VTS) was established for the Great Belt, and in 1996 a mandatory ship reporting system was developed for this passage. In 2004, the IMO approved the designation of the Baltic Sea as a particularly sensitive sea area.[118]

---

112 Schachte and Bernhardt, *supra* note 86, at p. 544.

113 *Convention Overview*, *supra* note 20.

114 *Ibid.* The Danish declaration, issued 16 November 2004 stated that "the exception from the transit passage regime provided for in article 35(c) of the Convention applies to the specific regime in the Danish straits (the Great Belt, the Little Belt and the Danish part of the Sound), which has developed on the basis of the Copenhagen Treaty of 1857. The present legal regime of the Danish straits will therefore remain unchanged."

115 *Passage Through the Great Belt (Finland v. Denmark)*, [1991] I.C.J. Reports 12 (Provisional Measures, Order of July 29). Some drilling rigs require almost 100 metres of clearance. U.S. aircraft carriers require a clearance of only 41 metres, and could have passed under the proposed bridge. Schachte and Bernhardt, *supra* note 86, at p. 528 n. 3.

116 T. A. Clingan. Jr., *The Law of the Sea: Ocean Law and Policy* (San Francisco: Austin and Winfield, 1994), pp. 118–121.

117 IMO Doc. Resolution A.579(14) (1985) and A.620(15) (1987).

118 See IMO Doc. MEPC 51/8/1.

The Turkish Straits[119]

The Turkish Straits consist of the Dardanelles, which connect the Aegean Sea to the Sea of Marmara, and the Bosporus, which connects the Sea of Marmara to the Black Sea. The total navigable length of the Straits from the entrance to the Dardanelles from the Aegean Sea to the exit of the Bosporus to the Black Sea is about 160 miles (257 km).

The Dardanelles are about 38 miles (61 km) long with a width ranging from a minimum of 3/4 mile (1.2 km) to a maximum of 4 miles (6.4 km). The Dardanelles are deep, averaging 55 metres and dropping to 91 metres at their deepest point. Despite two major currents, a surface current and a more saline undercurrent flowing in the opposite direction, the Dardanelles are not difficult to navigate because vessels can avoid the currents by staying in the middle. Numerous lights have been added to aid night navigation.

The Bosporus is narrow with abrupt and angular turns, winding generally in a northeasterly direction from the Sea of Marmara to the Black Sea. It is about 19 miles (31 km) long, and its width fluctuates from 750 metres to 2.25 miles (3.6 km) at its southern entrance. The depths in the main channel run from 36 to 124 metres. Unlike the Dardanelles, its strong currents can make navigation difficult and sometimes dangerous.

In 2000, between 100 and 150 ships passed through the Turkish Straits each day, including 10–15 oil tankers.[120] In the 1980s and early 1990s, some 10–30 collisions occurred annually, but this number was reduced to 2–4 annually after a traffic separation scheme was established for the Straits.[121] The pressure on these Straits will continue to increase with movement of hydrocarbons from the Caspian Sea.[122]

The Turkish Straits are one of the straits referred to in Article 35(c) of the LOS Convention as being governed by "long-standing international conventions" that are unaffected by the Convention's new rules. Transit through the Turkish Straits is currently controlled by Turkey, which exercises sovereign power in the Straits, but is governed by the provisions of

---

119 See generally, C. L. Rozakis and P. N. Stagos, *The Turkish Straits* (Dordrecht: Martinus Nijhoff, 1987); Űnlű, *supra* note 33.

120 Űnlű, *supra* note 33, at p. 58.

121 *Ibid.*, p. 60.

122 M. Hakki Casin, "The Security and Legal Aspects of Turkish Straits," in *Istanbul Straits Symposium*, *supra* note 1, at p. 86.

the Montreux Convention of 1936.[123] This Convention recognized and affirmed in Article 1 the principle of freedom of transit and navigation by sea in the Straits as a principle of international law. Despite this commitment and a provision in Article 28 that it "shall...continue without limit of time," the articles of the Convention place significant limitations on free passage. The Convention created different regimes for merchant vessels and warships. It further regulated transit based on when passage occurred – during time of war or time of peace. Finally, "time of war" was distinguished based on the belligerent or nonbelligerent status of Turkey.

Under the Montreux Convention, during times of peace both merchant vessels and warships enjoy freedom of transit and navigation in the Straits. Warships, however, must provide notice of their proposed transit at least eight days in advance of the trip and must communicate to a Turkish signal station when the journey begins (Article 13). Even in peacetime, vessels of war must begin passage only during daylight (Article 10) and must refrain from launching any aircraft they may be carrying (Article 15). Furthermore, the Convention limits the number of foreign naval vessels that can pass through the Straits at any one time to nine, weighing no more than 15,000 tons (Article 14), although the Black Sea nations may exceed this limit if their vessels pass through the Straits "singly, escorted by not more than two destroyers" (Article 11). Submarines can pass (but only on the surface) through the Straits to rejoin their base in the Black Sea, if they were constructed outside the Sea, and can transit outside for repair, if proper advance notification is given to the Turkish government. It has been asserted that the Montreux Convention as a practical matter prohibits aircraft carriers from transiting the Straits.[124] The aggregate

---

[123] *Convention regarding the Regime of Straits*, Montreux, 20 July 1936, League of Nations Treaty Series 173:213–241, reprinted in Rozakis and Stagos, *supra* note 119, at pp. 153–164. The parties to this convention are Bulgaria, France, Greece, Japan (with reservations), Romania, Turkey, USSR, United Kingdom, and Yugoslavia. The treaty permitted Turkey to refortify the straits in exchange for guarantees of free transit, subject to a number of conditions. See J. M. Van Dyke, "An Analysis of the Aegean Disputes Under International Law," *Ocean Development and International Law* 36 (2005): 63, 74, 79–83.

[124] D. Froman, "Kiev and the Montreux Convention: The Aircraft Carrier That Became a Cruiser to Squeeze through the Turkish Straits," *San Diego Law Review* 14 (1977): 681.

tonnage of non-Black Sea powers cannot exceed 45,000 tons at any one time and the vessels of such powers cannot remain in the Black Sea more than 21 days (Article 18).

The Montreux Convention thus gives Black Sea States particular rights not given to others and it is unique in giving Turkey the paramount role in enforcing the Treaty. Not only does Turkey supervise the passage of vessels of war through the Straits, but it is also charged under Article 18 with monitoring the total number of warships in the Black Sea and determining when the Sea is "filled."

In times of war, under Article 4, if Turkey is a nonbelligerent, merchant vessels can continue to enjoy freedom of transit and navigation in the straits. If Turkey is a belligerent, under Article 5, merchant ships not belonging to a country at war with Turkey also enjoy freedom of transit and navigation, provided that they enter the Straits only during the daytime and do not assist the enemy in any way. Under this provision, Turkey has an implied right to stop and search passing merchant vessels to assure that the vessels are not assisting the enemy.[125] Finally, Article 6 of the Convention allows Turkey to regulate merchant vessel passage if Turkey determines that it is "threatened with imminent danger of war."

If Turkey is a nonbelligerent, vessels of war of nonbelligerents continue to enjoy complete freedom of transit through the Straits, subject to the same conditions for passage during peacetime (Article 19). When Turkey is a belligerent, however, passage of all warships through the Straits is "left entirely to the discretion of the Turkish government" (Article 20).

Article 2 and Annex I of the Montreux Convention allow Turkey to levy some charges for general services on ships passing through the Strait, although Turkey has viewed the amount it can charge as inadequate to cover its actual expenses.[126] Turkey is allowed to levy charges on the tonnage of passing vessels for sanitary control services, lighthouses, light and channel buoys, fog sirens, and direction-finding stations. In 1994, with IMO approval, Turkey established a traffic separation scheme.[127] Turkey also promulgated that year, without complete IMO endorsement, the Turkish Straits Maritime Regulations, which established rules on ship

[125] Rozakis and Stagos, *supra* note 119, at p. 107.
[126] See generally, Űnlű, *supra* note 33.
[127] IMO Resolution A.19/827 (Nov. 24, 1994); Űnlű, *supra* note 33, at p. 64.

reporting and the use of pilots and tugs.[128] These rules were denounced by the Russian Federation as being contrary to "universally recognised provisions of the Law of the Sea by which no regulations issued by a coastal State may deny, hamper or impair the rights of freedom of passage through international straits" and also as being contrary to "the terms of the Montreux Convention, 1936."[129] The Legal Committee of the IMO noted that "a substantial number of States considered the Turkish regulations to be inconsistent with the Montreux Convention and the IMO rules and regulations" and decided that the IMO should look into this matter further.[130] Turkey refrained from applying some of the regulations during this period of review, but reissued the regulations in 1998 affirming most of the requirements.[131] In 2003, Turkey established a radar-based vessel control system at an initial cost of US$20 million.[132]

The Strait of Magellan

Since the conclusion of a boundary treaty in 1881 between Chile and Argentina, it has been established that Chile has sovereignty over the Strait of Magellan, which intersects the southern tip of South America.[133] The Strait spans 240 miles (386 km), measured by a straight east to west line. The total length of the Strait itself totals 311 miles, due to a bend around Brunswick peninsula, which accounts for its V-shape. The width of the

[128] Turkish Straits Maritime Traffic Scheme Regulations, adopted in 1994, amended in 1998, Official Gazette 6/11/1998 No: 23515. See G. Plant, "Navigation Regime in the Turkish Straits for Merchant Ships in Peacetime," *Marine Policy* 20 (1996): 15; G. Aybay and N. Oral, *Turkish Straits Maritime Traffic Scheme Regulations* (English Translation) (Istanbul: Aybay & Aybay, 1998).

[129] Scovazzi, *supra* note 29, at p. 347 (citing IMO Doc. MSC/63/7/15 of 24 January 1994).

[130] Űnlű, *supra* note 33, at p. 65.

[131] *Ibid.*, pp. 65–66.

[132] See website of Turkish Ministry of Transport, General Directorate of Coastal Safety and Salvage Administration, <www.coastalsafety.gov.tr>, 23 July 2006.

[133] Michael A. Morris, *The Strait of Magellan* (Dordrecht: Martinus Nijhoff, 1989), p. 76, referring to the 1881 boundary treaty between Chile and Argentina, which is reprinted in the Morris volume at pp. 205–207. A 1977 arbitration decision by a panel of the International Court of Justice concerning a dispute over the Beagle Channel held that "the 1881 treaty had given Chile exclusive control over the strait.... [and] that the waters of the strait were likewise Chilean since Chile controls both shores." *Ibid.*, p. 79.

Strait averages just over four miles (6.4 km), although the range varies from 22 miles (35 km) to about 1½ miles (2.4 km). Westerly winds are prevalent throughout the year and tidal currents tend to be strong and unpredictable.

Article V of the 1881 Treaty states that "[t]he Straits of Magellan shall be neutralized for ever, and free navigation assured to the flags of all nations." Because of this Treaty, the Strait of Magellan qualifies as one of the straits exempt from the rules promulgated in the LOS Convention because of Article 35(c). One commentator has interpreted the 1881 Boundary Treaty to say that "[t]here would seem to be no basic difference between the regime of transit as it exists now, based on the 1881 treaty, and that guaranteed in the 1982 Convention."[134] Another author, however, has stated that the appropriate regime governing this Strait "would appear to be innocent passage rather than transit passage," and that "Chilean authors have explicitly rejected the application of the transit passage regime to the Strait of Magellan."[135] The significance of this distinction would be that, under an innocent passage regime, Chile could require submarines to travel on the surface of the Strait, prohibit overflight, and prohibit noninnocent passage, including transport of, say, ultrahazardous substances.[136]

When Argentina ratified the LOS Convention on 1 December 1995, it submitted a declaration stating that the 1984 Peace and Friendship Treaty between Chile and Argentina "reaffirmed the validity of Article V of the Boundary Treaty of 1881 whereby the Strait of Magellan (Estrecho de Magallanes) is neutralized forever with free navigation assured for the flags of all nations."[137] Chile's declaration of 25 August 1997 contained

---

[134] Alexander, *supra* note 10, at p. 143.

[135] Morris, *supra* note 133, at p. 10. Morris gives two reasons to support the "innocent-passage" regime status in the Strait of Magellan. First, Article 38(1) of the LOS Convention exempts from transit passage straits formed by islands of a State and its mainland, and the configuration of the Strait of Magellan contains such geography. Second, "[b]ecause of the 1984 closing line drawn across the eastern mouth of the Strait of Magellan, the Atlantic side of the strait is fronted by an Argentine territorial sea and EEZ." *Ibid.*, p. 103.

[136] Another author agreeing that planes have no right of overflight over the Strait of Magellan and that submarines cannot navigate submerged is M. T. Infante, "Straits in Latin America: The Case of the Strait of Magellan," *Ocean Development and International Law* 26 (1995): 175, 183.

[137] *Convention Overview*, *supra* note 20.

similar language, explaining that both the 1881 Boundary Treaty and an 1873 unilateral Chilean declaration state that:

> the Strait of Magellan is neutralized forever with free navigation assured for the flags of all nations.... For its part the Argentine Republic undertook to maintain, at any time and in whatever circumstances, the right of ships of all flags to navigate expeditiously and without obstacles through its jurisdictional waters to and from the Strait of Magellan.[138]

The question of suspension of passage under the 1881 Treaty is unclear because the Treaty is vague, but some Chilean legal authorities have said that noninnocent passage may be suspended.[139] Even though Chile has never suspended passage in modern times,[140] if it sought to challenge a vessel transporting plutonium as noninnocent, for instance, it might have grounds to suspend passage or impose conditions.

The Strait of Gibraltar

Bound on the north by Spain and on the south by Morocco, the Strait of Gibraltar connects the Atlantic Ocean to the Mediterranean Sea. Thirty-six miles (58 km) long and eight miles (13 km) wide at its narrowest point, the Strait of Gibraltar is unquestionably one of the most important passages in the world's oceans, with an average of between 140[141] and 200[142] ships passing through the Strait each day. Many are tankers, which bring more than 200 million tons of oil through the Strait each year.[143] The Strait presents navigational challenges from changing and sometimes strong winds, periods of reduced visibility, and offshore shoals. A collision in 1979, for instance, cost 50 lives and spilled some 95,000 tons of crude oil.[144] In July 2000, two ferryboats collided near the Spanish port of Algeciras, killing five and injuring 17, caused by fog and the heavy summer traffic.[145] The Strait is now also used by Moroccans seeking a better life in Europe who

---

138 *Ibid.*

139 Morris, *supra* note 133, at pp. 103–104.

140 *Ibid.*

141 *Times Atlas of the Oceans, supra* note 89, at p. 151.

142 *Ibid.*, p. 157.

143 *Ibid.*

144 *Ibid.*

145 "Two Ferry Boats Collide in Gibraltar Strait," *ArabicNews.com*, 17 July 2000, <www.arabicnews.com/ansub/Daily/Day/000717/2000071734.html>, 25 October 2002.

attempt the crossing in small boats, leading frequently to accidents and drownings.[146] In June 2002, Abu Zubair al-Haili, a Saudi national thought to be one of the top 25 Al Qaeda leaders, was arrested in Morocco and accused of plotting to blow up U.S. and British warships in the Strait of Gibraltar.[147] The idea of connecting Spain with Morocco through an undersea railway tunnel, or even a bridge, remains under discussion, with various ideas being presented by futuristic thinkers.[148]

Both sides of the Strait of Gibraltar were controlled by Carthage between 573 B.C.E. and about 190 B.C.E. During this time, Carthage prohibited (by force) non-Carthaginian ships from passing through the Strait.[149] Again, between 711 A.D., when Muslim armies headed by Musa Ibn Nasayr and Tarik ibn Ziyad landed at Gibraltar Rock, and the early twelfth century, "[t]he Muslims completely controlled traffic in the strait for about 400 years, denying passage to all ships but their own."[150] Beginning in the twelfth century, trade increased from Western Europe into the Mediterranean region, and internal divisions among the Moors reduced their ability to limit passage through the Strait.[151] Ships thus began to pass through the Strait, but were sometimes "subjected to savage attacks by pirates based along the North African shores."[152] The town of Gibraltar was captured by a joint British-Dutch invasion in 1704 during the War of Spanish Succession,[153] and the British took advantage of this strategic location in its subsequent war with France and Spain, as well as during

---

[146] "27 Moroccan Nationals Missing in the Gibraltar Strait," *ArabicNews.com*, 18 December 2000, <www.arabicnews.com/ansub/Daily/Day/001218/2000121822.html>, 25 October 2002; "The Strait of Gibraltar: A New Wall of Shame," June 1994, <www.fecl.org/circular/2504.htm>, 25 October 2002.

[147] "Top Al Qaeda Leader in Custody," *CNN.com/U.S.*, 18 June 2002, <www.cnn.com/2002/US/06/187/alqaeda.arrest/>, 25 October 2002.

[148] See, e.g., International Tunnelling Association, *Gibraltar Strait Fixed Link*, Proceedings of 22–24 April 1999 Workshop in Rabat, Morocco, <www.ita-aites.org/gibraltar/gibindex.html>, 25 October 2002); Wikipedia, *Gibraltar Bridge*, <en.wikipedia.org/wiki/Gibraltar_Bridge>, 30 March 2008.

[149] Truver, *supra* note 16, at pp. 160–161.

[150] *Ibid.*, p. 162.

[151] *Ibid.*

[152] *Ibid.*

[153] *Ibid.*, p. 163.

the later Napoleonic Wars and the two World Wars.[154] In 1942, allied control of the Strait permitted the invasion of North Africa.[155]

Spain sought to recapture Gibraltar from Britain on several occasions, including the Great Siege of Gibraltar in 1779–1783, when it imposed a blockade to prevent any supplies going to the British garrison.[156] On 13 March 1780, Spain issued the Ordinance Relating to Neutral Navigation which allowed neutral merchant ships to pass through the Strait if they had their papers and cargoes in good order, kept to prescribed sea lanes, and avoided the Gibraltar area.[157] These requirements have been described by one commentator "as completely legal measures under the customary rules of naval war at the time," and that "[e]ven in the middle of the twentieth century, such action may be required by extreme circumstances and still remain within the bounds of legality."[158] Because of the continued tension between Britain and Spain, each required the ships of the other nation to show their flag when passing through the Strait until 1864 when Spanish shore batteries sank the British schooner *The Mermaid*, whose ensign was not clearly visible from shore because of a serious storm and high seas. After the Spanish rejected Britain's subsequent protests, the dispute was submitted to a mixed claims tribunal, which in 1869 awarded 3,866 pounds to the British owners of the ship.[159] The two countries agreed to drop the requirements that passing ships show their ensigns, but the agreement "made no mention of the customary regime of passage through straits," and "no special significance can be attributed to the Declaration of 1865 regarding the right of passage through the Strait of Gibraltar."[160]

Although this issue is not entirely clear, the right of passage through the Strait of Gibraltar was apparently not governed by any special regime established by treaty prior to the 1982 LOS Convention. Some writers

---

[154] *Ibid.*, p. 5.

[155] *Ibid.*, p. 6.

[156] *Ibid.*, p. 168.

[157] *Ibid.*, p. 169, citing J. D. Gonzalez Campos, "Navegacion por el Mar Territorial, Incluidos los Estrechos," in *La Actual Revision del Derecho del Mar-Una Perspectiva Espanola* (Madrid: Ministerio de Asuntos Exteriores, 1974), p. 396; see also, Bruel, *supra* note 87, Vol. 2, at p. 166.

[158] Truver, *supra* note 16, at p. 169.

[159] *Ibid.*, p. 170.

[160] *Ibid.*, p. 171.

have asserted that the 1904 Declaration between Great Britain and France respecting Egypt and Morocco[161] guaranteed freedom of navigation through the Strait, but others have argued that the reference in this document to "free passage" simply protects whatever navigational rights existed through any strait,[162] which was the regime of innocent passage of all vessels.[163] Alexander has pointed out that the regime of nonsuspendable innocent passage did not emerge until the 1949 *Corfu Channel Case*[164] and that Spain has argued that the 1904 Declaration did not include freedom of overflight.[165] The United States did fly its planes from the United Kingdom over the Strait of Gibraltar in April 1986 when it bombed Libya, claiming the right to do so based on the transit passage regime established in the 1982 LOS Convention.[166]

The division of waters in the Gibraltar Strait is particularly challenging because Spain disputes the legitimacy of the United Kingdom's 6.5 square kilometres enclave at the Port of Gibraltar, and Morocco likewise disputes the legitimacy of Spain's enclaves in North Africa along the Moroccan coast. One commentator has explained that this patchwork of disputed land claims on both sides of the Strait has "brought the possibility of delimiting the waters of the Strait to a stalemate."[167]

In 1760,[168] Spain claimed a six nautical mile territorial sea, which "placed the strait within Spanish territorial jurisdiction between the Spanish mainland and Ceuta [a Spanish enclave in North Africa], [but] Spain apparently never attempted to close the strait to any ship movements,

---

161 *Declaration Between Great Britain and France Respecting Egypt and Morocco, together with the Secret Articles*, London, 8 April 1904. The text of Article 7 of this agreement is found in Scovazzi, *supra* note 29, at p. 349, and in Truver, *supra* note 16, at p. 256.

162 Truver, *supra* note 16, at p. 179; J. A. Yturriaga, "Navigation Through the Territorial Sea Including Straits Used for International Navigation," in L. M. Alexander and T. A. Clingan, Jr., eds, *Hazards of Maritime Transit* (Cambridge, Mass: Ballinger, 1973), pp. 85–91.

163 Alexander, *supra* note 10, at p. 143.

164 *Corfu Channel Case (United Kingdom v. Albania)*, [1949] I.C.J. Reports 4 (April 9); see *supra* text at notes 11–16.

165 Alexander, *supra* note 10, at p. 144.

166 *Ibid.*

167 F. A. Ahnish, *The International Law of Maritime Boundaries and the Practice of States in the Mediterranean Sea* (Oxford: Clarendon Press, 1993), p. 288.

168 Truver, *supra* note 16, at p. 171.

either military or commercial."[169] With international acceptance of a 12 nautical mile territorial sea in the 1982 LOS Convention,[170] the Strait now falls almost entirely within Spanish and Moroccan territorial waters, except for a disputed portion of the northeastern section of the Strait, which the British claim by virtue of the dependency it has maintained at the port of Gibraltar since 1704.[171] Spain has been trying to negotiate with the British for joint sovereignty over Gibraltar[172] and, in addition, has taken the position that the 1713 cession[173] did not transfer any territorial waters to Britain.[174] Britain currently argues that the "cession of a territory automatically carries the cession of the appurtenant territorial waters,"[175] but in previous generations its officers recognized that the Treaty language was ambiguous and that any claim of offshore waters would be firmly opposed by Spain.[176]

On the southern side of the Strait, Morocco has drawn straight baselines along its coast that may not be justified based on the language of Article 7 of the LOS Convention.[177] The baselines "push[] the [median] line

[169] *Ibid.*, p. 169.

[170] LOS Convention, *supra* note 2, at Art. 3.

[171] In Article X of the 1713 Treaty of Utrecht, Spain yielded to Britain, "the full and entire propriety of the town and castle of Gibraltar, together with the port, fortifications, and forts thereunto belonging…without any territorial jurisdiction." See Ahnish, *supra* note 167, at p. 288. The United Kingdom removed its troops from Gibraltar in 1991, but has maintained it as a dependency. "Gibraltar," in *The Columbia Encyclopedia*, 6th ed. (New York: Columbia University Press, 2001), <www.bartleby.com/65/gi/Gibralta.html>, 25 October 2002.

[172] See generally, S. J. Lincoln, "Note, The Legal Status of Gibraltar: Whose Rock Is It Anyway?" *Fordham International Law Journal* 18 (1994): 285. The Gibraltarian government rejected a 1997 Spanish proposal for joint British-Spanish sovereignty. *Gibraltar*, *supra* note 171.

[173] Treaty of Utrecht, *supra* note 171.

[174] Ahnish, *supra* note 167, at pp. 290–291, citing declarations made by Spain when it signed the 1958 Convention on the Territorial Sea and the 1982 LOS Convention. See also, Truver, *supra* note 16, at pp. 164–167.

[175] Ahnish, *supra* note 167, at p. 291.

[176] *Ibid.*, p. 291.

[177] Article 7(1) of the LOS Convention, *supra* note 2, permits countries to draw straight baselines only if their "coastline is deeply indented and cut into, or if there is a fringe of islands along the coast in its immediate vicinity." If such lines are drawn, they "must not depart to any appreciable extent from the general direction of the coast." *Ibid.*, Art. 7(3).

[claimed by Morocco] further north than a true median line (measured from the low-water line) to a marked degree, if one takes into account the narrowness of the area involved."[178] More significantly, Morocco's claimed line ignores the Spanish claimed land territory at the eastern end of the southern side of the Strait – the peninsula of Ceuta and the uninhabited islet Perejil (Parsley), which Spain retained after Morocco became independent in 1956.[179] Ceuta was first occupied by Spain in 1580[180] and is now occupied by 70,000 Spaniards, many of Moroccan descent. Morocco has long complained that Spain's retention of this peninsula is a vestige of colonial occupation, arguing that Ceuta is part of Morocco's "national territory,"[181] and that Spain's occupation cannot be allowed to continue indefinitely.[182] Spain has recently been obliged to build four metre high steel walls around its Moroccan enclaves to block African refugees seeking entry into Europe.[183]

In the summer of 2002, Moroccan troops landed on the half mile wide Isla Perejil claimed by Spain 200 metres offshore of the southern coast and then withdrew after Spanish troops descended on the rock.[184] Spain claims to have controlled the island since 1668, but has not maintained a permanent presence there for four decades. In recent years it has been inhabited only by goats. After mediation facilitated by the United States, Spain agreed to remove its forces and both sides agreed that the islet would be returned to its previous demilitarized, unoccupied status.[185] One commentator has contended that "[i]t is unreasonable to argue... that the rock of Perejil... should be entitled to a full belt of territorial sea" because

---

[178] Ahnish, *supra* note 167, at p. 295.

[179] M. Slackman, "Spain Pulls Troops from Disputed Islet," *Honolulu Advertiser*, July 21, 2002, p. A23.

[180] Truver, *supra* note 16, at p. 162.

[181] Ahnish, *supra* note 167, at p. 278, quoting the statement of the Moroccan representative to the Second Committee, UNCLOS III, *Official Records*, pp. iv, 75–76.

[182] Slackman, *supra* note 179.

[183] V. Short, "Spain Strengthens Borders Against African Refugees," World Socialist Website, 2 June 1999, <www.wsws.org/articles/1999/jun1999/spai-j02.shtml>, 25 October 2002.

[184] J. Socolovsky, "Spain Offers Morocco Deal to Pull Troops from Island," *Honolulu Advertiser*, 18 July 2002, p. A12. In Arabic, the islet is called "Leila," which means "Night."

[185] Reuters, "Spain and Morocco Settle Dispute Over Occupied Isle," Honolulu Star-Bulletin, 21 July 2002, p. A6.

"it is a small, uninhabited rock and situated only 200 metres from the coast of Morocco."[186]

During the negotiations that produced the 1982 LOS Convention, both Spain and Morocco argued that the rule of innocent passage should govern navigation through all straits encompassed by expanded territorial seas.[187] They argued further that the regime of innocent passage should apply only to merchant vessels and that warships and submarines should be subject to regulation by coastal States. Spain had previously taken the view that the regime of nonsuspendable innocent passage applied in the Strait of Gibraltar,[188] and it viewed the concept of transit passage that was emerging during the UNCLOS negotiations as "inherently non-innocent."[189]

Because of these concerns, Spain introduced amendments in 1978 to the working draft text of the Convention that, if adopted, would have provided that aircraft do not have the right of transit passage; would have said explicitly that ships exercising transit passage could not engage in acts of propaganda, collect information, or interfere with coastal State communications; would have increased coastal State enforcement powers; would have allowed coastal States to regulate navigational aids, cables, and pipelines; and would have required ships passing through straits to have adequate insurance to cover any loss or damage they might cause and have required flag States to ensure prompt and adequate compensation for such losses.[190] At this same 1978 session, Morocco submitted "informal suggestions" that would have prohibited ships and aircraft from any use of weapons, the taking off or landing of aircraft from ships, hydrographic surveys or other research activities, deliberate acts of pollution, all fishing activities, intelligence-gathering by aircraft, and any interference with

---

[186] Ahnish, *supra* note 167, at p. 282.

[187] For a description of Spain's initiatives during the negotiations in the 1970s, see J. A. de Yturriaga, *Straits Used for International Navigation – A Spanish Perspective* (Dordrecht: Martinus Nijhoff, 1991), pp. 68–162.

[188] *Ibid.*, p. 49, quoting from a 1971 speech made by Spain's Minister for Foreign Affairs, Gregorio Lopez-Bravo.

[189] Truver, *supra* note 16, at p. 11, citing a private letter to the author from Jose Manuel Lacleta, a member of the Spanish delegation, dated 11 May 1976.

[190] K. L. Koh, *Straits in International Navigation* (London: Oceana Publications, 1982), pp. 145–146, citing Second Committee, Informal C.2/Informal Meeting/4 of 26 April 1978, in R. Platzoder, *Dokumente der Dritten Seerechtskonferenz der Vereinten Nationen – Genfer Session 1978*, Vol. 3 (July 1978), p. 929.

coastal communications during transit passage.[191] Morocco's proposal would also have required ships to maintain radio contact during their passage and to inform coastal authorities of any damage, unforeseen stop, or other change required by *force majeure*.[192] Morocco supported Spain's suggestions regarding insurance, liability, and compensation for damage.[193] These proposals were rejected by the maritime nations, who argued that "the question of straits had been fully debated and the compromise reached should not be reopened."[194] Nonetheless, the proposals illustrate the views of the nations that border on the Gibraltar Strait and their unhappiness about the transit passage regime that emerged in the LOS Convention.

Maritime transit through the Gibraltar Strait has been free and unimpeded for all vessels during most recent periods of history, although at times this free transit has had to be enforced with military might.[195] Since the drafting of the LOS Convention, no serious attempts have been made by Spain or Morocco to limit passage through the Strait of Gibraltar, but Spain did issue a declaration when ratifying the Convention saying that it understands that the straits regime in the Convention "is compatible with the right of the bordering State to enact and enforce in straits used for international navigation its own regulations, provided that such regulations do not interfere with the right of transit passage."[196] Spain's declaration also stated that the requirement in Article 39(3)(a) that government aircraft exercising transit passage will "normally" comply with the Rules of the Air established by the International Civil Aviation Organization means that they will do so "except for force majeure or serious difficulty."[197] In

[191] *Ibid.*, p. 147, citing Second Committee, Informal Meeting C.2/Informal Meeting/4 of 28 April 1978, in Platzoder, *ibid.*, p. 959.

[192] *Ibid.*

[193] *Ibid.*, p. 148.

[194] *Ibid.*, p. 145, citing Statement of Representative of United States of America of 17 April 1978, UNCLOS III, *Official Records*, Vol. 9 (United Nations, 1980), p. 130.

[195] See Truver, *supra* note 16, at pp. 178–181.

[196] Spanish Declaration, issued 15 January 1997, in *Convention Overview*, *supra* note 20; see generally, de Yturriaga, *supra* note 187, at pp. 303–304.

[197] *Ibid.* With regard to its enforcement powers, Spain's declaration interpreted Article 221 (Measures to avoid pollution arising from maritime casualties) "as not depriving the coastal State of a strait used for international navigation of its powers, recognized by international law, to intervene in the case of casualties referred to in that article," and further said that "article 233 must be interpreted, in any case, in conjunction with the provisions of article 34." Professor Scovazzi has observed that Spain's declaration

1996, the IMO approved a mandatory ship reporting system for the Strait of Gibraltar.[198] Spain and the United Kingdom both ratified the LOS Convention in 1997 (including conflicting declarations on the status of Gibraltar); Morocco ratified the Convention in 2007.[199] In 1991, a Spanish commentator wrote that Spain was "the ideal 'persistent objector'" seeking to insist on a regime of nonsuspendable innocent passage and to oppose "the emerging customary rule on transit passage through straits used for international navigation,"[200] but "political" realities later led to Spain's ratification of the Convention, which weakens Spain's position "despite the interpretive declaration attached to it."[201]

The Strait of Hormuz[202]

One of the most important waterways in the world, economically, politically, and strategically, the Strait of Hormuz connects the Persian Gulf to the Gulf of Oman. The Strait is about 104 miles (167 km) long at its median point. Its width varies from about 52.5 nautical miles (nm) (97.3 km) to 20.75 nm (38.4 km). With the extension of the territorial sea to 12 nm, the Strait falls within the overlapping Iranian and Omani territorial seas.

As of 1978, Iran and Oman were maintaining unimpeded transit through the Strait by means of the Iranian-Omani Joint Patrol of the Strait of Hormuz.[203] These countries have had disputes over islands and boundary delimitations and the area in general has been an area of international tension and conflict. No treaty governs this Strait. From the perspective of the maritime powers it is the classic international strait through which transit must be permitted without interruption. During the final negotiating session of UNCLOS in 1982, however, Iran stated that it "could not give an unconditional guarantee of freedom of navigation"

"gives a meaning of the wording of articles 39(3) and 42(1)(b) which is different from their ordinary sense." Scovazzi, *supra* note 29, at p. 341.

198 IMO Resolution MSC.63(67) (3 December 1996).

199 *Convention Overview, supra* note 20.

200 de Yturriaga, *supra* note 187, at p. 329.

201 *Ibid.*, p. 330.

202 See generally, R. K. Ramazani, *The Persian Gulf and the Strait of Hormuz* (Dordrecht: Martinus Nijhoff, 1979).

203 See generally, *ibid.*, pp. 72–88.

and would "guarantee passage only to vessels that did not pose a threat to its security."[204] Iran also issued a "declaration of understanding" at the end of the negotiations in 1982 that the right of transit passage through international straits was a new international norm – the "product of *quid pro quo* which [does] not necessarily purport to codify the existing customs or established usage (practice) regarded as having an obligatory character" and hence that "only States parties to the Law of the Sea Convention shall be entitled to benefit from the contractual rights created therein."[205] At the same time, Oman issued an "understanding" that the transit passage regime "does not preclude a coastal State from taking such appropriate measures as are necessary to protect its interest of peace and security."[206]

The Strait of Bab al-Mandeb[207]

Linking the Gulf of Aden and the Red Sea, the Strait of Bab al-Mandeb is about 14.5 miles (23 km) wide at its narrowest point and is bordered by Yemen, Djibouti and Ethiopia. All these littoral States have claimed a 12 nm territorial sea that precludes any area within the Strait from being high seas. Because no specific international agreement governs the Strait of Bab al-Mandeb, the Strait has been subject to the general regime of international straits, which, until the entry into force of the 1982 LOS Convention, was freedom of navigation and overflight in the high seas zone and nonsuspendable innocent passage in the territorial seas for all ships of commerce and war.

Because the Strait of Bab al-Mandeb fits within the definition of an international strait by linking one area of high seas or an exclusive economic zone with another such area[208] – the Red Sea and the Gulf of Aden – the right of transit passage through international straits in Article 38 of the LOS Convention currently governs transit through the Strait. This liberal regime was also recognized in the 1975 Memorandum of Agreement between the Government of Israel and the United States: United

[204] *Straits Legislative History, supra* note 5, at p. 138.

[205] *Ibid.*, p. 155.

[206] *Ibid.*

[207] See generally, R. Lapidoth-Eschelbacher, *The Red Sea and the Gulf of Aden* (Dordrecht: Martinus Nijhoff, 1982).

[208] LOS Convention, *supra* note 2, at Art. 37.

States-Israeli Assurances[209] and in a 1978 unilateral declaration by the government of the People's Democratic Republic of Yemen (Southern Yemen).[210] In the closing negotiating session of UNCLOS in 1982, however, the Yemen Arab Republic issued an "understanding" that reaffirmed its sovereignty over its territorial waters and asserted that "nuclear-powered craft, as well as warships and warplanes in general, must obtain the prior agreement of the Yemen Arab Republic before passing through its territorial waters, in accordance with the established norm of general international law relating to national sovereignty."[211]

The Strait of Dover/Pas de Calais[212]

Connecting the North Sea to the English Channel, the Strait of Dover/Pas de Calais historically has been open to ships. At its narrowest, the Strait is only 18 nautical miles wide. Prior to the international acceptance of the 12 nautical mile territorial sea, the United Kingdom had not claimed any distance greater than three miles. Hence, even though the French adopted a 12 nautical mile territorial sea in 1971, the Strait had a sufficient high seas route available for free navigation.[213]

Now, with the 12 nautical mile territorial sea having been generally accepted, the narrowest portion of the Strait would fall entirely under French and British territorial jurisdiction, but both countries have indicated that they accept the LOS Convention regime of transit passage through the Strait of Dover. Both Britain and France have a great stake in free navigation. Britain, in fact, introduced the concept of transit passage.[214] France, in declaring a 12 nautical mile territorial sea in 1971, foresaw the need to ensure free navigation "where the distance between the baselines of the French coasts and the baselines of the coasts of an opposite foreign

[209] 14 *I.L.M.* 1468 (1975).

[210] Lapidoth-Eschelbacher, *supra* note 207, at p. 149, quoting from U.N. document NV/78/63 (12 July 1978).

[211] *Straits Legislative History*, *supra* note 5, at p. 157.

[212] See generally, L. Cuyvers, The Strait of Dover (Dordrecht: Martinus Nijhoff, 1996).

[213] *Ibid.*, pp. 53–54.

[214] *Ibid.*, p. 54, citing *Draft Articles on the Territorial Sea and Straits*, 3 July 1974 (submitted to Committee 11 by the United Kingdom), *Official Records*, Vol. 3, UNCLOS III, U.N. Doc. A/Conf.62/C.2/L.3 (1974).

state is equal to – or less than – 24 miles, or does not allow any longer the existence of a zone of high seas sufficient for navigation."

Because of the density of traffic in the Strait of Dover,[215] vessel traffic has been managed for the last 150 years.[216] A formal traffic separation scheme was adopted in 1972, approved by the IMO (then the Intergovernmental Maritime Consultative Organization), which is administered jointly by France and the United Kingdom.[217] This approach has established a high degree of cooperation and sharply reduced the number of collisions in the Strait. An under-keel clearance requirement of 5.2 metres has also been adopted for this Strait.[218] In 1998, the IMO approved the establishment of a ship reporting system,[219] and the following year, a mandatory ship reporting system was established for vessels more than 300 gross tons.[220]

Strenuous efforts have also been undertaken to reduce pollution from vessels in the Strait, but these efforts have been notably less successful.[221] It has proved to be extremely difficult to enforce pollution control regulations on vessels and the incentives on vessels to comply are limited.

---

[215] It has been estimated that:

> In 2001, as many as 120,000 vessels passed through [the Dover Straits] and in addition there were 74,000 movements by ferries across the strait, carrying 21 million passengers. That year, 654 incidents were noted by the Dover Coastguards: 193 persons were rescued and 21 lost their lives.

D. H. Anderson, "The Legal Regime of the Straits Around Great Britain," in *Istanbul Straits Symposium, supra* note 1, at pp. 20, 26.

[216] Cuyvers, *supra* note 212, at pp. 62–77.

[217] See *ibid.*, p. 71 (map); Anderson, *supra* note 215, at p. 27.

[218] Alexander, *supra* note 10, at p. 113.

[219] IMO Resolution MSC.85(70), 3 December 1998.

[220] Anderson, *supra* note 215, at p. 28.

[221] *Ibid.*, pp. 93, 94–96. One notable effort is the Bonn Agreement for Cooperation in Dealing with Pollution of the North Sea by Oil, which was signed 9 June 1969 by Belgium, Denmark, France, Germany, the Netherlands, Norway, Sweden, and the United Kingdom. *Ibid.*, p. 93, citing 11 *I.L.M.* 262 (1972). Under this regime, a zone of joint responsibility was established for the Strait of Dover, which was allocated to Belgium, Denmark, France, Germany, the Netherlands, Norway, Sweden, and the United Kingdom. Each State was given the responsibility of monitoring this strait and assessing the movements of oil. After 1979, the contracting parties agreed that this treaty should apply to other hazardous materials as well as to oil. In 1983, the agreement was formally replaced by the Bonn Convention for International Cooperation in Dealing with Pollution of the North Sea by Oil and Other Harmful Substances. *Ibid.*, pp. 93–94.

One initiative that has been given increased recent attention has been to establish port State control. This approach, authorized in Article 218 of the LOS Convention, gives ports a responsibility to monitor vessels as they arrive. It is designed to supplement, or even replace, the flag-State enforcement system, which has proved to be inadequate because of the use of flags of convenience.

Two weeks after the *Amoco Cadiz* disaster in 1978, the countries of Belgium, Denmark, France, West Germany, the Netherlands, Norway, Sweden and the United Kingdom signed the Memorandum of Understanding between Certain Maritime Authorities on the Maintenance of Standards on Merchant Ships.[222] Each signatory agreed to harmonize its procedures to inspect ships in its ports. The actual inspections were not, however, carried out with the vigour anticipated and many lapses continued to occur.[223]

Another step taken was the Paris Memorandum of Understanding on Port State Control,[224] which was approved in 1982 and has been supported by 14 Western European countries. This understanding requires each country to inspect at least 25 per cent of the foreign vessels that visit its ports. It also establishes a commission to monitor the operations of each country and to facilitate achievement of the goals of the Memorandum. Each country is now vigorous in its inspection programmes. This approach has been successful in uncovering deficiencies and in encouraging vessel owners to maintain their ships in better condition.[225]

The Strait of Malacca

The Strait of Malacca is critical for international shipping, linking the Pacific and Indian Oceans, and it is a major artery for the transport of oil and other commodities.[226] Some 150–200 ships pass through the Strait each day, or more than 50,000 each year.[227] The Strait of Malacca

---

222 *Ibid.*, p. 103. This agreement was signed in The Hague on 2 March 1978.

223 *Ibid.*, pp. 103–104.

224 *Paris Memorandum of Understanding on Port State Control*, Paris, 26 January 1982, <www.parismou.org>, 10 March 2008.

225 *Ibid.*, pp. 107, 108–115.

226 M. Leifer, *Malacca, Singapore, and Indonesia* (Dordrecht: Martinus Nijhoff, 1978), p. 52.

227 Alexander, *supra* note 10, at p. 127.

is dangerous for shipping because it is quite shallow (requiring regular dredging), the water level changes with the tides, and the seabed shifts, creating serious risks of grounding.[228] Danger from collisions also exists because the waterway is often congested and the ships' speed makes it difficult for them to stop quickly.[229] One report stated that 476 accidents occurred between 1978 and 1994.[230] Another report stated that 500 accidents had occurred in the six years ending in 1993.[231] Accidents involving oil tankers include the *Showa Maru* (1975), *Diego Silang* (1976), *Nagasaki Spirit* (1992), and *Maersk Navigator* (1993).[232]

The waters of the Strait of Malacca are divided among the three straits States – Singapore, Malaysia and Indonesia. All three have a common interest in safety and protection of the coastal environment, but Singapore's overriding interest has always been in freedom of navigation and encouraging shippers to use the Strait and its port facilities. Japan, a major user of the Strait, conducted and paid for a number of hydrographic studies to improve safety, and has been vitally concerned with keeping the strait open for its supertankers. By the end of 1975, a series of accidents, including the *Showa Maru* tanker incident, called the "*Torrey Canyon* of the East," had increased the safety and environmental concerns, and Malaysia and Indonesia asserted their right to control the Straits at the negotiations that produced the 1982 LOS Convention. A safety agreement was signed in Manila on 24 February 1977 during a meeting of the Association of Southeast Asian Nations (ASEAN). The Agreement included a 3.5 metre under-keel clearance requirement[233] and a traffic separation scheme for "three specified critical areas of the Straits

---

[228] Leifer, *supra* note 226, at pp. 55, 56. One notable grounding was the 244,000 ton Japanese tanker *Showa Maru*, which spilled 844,000 gallons of crude oil into the Strait of Singapore in January 1975. Alexander, *supra* note 10, at p. 113. See generally, "The Straits of Malacca: A Profile," in *Straits of Malacca, supra* note 1, at p. 3.

[229] Leifer, *supra* note 226, at p. 53.

[230] Malacca Straits Research and Development Center, <www.fsas.upm.edu.my/~masdec/web/straits.html>, 23 July 2006.

[231] Hamzah bin Ahmad, *supra* note 40, at p. 127, citing a report from Oil Companies International Forum, submitted to the IMO Sub-Committee on Safety of Navigation, 39th Session, 29 June 1993.

[232] *Ibid.*

[233] This 3.5 metre under-keel clearance requirement was approved by the IMO Assembly on 19 November 1981 in Resolution A.375(X). See W. Awang bin Wan Yaacob, *supra* note 52, at pp. 18–19.

of Malacca and Singapore, namely in the One Fathom Bank area, the Main Streets and Phillips Channel, and off Horsburg Lighthouse."[234] The traffic separation scheme was amended in 1981 and 1998. The 3.5 metre under-keel clearance requirement has the effect of limiting fully loaded tankers to about 230,000 dwt and requiring larger tankers to go through the Sunda or Lombok Straits in Indonesia, which, respectively, add 600 and 1,000 nautical miles to the trip between the Middle East and Japan.[235] In 1981, a Fund was established to provide compensation in the event of a spill, supported at present solely by Japan without contributions from other user States, and administered by Japan, Malaysia, and Singapore.[236] A mandatory ship reporting system was adopted by the IMO in 1998 for ships passing through the Malacca and Singapore Straits.[237]

In recent years, the countries bordering on the Straits have installed sophisticated onshore and offshore navigational and safety aids to prevent collisions and pollution. Malaysia has invested RM 52 million to install 256 navigational aids and RM 100 million for a Vessel Traffic Management and Information Service (VTMIS).[238] Singapore has, for instance, recently installed radar-based vessel monitoring system, at a cost of Singapore$40 million (with annual maintenance costs of about Singapore$1.5 million)

---

[234] *Tri-Partite Agreement of 24 February 1977*, reprinted in K. Kantaatmadja, "Various Problems and Arrangements in the Malacca Straits (An Indonesian Perpective)," in J. M. Van Dyke, L. M. Alexander and J. R. Morgan, eds, *International Navigation: Rocks and Shoals Ahead?* (Honolulu: Law of the Sea Institute, 1988), pp. 165, 168.

[235] Leifer, *supra* note 226, at pp. 63, 69, 72, 205; Alexander, *supra* note 10, at p. 113. During the final negotiating session of the Third U.N. Conference on the Law of the Sea in 1982, Indonesia, Malaysia and Singapore presented a joint statement to the conference stating that in their view their enforcement of the requirement that vessels maintain a 3.5-metre keel clearance in the Malacca-Singapore Strait did not constitute an interference with the right of transit passage in violation of Article 42(2) or 44 of the LOS Convention. *Straits Legislative History*, *supra* note 5, at p. 144.

[236] The Malacca Straits Council, funded by Japanese businesses, financed the installation of 40 navigational aids between 1968 and 1988. T. A. Grigalunas, Y-T. Chang, and J. Opaluch, "Sustainable Financing for Controlling Transboundary Pollution by Shipping in the Malacca Straits – Options and Implications," *International Journal of Maritime Economics* 2 (2000): 331, 337. In 1997, the funding for this organization was described as "inadequate, and is not all encompassing." W. Awang bin Wan Yaacob, *supra* note 52, at p. 20.

[237] IMO Resolution MSC.73(69), 29 May 1998.

[238] Malacca Straits Research and Development Centre website, <www.fsas.upm.edu.my/~masdec/web/straits.html>, 23 July 2006.

and spends Singapore$3 million annually to operate five lighthouses in the Strait.

The IMO has been trying to assist, developing the Marine Electronic Highway Project, which was initiated in 2001 in the Straits of Malacca and Singapore[239] to guide ships through the Straits even during hazardous fog conditions. This project is expensive – costing US$15 million for a 100 kilometre stretch of the Strait of Malacca – and it remains unclear whether users of this system can be charged a fee under Article 26(2) of the LOS Convention.[240]

Historically, charges have been imposed on ships passing through the Malacca Straits. When the Portuguese controlled the Straits, between 1511 and 1641, they imposed a toll on vessels.[241] The Dutch took over the Strait in 1641 and carefully controlled passage.[242] Not until 1824, when the Dutch signed the Treaty of London with the United Kingdom, was the right of passage for vessels of all nations established.[243]

Both Malaysia and Indonesia have asserted that the Straits are part of their territorial seas[244] and that "the Straits of Malacca and Singapore are not international straits."[245] The earlier position of Indonesia and Malaysia has been that "the regime of innocent passage should obtain in straits used for international navigation that have been assimilated either by territorial or internal waters," such as the Strait of Malacca.[246] The major maritime powers objected to this position as too restrictive, and, as noted earlier, the LOS Convention adopted the transit passage regime through international straits to ensure that straits would be open to navigation.

---

239 IMO Council, C93/6 at 18.

240 See *supra* text at notes 51–71.

241 K. E. Shaw, *The Straits of Malacca* (Singapore: University of Education Press, 1973), pp. 89–91.

242 *Ibid.*

243 *Ibid.*; the text of the 1824 Treaty of London can be found in Wikipedia, *Anglo-Dutch, Treaty of 1824* <en.wikipedia.org/wiki/Anglo-Dutch_Treaty_of_1824>, 30 March 2008.

244 Leifer, *supra* note 226, at p. 91.

245 *Joint Statement of the Governments of Indonesia, Malaysia, and Singapore, 16 November 1971*, reprinted in G. Knight and H. Chiu, *The International Law of the Sea* (London: Elsevier Applied Science, 1991), p. 294; in Kantaatmadja, *supra* note 234, at p. 168; and in Awang bin Wan Yaacob, *supra* note 52, at pp. 15, 16.

246 Leifer, *supra* note 226, at p. 88.

The Malacca Strait has been generally open to all international transit, but Singapore and Indonesia opposed in 1993 the passage of a Japanese plutonium ship through the Strait because of the danger of collisions and piracy.[247] Malaysia developed a plan to escort the ship through the Strait, if that route were taken,[248] but also threatened to block passage as a threat to its national security.[249] The ship did not pass through the Strait, and instead went south around Australia.[250]

The Lombok Strait and Archipelagic Waters

An archipelagic State enjoys a special status under the LOS Convention.[251] The breadth of the territorial sea (Article 48) of such a State is measured from straight baselines around the islands under the rules articulated in Article 47. The waters inside such baselines are archipelagic waters (Article 49) and internal waters (Article 50).[252] Archipelagic States are required to designate archipelagic sea lanes, through which the vessels of all States can exercise the right of archipelagic sea lanes passage, which is similar to the right of transit passage through international straits.[253] Vessels also have a right of innocent passage through archipelagic waters (Article 52), subject to specific restrictions (Article 53).

---

247 See *supra* note 32.

248 Agence France Presse, 10 November 1992.

249 "Malaysia May Cite Security Laws to Block Japanese Plutonium Ship," *UPI Business and Financial Wire*, 24 September 1992.

250 See generally, Van Dyke, *supra* note 32.

251 LOS Convention, *supra* note 2, at Article 46(b) defines an archipelago as "a group of islands, including parts of islands, interconnecting waters and other natural features which are so closely interrelated that such islands, waters and other natural features form an intrinsic geographical, economic, and political entity, or which historically have been regarded as such." Article 46(a) defines an archipelagic State as "a State constituted wholly by one or more archipelagos and may include other islands."

252 *Ibid.*, Arts. 9, 10, and 11.

253 *Ibid.*, Art. 53; see generally, Van Dyke, et al., *supra* note 234. When the Philippines signed the LOS Convention in 1982 and ratified in 1984, it issued a declaration that appears to be inconsistent with the language of the Convention, saying that: "The concept of archipelagic waters is similar to the concept of internal waters under the Constitution of the Philippines, and removes straits connecting these water with the economic zone or high sea from the rights of foreign vessels to transit passage for international navigation." *Convention Overview, supra* note 20.

The Lombok Strait passes between the Indonesia islands of Lombok and Bali. It is an alternative route to the Strait of Malacca and, unlike that Strait, is easily navigable. Japanese supertankers use the Lombok route extensively because it is deep, even though the distance required by this route is longer than through the Malacca Strait.[254]

Indonesia considers the Lombok Strait to be part of its archipelagic waters.[255] Even if Indonesia did not formally designate this Strait as an archipelagic sea lanes, it would be almost automatically in this category under Article 53(12) of the LOS Convention, which says that "[i]f an archipelagic State does not designate sea lanes…, the right of archipelagic sea lanes passage may be exercised through the routes normally used for international navigation."

Indonesia expressed its strong preference for the Japanese plutonium ship in 1993 to avoid its archipelagic waters, but also expressed concern that it did not have the power to prohibit the ship from passage through its sea lanes.[256] Indonesia offered protection to the vessel if it did pass through its waters.[257]

The Situation in the Aegean Sea

Greece has declared territorial seas of only six nautical miles around its islands in the Aegean Sea. If it were ever to claim zones of 12 nautical miles around the islands in this crowded sea, it would be necessary to determine whether the right to transit passage would exist in every passageway or only in certain routes.[258] One would think that this right of transit passage would exist at least for the major shipping routes leading from the Turkish Straits into the Mediterranean. However, one Greek scholar has suggested that "[i]t would be reasonable to assume" that "the narrows between the Kos and Astipalaia islands, Amorgos and Kalimnos, Naxos and Patmos, [and] Mikonos and Ikaria," which he characterized as "borderline cases," "fall short of the definition of straits used for international navigation, and

254 Leifer, *supra* note 226, at p. 80.

255 *Ibid.*, pp. 91–92.

256 "Malaysia Reports Concerns over Plutonium Shipment," *Kyodo News Service*, 16 October 1992.

257 L. Makabenta, "South-East Asia: Fear of Japan's Plutonium Shipment," *Interpress Service International News*, 18 September 1992; Van Dyke, *supra* note 32, at p. 412.

258 See Van Dyke, *supra* note 123, at pp. 83–85, 91–94.

consequently would be subject to the more restrictive, innocent passage regime."[259] These "borderline" "narrows" are, in fact, the major and most logical route to get from the Turkish Strait into the eastern Mediterranean and the many ports in the Middle East. But opinions are decidedly mixed on this topic. Commentators neutral to the Aegean region have observed that "minor" straits, including perhaps those in the Aegean that connect an EEZ or high seas area with a territorial sea, may be governed by the regime of "nonsuspendable innocent passage,"[260] which differs from transit passage because it does not allow submarines to pass submerged, nor does it guarantee overflight rights of airplanes.[261]

It is also unclear whether the right of transit passage would apply, for instance, to ships leaving the port of Izmir in Turkey and going to another Turkish port along the Turkish coast to the north or south. The second sentence of Article 38(2) says that transit passage includes "passage through the strait for the purpose of entering, leaving or returning from a State bordering the strait," but what if the passage is from one port to another in the same State?[262] Such a vessel would not be passing from one area

[259] Politakis, *supra* note 36, at p. 301 (citing for support Nandan and Anderson, *supra* note 3); Politakis also acknowledged that the counter-argument can be made, i.e., "that all the above-mentioned straits should rather be regarded as organically interconnected, forming continuous maritime routes linking the Mediterranean with the northern Aegean, and thus subject to the transit passage rules." *Ibid.*, p. 302.

[260] See, e.g., E. Franckx, "The Work of the International Law Association's Committee on Coastal State Jurisdiction Relating to Marine Pollution and Its Implications for the Aegean Sea," in B. Öztürk, ed., *The Aegean Sea* (Istanbul: Turkish Marine Research Foundation, 2000), pp. 221, 234 (citing T. Treves, "Navigation," in R. J. Dupuy and D. Vignes, *A Handbook on the New Law of the Sea*, Vol. 2 (Dordrecht: Martinus Nijhoff, 1991), pp. 835, 970–976).

[261] Compare the LOS Convention, *supra* note 2, Art. 39(1)(c), permitting submarines to transit in "their normal modes of continuous and expeditious transit" during transit passage through international straits, with Art. 20, which requires submarines to surface when exercising innocent passage.

[262] LOS Convention, *supra* note 2, at Art. 37. See, e.g., Politakis, *supra* note 36, at p. 302: "It is important to note also that although a large interpretation of Arts. 37–38 and Arts. 53–54 might bring some Aegean straits within the ambit of transit passage, it is not at all clear whether ships heading to or departing from Turkish ports on the Aegean coast, such as that of Izmir, could be equally considered as engaged in transit passage.... [T]he traditional innocent passage regime would still apply to ships entering or clearing certain Turkish ports on the Aegean coast." Article 45 indicates that the regime of nonsuspendable innocent passage would apply.

of high seas or EEZ into another[263] and would not be involved in international navigation in the normal sense of that term. A final unresolved issue is what the passage rights through the Aegean would be in times of war. An Italian scholar has written that "the status of international straits in time of war has never been completely clarified."[264]

Greece opposed the concept of transit passage through international straits when this notion was being developed in the negotiations that led to the 1982 LOS Convention.[265] When it signed the Convention in December 1982, Greece made the following declaration:

> The present declaration concerns the provisions of Part III "on straits used for international navigation" and more especially the application in practice of articles 36, 38, 41 and 42 of the Convention on the Law of the Sea.
>
> In areas where there are numerous spread out islands that form a great number of alternative straits which serve in fact on and the same route of international navigation, it is the understanding of Greece, that the coastal State concerned has the responsibility to designate the route or routes, in the said alternative straits, through which ships and aircrafts of third countries could pass under transit passage regime, in such a way as on the one hand the requirements of international navigation and overflight are satisfied, and on the other hand the minimum security requirements of both the ships and the aircrafts in transit as well as those of the coastal State are fulfilled.[266]

This statement, repeated when Greece ratified the Convention in 1995, raises several issues, which would become particularly acute if Greece should ever extend its Aegean territorial seas to 12 nautical miles. Greece has asserted the right to designate those straits that international shipping

---

263 One Greek scholar has said that if Greece expands its territorial sea from six to 12 nautical miles, "it would be no longer possible for Turkish warships stationing at Izmir to join the high seas without first passing through Greek territorial waters, and thus subject to the regime of innocent passage." Politakis, *supra* note 36, at p. 295.

264 N. Ronzitti, *The Law of Naval Warfare* (Cambridge: Grotius Publications, 1993), p. 14.

265 A. Strati, "Greek Shipping Interests and the UN Convention on the Law of the Sea," in T. C. Kariotis, ed., *Greece and the Law of the Sea* (The Hague: Martinus Nijhoff, 1997), p. 279. See also, Strati, *supra* note 36, at p. 92, "where Greece's ocean resource and national security interests conflicted with its shipping interests, the former took precedence."

266 *Convention Overview*, *supra* note 20. Turkey protested this statement after it was originally made, and also when it was repeated later. See UNCLOS III, 17 *Official Records*, Part B, U.N. Doc. A/Conf.62/WS/34, p. 226, and 30 *Law of the Sea Bulletin* (1996).

(and aircraft) can utilize, but the United States and other maritime powers have argued that the transit passage right applies to every strait and that no rights of designation exist. Some commentators have speculated that Greece would like to "prevent Turkish aircraft from flying through straits near the Greek mainland, particularly the Kea Strait southeast of Athens."[267] The Kea Strait may not to be subject to transit passage under the regime established by the LOS Convention in any event because of the "Messina exception" in Article 38(1), which says that "transit passage shall not apply if there exists seaward of the island a route through the high seas or through an exclusive economic zone of similar convenience with respect to navigational and hydrographical characteristics."[268]

Other Important Straits

Other straits that provide important passageways include, the Strait of Tiran (leading into the Gulf of Aqaba),[269] the Strait of Jubal (at the southern end of the Gulf of Suez), the Northwest Passage through Canada's Arctic region,[270] the Kerch Strait (connecting the Sea of Azov with the Black Sea), and the Kurile Strait. The crowded seas of Northeast Asia are full of important straits and the countries of this region have limited their territorial seas claims around the straits to promote free navigation through them. Japan, which asserts a 12 nautical mile territorial sea in general, claims only a three nautical mile territorial sea in the Soya Strait, the Tsugaru Strait, the eastern and western channels of the Tsushima

---

[267] D. S. Saltzman, "A Legal Survey of the Aegean Issues of Dispute and Prospects for a Non-Judicial Multidisciplinary Solution," in B. Öztürk, ed., *The Aegean Sea* (Istanbul: Turkish Marine Research Foundation, 2000), pp. 179, 187 n. 65.

[268] One U.S. author has stated that the language in Article 38(1) of the LOS Convention "precisely describes the Kea Strait case." Lt.Col. Michael N. Schmitt, "Aegean Angst – The Greek Turkish Dispute," *Naval War College Review* 49(3) (1996), pp. 42, 51. See *supra* text at notes 72–85.

[269] When it ratified the LOS Convention on 26 August 1983, Egypt issued a declaration stating that the 1979 Peace Treaty between Egypt and Israel was consistent with Part III of the LOS Convention and that the Strait of Tiran was governed by Part III. *Convention Overview*, *supra* note 20.

[270] Canada claims that the waters in the Northwest Passage are internal waters pursuant to a historic waters claim of Canada to these waters, but most of the rest of the world views it as an international strait. D. Ballingrud, "Warming May Open Northwest Passage Seaway," *Honolulu Star-Bulletin*, 3 December 2002, p. C6, col. 5.

Strait, and the Osumi Strait.[271] Both the Republic of Korea and Japan have limited their territorial sea claims around the land areas adjacent to the Korean Strait to three nautical miles in order to permit unimpeded passage through this area.[272]

Canals

Canals are considered to be internal waters under the law of the sea and, therefore, no right of innocent passage exists through the canal itself. Because canals are human-made and always present unique problems of management, maintenance, financial integrity, and control, greater regulation of passage is permitted through these waterways. Passage through canals is generally governed by international treaties, such as the treaty that led to the return of the Panama Canal to Panama,[273] which says that passage cannot regulated in a discriminatory manner, but does allow for restrictions based on safety concerns. Article 1 of the 1888 Convention of Constantinople says that the Suez Canal "shall always be free and open, in time of war as in time of peace, to every vessel of commerce or of war, without distinction of flag."[274] At least during the time the United States governed the Panama Canal, the obligation to keep the canal open for all ships was "somewhat less explicit."[275]

The Kiel Canal, a 53 mile waterway linking the Baltic and North Seas, which can accommodate vessels of eight metres draft, opened through

---

271 Japanese Law on the Territorial Sea No. 30 of May 2, 1977, listed in *National Legislation on the Territorial Sea, the Right of Innocent Passage and the Contiguous Zone* (U.N. Sales No. E.95.V.7, 1995), pp. 177–182; see also, U.S. Dept. of State, *Limits in the Sea No. 120, Straight Baseline and Territorial Sea Claims: Japan* (1998).

272 See, e.g., C.-H. Park, "The Korea Strait," in Van Dyke, et al., *supra* note 234, at p. 173. Professor Oxman has written that the same approach of claiming limited three nautical mile territorial seas utilized by Japan and Korea to allow navigational freedom has also been used by "Germany and Denmark, and by Denmark, Sweden and Finland." B. H. Oxman, "Applying the Law of the Sea in the Aegean Sea," in B. Öztürk and N. Algan, eds, *Problems of Regional Seas 2001* (Istanbul: Turkish Marine Research Foundation, 2001), pp. 266, 279.

273 *Treaty Concerning the Permanent Neutrality and Operation of the Panama Canal*, Washington, D.C., 7 September 1977, 16 *I.L.M.* 1040 (1977).

274 See M. McDougal and W. Burke, *The Public Order of the Oceans: A Contemporary International Law of the Sea* (New Haven: Yale University Press, 1962), p. 1075 n. 113.

275 *Ibid.*

Germany in 1895 for the purpose of providing access to the sea by Germany's navy and ensuring that its movement between the North Sea and the Baltic could not be cut off.[276] The Versailles Treaty transformed it into an international waterway, saying:

> The Kiel Canal and its approaches shall be maintained free and open to the vessels of commerce and of war of all nations at peace with Germany on terms of entire equality.[277]

In 1936, Germany declared that the Canal was not an international waterway and that foreign warships could pass through it only with advance authorization,[278] which led to protests from France and Czechoslovakia.[279] After World War II, no formal action was taken to internationalize the Canal, but it has been open to international transit.[280] In 1980, two German commentators said that in the opinion of "most authors in international law – the internationalization system of the Versailles Treaty for the Kiel Canal has been abandoned."[281]

The Panama Canal Regulations, implemented by Panama when it assumed control, explicitly allow Canal authorities to prohibit ultrahazardous cargoes. The regulation "Denial of Passage to Dangerous Vessel" states that: "The Canal authorities may deny any vessel passage through the Canal when the character or condition of the cargo…is such as to endanger the structures pertaining to the Canal, or which might render the vessel liable to obstruct the Canal."[282] Article III(1)(a) of the 1977

[276] Alexander, *supra* note 10, at p. 181 (citing R. Baxter, *The Law of International Waterways: With Particular Regard to Interoceanic Canals* (Cambridge, Mass: Harvard University Press, 1964), p. 164); Vitzthum, *supra* note 90, at p. 544.

[277] *Versailles Treaty*, 28 June 1919, Art. 380, <www.yale.edu/lawweb/avalon/imt/partxii.htm>, 30 March 2008; see R. Baxter, *The Law of International Waterways: With Particular Regard to Interoceanic Canals* (Cambridge, Mass: Harvard University Press, 1964), p. 172.

[278] *Ibid.*

[279] *Ibid.*

[280] Alexander, *supra* note 10, at p. 181.

[281] I. von Muench, "Commentary," in C.-H. Park, ed., *The Law of the Sea in the 1980s* (Honolulu: Law of the Sea Institute, 1983), pp. 602, 608; Guenther Jaenicke, "Comments," *ibid.*, p. 615.

[282] 35 C.F.R. sec. 103.2.

Treaty between the United States and Panama[283] permits Panama to adopt regulations that are "necessary for safe navigation and efficient, sanitary operation of the Canal." This Treaty emphasizes at several places the importance of keeping the Canal open for commerce and the necessity of providing sufficient security to ensure that it remains open.

Panama is entitled to demand full indemnity[284] and insurance (as the Suez Canal does), an environmental impact assessment, notification and consultation with shipping States, and the development of emergency response plans for shipments transiting the Canal. The Panama Canal Authority requires that notification be given 30 days in advance of the arrival of a vessel in Canal waters for all cargoes of fissionable materials, in order to obtain approval to transit such cargo.[285]

Conclusion

Professor Scovazzi has explained in his recent writings that the regime of transit passage through international straits is still in a period of evolution and that the rules found in the LOS Convention cannot yet be viewed as accepted customary international law in all respects. A few straits States, such as Iran and Turkey have not ratified the Convention, nor has one of the major maritime powers, the United States.[286] Although the Convention contains a series of provisions articulating the principles that apply to the regime of transit passage, many ambiguities remain. Some straits are exempt under Article 35(c) because they are "regulated in whole or in part by long-standing international conventions," and another group of straits are exempt under Article 38(1) because "there exists seaward of the island a route...of similar convenience with respect to navigational

---

283 *Treaty Concerning the Permanent Neutrality and Operation of the Panama Canal, supra* note 274.

284 See Article 139 of the Regulations on Navigation in Panama Canal Waters: "[V]essels carrying radioactive substances shall be required to provide current proof of financial responsibility and adequate provision for indemnity to the Republic of Panama, the Authority, or any agency thereof, covering public liability and loss as a result of accidents owing to radioactive cargo."

285 Regulations on Navigation in Panama Canal Waters, Article 137.

286 Convention Overview, *supra* note 20.

and hydrographical characteristics." Canals, although they serve a similar purpose as straits, are governed by a separate regime altogether. Judge David Anderson suggested during an international conference that we should examine the principles that govern navigational rights on international rivers for ideas that would logically apply to straits.[287] As environmental and security concerns become more focused, it is probable that strait States will take new initiatives to protect their coastal environments and populations. It is likely, therefore, that conflicts will continue between shipping and maritime powers and States bordering straits.

[287] Judge Anderson made this suggestion during discussions at the meeting that produced the *Istanbul Straits Symposium, supra* note 1, in Istanbul, 16 November 2002.

The Establishment of a Cooperative Mechanism for the Straits of Malacca and Singapore under Article 43 of the United Nations Convention on the Law of the Sea

Robert Beckman*

Introduction

In September 2007, after three years of discussions and negotiations, the three littoral States bordering the Straits of Malacca and Singapore (Indonesia, Malaysia and Singapore) formally launched a cooperative mechanism that is designed to enable user States and other stakeholders to cooperate with the littoral States in enhancing navigational safety and environmental protection in the Straits of Malacca and Singapore. The cooperative mechanism is intended to implement Article 43 of the 1982 United Nations Convention on the Law of the Sea (LOS Convention)[1] which provides that user States and States bordering a strait should by agreement cooperate with respect to navigational safety and environmental protection. This is the first time that Article 43 has been implemented for a strait used for international navigation. This contribution will trace the development of this cooperative mechanism and evaluate its significance.

---

* Associate Professor, Faculty of Law, National University of Singapore, Singapore.

[1] *United Nations Convention on the Law of the Sea*, Montego Bay, 10 December 1982, UN/Doc. A/CONF.62/122, 7 October 1982, 21 *I.L.M.* 1261 (1982) [hereinafter LOS Convention].

The Importance of the Straits of Malacca and Singapore

The Malacca Strait is 520 nautical miles in length. Its width varies from 200 nautical miles in the north to 11 nautical miles in the south. The lower half of the Malacca Strait is within the territorial sea (or archipelagic waters) of Indonesia and Malaysia. At its southern end it joins with the Singapore Strait, which is narrower and more difficult to navigate. The Singapore Strait is within the territorial sea (or archipelagic waters) of Singapore and Indonesia.

The littoral States of Indonesia, Malaysia and Singapore agreed in 1971 to adopt a common policy in the two Straits.[2] As a result, the two straits are treated as one at the International Maritime Organization (IMO) and are commonly referred to together as the Straits of Malacca and Singapore.

The Straits are one of the world's most important shipping lanes. They are the major route used for international navigation between the Indian Ocean and the South China Sea. Ships that pass through the Straits carry an estimated one quarter of the world's trade and one half of the world's oil. Approximately 11 million barrels oil transit the Straits each day, carrying a significant portion of the crude oil supplies of China, the Republic of Korea and Japan.[3]

Figures on vessel traffic through the Straits vary. According to a study conducted by the Secretariat of the IMO, approximately 60,000 vessels per year transit the Straits.[4] However, a study by the Ministry of Land, Infrastructure and Transport of Japan in cooperation with the Nippon Foundation found that in 2004 approximately 94,000 vessels of more than 100 gross tonnage passed through the Straits. (The discrepancy in the numbers may be a result of the size of the vessels being monitored, as only vessels of 300 gross tonnage and above are required to participate in the Straits' mandatory ship reporting system.) According to the Japanese study, the number of tankers passing through the Straits in 2004 was

---

[2] Joint Statement of the Foreign Ministers of the Republic of Indonesia, Malaysia and the Republic of Singapore, announced simultaneously by the three Governments on 16 November 1971.

[3] Profile of the Straits of Malacca and Singapore, Annex, Note by the Secretary-General, Protection of Vital Shipping Lanes, IMO Document C 93/15, 7 October 2004, Annex, p. 3.

[4] *Ibid.*

almost 23,000, which was 25 per cent of the total number of vessels and 47 per cent of the total tonnage of vessels passing through the Straits. The Japanese study concluded that if the current trend continues, approximately 150,000 vessels will pass through the Straits by 2020 and the share of traffic volume of tankers passing through the Straits in terms of tonnage will increase to 55 per cent.[5]

Past Cooperation Among the Three Littoral States

The three littoral States have been cooperating with regard to the Straits since the early 1970s. Three ministerial meetings of the three littoral States were held before the adoption of the LOS Convention in 1982.

The first ministerial meeting, held in October 1971, resulted in the Joint Statement of 16 November 1971. In the 1971 Joint Statement, the three governments indicated that they were not in agreement on the international legal status of the Straits. They did agree on several matters, including (1) that safety of navigation in the Straits was the responsibility of the three littoral States; (2) that there was a need for tripartite cooperation on the safety of navigation in the Straits; and (3) that a body for cooperation to coordinate efforts on the safety of navigation, comprised only of the three littoral States, should be established.[6]

The second ministerial meeting of the three littoral States took place in Singapore in February 1975. The issue of immediate concern was whether conditions should be placed on the passage through the Straits of very large crude carriers (VLCCs). The Joint Statement of 19 February 1975 included the following points: (1) an expert group should be appointed to study the problem of VLCCs and the possibility of measures to enhance safety of navigation; (2) a traffic separation scheme should be established in the Straits; (3) there should be consultation and cooperation on anti-pollution policy and on compensation for oil pollution damage; (4) a Council of Ministers should be established by the three littoral States on the safety of navigation and marine pollution that should meet annually;

[5] Y. Okanishi, "Emerging trends in vessel traffic through the Straits," Singapore Meeting on the Straits of Malacca and Singapore: Enhancing Safety, Security and Environmental Protection, 4–6 September 2007, Singapore.

[6] 1971 Joint Statement, *supra* note 2.

and (5) a Committee consisting of senior officials should be established to assist the Council of Ministers.[7]

The third ministerial meeting of the littoral States was held in February 1977. The ministers reviewed the report of a meeting of senior officials that had been held in Jakarta in December 1976 and adopted the recommendations in the report. The recommendations adopted included (1) vessels transiting the Straits maintain an under-keel-clearance of at least 3.5 meters at all times; (2) deep draught vessels be required to use a deep water route to be designated in part of the Singapore Strait; (3) a traffic separation scheme be established in three areas in the Straits; (4) navigational aids and facilities be improved for the effective implementation of the traffic separation scheme; and (5) a joint policy to deal with marine pollution be formulated.[8]

These adopted recommendations were submitted to the Inter-Governmental Maritime Consultative Organization (IMCO),[9] and on 14 November 1977 the IMCO Assembly adopted Resolution A.375(X) on Navigation through the Straits of Malacca and Singapore.[10] The resolution mandated a new routeing system for the Straits which included traffic separation schemes, deep water routes, and rules relating to under-keel-clearance for deep-draft vessels. It also endorsed the recommendation that all oil tankers navigating through the Straits be adequately covered by relevant insurance and compensation schemes. Annex V of the Resolution contained the "Rules for Vessels Navigating through the Straits of Malacca and Singapore."

The group of experts established in 1975 has come to be known as the Trilateral Technical Experts Group (TTEG). The TTEG meets regularly to coordinate policies relating to safety of navigation and environmental protection in the Straits. The Council of Ministers, however, was never established, and following 1976 no ministerial level meetings specifically dealing with cooperation in the Straits were held until August 2005.

---

[7] Joint Statement of the Foreign Ministers of the Republic of Indonesia, Malaysia and the Republic of Singapore, 19 February 1975, Singapore.

[8] Joint Statement of the Foreign Ministers of the Republic of Indonesia, Malaysia and the Republic of Singapore, 24 February 1977, Manila.

[9] IMCO is now known as the International Maritime Organization (IMO).

[10] IMO Assembly Resolution A.375(X), Navigation through the Straits of Malacca and Singapore, adopted on 14 November 1977.

As a result of close cooperation through the TTEG, the littoral States have recommended to the IMO the adoption of other measures designed to enhance the safety of navigation in the Straits. For example, in 1998 the three littoral States proposed to the IMO that a mandatory ship reporting system be adopted in the Straits. The Maritime Safety Committee (MSC) of the IMO, acting in accordance with Regulation V/8-1(h) of the International Convention for the Safety of Life at Sea (SOLAS),[11] adopted resolution MSC.73(69) on 19 May 1998, establishing the mandatory ship reporting system known as "STRAITREP" for the Straits.[12]

In 1998, the three littoral States also proposed to the MSC[13] that the traffic separation scheme for the Straits be amended to facilitate the safe and efficient flow of traffic through the Straits. The MSC approved the amendment of the traffic separation scheme so that it extended for more than 260 nautical miles, from One Fathom Bank, off Port Klang, the main port for Kuala Lumpur, through the southern half of the Malacca Strait and the entire Singapore Strait, to the Horsburgh lighthouse area in the South China Sea. The Horsburgh lighthouse is 40 nautical miles from Singapore, at the eastern approach to the Singapore Strait. The measures also amended the traffic separation scheme in the Singapore Strait by creating three precautionary areas.[14]

---

11 Adopted on 1 November 1974, entered into force on 25 May 1980. The text with amendments of Chapter V of SOLAS is available online at U.S. Coast Guard Navigation Center, <www.navcen.uscg.gov./marcomms/imo/SOLAS_V.pdf>, 28 September 2007.

12 IMO Resolution MSC.73(69), Mandatory Ship Reporting Systems, adopted on 19 May 1998, Annex I, Description of the Mandatory Ship Reporting System in the Straits of Malacca and Singapore.

13 IMO Assembly Resolution A.858(20) of 27 November 1997 sets out the procedures for the adoption and amendment of traffic separation schemes and other routeing measures, and authorizes the MSC to act on behalf of the IMO in approving such measures.

14 COLREG.2/Circ. 44, dated 26 May 1998; see also R. C. Beckman, "Enhancing the Safety of Navigation," *Port of Singapore Magazine* Issue 1 (1999): pp. 26–29.

The LOS Convention Regime on Straits used for International Navigation

The Straits are governed by Part III of the LOS Convention[15] on straits used for international navigation. Pursuant to Part III of the LOS Convention, commercial and military vessels of all States have the right of transit passage through the Straits and such right cannot be impeded or suspended.[16] Ships exercising transit passage through the Straits are bound under the LOS Convention to comply with generally accepted international regulations, procedures and practices governing safety at sea and pollution from vessels.[17] The rights of the littoral States to regulate ships exercising transit passage respecting safety and ship-source pollution are severely restricted. Littoral States may only establish sea lanes and traffic separation schemes and "give effect to" applicable international regulations regarding the discharge of oil, oily wastes, and other noxious substances.[18] In effect, this means that the power to adopt laws and regulations governing vessels exercising the right of transit passage through the Straits rests with the IMO rather than with the littoral States.

Background on Article 43 of the LOS Convention

The regime governing straits used for international navigation in the LOS Convention contains one provision that was intended to ensure that, while user States were able to enjoy the benefits of transit passage, they were also expected to share some of the burdens to ensure safety of and to control pollution in such straits. That provision is Article 43, which places an obligation on user States to cooperate with States bordering straits used for international navigation with regard to navigational safety and vessel-source pollution. Article 43 reads as follows:

> User States and States bordering a strait should by agreement cooperate:
> (a) in the establishment and maintenance in a strait of necessary navigational and safety aids or other improvements in aid of international navigation; and
> (b) for the prevention, reduction and control of pollution from ships.

---

[15] LOS Convention, *supra* note 1.
[16] *Ibid.*, Arts. 38, 44.
[17] *Ibid.*, Art. 39.
[18] *Ibid.*, Art. 42.

Unfortunately, with the exception of Japan, user States did not show any interest in cooperative arrangements under Article 43 for the Straits of Malacca and Singapore. User States willingly accepted the benefits of the straits used for international navigation regime, but they were not willing to share the burdens. Consequently, Indonesia, Malaysia and Singapore have had to bear the financial burden, with assistance from Japan, of ensuring navigational safety in the Straits. They have also had to bear the burden of dealing with oil pollution incidents and coping with illegal discharges of oil and oily wastes in the Straits.

Article 43 was a compromise provision which, as already noted, sought to balance the interests of the littoral States and user States with regard to the safety of navigation and environmental protection.[19] The use of the word "should" in the chapeau of Article 43 indicates that the provision was intended to encourage cooperation between user States and States bordering the strait for the specific purposes set out in paragraphs (a) and (b) of the article. The cooperation is to be by "agreement," but the word agreement is sufficiently vague to include either formal written agreements or less formal arrangements. The provision does not define "user States" but common sense dictates that user States would include States whose vessels use such straits and States who otherwise benefit from the passage of vessels through such straits. Article 43 does not mention the IMO, but this omission would not foreclose a role for the IMO in the process leading to cooperative arrangements.

Commentators have suggested that Article 43 was an attempt to meet the concerns of States bordering straits of having to bear the burden of establishing and maintaining aids to navigation and the risk of pollution of their marine and coastal environment from ships transiting straits, while receiving little benefit since most of the vessels transiting the straits used for international navigation would not stop in the State bordering the strait.[20] According to leading commentators, Article 43 was drafted with

---

[19] On the legislative history of Part III, see *Straits Used for International Navigation, Legislative History of Part III of the United Nations Convention on the Law of the Sea*, Volume II (New York: U.N. Division of Ocean Affairs and the Law of the Sea, 1992); S. N. Nandan and D. H. Anderson, "Straits Used for International Navigation: A Commentary on Part III of the United Nations Convention on the Law of the Sea 1982," *British Yearbook of International Law* 60 (1989): 159–204; and S. N. Nandan and S. Rosenne, eds, *United Nations Convention on the Law of the Sea 1982: A Commentary*, Volume II (Dordrecht: Martinus Nijhoff Publishers, 1993).

[20] Nandan and Rosenne, *ibid.*, p. 383.

straits such as the Straits of Malacca and Singapore in mind,[21] and one of the purposes of the provision was to ensure that States bordering such straits do not unilaterally impose charges on passing vessels.[22]

Commentators have also pointed out that States bordering straits used for international navigation are under no express duty under the LOS Convention to establish a system of navigational and safety aids in straits.[23] The only duty on States bordering such straits is to give appropriate publicity to any danger to navigation within the strait of which they have knowledge.[24] However, SOLAS, Chapter V, Regulation 13, provides that each Contracting Government

> undertakes to provide, as it deems practical and necessary either individually or in cooperation with other Contracting Governments, such aids to navigation as the volume of traffic justifies and the degree of risk requires.[25]

Furthermore, Regulation 13 provides that when establishing such aids to navigation, Contracting Governments undertake to take into account international recommendations and guidelines.

It can be concluded that Article 43 is intended to promote and foster cooperation between States bordering straits and user States to enhance navigational safety and environmental protection in straits used for international navigation. This is especially true in straits where user States are the major beneficiaries of safe passage through the straits, and the burdens and risks borne by the States bordering the straits far exceed the benefits they derive from passage of ships through the straits.

Efforts to Implement Article 43 in the Straits

In July 1994, agreement was reached at the United Nations on the Agreement relating to the implementation of Part XI of the United Nations Convention on the Law of the Sea of 10 December 1982 (1994 Imple-

---

[21] Nandan and Anderson, *supra* note 19, at p. 193.

[22] S. N. Nandan, "The Provisions on Straits Used for International Navigation in the 1982 United Nations Convention on the Law of the Sea," *Singapore Journal of International & Comparative Law* 2 (1998): 393–399 at 397.

[23] Nandan and Rosenne, *supra* note 19, at p. 383.

[24] LOS Convention, *supra* note 1, at Art. 44.

[25] SOLAS, *supra* note 11.

mentation Agreement).[26] The 1994 Implementation Agreement addressed the concerns of the industrialized countries about the provisions in Part XI of the LOS Convention on deep sea mining in areas beyond the limits of national jurisdiction. The 1994 Implementation Agreement ensured that the LOS Convention would become universally accepted and that its provisions, including Part III on straits used for international navigation, would be accepted by all States as legally binding.[27]

Indonesia, Malaysia and Singapore realized that this development presented them with an opportunity to attempt to implement Article 43 in the Straits of Malacca and Singapore. This resulted in discussions and papers at international conferences and workshops on cooperative arrangements to improve navigational safety and prevent pollution in the Straits, including the following:

- 1994 Kuala Lumpur Workshop on the Strait of Malacca / 1995 Malaysian Institute of Maritime Affairs International Conference on the Strait of Malacca: Meeting the Challenges of the 21st Century[28]
- 1996 International Maritime Organization / Institute of Policy Studies Conference in Singapore: Navigational Safety and the Control of Pollution in the Straits of Malacca and Singapore – Modalities of International Cooperation[29]
- 1996 Regional Conference on Sustainable Financing Mechanisms for the Prevention of Marine Pollution: Public Sector-Private Sector Partnerships[30]

---

26 *Agreement relating to the implementation of Part XI of the United Nations Convention on the Law of the Sea of 10 December 1982*, New York, 28 July 1994, U.N. General Assembly A/RES/48/263, 33 I.L.M. 1309.

27 In July 1994, the 63 States who were parties to the LOS Convention did not include any of the major industrialized countries. As of 31 May 2007, 155 States are parties to the LOS Convention, including most of the industrialized countries.

28 Selected papers from these workshops were published in H. Ahmad, ed., *The Straits of Malacca: International Co-operation in Trade, Funding and Navigational Safety* (Kuala Lumpur: Maritime Institute of Malaysia, 1997).

29 The papers presented at the conference and the rapporteur's reports summarizing the points made in the papers and the discussions are published in the *Singapore Journal of International & Comparative Law* (1998), No. 2.

30 The regional conference was organized by the GEF/UNDP/IMO Programme for the Prevention and Management of Marine Pollution in the East Asian Seas. It is noteworthy because one of the papers was co-authored by B. A. Hamzah and M. Nizam Basiron

- 1999 International Maritime Organization/Institute of Policy Studies Conference in Singapore: Towards Implementation of Article 43 of the LOS Convention for the Straits of Malacca and Singapore[31]
- 1999 International Conference on the Straits of Malacca: Towards Sustainable Development of the Straits of Malacca (Malacca, Malaysia)[32]

The International Maritime Organization (IMO) co-organized conferences in Singapore in 1996 and 1999 with the Institute of Policy Studies (IPS). IPS was headed by Ambassador-at-Large Professor Tommy Koh, who had served as President of the Third United Nations Conference on the Law of the Sea (UNCLOS III), which negotiated the LOS Convention. The 1999 IMO/IPS Conference made significant progress toward reaching a consensus on how to implement Article 43. In summarizing the discussions of the 1999 Conference, the Chair, Professor Tommy Koh, listed 12 points of convergence, including the following:

> 6. The initiative for implementing Article 43 should be taken by the three littoral States, but that in doing so, the three littoral States should cooperate with the International Maritime Organization.
>
> 9. With respect to the implementation of Article 43, 'user States' should cooperate with the littoral States to do what the 1982 Convention enjoins them to do and what equity compels them to do.
>
> 11. In implementing Article 43, we must take into account the equities on both sides, including the fact that most of the Straits of Malacca and Singapore are within the territorial sovereignty of the coastal States, and the fact that user States have legitimate rights and interests in the Straits.[33]

---

of the Maritime Institute of Malaysia, and was entitled "The Straits of Malacca: Some Funding Proposals." This paper was also published separately as a MIMA paper.

[31] The papers presented at the conference and the rapporteur's reports summarizing the points made in the papers and the discussions are published in a Special Feature volume of the *Singapore Journal of International & Comparative Law* (1999), No. 3, on Implementing Article 43 of the LOS Convention in the Straits of Malacca and Singapore.

[32] M. Shariff, F. M. Yusoff, N. Gopinath, H. M. Ibrahim and R. A. Nik Mustapha, eds, *Towards Sustainable Management of the Straits of Malacca: Proceedings of the International Conference on the Straits of Malacca, 19–22 April, 1999, Malacca, Malaysia* (Selangor: Malacca Straits Research and Development Centre (MASDEC), 2000).

[33] R. C. Beckman, "Towards Implementation of UNCLOS Article 43 for the Straits of Malacca and Singapore – Rapporteur's Report on the 1999 IPS/IMO Conference on

The Rapporteur-General for the 1999 Conference, Robert Beckman, stated in his summary that among the matters on which there is now a better understanding are the following:

> 6. The three straits States concerned must decide amongst themselves on the nature and extent of assistance they seek and the modalities they wish to establish for such cooperation, and that the initiative for such cooperation must come from the straits States.
>
> 7. That a consensus is emerging on certain principles that should govern any funding mechanism under Article 43, and that some form of funding mechanism to implement Article 43 with respect to the Straits of Malacca and Singapore is inevitable.[34]

It was agreed at the 1999 Conference that the littoral States had to decide amongst themselves on the nature and extent of assistance they should seek and the modalities they wish to establish for cooperation. The only suggestion on how to proceed was made by Ambassador Hasjim Djalal of Indonesia:

> In conclusion, I would suggest that the Tripartite Committee of Senior Officials of the Straits of Malacca and Singapore be convened immediately to revive the discussion on the implementation of Article 43, either through informal voluntary cooperation with the users of the straits or through a much more formal mechanism through the IMO or through other conferences.[35]

The three littoral States did not take up Ambassador Djalal's suggestion. From 1999 to 2005 there was little action or progress relating to burden-sharing arrangements and the implementation of Article 43. No meeting of senior officials was held; no follow-up conference was held; and no discussions among the littoral States took place. The issue of Article 43 was put on the back-burner as a result of the turbulence facing the region arising from the Asian financial crisis from 1997 to 2002, the political upheavals in Indonesia after the fall of President Suharto in 1998, and the 2003 Asian severe acute respiratory syndrome (SARS) crisis. During this

---

the Straits of Malacca and Singapore," *Singapore Journal of International & Comparative Law* 3 (1999): 253–292, at 285–286.

[34] *Ibid.*, pp. 290–291.

[35] H. Djalal, "Funding and Managing International Partnership for the Malacca and Singapore Straits Consonant with Article 43 of UNCLOS, 1982," *Singapore Journal of International & Comparative Law* 3 (1999): 457–469, at 469.

period, Japan continued to be only user State that willingly accepted the burdens as well as the benefits of the regime of transit passage through the Straits.

Events Leading to a New Spirit of Cooperation in 2005

After the 11 September 2001 incident in the United States there was increased concern in the international shipping community about maritime terrorism. Maritime security became a high priority at the IMO, and IMO adopted measures to enhance the security of ships and ports.[36]

As security in ports was enhanced, concern increased about the vulnerability to terrorist attacks of vital sea lanes such as the Straits of Malacca and Singapore. The Indian Navy reportedly escorted vulnerable vessels through the Straits at the request of the United States.[37] The United States took various initiatives with coalitions of willing partners, including the Container Security Initiative (CSI)[38] and the Proliferation Security Initiative (PSI),[39] the latter of which raised some concerns in Asia about whether the LOS Convention provisions on sovereignty and jurisdiction would be respected. A remark by a U.S. Admiral about the establishment

---

[36] A new comprehensive security regime for international shipping was adopted by the IMO in December 2002 and entered into force on 1 July 2004. The mandatory security measures include a number of amendments to SOLAS 1974. They include the new International Ship and Port Facility Security Code (ISPS Code), which contains detailed security-related requirements for governments, port authorities and shipping companies in a mandatory section (Part A), together with a series of guidelines about how to meet these requirements in a second, non-mandatory section (Part B).

[37] V. Sakhuja, "Challenging Pirates in Malacca Straits," Society for the Study of Peace and Conflict, Opinion/ Analysis, Article No. 6, September 20, 2004, <www.sspconline.org>, 13 September 2007.

[38] The Container Security Initiative (CSI) was launched in 2002 by the U.S. Bureau of Customs and Border Protection (CBP), an agency of the Department of Homeland Security. Its purpose was to increase security for container cargo shipped to the United States. See <www.cbp.gov/xp/cgov/border_security/international_activities/>, 13 September 2007.

[39] The Proliferation Security Initiative (PSI) was launched by U.S. President George W. Bush in May 2003. It is aimed at stopping shipments of weapons of mass destruction (WMD), their delivery systems, and related materials. See <usinfo.state.gov/products/pubs/proliferation/>, 13 September 2007.

of a Regional Maritime Security Initiative (RMSI) in Southeast Asia was met with suspicion and hostility in Indonesia and Malaysia.[40] At the same time, the number of incidents of piracy and armed robbery against ships in the Straits of Malacca increased.[41] This resulted in the Malacca Strait being declared a war-risk zone for the purposes of marine insurance by Lloyd's.[42] Other developments which did not go unnoticed in the Southeast Asia were the steps taken by China to develop a blue-water navy,[43] the announcement by India that its maritime security interests extend to the South China Sea,[44] and proposals in Japan for its Coast Guard to undertake patrols to safeguard sea lanes vital to its maritime interests.[45]

In addition, Japan indicated in 2004 through informal channels that it would no longer be able to bear all of the costs of maintaining and replacing aids to navigation in the Straits and that assistance from other user States would be necessary. As already noted, for many years Japan had been the only user State to contribute to navigational safety and environmental protection in the Straits. Through the Malacca Straits Council of Japan, it had contributed 13.2 billion yen (US$113 million)

---

40 C. S. Kuppuswamy, "Straits of Malacca: Security Implications," South Asia Analysis Group, Paper No. 1033, 18 June 2004. See <www.saag.org/papers11/paper1033.html>, 13 September 2007.

41 Of the total number of incidents of piracy and armed robbery against ships reported to the ICC International Maritime Bureau (IMB) Piracy Reporting Centre in Malaysia in calendar year 2004, eight were in the Singapore Straits, 37 were in the Malacca Straits, and 93 were in Indonesia. The IMB reported that the number of attacks in the Malacca Straits was the second highest since the IMB began compiling statistics in 1991, and that many on the attacks in the Malacca Straits were serious and involved vessels being fired upon and crew kidnapped for ransom. ICC International Maritime Bureau, "Piracy and Armed Robbery against Ships, Annual Report, 1 January–31 December 2004" (Kuala Lumpur: ICC International Maritime Bureau, 2004).

42 In June 2005 the Joint War Committee (JWC) section of Lloyd's Market Association in London declared the Malacca Strait a war-risk zone. The decision to add the Straits of Malacca to JWC's list was taken following recommendations by a private defence consultant, Aegis Defence Services, which carried out risk assessments on the area and suggested that it was a potential site for a terrorist attack.

43 "Blue-Water Ambitions," *Asiaweek* 26(11) 24 March 2000, <www.asiaweek.com/asiaweek/magazine/2000/0324/nat.2china.navy.html>, 13 September 2007.

44 "India Challenges China in the South China Sea," *Asia Times Online*, 27 April 2000, <www.atimes.com/ind-pak/BD27Df01.html>, 13 September 2007.

45 "Japan to begin Asian piracy patrols," CNN.com, 6 June 2001, <archives.cnn.com/2001/WORLD/asiapcf/east/06/06/japan.sea.piracy/index.html>, 28 September 2007.

from 1968 to 2005. Japan's indication that it would no longer be able to bear this cost alone created pressure to establish a framework for cooperation under Article 43.

The final impetus for action came from leadership of the Secretary-General of the IMO, Mr. Efthimios E. Mitropolous, with the support of the IMO Council. At its 92nd session in June 2004, the IMO Council addressed concerns about the security of vital shipping lanes and requested the Secretary-General to work on the issue in collaboration with the parties concerned.[46] The IMO Secretariat undertook a study of vital shipping lanes and gave special consideration to the Straits of Malacca and Singapore. At the IMO Council's 93rd session in October 2004, the Secretary-General submitted a document that analyzed the key issues involved in protecting vital shipping lanes against terrorist attacks.[47] The document included a profile of the Straits of Malacca and Singapore. The IMO Council agreed that the organization had and could play a role in the protection of shipping lanes of strategic importance and significance.

In March 2005, the IMO Secretary-General delivered the keynote address at the ASEAN Regional Forum on Confidence Building Measures in Singapore. The Secretary-General stated that, as part of his efforts to push forward collaboration on maritime security in the region, he had initiated a round of consultations with the three littoral States and selected user States. He referred to the planned IMO-sponsored meeting to be hosted by Indonesia in September 2005 and stated that this gathering would provide a unique opportunity for the littoral States, user States, and other stakeholders to work to enhance the safety of navigation, environmental protection and overall security in the Straits.[48]

---

[46] IMO, "Council – 92nd session: 21–25 June 2004," <www.imo.org/Safety/mainframe.asp?topic_id=114&doc_id=3580>, 28 September 2007.

[47] Note by the Secretary General, Protection of Vital Shipping Lanes, IMO Doc. C93/15, 7 October 2004. Document includes Annex, "Profile of the Straits of Malacca and Singapore."

[48] E. E. Mitropolous, "Regional Co-operation in Maritime Security," ASEAN Regional Forum Conference, 2 March 2005, at IMO, Newsroom, Speeches of the Secretary-General, <www.imo.org>, 13 September 2007.

The Batam Joint Ministerial Statement, August 2005

In August 2005, one month before the IMO-sponsored meeting in Jakarta, the foreign ministers of the three littoral States met in Batam, Indonesia, to discuss matters pertaining to the safety of navigation, environmental protection and maritime security in the Straits. This was the 4th trilateral ministerial meeting of the three littoral States on matters relating to the Straits of Malacca and Singapore and the first since 1977.

The Batam Joint Statement[49] sets out the views of Indonesia, Malaysia and Singapore on the respective roles of the littoral States, user States, and international agencies such as the IMO in enhancing safety, security, and environmental protection in the Straits. The Joint Statement reaffirms that the three littoral States have sovereignty and sovereign rights in the Straits and the primary responsibility for the safety of navigation, environmental protection and maritime security in the Straits. At the same time, the Statement acknowledges the interests of user States and the relevant international agencies (e.g., the IMO), and the role they could play in respect of the Straits. It also states that any cooperative measures taken in the Straits must be in conformity with international law, including the LOS Convention.

With respect to the possible roles of user States and others in cooperative arrangements, the Joint Statement provides that, bearing in mind the responsibility and burden of littoral States and the interests of user States, the ministers welcome the assistance of the user States, relevant international agencies, and the shipping community in the areas of capacity building, training and technology transfer, and other forms of assistance in accordance with the LOS Convention.

The Batam Joint Statement recognizes the importance of the tripartite ministerial meeting in providing the overall framework for cooperation. The ministers agreed that the ministers and the senior officials should meet on a more regular basis to address relevant issues in a timely manner. Such meetings may include, if necessary, representatives of other relevant agencies of the respective littoral States. This wording is important since it acknowledges that some types of cooperation are beyond the

---

[49] *The Batam Statement of the 4th Tripartite Ministerial Meeting on the Straits of Malacca and Singapore*, Batam, Indonesia, 1–2 August 2005, IMO Doc. IMO/SGP 1/INF.3, Annex 3.

competence of any one ministry and require coordination between government agencies.

In the Batam Statement, the three littoral States agreed amongst themselves on principles for cooperation in advance of the discussions that were to take place at the IMO-sponsored meeting in Jakarta with the IMO, user States and the international shipping community. The agreed principles for cooperation included (1) that the littoral States have sovereignty and sovereign rights in the Straits; (2) that the littoral States have primary responsibility over the safety, environmental protection and security of the Straits; and (3) that any measures adopted or taken on theses matters must be in accordance with international law, including the LOS Convention.

2005 IMO Meeting and the Jakarta Statement, September 2005

The IMO-sponsored meeting in Jakarta was held on 7 and 8 September 2005 and was attended by over 30 States, as well as international organizations and international non-government organizations. The statement from the meeting, known as the Jakarta Statement,[50] identified several points of agreement relating to safety and environmental protection in the Straits. First, the Statement indicated that it was agreed that the work of the Tripartite Technical Experts Group (TTEG) on safety of navigation should continue and be supported and encouraged. Second, it was agreed that the three littoral States should establish a mechanism for meetings on a regular basis with user States, the shipping industry and other stakeholders, to discuss the issues of safety, security and environmental protection in the Straits and to facilitate cooperation, including exploring the possible options for burden-sharing. Third, it was agreed that the three littoral States should establish and enhance mechanisms for information exchange within and between States so as to enhance maritime domain awareness and to contribute to the enhancement of cooperative measures. It was agreed that where possible, these mechanisms should build on existing arrangements such as the TTEG mechanism.

[50] *Jakarta Statement on Enhancement of Safety, Security and Environmental Protection in the Straits of Malacca and Singapore*, Jakarta, 8 September 2005, IMO Doc. IMO/JKT 1/2.

2006 IMO Meeting and the Kuala Lumpur Statement, September 2006

A follow-up meeting was held from 18 to 20 September 2006 in Kuala Lumpur, Malaysia. At the Kuala Lumpur meeting representatives from the three littoral States presented papers and made presentations outlining the cooperative mechanism. The proposed cooperative mechanism consists of three elements: specific projects, a cooperation forum, and a fund.

First, the three littoral States proposed six specific projects to enhance safety and environmental protection and requested user States to cooperate directly in these projects. Under the coordination of the littoral States' TTEG, the following six projects were agreed upon: (1) the removal of wrecks in the traffic separation scheme (TSS); (2) cooperation and capacity-building in relation to the Protocol on Preparedness, Response and Co-operation to Pollution Incidents by Hazardous and Noxious Substances, 2000 (OPRC-HNS Protocol);[51] (3) a demonstration project of class B automatic identification systems (AIS) transponder on small ships; (4) setting up tide, current and wind measurement systems; (5) replacement and maintenance of aids to navigation; and (6) replacement of aids to navigation damaged in the 2004 tsunami. One of the most important developments at the Kuala Lumpur meeting was that two major user States, China and the United States, expressed interest in participating in three of the proposed projects.

Second, the three littoral States proposed the establishment of a Cooperation Forum to promote open dialogue and discussions between the littoral States, user States and other interested parties. The Cooperation Forum is to be the main avenue for user States and other interested parties to meet and cooperate with the littoral States on improving safety and environmental protection in the Straits. Third, the three littoral States agreed to establish an Aids to Navigation Fund that would enable user States and other stakeholders to make voluntary contributions to enhance navigational safety and environmental protection. Although the Cooperation Forum and the Aids to Navigation Fund were outlined as integral parts of the cooperative mechanism, few details were provided on how they would be established and how they would operate.

---

[51] *Protocol on Preparedness, Response and Co-operation to Pollution Incidents by Hazardous and Noxious Substances, 2000*, 15 March 2000, Australian Treaties Library, [2003] ATNIF 9 [hereinafter (OPRC-HNS Protocol)].

The formal document which emerged from the meeting was the Kuala Lumpur Statement.[52] The Statement sets out several points of agreement with respect to the cooperative mechanism, including the following points. First, it was agreed that the cooperative mechanism is intended to promote dialogue and facilitate close cooperation between the littoral States, user States, the shipping industry, and other stakeholders. Second, it was agreed that the six projects presented by the three littoral States for enhancing safety of environmental protection should be supported. Third, it was agreed that the littoral States, user States, the shipping industry, and other stakeholders should cooperate towards the establishment of a mechanism for voluntary funding of the proposed projects and for the maintenance and renewal of the aids to navigation in the Straits. It was also agreed that the IMO should act to facilitate a follow-up meeting and to attract sponsors for the proposed projects and contributors for fund.

2007 IMO Meeting and the Singapore Statement, September 2007

A third and final meeting on Enhancement of Safety, Security and Environmental Protection in the Straits of Malacca was convened by the IMO and the Government of Singapore from 4–6 September 2007. The purpose of the meeting was to provide a follow-up forum to build upon the outcomes of the Jakarta and Kuala Lumpur meetings. Indonesia, Malaysia and Singapore formally launched the cooperative mechanism announced at the Kuala Lumpur meeting and provided more details on the three components. They also briefed participants on the progress made to secure sponsors for the six proposed projects. Participants were also updated on the measures being taken by the defence forces of the three littoral States to enhance security in the Straits. The Singapore meeting was attended by representatives from 50 States and 17 maritime-related organizations.

The product of the meeting was the Singapore Statement,[53] which affirmed that the littoral States have sovereignty, jurisdiction and territorial integrity over the Straits and primary responsibility for safety, security and

[52] *Kuala Lumpur Statement on Enhancement of Safety, Security and Environmental Protection in the Straits of Malacca and Singapore*, Kuala Lumpur, 20 September 2006, IMO Doc. IMO/KUL 1/4.

[53] *Singapore Statement on Enhancement of Safety, Security and Environmental Protection in the Straits of Malacca and Singapore*, Singapore, 6 September 2007, IMO Doc.

environmental protection in the Straits. At the same time, the Statement acknowledges the role of the IMO, user States, shipping industry, and other stakeholders in cooperating with the littoral States in promoting and enhancing safety of navigation and environmental protection, and in ensuring the uninterrupted flow of traffic in the Straits. The Statement invites the IMO to participate in the cooperative mechanism, to continue to cooperate with the littoral States, and to provide assistance in attracting sponsors for the projects and contributors to the Aids to Navigation Fund. The Statement makes it clear, however, that the cooperative mechanism initiative is being led by the littoral States and that the IMO's role is secondary.

The Singapore Statement also makes it clear that the cooperative mechanism builds upon and enhances existing cooperative arrangements. This is important in the regional context, since Indonesia usually favours building upon existing cooperative mechanisms rather than developing new mechanisms. Therefore, because the TTEG on the Safety of Navigation had worked successfully for years, the littoral States decided to build upon it and enhance its functions.

The Singapore meeting did not establish anything new. Rather, it clarified and solidified what had been agreed to in the previous meetings and formally launched the cooperative mechanism. In particular, the littoral States provided more detail on the cooperative mechanism and publicly announced that user States had come forward to support five of the six projects and the Aids to Navigation Fund. Support for one or more of the projects came from China, Japan, Korea and the United States. In addition, Korea and the United Arab Emirates announced that they would contribute to the Aids to Navigation Fund. Several other States, including Germany, announced that they would try to find ways to participate in the projects.

The lack of support from private industry and other stakeholders for the Aids to Navigation Fund was a disappointment. The IMO Secretary-General stated that the main challenge in this regard is how to sensitize the interested parties.[54] However, the one exception, as expected, was

IMO/SGP 1/WP.1, <www.mpa.gov.sg/infocentre/newsreleases/2007/nr070906.htm>, 25 November 2007.

[54] Opening Address by Mr. E. E. Mitropoulos, Secretary-General, International Maritime Organization, Singapore Meeting on the Straits of Malacca and Singapore: Enhancing

the Nippon Foundation of Japan. Consistent with its past generosity, it announced that it would be prepared to contribute to the Aids to Navigation Fund by providing up to a third of the costs of maintaining and repairing the aids to navigation during the first five-year period.

Historic Breakthrough of Great Significance?

In his opening address at the Singapore meeting, the IMO Secretary-General stated that he viewed the mechanism as "an historic breakthrough of great significance" and that the cooperative mechanism constituted "the first attempt to put in place a formal framework to promote implementation of the both the spirit and intent of Article 43 of the United Nations Law of the Sea Convention."[55]

The cooperative mechanism devised by the littoral States in consultation with the IMO and user States is indeed an historic breakthrough. It addresses (or side-steps) all of the concerns raised in the conferences and workshops held from 1994 to 1999 on the implementation of Article 43 in the Straits. The cooperative mechanism is open to user States without attempting to define the meaning of user States. The cooperative mechanism goes beyond Article 43 in two ways. First, it clarifies that the IMO has a role in the mechanism. Second, it calls for participation by private entities, including the shipping industry and other stakeholders.

User States and other stakeholders are likely to participate in the Cooperation Forum since it will enable them to exchange ideas with the littoral States on how to enhance safety, security and environmental protection in the Straits. The International Association of Independent Tanker Owners (INTERTANKO) was the first to recognize the potential of the Cooperation Forum. At the 2007 Singapore meeting, it stated that it considers its involvement in the Cooperation Forum as a key opportunity and that it is eager to contribute its time and resources to the Cooperation Forum and its future work by acting as a link between the littoral States, user States and the tanker industry.[56] The Cooperation Forum will also enable user States and other stakeholders to work together with the littoral States

Safety, Security and Environmental Protection, Singapore, 4 September 2007, p. 5, IMO Doc. IMO/SGP 1/INF.7.

[55] *Ibid.*, p. 4.

[56] Presentation by Mr. T. Wilkins, Regional Manager Asia-Pacific, International Association of Independent Tanker Owners (INTERTANKO), Singapore Meeting on the Straits

to identify new projects and new cooperative measures to enhance safety, security and environmental protection in the Straits.

User States and other stakeholders who contribute to the Aids to Navigation Fund will also have a voice in the management and operation of the Fund. A committee comprising the littoral States, contributing user States and other contributors will be established to manage and operate the Fund. User States who contribute to specific projects are to sit as members of the Project Coordination Committee. These mechanisms will ensure that user States and other stakeholders who cooperate with the littoral States will be given a voice in both the projects and the Fund as well as in the Cooperation Forum.

The cooperative mechanism is likely to work and be successful for three reasons. First, the mechanism is initiated and driven by the littoral States, recognizes the sovereignty and jurisdiction of the littoral States in the Straits, and is consistent with international law, including the LOS Convention. Second, it recognizes that the IMO, user States and other stakeholders have a role in cooperating with the littoral States and gives to those who are cooperating a voice in the management and operation of the mechanisms. Third, because it is inclusive, open and flexible, it will be possible to adapt the cooperative mechanism to meet changing circumstances.

What Next?

Indonesia, Malaysia and Singapore will be taking steps in the next several months to bring the three components of the cooperative mechanism into effect. The inaugural meetings of the three components of the cooperative mechanism—the Cooperation Forum, the Project Coordination Committee and Aids to Navigation Fund—will be organized in 2008.

Once the cooperative mechanism is established, it is expected that the littoral States, in consultation with user States and other stakeholders, will identify additional projects to enhance safety and environmental protection in the Straits. Five of the six projects that have been approved relate to the safety of navigation. Additional projects that might be considered to enhancement environmental protection in the Straits include:

---

of Malacca and Singapore: Enhancing Safety, Security and Environmental Protection, 4–6 September 2007, Singapore, Session 6, 5 September 2007.

1. Assistance to littoral States in ratifying and effectively implementing all of the annexes to the International Convention for the Prevention of Pollution from Ships (MARPOL 73/78),[57] including the provision of adequate reception facilities in major ports in the littoral States;
2. Cooperative measures to combat intentional illegal discharges of oil or other hazardous and noxious substances in the Straits;
3. Cooperative measures to develop contingency plans to deal with a major transboundary oil pollution incident, including whether it is desirable to have all three littoral States become parties to the International Convention on Oil Pollution Preparedness, Response and Co-operation, 1990 (OPRC 1990),[58] the 2000 OPRC-HNS Protocol,[59] and the Protocol of 1992 to amend the International Convention on the Establishment of an International Fund for Compensation for Oil Pollution Damage, 1971 (Fund 1992);[60] and
4. Cooperative measures to ratify and effectively implement other IMO conventions on ship-source pollution, including the International Convention on Civil Liability for Bunker Oil Pollution Damage, 2001 (Bunkers Convention)[61] and the International Convention for the Control and Management of Ships' Ballast Water and Sediments, 2004 (Ballast Water Convention).[62]

---

[57] *International Convention for the Prevention of Pollution from Ships*, London, 2 November 1973, 1340 *U.N.T.S.* 184, as amended by *Protocol of 1978 relating to the International Convention for the Prevention of Pollution from Ships of 1973*, 17 February 1978, 1340 U.N.T.S. 61.

[58] *International Convention on Oil Pollution Preparedness, Response and Co-operation, 1990*, Australian Treaties Library, [1995] ATS 12.

[59] OPRC-HNS Protocol, *supra* note 51.

[60] *Protocol of 1992 to amend the International Convention on the Establishment of an International Fund for Compensation for Oil Pollution Damage, 1971*, Australian Treaties Library, [1996] ATS 3.

[61] *International Convention on Civil Liability for Bunker Oil Pollution Damage, 2001*, London, 23 March 2001, Australian Treaties Library, [2002] ATNIA 25.

[62] *International Convention for the Control and Management of Ships' Ballast Water and Sediments, 2004*, London, 13 February 2004, Australian Treaties Library, [2005] ATNIF 18.

How was the Historic Breakthrough Achieved?

Ambassador Tommy Koh, the former President of the UNCLOS III and Chair of the 1996 and 1999 IMO/IPS Conferences on the Straits, chaired the closing session of the 2007 Singapore Conference. His comments included the following statements:

> I was... very happy when [IMO] Secretary-General Mitropoulos launched the "Protection of Vital Shipping Lanes" initiative, the first of which concentrated on the Straits of Malacca and Singapore. Today, owing to the commitment of the IMO, the tireless efforts of the three littoral States over the past three years and the willingness of the user States and shipping industry to contribute to ensure the safety, security and cleanliness of the Straits, we have finally been able to implement Article 43 of the LOS Convention. This is indeed a historical breakthrough.
>
> On this momentous occasion, allow me to share with you 5 lessons that I have learnt from this process.
>
> **Lesson 1:** The three littoral States have been able to work together. Indeed, we have been cooperating on improving navigational safety and environmental protection in the Straits since the 1970s, through the Tripartite Technical Experts Group (TTEG).
>
> **Lesson 2:** The three littoral States were committed to upholding and applying applicable international law in the Straits of Malacca and Singapore.
>
> **Lesson 3:** The three littoral States share a mindset that the Cooperative Mechanism should be open and inclusive, one that acknowledges the legitimacy and rights of user States while still upholding the sovereignty and territorial integrity of the littoral States.
>
> **Lesson 4:** The IMO has played a vital role in the process. Without the commitment of the IMO, we would not be here today celebrating this successful outcome.
>
> **Lesson 5:** There is a convergence of interest between littoral States and user States in enhancing the navigational safety and environmental protection of the Straits.

I agree with Ambassador Koh's points. The convergence of events in 2004 and 2005 and the IMO Protection of Vital Shipping Lanes initiative pressured the littoral States to work together. The tripartite ministerial meeting in Batam in August 2005 was critical. Once the three littoral States agreed amongst themselves on the principles that would form the basis for discussions with the IMO, user States and other stakeholders, a framework for negotiations was in place. It was then simply a matter of hard work by the officials of the three littoral States. Numerous meetings were held between the officials of the littoral States between the annual IMO meetings to work out the details of the cooperative mechanism. The key fact is that although the process was initiated by the IMO, it was taken over and driven by the three littoral States, with the IMO playing only an advisory role. As the littoral States reached agreement among themselves on the details of the cooperative mechanism, they consulted informally with the IMO and with key user States. This enabled the making of minor adjustments to the mechanism in order to ensure support from key user States.

If there was a weakness in this process, it was that the process was primarily State-driven, making it difficult for the littoral States to consult with non-State stakeholders such as shipping companies and shipping organizations. Because of its past record of generous support and ongoing activities to maintain and replace aids to navigation in the Straits, the Nippon Foundation was likely kept apprised of developments. However, because other stakeholders were not brought into the discussions, they are not likely to feel a sense of ownership in it. It was, therefore, not surprising that at the Singapore meeting the only non-government stakeholder to come forward and pledge funding for the Aids to Navigation Fund was the Nippon Foundation.

One of the major tasks that now faces the littoral States and the IMO is to convince private shipping companies and other stakeholders that it is in their interests to take an active part in the cooperative mechanism and make voluntary contributions to the Aids to Navigation Fund. Some companies may be willing to do so if important persons in the international maritime community, such as the IMO Secretary-General, appeal to their sense of corporate social responsibility. Another factor which should encourage more private companies to make voluntary contributions is that they will be able to participate in the Cooperation Forum. If shipping companies and representatives of the shipping industry are invited to engage in a dialogue with the littoral States in the Cooperation Forum on how to enhance safety, security and environmental protection in the Straits, and their views are taken into account by the littoral States, they should be more willing to make voluntary contributions to the Aids to Navigation Fund.

A major difficulty in getting private industry to make voluntary contributions is that they fear that their participation will establish a precedent that will be followed in other straits used for international navigation. The Straits of Malacca and Singapore was selected by the IMO as the "first" of several vital shipping lanes that are in need of protection and cooperative measures. However, private industry should be somewhat assured by the fact that it took more than ten years and a convergence of events to attain the historic breakthrough in the Straits of Malacca and Singapore. If it proves as difficult to implement Article 43 in other straits used for international navigation, the shipping companies should have little fear that the cooperative mechanism for the Straits of Malacca and Singapore will open the flood gates. Also, since the cooperative mechanism is voluntary, it will not work as a precedent for other straits used for international navigation unless major user States and shipping companies are willing to voluntarily cooperate with the States bordering the strait.

Whither Cooperation to Enhance Maritime Security?

It should be noted that the cooperative mechanism that has been established for the Straits is limited to measures to enhance safety and environmental protection and does not include measures to enhance maritime security in the Straits. This is consistent with the wording of Article 43 of the LOS Convention as there is no mention of security in the article.

However, the IMO commenced the Protection of Vital Shipping Lanes initiative because of the concern for security of the shipping lanes. Also, the major reason that the littoral States took steps to enhance their cooperation on Straits issues in 2005 was because of concerns with maritime security after the September 11 incident and because of the risk to international shipping from piracy and armed robbery against ships.

The 2005 Batam Statement contains several paragraphs which suggest that the process was intended by the ministers to be used as a framework for cooperation to enhance security in the Straits, as well as safety and environmental protection. The 2005 Batam Statement provides that the ministers agreed to establish a TTEG on maritime security. However, in the following discussions the littoral States seemed to have second thoughts about including security in the cooperative mechanism. The TTEG on maritime security was mentioned in the 2006 Kuala Lumpur Statement, but there is no mention of it in the 2007 Singapore Statement. It appears that the littoral States decided against establishing a TTEG on maritime security. Instead, they have left cooperative measures to enhance maritime security to the defence forces of the littoral States. The 2007 Singapore Statement commends

> the joint efforts of the armed forces of the three littoral States in contributing to the security of the Straits, through the Malacca Straits Sea Patrols and the "Eyes in the Sky" maritime air patrols, as formalized in the signing of the Malacca Straits Patrols Standard Operating Procedures on 21 April 2006

Thus, the new cooperative mechanism for the Straits appears to be limited, as least for the time being, to the enhancement of safety and environmental protection in the Straits. Cooperation to enhance maritime security will also continue to be enhanced, but by the defence forces at the operational level and by the foreign affairs and defence ministries in other forums.

Conclusion

The cooperative mechanism established by Indonesia, Malaysia and Singapore in the Straits of Malacca and Singapore is the first attempt by the international community to implement Article 43 of the LOS Convention. The cooperative mechanism as set out in the 2007 Singapore Statement is consistent with the letter and spirit of Article 43 and with Part III on straits used for international navigation. The cooperative mechanism is likely to succeed because it recognizes the sovereignty and territorial integrity of the littoral States, while at the same time acknowledging the interests of user States, the IMO and the international community. It is also likely to succeed because it establishes a voluntary scheme that gives a voice to the stakeholders and avoids the controversial issue of tolls or charges.

If it is successful, the cooperative mechanism will enhance cooperation between the littoral States and the IMO, user States and other stakeholders in order to benefit the marine environment and international shipping, without threatening or undermining the sovereignty and territorial integrity of the littoral States. The same principles and scheme can be used to develop cooperative mechanisms to enhance maritime security in the Straits.

The most important result of the establishment of the cooperative mechanism for the Straits of Malacca and Singapore is that it is likely to enhance cooperation and build trust among the three littoral States themselves. The important lesson that Indonesia, Malaysia and Singapore should have learned over the past years is that, when they identify their common interests and work together to accomplish them, they can achieve far more than when they act unilaterally. It is hoped that the three littoral States will use this experience as a model for further cooperative efforts to protect and preserve the marine and coastal environment in Southeast Asia.

The Compulsory Pilotage Regime in the Torres Strait – A "Melting Pot" of Operational, Legal, and Political Considerations

Sam Bateman*

Introduction

On 6 October 2006, Australia introduced compulsory pilotage for the Torres Strait and Great North East Channel.[1] The shipping route through this area constitutes a strait used for international navigation to which the transit passage regime in Part III of the 1982 United Nations Convention on the Law of the Sea (LOS Convention) applies.[2] The route passes through the Particularly Sensitive Sea Area (PSSA) designated and approved by the International Maritime Organization (IMO) in 2005.[3] Thus the area is one that needs special protection through action by

* Professorial Research Fellow, Australian National Centre for Ocean Resources and Security, University of Wollongong, Australia, and Senior Fellow, Maritime Security Programme, S. Rajaratnam School of International Studies, Nanyang Technological University, Singapore. The author is most grateful to Mr. Paul Nelson, Manager-Environment Protection, Australian Maritime Safety Authority, for the provision of the map at Figure 1 and the data in Table 1.

1 Australian Marine Notice 8/2006 and associated Marine Orders Part 54.

2 *United Nations Convention on the Law of the Sea*, Montego Bay, 10 December 1982, UN/Doc.A/CONF.62/122, 7 October 1982, 21 *I.L.M.* 1261 (1982) [hereinafter LOS Convention].

3 *Designation of the Torres Strait as an Extension of the Great Barrier Reef Particularly Sensitive Sea Area*, adopted by IMO Resolution MEPC.133(53), 22 July 2005, available in *Report of the Marine Environment Protection Committee on its Fifty-Third Session,* Annex 21, MEPC 53/24/Add.2 (1 August 2005).

the IMO because of its significance for recognized ecological or socio-economic or scientific reasons, and which may be vulnerable to damage by international maritime activities.[4]

This issue was hotly debated at the IMO, and the compulsory pilotage regime has been formally protested by the United States and Singapore.[5] However, Australia and Papua New Guinea (PNG), as the two littoral States, believe that compulsory pilotage is necessary to protect sensitive marine habitats in the Torres Strait and is in accordance with international law. They consider it an appropriate and necessary associated protective measure (APM) for the PSSA to control maritime traffic in the area and to reduce the risks of accidental pollution by ships. At the heart of the matter are issues related to what Douglas Johnston called the "greening" of the law of the sea,[6] and which others see as "creeping jurisdiction."[7]

Like most law of the sea issues with operational and international implications, the pros and cons of compulsory pilotage in the Torres Strait cannot be seen in legal terms alone. There are a host of operational and political factors to be considered as well. Legal considerations relate to whether the action is justifiable in international and domestic law. Political factors arise in assessing the costs and benefits of the action in political terms both domestically and internationally, and operational considerations cover assessments of what might be technically, operationally or administratively feasible and necessary. When all these factors are taken into account, there is no certainty that the legal position will carry the day. Ultimately governments will do what they think is necessary in the national interest; what they think they can get away with; and what brings them either the most political kudos or has the least political costs.

---

[4] For the IMO definition of a PSSA, see *Revised Guidelines for the Identification and Designation of Particularly Sensitive Sea Areas*, adopted by IMO Assembly Resolution A.982(24), 1 December 2005 [hereinafter Revised Guidelines].

[5] R. C. Beckman, "PSSAs and Transit Passage – Australia's Pilotage System in the Torres Strait Challenges the IMO and UNCLOS," *Ocean Development and International Law* 38(4) (2007), p. 337.

[6] D. M. Johnston, "The challenge of international ocean governance – Institutional, ethical and conceptual dilemmas," in D. R. Rothwell and D. L. VanderZwaag, eds, *Towards Principled Oceans Governance – Australian and Canadian Approaches and Challenges* (London and New York: Routledge, 2006), p. 379.

[7] Beckman, *supra* note 5, p. 349.

This essay addresses the mix of legal, operational and political considerations with the Torres Strait pilotage regime. While the regime has been introduced into Australian domestic law, doubts have been raised about whether the regime is in accordance with international law, particularly Part III of the LOS Convention.[8] Ambassador Tommy Koh of Singapore, who was deeply involved in negotiating the LOS Convention, considers that Australia's case is "weak," and sets "an unfortunate precedent."[9] On the other hand, there are many political and operational factors that suggest that the Australian government had no alternative but to introduce the system of compulsory pilotage for the Torres Strait.

Operational Considerations

Navigational Issues

Geographically, the Torres Strait comprises the waters lying between Cape York Peninsula in the extreme north of Australia and Papua New Guinea. These waters are shallow, usually no deeper than about 13 metres, and strewn with numerous islands, small islets, reefs and shoals. The northern half of the Strait is only navigable by vessels with a very shallow draught, and deep draught vessels are restricted to using narrow channels in the southern part of the Strait between the various islands off Cape York, principally the Prince of Wales Channel immediately north of Hammond Island.

Navigation in the Torres Strait is extremely hazardous. Apart from the complex topography of the area and relatively shallow waters, the tidal regime in the Strait is highly complex and variable.[10] Tidal streams can exceed seven knots. This is an area of confluence between two major ocean systems. Tides in the Coral Sea and Pacific Ocean are semidiurnal while tides in the Indian Ocean are diurnal, causing tides in the Strait to

---

8 The most cogent criticisms have been raised by Beckman, *ibid.*, pp. 325–357.

9 T. Koh, "UN Convention on the Law of the Sea – A passage to maritime safety," *Straits Times Interactive*, 28 February 2007.

10 R. Babbage, "The Strategic Significance of Torres Strait," *Canberra Papers on Strategy and Defence* No. 61 (Canberra: Strategic and Defence Studies Centre, Australian National University, 1990), p. 3.

fall into no coherent pattern.[11] The *Australian Seafarers Handbook* warns vessels planning a passage through the Torres Strait that "[A]ccurate calculations and local knowledge are necessary to establish a tidal window for any particular day and draught."[12] Currents are also very strong, and visibility can be impaired by flash squalls and storms. Radar propagation is frequently degraded throughout the year due to either heavy precipitation during the "wet" season, or sea mist caused by the strong, gusty southeast trade winds in the "dry" season.[13] All in all, it is one of the most dangerous stretches of water in the world routinely used by large vessels.

International shipping passing through the Torres Strait uses the Prince of Wales Channel. This passage is narrow, being only 800 metres wide at its narrowest point,[14] and subject to very strong tidal streams. Most ships are bound to or from Australian ports and use the Inner Route of the Great Barrier Reef. However, ships bound to and from South Pacific ports use the Great North East Channel. As shown in Figure 1, ships using the Inner Route pass southeastwards off Cape York while the vessels using the latter channel proceed northeastwards towards Bramble Cay to enter the Coral Sea through Bligh Entrance. It is these latter vessels to which the straits' transit passage regime in the LOS Convention applies. Ships using the Inner Route pass through Australia's internal waters. This means there are no law of the sea implications, and the passage of these ships does not constitute transit passage within the meaning of Article 38(2) of the LOS Convention.[15]

Table 1 shows the number of piloted transits of the Great Barrier Reef Inner Route and Great North East Channel by flag for the year ending 30 September 2007. As the compulsory pilotage regime for the Torres Strait commenced in early October 2006, Figure 1 includes all transits and thus shows the major users of these routes. There were roughly three transits per day of the Great North East Channel while nearly nine ships

[11] S. B. Kaye, *The Torres Strait* (The Hague: Martinus Nijhoff Publishers, 1997), p. 4.

[12] Commonwealth of Australia, *Australian Seafarers Handbook* (Wollongong: Australian Hydrographic Services, 2007), p. 111.

[13] RAN Hydrographer, "The Meteorological Effects on Maritime Operations conducted in the Torres Strait Region," in Babbage, *supra* note 10, pp. 217–218.

[14] *Routeing of Ships, Ship Reporting and Related Matters – Torres Strait PSSA Associated Protective Measure – Compulsory Pilotage*, submitted by Australia and Papua New Guinea, 22 March 2004, IMO Doc. NAV 50/3, para. 2.2.5.

[15] Kaye, *supra* note 11, at p. 125.

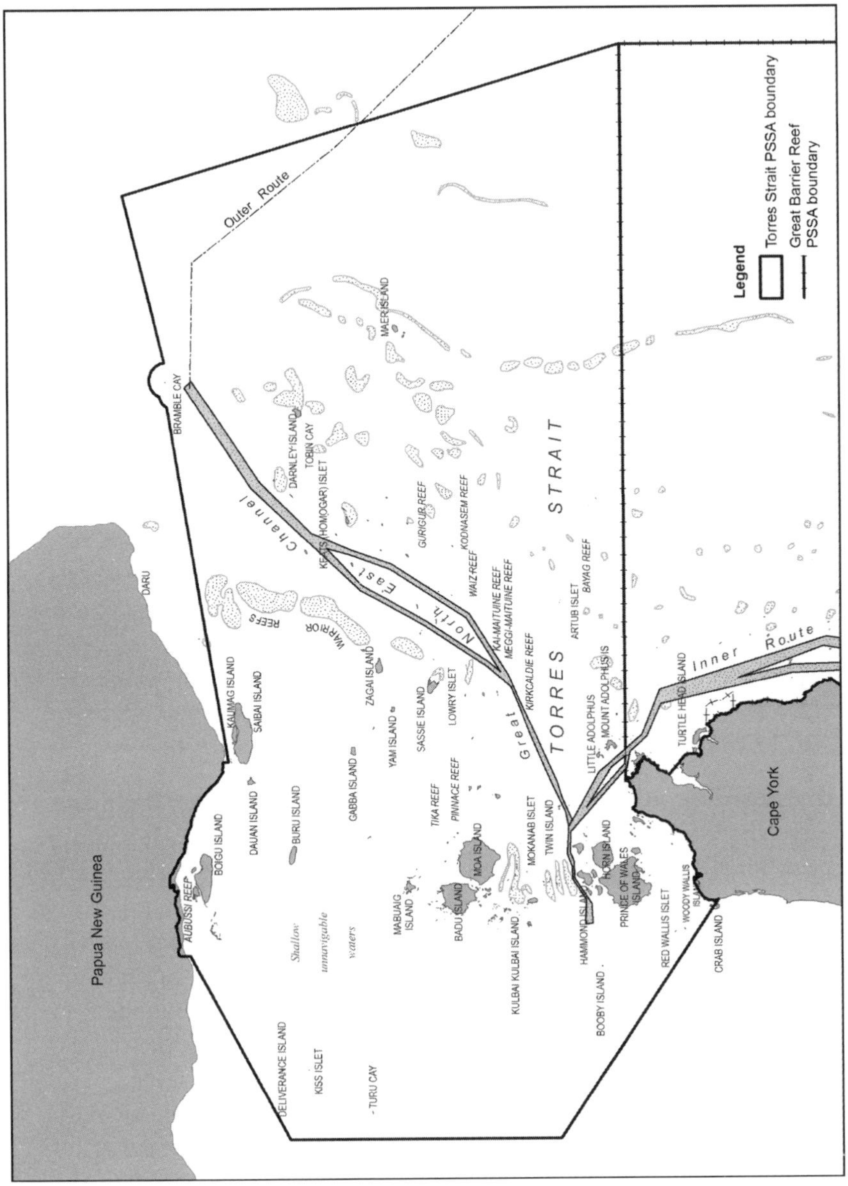

Source: Australian Maritime Safety Authority

Figure 1. Torres Strait shipping routes and PSSA

Table 1. Piloted transits of Great Barrier Reef (GBR) Inner Route and Great North East Channel (GNEC) year ending 30 September 2007

Flag	GBR Inner Route	GNEC	Total
Antigua & Barbuda	43	22	65
Australia	235	12	247
Bahamas	68	71	139
China	22	13	35
Cyprus	52	24	76
Denmark	89	22	111
France	4	45	49
Germany	51	9	60
Greece	56	16	72
Hong Kong China	186	107	293
India	53	14	67
Isle of Man UK	9	47	56
Liberia	142	43	185
Malaysia	61	19	80
Marshall Islands	87	41	128
Netherlands	50	11	61
Panama	535	297	832
Singapore	190	94	284
United Kingdom	74	6	80
Vietnam	1	16	17
Other	175	75	250
TOTAL	2183	1004	3187

Source: Australian Maritime Safety Authority

per day passed through the Prince of Wales Channel. The main flag States that used the Great North East Channel are Panama (297 transits during the period in Table 1), Hong Kong (107) and Singapore (94). Three U.S. flag vessels transited this channel during the year.

While the Prince of Wales Channel is the most dangerous part of Torres Strait used for international navigation, navigation through the Great North East Channel is also hazardous. The channel is bounded on the western side by the Dungeness and Warrior Reefs and on the eastern and southeastern side by a mass of coral reefs and islands. In several parts of the channel, ships must navigate within one nautical mile of reefs and others dangers. The channel has a minimum depth of 12.2–12.8 metres and is subject to strong tidal flows.[16] These flows, along with the

[16] *Ibid.*, p. 15.

currents through the Strait, are mostly across the main channel, and this means that ships passing along the Great North East Channel must navigate with extreme caution. Unless ships exercise great vigilance, they can easily be swept off their intended route. The channel becomes even more dangerous at night or in poor visibility when the reefs and other dangers are not visible. Torres Strait waters are also extensively used by fishing and recreational vessels, and between 1985 and 2003, there were twelve collisions between trading vessels and fishing vessels caused in most instances by a failure to keep a proper lookout on both vessels.[17]

There can be no doubt that the use of a pilot markedly reduces the risks of a marine accident. Merchant ships operate with small crews and navigating safely through the Torres Strait via the Prince of Wales and Great North East channels is a most demanding task for the small staff on the bridge of a large vessel. Research undertaken on behalf of Australia by representatives of the U.S. and Canadian coast guards found that compulsory pilotage in the Torres Strait could reduce the risks of groundings by between 45 per cent and 57 per cent and collisions by between 57 per cent and 67 per cent depending on the specific location within the strait.[18] Other analysis by Det Norske Veritas in 2001 indicated that compulsory pilotage would reduce these risks by 35 per cent.[19] These are significant reductions in risk that cannot be ignored.

A pilot service has been available for the Torres Strait and Great Barrier Reef for well over one hundred years.[20] An essential link in the chain of safety for navigation through the Torres Strait is provided by the pilot,

> [w]hose local knowledge and experience in reef waters enables him to relate charted information and nav-aids to the practical function of guiding a ship safely through the area. A pilot knows how reefs, islands, rocks and shoals appear at all stages of time and tide, how currents change with each passing mile. His knowledge assists him to pass dangers at close range; to meet other ships in tight areas and yet pass in safety.[21]

[17] *Australian Seafarers Handbook, supra* note 12, at p. 109.

[18] J. Roberts, *Marine Environment Protection and Biodiversity Conservation – The Application and Future Development of the IMO's Particularly Sensitive Sea Area Concept* (Berlin: Springer, 2007), p. 151.

[19] IMO NAV 50/3, *supra* note 14, at para. 5.8.

[20] The first pilot licenses for the area were issued in 1884. J. C. H. Foley, *Reef Pilots – The History of the Queensland Coast ant Torres Strait Pilot Service* (Sydney: Banks Bros. & Street, 1982), p. 34.

[21] *Ibid.*, pp. xviii–xix.

Pollution Concerns

The Torres Strait represents an extremely sensitive marine habitat that could easily be severely damaged as a result of a marine accident. This was recognized by the IMO when the area was declared a PSSA. The impact of crude oil on mangroves in the Strait was still evident some 27 years after a major spill in the Strait as a result of the grounding of the tanker *Oceanic Grandeur*.[22] All the materials listed in the International Maritime Dangerous Goods (IMDG) Code are carried through the Torres Strait, either in containers or as bulk cargoes.[23] They include crude oil, petroleum products, bulk fertilizer and mineral concentrates. Most of these substances, as well as bunker fuel oil, would have a great impact on the marine environment if they were released as a consequence of a marine accident. Due to the strength of the tidal streams and currents in the area, containment and clean-up arrangements would be difficult and the adverse effects could quickly spread across numerous islands and reefs in the Strait.

As a result of concerns over the risks of pollution damage to the environmentally sensitive Great Barrier Reef, Australia applied initially to the IMO to have the Great Barrier Reef identified as a PSSA. This was approved in 1990 along with a recommendation that IMO Member States should inform ships flying their flags to comply with the system of pilotage introduced by Australia.[24] That system became a compulsory one, and this has been accepted without challenge by other countries.

With regard to the Torres Strait, the IMO had earlier adopted a resolution promoting voluntary pilotage in the Strait.[25] This was extended further with a 1991 resolution, superseding the earlier one, recommending that certain classes of vessel use a pilot when passing through the Torres Strait and Great North East Channel.[26] While these recommendatory regimes were initially reasonably successful, non-compliance has increased signifi-

[22] IMO NAV 50/3, *supra* note 14, at para. 4.3.

[23] *Ibid.*, para. 2.3.1.

[24] *Protection of the Great Barrier Reef Region*, IMO Resolution MEPC.45(30), adopted 16 November 1990.

[25] *Use of Pilotage Services in the Torres Strait and Great Barrier Reef Area*, IMO Resolution A.619(15), adopted 16 November 1987.

[26] *Use of Pilotage Services in the Torres Strait and the Great North East Channel*, IMO Resolution A.710(17), adopted November 1991.

cantly. Data from 1995 and 2001 shows that while 70 per cent of vessels on eastbound voyages were taking a pilot in 1995, this figure had fallen to 32 per cent by 2001.[27] Similar figures for westbound voyages were 55 per cent and 38.5 per cent. These figures equate to about 500 un-piloted transits each year.[28] As a consequence, Australia and PNG agreed that the risks of a major shipping incident in the Strait leading to serious pollution of the area were unacceptably high.

As a result of these concerns, Australia and PNG jointly proposed an extension to the existing Great Barrier Reef PSSA to include the waters of the Torres Strait (see Figure 1). This was approved in July 2005 through a resolution regarding governments informing ships flying their flags to comply with the system of pilotage introduced by Australia.[29]

Pilotage System

Australia subsequently issued the regulations establishing the compulsory pilotage regime for the Torres Strait and Great North East Channel. These regulations recognise the principle of sovereign immunity for warships and government vessels not employed on commercial service. They also include a system of pilotage exemption for masters of ships that use the Torres Strait on a regular basis. Other countries and international shipping organizations, including INTERTANKO and the International Chamber of Shipping, protested these regulations at the 55th Session of IMO's Marine Environment Protection Committee held in August 2006.[30]

Australia has put in place measures to ensure that ships approaching the Torres Strait are notified well in advance of their approach of the need to take on a pilot when transiting the Torres Strait. Ships planning to enter Australia's exclusive economic zone (EEZ) are required to report their intentions and are tracked using the Australian Maritime Information System (AMIS) managed by the Border Protection Command. Their movements are then monitored within the Torres Strait and Great Barrier

---

[27] *Torres Strait PSSA Associated Protective Measures – Compulsory Pilotage*, submitted by Australia and Papua New Guinea, IMO Doc. LEG 89/15 (24 August 2004), para. 6.

[28] Roberts, *supra* note 18, at p. 156.

[29] IMO Resolution MEPC.133(53), *supra* note 3.

[30] *Report of the Marine Environment Protection Committee on its Fifty-fifth Session*, IMO Doc. 55/23 (16 October 2006), especially at pp. 52–54.

Reef by REEFCENTRE, which operates REEFVTS, the vessel traffic and information system for these shipping routes.[31]

As a vessel approaches the Torres Strait, it is interrogated by automatic identification system (AIS) shore stations and tracked by shore-based radar. Within the vicinity of the Prince of Wales Channel, it will also be identified by remotely operated video cameras. Should a vessel not take a pilot and fail to identify itself, it will be positively identified by surveillance aircraft and subject to legal proceedings when it next enters an Australian port. No attempt will be made to physically enforce the compulsory pilotage regime by denying passage.

Advice from the Australian Maritime Safety Authority (AMSA) is that since the introduction of compulsory pilotage, there has been one hundred per cent compliance with the regime. One ship transited the Great North East Channel without a pilot in the very first days of the new requirement for compulsory pilotage. As that vessel followed all the appropriate procedures for requesting a pilot but one was not available, and the vessel continued on passage without a pilot, AMSA does not regard this incident as a breach of the regulations and will not take action against the ship if she were to enter any Australian port in the future. Despite the international protests, compulsory pilotage is achieving its objective of improved protection for the sensitive and pristine marine habitats of the Torres Strait and adjacent areas.

Legal Issues

Torres Strait Treaty

The Torres Strait Treaty between Australia and PNG establishes sovereignty over islands in the Strait and a system of agreed maritime boundaries.[32] It is a complicated treaty creating territorial sea enclaves, non-coincident

[31] REEFREP is a mandatory ship reporting system adopted by the IMO in 1996. In December 2004, REEFREP was enhanced with the introduction of a vessels traffic service and became the Great Barrier Reef and Torres Strait Vessel Traffic Service (REEFVTS). *Australian Seafarers Handbook, supra* note 12, at p. 157.

[32] *Treaty between Australia and the Independent State of Papua New Guinea concerning the Maritime Boundaries in the Area between the Two Countries, including the Area known as Torres Strait, and Related Matters,* Sydney, 18 December 1978, 1985 *Australian Treaty*

seabed and water column boundaries, and a large protected zone with extensive management arrangements. The principal purpose of the protected zone is to acknowledge and protect the traditional way of life and livelihood of the traditional inhabitants, including fishing and free movement. Generally the Strait is an area of high marine biodiversity with sensitive marine habitats and extensive fishing activity both commercial and subsistence.

Article 7 of the Torres Strait Treaty addresses freedoms of navigation and overflight. It states, *inter alia*, that:

> ...a regime of passage over routes used for international navigation in the area between the two countries, including the area known as Torres Strait, shall apply in respect of vessels that is no more restrictive of passage than the regime of transit passage through straits used for international navigation described in Articles 34 to 44 inclusive [of the UN document that later became the LOS Convention].[33]

Jurisdiction

Jurisdiction in the Torres Strait is extremely complex. The Torres Strait Treaty established a seabed boundary between Australia and PNG, but there is a non-coincident fisheries jurisdiction line that loops north to enclose the inhabited Australian islands of Boigu (Talbot Island in Figure 1), Dauan and Saibai that are adjacent to the PNG coast.[34] The territorial seas around islands in the Strait have varying widths depending on their location in the strait. The Torres Strait Treaty provides that the territorial sea of the Australian islands north of the seabed boundary is restricted to three nautical miles.[35] Australia extended its territorial sea from three to twelve nautical miles in 1990,[36] while PNG took this step in 1977.[37] However, both countries are restricted under the Torres Strait Treaty

Series No. 5, entered into force on 15 February 1985 [hereinafter Torres Strait Treaty]; reproduced in Kaye, *supra* note 11, at Appendix A.

33 *Ibid.*, Art. 7(6).

34 B. R. Opeskin and D. R. Rothwell, "Australia's Territorial Sea: International and Federal Implications of Its Extension to 12 Miles," *Ocean Development and International Law* 22 (1991), p. 399.

35 Torres Strait Treaty, *supra* note 32, at Art. 3(2).

36 Opeskin and Rothwell, *supra* note 34, at p. 395.

37 *Ibid.*, p. 400.

from extending their territorial sea beyond three nautical miles in parts of the Strait.[38]

Australian jurisdiction within the Strait is complicated by the fact that Australia has a federal system of government and a large part of Australia's maritime area of jurisdiction in the Strait comprises coastal waters of the State of Queensland. *Coastal waters* in the Australian context are defined as a belt of water between the limits of the Australian states and the Northern Territory and a line three nautical miles seaward of the territorial sea baseline.[39] Effectively they include internal waters within territorial sea baselines and the first three miles of territorial sea. This limit to state jurisdiction over the territorial sea was established as part of the Offshore Constitutional Settlement negotiated between the Commonwealth of Australia, the six Australian states and the Northern Territory in 1979 when it was agreed that if the width of Australia's territorial sea were ever declared to be greater than three nautical miles, the coastal waters of the states and the Northern Territory would remain fixed at three nautical miles.[40]

Australia proclaimed territorial sea baselines in 1983 and these were amended slightly in 1987 to take account of certain historic bays in South Australia.[41] This proclamation was revoked in 2006 when a revised set of baselines was proclaimed.[42] Most of these baselines are normal baselines drawn in accordance with Article 5 of the LOS Convention, while the remainder comprises straight baselines drawn across bays and river mouths, and around fringing islands, in accordance with Articles 7, 9, 10 and 14 of the LOS Convention. Opeskin has noted that the baselines proclaimed in 1983 and 1987 "generally reflect a cautious approach to the rules, such

[38] Torres Strait Treaty, *supra* note 32, at Art. 3.

[39] See Geoscience Australia, <www.ga.gov.au/nmd/mapping/marbound/bndrs.jsp>, updated 6 June 2006.

[40] Opeskin and Rothwell, *supra* note 34, at pp. 408–410.

[41] B. R. Opeskin, "The Law of the Sea," in S. Blay, R. Piotrowicz and B. M. Tsamenyi, eds, *Public International Law – An Australian Perspective* (Melbourne: Oxford University Press, 1997), p. 332.

[42] Commonwealth of Australia, *Seas and Submerged Lands (Territorial Sea Baseline) Proclamation 2006*, Federal Register of Legislative Instruments F2006L00525, 15 February 2006.

as in relation to the Great Barrier Reef."[43] While a less cautious approach could extend Australia's zones of maritime jurisdiction, it would also mean a significant increase in the areas of coastal waters under the jurisdiction of the states rather than the Commonwealth.

Australia employs a system of straight baselines in the Torres Strait.[44] These were part of the 1983 Proclamation but have been retained unchanged in the 2006 Proclamation. The baselines enclose as fringing islands the main islands lying to the north and northeast of Cape York as well as the fringing islands down the east coast of Cape York. After the declaration of the 12 nautical mile territorial sea, Australia might have been able to extend its system of straight baselines in the Torres Strait to include islands such as Murray, Darnley and Stephens lying in the eastern approaches to the Strait. However, it has chosen not to do so, presumably because much of the area would then have become coastal waters of Queensland rather than extended territorial sea under Commonwealth jurisdiction. The actual area of any additional waters brought under Australian jurisdiction, including any additional EEZ, would have been relatively small.[45]

Transit Passage

The transit passage regime for straits used for international navigation set out in Section 2 of Part III of the LOS Convention applies to the Torres Strait. Introduction of this regime overcame the difficulty that many straits, which had previously been high seas, became territorial seas when the maximum width of the territorial sea was extended in the LOS Convention to 12 nautical miles. Recognition of the broader territorial sea had a considerable impact upon navigation through straits used for international navigation. It has been estimated that there were 153 straits around the world with breadths of between 6 and 24 nautical miles that had previously included high seas "corridors" but became enclosed within the territorial sea of one or more coastal States with acceptance of the

---

[43] Opeskin, *supra* note 41, at p. 332.

[44] H. Burmester, "Review of the Strategic Significance of the Torres Strait: Legal Issues," in Babbage, *supra* note 10, at Figure 1, p. 305.

[45] A map showing the effect of extending Australia's territorial sea to 12 nautical miles in the Torres Strait region is available in Babbage, *ibid.*, Figure 8, p. 45.

12 mile limit.[46] Without the transit passage regime, only innocent passage would have been available through these straits. This is a more restrictive regime not available to aircraft or submerged submarines, and it can be suspended in certain circumstances by a coastal State.

Transit passage is defined in Article 38(2) of the LOS Convention as the exercise of the freedom of navigation and overflight by ships and aircraft through a strait used for international navigation "between one part of the high seas or an exclusive economic zone and another part of the high seas or exclusive economic zone." Passage must be "continuous and expeditious," and Article 42(2) of the LOS Convention states that the laws and regulations of States bordering straits shall not "in their application have the practical effect of denying, hampering or impairing the right of transit passage."

There is some doubt in the literature as to the extent to which transit passage is part of customary international law. Scovazzi has noted that "in international practice the transit passage regime has been the subject of a series of exceptions, reservations, declarations, qualifications, attenuations' and "[i]t is therefore possible to argue that the LOS Convention transit passage regime is still far from fully corresponding to present customary international law."[47] Churchill and Lowe agree that a general right of transit passage is not yet established in customary international law,[48] and as two American authors have observed that, "the transit passage regime has not been accepted universally; and some States have sought to impose restrictions on the use of these straits inconsistent with the terms of the Convention."[49]

Indeed the right of transit passage may be increasingly qualified in the future by the growing trend among coastal States to introduce measures

---

[46] L. M. Alexander, "Exceptions to the Transit Passage Regime: Straits with Routes of 'Similar Convenience'," *Ocean Development and International Law* 18(4) (1987), pp. 480–481.

[47] T. Scovazzi, "Management Regimes and Responsibility for International Straits," in H. Ahmad, ed., *The Straits of Malacca – International Cooperation in Trade, Funding and Navigational Safety* (Petaling Jaya: Pelanduk Publications, 1997), p. 344.

[48] R. R. Churchill and A. V. Lowe, *The Law of the Sea,* 3rd ed. (Manchester: Juris Publishing, Manchester University Press, 1999), p. 113.

[49] G. V. Galdorisi and K. R. Vienna, *Beyond the Law of the Sea – New Directions for U.S. Oceans Policy* (Westport: Praeger, 1997), p. 147.

for the protection of the marine environment which impact upon navigation.[50] This reflects both the higher shipping traffic and increased carriage of cargoes potentially hazardous to the marine environment in the case of an accident, as well as growing concern for the health of the waterways of the world. Inevitably there must be some impact on the navigational rights and freedoms that may have been customary in the past.[51]

Australia has conceded that the LOS Convention transit passage regime applies to the Torres Strait. While large parts of the route through the Prince of Wales Channel and the Great North East Channel pass through Australia's internal waters enclosed within territorial sea baselines, these are straight baselines and thus subject to Article 8(2) of the LOS Convention which provides for a right of innocent passage through internal waters where the use of straight baselines has the effect of enclosing as internal waters "areas which had not previously been considered as such." The same language is used in Article 35(a) of the LOS Convention, which has the effect of bringing such waters within the scope of the transit passage regime.

It may well be that Australia was premature in conceding the application of the transit passage regime to the Torres Strait. An argument might have been made that the enclosed waters lying between the many islands between Mabuiag Island and Cape York have always been considered internal waters of Australia. They are very confined waters that are of great significance to the indigenous inhabitants of the region who have a strong cultural affinity with them along with a deep concern for the protection of the marine environment. Just as marine environmental concerns have increased in recent decades, so also has the appreciation of the rights of indigenous peoples.

It is important to note that as far as international shipping is concerned, the transit passage regime applies only to the route from the Arafura Sea through Torres Strait via the Great North East Channel to the Coral Sea. This is the only route that meets the criteria of one used for international navigation between one part of an EEZ or the high seas and another. While

---

50 Scovazzi, *supra* note 47, p. 332.

51 S. Bateman, D. R. Rothwell and D. VanderZwaag, "Navigational Rights and Freedoms in the New Millennium: Dealing with 20th Century Controversies and 21st Century Challenges," in D. R. Rothwell and S. Bateman, eds, *Navigational Rights and Freedoms and the New Law of the Sea* (The Hague: Martinus Nijhoff, 2000), pp. 314–335.

there are other passages through the Torres Strait, these are all shallow, narrow and poorly charted. They are certainly not used for international navigation. Due to the depth of water in its channels, submerged transit of the Strait is out of the question.

Arguments For and Against

The main legal arguments against compulsory pilotage in the Torres Strait are that the IMO did not specifically approve it; it has the practical effect of "denying, hampering or impairing the right of transit passage" and is thus contrary to the LOS Convention; and it establishes a precedent that, if adopted by other countries adjacent to a strait used for international navigation, would constitute a very significant impairment of the freedom of navigation. Australia strongly refutes these arguments. Firstly, it notes that the IMO endorsed the regime when it recommended that governments should "inform ships flying their flag that they should act in accordance with Australia's system of pilotage for merchant ships 70 m in length and over or oil tankers, chemical tankers, and gas carriers, irrespective of size."[52] This language is identical to that used by the IMO when it recommended that ships act in accordance with Australia's system of pilotage for the Inner Route of the Great Barrier Reef. Australia also notes that it is not in the nature of the IMO to formally approve traffic management schemes but rather to recommend their acceptance.

Secondly, Australia does not accept that compulsory pilotage amounts to "denying, hampering or impairing" passage through the Torres Strait. It argues in fact that the measure, by facilitating the safe passage of vessels, actually enhances transit passage through the Strait.[53] The pilotage arrangements made by Australia ensure that the passage of a ship will not be denied, hampered or impaired by the need to take a pilot. The regime is aimed solely at enhancing safe navigation and protection of the marine environment. It is a commercial system with pilotage revenues going to a private company rather than a government agency. It is a commercial cost and not a fee for transit. While Australia has made certain guarantees relating to the availability of a pilot, in the event that one was not

[52] IMO Resolution MEPC.133(53), *supra* note 3.

[53] *Torres Strait PSSA Associated Protective Measure – Compulsory Pilotage*, submitted by Australia and Papua New Guinea, IMO Doc. LEG 89/15, 24 August 2004.

available and the transiting vessel had taken all the appropriate actions to request a pilot and report its transit, this would be accepted as a defence to any subsequent charge.

The argument that compulsory pilotage is contrary to the LOS Convention draws also on the language of Articles 41 and 42 of the LOS Convention. Beckman contends that a coastal State "may only adopt laws and regulations designating sea lanes and traffic separation schemes, as provided in Article 41," and that "coastal states may only adopt laws and regulations that give effect to applicable international regulations regarding the discharge of oil, oily wastes, and other noxious substances."[54] However, he has introduced the word "only" into these limitations and this may be an overly restrictive reading of these two articles.

Lastly, Australia does not accept that its arrangements in the Torres Strait are a precedent for other straits used for international navigation. Every strait is different with regard to the quantity of shipping traffic, the extent of navigational dangers, depth of water, the sensitivity of the marine environment, and the feasibility of a pilotage system. Pilotage in the Torres Strait has a long history and until comparatively recently, most ships without good local knowledge acted prudently and were prepared to take a pilot. Paradoxically, the number of ships complying with the voluntary pilotage regime began falling at a time when greater concern was being expressed about the human element as a cause of marine accidents.

The question might be asked about how Australia would respond to an attempt by Indonesia to introduce compulsory pilotage in the Sunda or Lombok Straits, which are vital waterways for Australian trade. However, these straits are wide and relatively clear of navigational dangers. They are nowhere near as dangerous or as demanding for navigation as the Torres Strait is, and have no history of pilotage being available.

The Torres Strait is one of the most hazardous and navigationally difficult stretches of water in the world routinely used by international shipping. The level of shipping traffic through the Great North East Channel is not high, and it is administratively and operationally feasible to provide a pilot without delaying passage. Most importantly, the Torres Strait has been approved by the IMO as a PSSA for which special mandatory measures to preserve and protect the marine environment are required.[55] Australia's

[54] Beckman, *supra* note 5, at p. 344.
[55] LOS Convention, *supra* note 2, at Art. 211(6)(a).

scheme is not a direct application of compulsory pilotage to a strait. It is a necessary and proportionate measure to protect an approved PSSA. Any other country or countries seeking to use the Torres Strait precedent would first have to go through the step of having the strait, or the waters of which it is a part, approved as a PSSA by the IMO.

Environmental Norms

The LOS Convention was formulated in a period when there was less concern for the health of the marine environment than there is at present and modern international environmental law was underdeveloped. Norms and principles for the preservation and protection of the marine environment have multiplied exponentially over the last two decades. Many imperatives for change have been associated with increased concern about threats to the marine environment posed by ships and shipping operations arising as a result of highly publicized marine accidents.[56] The navigational regimes in the LOS Convention provide an example of the relatively lower level of concern for the marine environment than was current in the 1970s. The regimes of straits transit passage and archipelagic sea lanes (ASL) passage apply to "all ships and aircraft," and there is no clear right of the coastal or archipelagic State to prevent the passage of a vessel that might be perceived to be a serious threat to the marine environment.

Chircop had addressed the issue in the context of a ship in distress seeking a place of refuge in a strait used for international navigation concluding that:

> While respecting the intention of the UNCLOS III negotiators to protect freedom of navigation through straits, one should be wary of applying too restrictive an interpretation that might not permit the coastal state to intervene to prevent a casualty from harming vital interests. It is possible to argue that international straits are not exempted from the right of protection of the coastal state under general international law and the precautionary principle under international law apply.[57]

---

[56] Bateman, et al., *supra* note 51, at p. 317.

[57] A. Chircop, "Law of the Sea and International Environmental Law Considerations for Places of Refuge for Ships in Need of Assistance," in A. Chircop and O. Linden, eds, *Places of Refuge for Ships – Emerging Environmental Concerns of a Maritime Custom* (Leiden: Martinus Nijhoff Publishers, 2006), p. 246.

This point could be a response to the argument that Australia, by introducing compulsory pilotage in the Torres Strait, has exceeded its jurisdiction allowed by Article 34(2) of the LOS Convention.[58] That article states that the jurisdiction of States bordering the straits is exercised subject to Part III of the LOS Convention and other rules of international law. The latter could include the principle of protection and the precautionary principle.[59]

All State parties to the LOS Convention have a general obligation to protect and preserve the marine environment.[60] However, since the LOS Convention was drafted nearly thirty years ago, the major developments with the law of the sea have arisen from increased awareness of the importance of fulfilling this obligation. This is reflected in the "steady succession of new global instruments with an oceanic orientation."[61] The prevention of accidental ship-sourced marine pollution has been the focus of many of the new instruments. Much has been done to achieve this goal but as the latest edition of the *Gard Handbook on Protection of the Marine Environment* notes:

> ...there is no room for complacency. Due to the regularity of maritime accidents involving marine pollution, the UN General Assembly reiterated its concern about the degradation of the marine environment from ships once again in 2000. Although the interrelationship of "safer ships and cleaner oceans" is now better understood and widely accepted, there will now have to be better attention given to the weak links in the system: insufficient attention on prevention and human error. Almost all accidents are caused by one or both of these.[62]

There are now many examples of how this increased concern has impacted on the law of the sea, including the freedoms of navigation. The *Oil Pollution Act* 1990 (OPA90)[63] introduced in the United States following the *Exxon Valdez* disaster introduces several restrictions on the freedoms of navigation in the EEZ, including the requirement for double hulls for all

---

[58] Beckman, *supra* note 5, at p. 343.

[59] Chircop, *supra* note 57, footnote 76, at p. 246.

[60] LOS Convention, *supra* note 2, at Art. 192.

[61] Johnston, *supra* note 6, at p. 378.

[62] E. Gold, *Gard Handbook on Protection of the Marine Environment,* 3rd ed. (Arendal: Gard AS, 2006), p. 155.

[63] 33 USC §2701–2761 (1990).

new tank vessels and ocean-going barges.[64] Both Indonesia and Singapore backed Malaysia's insistence that Japanese plutonium shipments should not be routed through the Malacca Strait for fear of the environmental risks involved.[65] The concern is also evident in the Revised Guidelines for the Identification and Designation of PSSAs adopted by the IMO on 1 December 2005.[66] These guidelines acknowledge that with the increase in global trade, shipping activities are also increasing with greater potential for adverse effects and damage to the marine environment. The guidelines are far more detailed and "liberal" in their approach than Article 211(6) of the LOS Convention reflecting the more sophisticated and comprehensive scientific understanding of the dangers posed by ships to the marine environment than was the case when the LOS Convention was being negotiated.[67]

Political Factors

Domestic Issues

Australia has been very active over the years in promoting measures to preserve and protect the marine environment, particularly from ship-source marine pollution, and to conserve marine living resources. For example, Australia has taken a strong stand at the IMO with developing international instruments to prevent the introduction of foreign organisms through discharge of ballast water and to provide compensation for oil-spill damage from ships other than oil tankers. Reasons for the strong Australian position on the marine environment are not hard to find.

First, Australia has a very large EEZ for which there is a clear obligation and duty to preserve and protect the marine environment. Furthermore, the Australian marine environment is relatively pristine and includes six

[64] Gold, *supra* note 62, at p. 357.

[65] J. M. Van Dyke, "Sea Shipment of Japanese Plutonium under International Law," *Ocean Development and International Law* 24 (1993), pp. 399–403.

[66] Revised Guidelines, *supra* note 4.

[67] *West European Particularly Sensitive Sea Area (PSSA) – Comments made by the Division for Ocean Affairs and the Law of the Sea of the United Nations (DOALAS) in connection with issues raised in document LEG 87/16/1*, IMO Doc. LEG 87/16/1, Annex 7 (23 October 2003).

marine areas currently on the World Cultural and National Heritage List – the Great Barrier Reef, the Lord Howe Island Group, Shark Bay, Fraser Island, Heard and McDonald Islands, and Macquarie Island.[68] Two of the eleven PSSAs that have been designated to date around the world lie within Australia's EEZ. Australia takes its responsibilities towards the marine environment very strongly, and the threat to the Torres Strait environment posed by a large vessel navigating in the area unpiloted was unacceptably high for the Australian government.

Secondly, Australia is a major global shipping nation, ranking high in terms of seaborne trade measured in ton-miles. This is because Australia's exports are mainly bulk ores and grain. Thus, Australia has a high demand for the services of bulk carriers, which are generally regarded as the class of ship posing relatively higher risks to the marine environment than other classes of vessel. Dry bulk carriers figure prominently in marine accident statistics and are more likely to operate with sub-standard maintenance and crews. They are "low-tech" ships often with low value cargoes. Bulk carriers figured prominently in the concerns of inquiries, such as the Donaldson Report in the United Kingdom in 1994,[69] and the earlier *Ships of Shame* report in Australia[70] (Australian Government, 1992) that addressed problems of sub-standard shipping. Another serious threat to the marine environment is posed by the tankers carrying large quantities of crude oil and petroleum products around the Australian coast.

Lastly and perhaps due to its "beach culture," the Australian community has a strongly developed consciousness of the importance of a clean marine environment. Shipping accidents in Australian waters leading to actual or potential marine pollution attract considerable media attention and public uproar. Huge public outrage would result from a major shipping accident in the Torres Strait, and the Australian government would be severely criticized if it could be accused of not having taken all necessary precautions to prevent such an accident. There have been periodic calls over the years for shipping traffic to be denied access to the Inner Route

---

[68] See World Heritage List, <whc.unesco.org/pg.cfm?cid=31>, 4 March 2008.

[69] United Kingdom Government, *Safer Ships, Cleaner Seas* (The Donaldson Report), CMD 2569 (London: Her Majesty's Stationery Office, 1994).

[70] Australian Government, House of Representatives Standing Committee on Transport, Communications and Infrastructure, *Ships of Shame, Inquiry into Ship Safety* (Canberra: Australian Government Publishing Service, 1992).

of the Great Barrier Reef.[71] It is not surprising that Australia has a long history of taking strong measures in relation to pilotage in the waters of the Great Barrier Reef and Torres Strait.

In an example of the type of accident that could occur in the Torres Strait and the public uproar that would result, in 2000, the 22,000 tonne Malaysian-registered container ship MV *Bunga Teratai Satu* ran aground in the marine park on Sudbury Reef off Cairns.[72] It was on the Inner Route of the Great Barrier Reef but outside the compulsory piloted areas when the accident occurred. The ship carried 1,200 tonnes of fuel oil. It was eventually refloated without loss of fuel or cargo but about 1,500 square metres of the reef was destroyed by the impact of the vessel, and the grounding also resulted in contamination from toxic tributyltin (TBT), a component of the ship's anti-foulant paint. The ship owners were fined AU$400,000, after previously having agreed to pay clean-up costs of AU$1 million. The incident generated considerable public interest,[73] and led to the *Great Barrier Reef Marine Park Amendment Act 2001* that redefined the compulsory pilotage area in the Great Barrier Reef Marine Park.

A recent high profile accident in Australian waters involved the large bulk carrier *Pasha Bulker* which went aground on Nobby's Beach near the port of Newcastle on 8 June 2007 during a severe storm. This was a very public event as it occurred on a popular surfing beach in the inner suburbs of a major city. Fortunately the vessel was later refloated without causing any major environmental damage, but initial inquiry results reveal that the accident was a consequence of human errors with the ship's master having made several poor decisions.[74]

---

71 BBC News, "No bans on Barrier Reef ships," 6 November 2000, <news.bbc.co.uk/1/hi/world/asia-pacific/1009833.stm>, 4 March 2008.

72 The information in this paragraph is from Parliament of Australia, Parliamentary Library, *Great Barrier Reef Marine Park Amendment Bill 2001*, Bills Digest No. 128 2000–01 (17 May 2001), <www.aph.gov.au/library/Pubs/bd/2000-01/01BD128.htm>, 4 March 2008.

73 Australian Government, Department of Transport and Regional Services, *Annual Report 2000–2001,* Australian Transport Safety Bureau, <www.dotars.gov.au/department/annual_report/2000_2001/P3-3.aspx#_2>, 13 July 2007.

74 "Pasha ship master blamed but no charges," *National Nine News,* 5 December 2007, <news.ninemsn.com.au/article.aspx?id=272035>, 11 December 2007.

International Issues

Beckman has speculated on the possibility that one or more maritime States might challenge Australia's regulations on compulsory pilotage before an international court or arbitral tribunal.[75] However, he rightly calls into question whether any State would believe that the potential benefits of a legal challenge outweigh the costs. The United States is the principal critic of compulsory pilotage in the Torres Strait but is unable to challenge Australia in an international court or tribunal while not a party to the LOS Convention. The United States may choose to support a challenge by another State that is a party to the LOS Convention. Singapore could be a likely contender in this regard. Singapore has also been a leading critic of compulsory pilotage, but it has a strong bilateral relationship with Australia from which it gains a lot, including extensive access to defence training facilities in Australia. It may be reluctant to jeopardize any aspect of that relationship for what, at best, is a matter of principle.

Despite Beckman's conclusion that the chances are good that a challenge would succeed, there is no certainty of this outcome. It would be a messy challenge with environmental issues looming large. The environmental credentials of the parties may well be called into question. In this regard, Singapore as a challenger would fare badly. Singapore has no EEZ of its own and has been frequently criticized by its neighbours, Malaysia and Indonesia, for lack of marine environmental awareness, most notably on the issues of land reclamation and sand mining. When it comes to a balance between the protection of the marine environment and the freedom of navigation, Singapore comes down on the side of the latter.[76] Singapore claims to have encouraged ships flying its flag to conform to the compulsory pilotage regime in the Torres Strait,[77] but was not prepared to take any further action to ensure that ships did conform.

In stark contrast with Singapore, Australia has an impeccable record with regard to the preservation and protection of the marine environment. It has taken a leading role at the IMO on issues such as ballast

[75] Beckman, *supra* note 5, at pp. 348–349.

[76] M. Hand, "Singapore urges balance in green issues," *Lloyd's List*, 22 November 2007.

[77] Chan Beng Seng, Director (Sea Transport), Ministry of Transport, Singapore, "Compulsory Pilotage in Torres Strait – Aussie view of UNCLOS, IMO positions incorrect," *Straits Times Forum*, 28 December 2006.

water management and liability for pollution damage from bunker spills. Australia has also been prominent in moves over the years to develop the international regimes for PSSAs and marine protected areas (MPAs). As well as the two PSSAs designated in littoral waters, Australia has an extensive estate of 14 MPAs that are Commonwealth reserves under the *Environment Protection and Biodiversity Conservation Act 1999* (EPBC Act),[78] as well as additional marine parks managed by state governments. The proposed South Tasman Rise Commonwealth Marine Reserve (CMR) covers 27,704 square kilometres of EEZ in the deep ocean to the southeast of Tasmania, including a section of the mid-continental slope of the South Tasman Rise in 1,200 to 3,000 metres depth. Its southern edge follows the boundary of the Australian EEZ.

During negotiations on the LOS Convention at the Third United Nations Conference on the Law of the Sea, Australia was among the maritime States that were strong supporters of the freedom of navigation. Australia is ringed from the northeast to the northwest by archipelagos and is dependent both strategically and economically upon the free movement of shipping through those archipelagos. It has been suggested that Australia, in introducing compulsory pilotage in the Torres Strait, failed to consider the strategic implications and what might be its reaction if, for example, Indonesia introduced a similar system of pilotage in one or more of its archipelagic straits.[79] However, as noted above, all straits have different characteristics and the Torres Strait is significantly more dangerous than the Indonesian straits.

Conclusion

While the legal status of the compulsory pilotage regime in the Torres Strait was not clarified at the IMO Legal Committee, there was almost equal support for the opposing views. Only a formal determination through an international dispute settlement procedure will decide whether the measure

[78] For a map of current MPAs in Australia, see Australian Government, Department of the Environment and Water Resources, <www.environment.gov.au/coasts/mpa/>, 4 March 2008.

[79] Ambassador Tommy Koh speaking at the Maritime Policy Forum held at the S. Rajaratnam School of International Studies, Singapore, 10 May 2007.

is in accordance with the international law of the sea. This might arise, for example, through a legal challenge by the flag State of a vessel against which Australia had taken enforcement action for not taking pilot through the Great North East Channel of the Torres Strait.[80]

Despite flag States encouraging ships flying their flag to take a pilot through the Strait, it is clear that the recommendatory regime for pilotage was not working, and with all other feasible vessel traffic services (VTS) arrangements in place, something more was required to preserve and protect the environment of the Torres Strait. In political terms, the domestic position in Australia with regard to compulsory pilotage through the Torres Strait is very clear. Both the Commonwealth government and the Queensland state government are strongly committed to the protection and preservation of the marine environment, and this position is equally strongly supported by the Australian public. A shipping accident in the Torres Strait leading to major environmental damage would have major political repercussions if the governments concerned had not taken all possible precautions against such an accident. There was a change of Government in Australia in November 2007, and politically, the new Australian Labor Party government will be less prepared to compromise on environmental issues than its conservative predecessor.

There may be some international criticism of Australia's position, but the Australian government will robustly defend its position. In the final analysis, Australia's position on PSSAs and associated measures, as well as on marine environmental protection more generally, are only reflective of the markedly increased international concern for the protection and preservation of the marine environment that has become apparent over the past twenty years or so. This concern was very clear in the 1998 Report by the Independent World Commission on the Oceans, entitled *The Ocean – Our Future*,[81] and continues to be evident in forums such as the World Summit on Sustainable Development in 2002 and the on-going United Nations Open-ended Informal Consultative Process on Oceans and the Law of the Sea. Greater concern for the marine environment is the most significant development in oceans management and policy that has

---

80 J. Roberts, "Compulsory Pilotage in International Straits: The Torres Strait PSSA Proposal," *Ocean Development and International Law* 37 (2006), p. 107.

81 Independent World Commission on the Oceans, *The Ocean – Our Future* (Cambridge: Cambridge University Press, 1998).

occurred over the last two decades. As a consequence of this concern, the freedoms of navigation are no longer the same as those that prevailed in the more laissez-faire environment of the 1970s when the LOS Convention was being negotiated. We see this in measures restricting freedom of navigation in various parts of the world, including in Europe and North America. This is all part of the "greening" of the law of the sea mentioned in the introduction to this essay.

Introducing compulsory pilotage in the Torres Strait was both a clear and necessary decision for Australia. The risks of a marine accident in the Strait involving an unpiloted ship were unacceptably high and there was a clear need to enhance navigational safety. The domestic political costs of not pursuing all possible measures were high for the Australian government and the international political costs, to the extent that they might outweigh the benefits, were acceptable. Lastly, sufficient doubt surrounds whether or not the measure could be justified legally. Another State may take Australia to international arbitration on the issue or seek an advisory opinion but for the reasons discussed in this essay, particularly the operational and political ones, that action would seem unlikely.

Northern Sea Route Navigation: The Last Frontier?

Edgar Gold*

Author's Note

This essay is part of a tribute to honour the memory of the late Professor Douglas Millar Johnston. Douglas Johnston was a remarkable human being, who touched many lives with his powerful intellect, extraordinary academic abilities, innovative approach to legal and other problems, and sparkling sense of humour. He was also my former teacher, colleague and friend for over 35 years. He had a profound influence on my own career from the very beginning of his appointment at Dalhousie University when I became his research associate and we created the framework for one of the most important marine law and policy institutes in the world. Over the next decades we worked, collaborated, and conspired together, not only on many Canadian projects, but also on numerous new and innovative programmes, in many other parts of the world. These are all successful milestones that not only reflect Douglas Johnston's abilities in so many ways, but will be a permanent testimonial to a man who never shrank from venturing beyond the traditional borders of law and policy.

During his long and distinguished career Douglas Johnston had been interested in and written about the various legal aspects related to the Arctic from time to time. In fact, in his last years he had become actively involved in a Canadian Arctic project and had expressed the view that Arctic development and navigation needed to be examined, especially in the Canadian context, in a much more systematic way.[1] He planned to

* Adjunct Professor, Marine and Shipping Law Unit, University of Queensland, Brisbane, Australia.

[1] See: note 68 *infra.*

do this but ran out of time. Accordingly, it is most appropriate to include this subject in a publication that honours Douglas Johnston.

Introduction

This contribution will describe and examine some of the work of two important international projects that have carried out detailed analyses of the future of Arctic navigation during the period 1993–2005. Much of this work has concentrated principally on the Northern Sea Route (NSR), i.e., the navigational path that traverses Russian Arctic waters (Figure 1). Regrettably, even though invited, Canada chose not to participate officially in these projects, although a number of Canadian researchers did. Nevertheless, many of the conclusions of these projects have important implications for Canadian Arctic navigation and development.

The starting point for the first project was the announcement by President Gorbachev of the Soviet Union in 1992 that the previously closed waters of the Russian Arctic regions were to be opened up for international navigation. Further, an international Northern Sea Route, which would provide a new navigational route from the Northern Pacific Ocean to the Arctic Ocean, should be established as soon as possible. This resulted in the first comprehensive, scientific, technical and legal examination of polar commercial navigation ever undertaken at the international level.

The six-year (1993–1999) multi-million dollar project entitled "International Northern Sea Route Programme" (INSROP) was jointly funded by three principal research institutions in Norway, Russia and Japan, and involved participation by researchers from over 20 States.[2] One of INSROP's main conclusions was that regular, international, commercial trans-Arctic navigation was limited by a number of questions related to the availability of insurance risk coverage, as well as numerous technical problems. On the other hand, it also led to the conclusion that limited NSR navigation providing access to the resources of the Russian Arctic was

[2] C. L. Ragner and R. D. Brubaker, *International Northern Sea Route Programme (1993–1999) Draft Report to the Arctic Council* (Oslo: Fridtjof Nansen Institute, 2008).

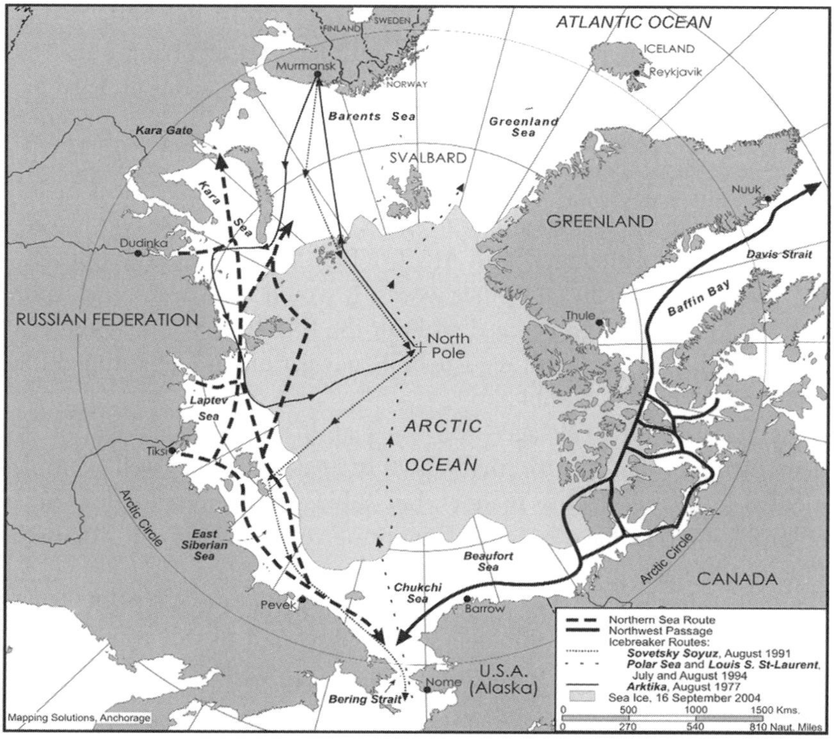

Source: Courtesy of Dr. Lawson Brigham, AMSA, Alaska.

Figure 1. Arctic navigational routes

becoming much more feasible as it could build on the extensive Russian navigational experience in the region.

The next few years were utilized in digesting, analyzing and further examining the vast knowledge base, consisting of literally hundreds of publications and other information, amassed by INSROP. By 2001, the continuing global mineral resource shortages had resulted in renewed interest in accessing Russian Arctic resources, especially by the European Union. This led directly to the second major research project, this time with significant funding from the European Union and led by a Finnish research institution, but with participation from a large conglomerate of private and public sector institutions as well as numerous international researchers. This was the Arctic Operational Platform (ARCOP) project, which concluded in late 2005.

There were distinct differences between the INSROP and ARCOP projects. INSROP was principally an academic exercise, involving numerous researchers who developed scientific, environmental, technical, legal and economic data for the NSR as a transit route, as well as an access route to Russian Arctic region resources. On the other hand, ARCOP developed a more practical data base required for actual economic investment mainly in the Russian Arctic/Barents Sea region. Nevertheless, the two projects were quite complementary, with ARCOP building on INSROP, but moving several steps further. In other words, it would have been much more difficult for ARCOP to proceed without the INSROP base that explored the possibility of extended, regular NSR navigation and determined that such navigation was feasible.

As a result, ARCOP accepted the viability of such navigation by demonstrating that it can actually be further developed as well as implemented in a cost-effective manner, but subject to overcoming a number of problems that have been and continue to be extensively discussed. One of the main problems is the provision of adequate risk coverage for the high-value vessels that are expected to operate on the NSR. These and a number of other problems will be discussed in this contribution. Finally, further ongoing research in this area, as well as the implications for Canada, will be summarized.

The International Northern Sea Route Programme (INSROP)

Background

As already indicated, INSROP was a major multi-disciplinary and multilateral research programme that examined the feasibility of the navigational use of Arctic waters from various aspects: environmental, technical, political, commercial and legal, during the period 1993–1998. The multi-funded programme was coordinated by the Fridtjof Nansen Institute (FNI) of Oslo, Norway, in cooperation with the Ship and Ocean Foundation (SOF) of Tokyo, Japan, and the Central Marine Research and Design Institute (CNIIMF) of St. Petersburg, Russia. The principal purpose of INSROP was to build a scientific knowledge base able to provide an adequate foundation for long-term planning by governments and private industry for the promotion of rational decision-making concerning the use of the Northern Sea Route for maritime transit and regional development.

INSROP's overall research programme consisted of four principal sectors: (1) natural conditions and ice navigation; (2) environmental factors; (3) NSR trade and commercial shipping aspects; and, (4) political, legal and strategic factors. These sectors consisted of approximately 70 sub-projects that were designed to examine specific aspects in considerable detail. The numerous working and discussion papers setting out the work of these sub-projects form a systematic knowledge and research base on the Northern Sea Route.[3] An important aspect of INSROP has also been its inter-disciplinary approach. It was understood at the beginning that INSROP was not simply a thorough scientific exercise, but that its practical implications would eventually lead to significant decision-making on various levels in many areas.

INSROP: NSR Navigation Risk Assessment

One of INSROP's principal research areas was an examination of the feasibility of providing the projected shipping services with adequate risk coverage. There is full agreement that NSR navigation and development could not proceed if the international marine insurance market could or would not provide the required risk coverage. Accordingly, marine insurance coverage was considered to be an essential foundation for the Northern Sea Route. Such a route would have no viability without commercial users, and users cannot operate vessels without adequate insurance coverage.

The marine insurance research group set up within INSROP realized from the start that they were dealing with uncharted territory. Not only did the ice-infested waters of the Northern Sea Route involve new marine risks, but the route also involved the Russian Federation, which had only recently emerged from its lengthy isolation and, as a result, would present private industry with a large number of unknown factors. In addition, it was also known that the marine insurance industry, although immensely practical, responsive and innovative, had little experience in reacting to challenges that would be theoretical and hypothetical. Although the industry had some limited experience with Russian operations in northern

[3] This work is reflected in some 170 substantial Working Papers and other publications. See the INSROP Website, <www.fni.no/insrop>, for further details. A summary of INSROP's conclusions was also published. See W. Østreng, ed., *The Natural and Societal Challenges of the Northern Sea Route* (Dordrecht/Boston/London: Kluwer, 1999).

waters, the idea of using the NSR for regular international transit, possibly throughout the year, was entirely new.

As a result, it was decided that this research area would have to consist not only of developing an adequate risk coverage knowledge base for INSROP itself, but also have to determine what information the insurance sector would need in order to make rational risk coverage decisions. This would require contacts with most of the major insurance markets and, at the same time, provide decision-makers in such markets with basic information about INSROP. In retrospect, the latter factor would become of critical importance for the sub-project specifically and INSROP generally.

In order to carry out its work the research group established the following research objectives:

- Northern Sea Route Overview: to gain an overview of the NSR region's natural conditions, ice navigation, environmental factors, trade and commercial shipping aspects, and political, legal and strategic factors;
- Northern Sea Route History: to examine the history, longer term statistical results, and actuarial records of vessels and cargoes operating in the region;
- Northern Sea Route Marine Insurance Interests: to identify and establish contact with hull and machinery, cargo, and protection and indemnity (P&I) underwriters and the principal marine insurance brokers who may have an interest in the projected INSROP "market"; and
- Northern Sea Route Shipping and Ancillary Interests: to identify and establish contact with shipping companies, ship construction companies, marine surveyors, shipping agents, maritime law firms and other parties who may have operational experience and/or an interest in the NSR;
- Northern Sea Route's Initial Knowledge Base: to develop an initial knowledge base, drawn from the above work, to provide INSROP generally, and the various marine insurance, shipping and ancillary interests specifically, with an initial feasibility study on the provision of risk coverage.

INSROP Tokyo Symposium 1995

The culmination of the first stage of project research was the first INSROP Symposium, held in Tokyo in October 1995. The meeting, coordinated by the Ship and Ocean Foundation, brought the bulk of the various INSROP research groups together for the first time. This was probably the very first international meeting to discuss NSR navigation in an organized, comprehensive way. As a result, many of the Tokyo conclusions are as valid in 2008 as they were in 1995. The five-day meeting attracted researchers from the three principal INSROP supporting States: Japan, Norway and Russia, as well as researchers from Finland, Canada, Germany, the United Kingdom, and the United States. Some 60 papers were presented at sessions divided into: Natural Conditions and Ice Navigation; Environmental Factors; Trade and Commercial Shipping Aspects; and Political and Legal Aspects. This research reflected the work carried out by INSROP during its first two years and consisted of the first systematic compilation of a knowledge base on the subject.[4]

The Symposium was also presented with the initial conclusions of the first INSROP experimental Northern Seas Route voyage, planned and carried out by the Japanese INSROP research group, with some assistance from other researchers, in the summer of 1995.[5] This voyage took a Russian-flag, 15,000 DWT,[6] SA-15 type, ice-strengthened cargo vessel on a commercial voyage along the NSR from Yokohama to northern Norway in 21 days, including the time spent on ice and other scientific experiments. Although the voyage was principally undertaken for a large number of scientific experiments, it also provided evidence that the NSR was navigable without icebreaker assistance in the summer season and that the route would significantly reduce the normal Far East-to-Europe transit time.

Other papers presented at the Tokyo Symposium concentrated heavily on scientific aspects related to ship design for ice navigation, ice-breaking criteria and ice conditions. It was concluded initially that the more severe ice conditions will continue to require icebreaker assistance and that, even in the summer season, vessels without ice-strengthened hulls are at a greater risk

---

[4] INSROP, *INSROP Symposium Tokyo '95: Proceedings* (Tokyo/Oslo: SOF/INSROP, 1995) [hereinafter *Tokyo Symposium Proceedings*].

[5] H. Yamaguchi, "NSR Sea Trial Test," *ibid.*

[6] Deadweight Tons.

than strengthened vessels.[7] The Symposium's environmental sessions provided significant evidence that the Arctic environment is especially vulnerable to oil and chemical spills and that environmental factors, conceptual design and environmental impact assessment must underlie navigation in the region.[8] Russian researchers indicated that the ecological safety of navigation, as well as traditional national security, on the Northern Sea Route was being given a very high profile by the Russian government. This was also further confirmed in another, later INSROP study emphasizing the importance of international environmental security that would not permit Western operations in Russia to sidestep strict environmental requirements.[9] This confirmed the commercial and insurance implications already referred to above. It is certain that some new or special arrangements for environmental liabilities would have to be made if the NSR is to be used regularly by vessels with pollution potential.

Trade and commercial shipping aspects were also discussed at the Tokyo Symposium. However, the discussion here was more fragmented. One direction taken was the Russian interest in the development of the Russian "North" that had been lagging and, in many cases, drastically reduced since the collapse of the Soviet Union. The thrust was for foreign investment that would access the abundant natural resources in the region and, at the same time, lead to regional, infrastructural development.[10] Evidence of this development was provided by presentations from a Finnish group that had been engaged in oil and gas transportation from the region for the previous three years.[11] As will be seen below, this Finnish experience undoubtedly became a valuable precedent for later NSR research as well as actual commercial development and investment and trade in the area. Nevertheless, the

---

7 "Technical Sessions Theme I and Theme I-2," in *Tokyo Symposium Proceedings*, *supra* note 4.

8 "Technical Session Theme II," *ibid.*

9 A. Y. Roginko, *The NSR in the Context of Arctic Military and Ecological (Environmental) Security*. INSROP Working Paper No. 13–1995, IV.2.1 (Oslo: FNI/INSROP, 1995).

10 A. Granberg, "The Significance of the NSR for Regional Development in Arctic Areas of Russia," in *Tokyo Symposium Proceedings*, *supra* note 4; S. Hoifodt, "The Northern Sea Route and the Regional Development in the North," in *Tokyo Symposium Proceedings*, *supra* note 4.

11 M. Niini, "Experiences of Three Years of Oil Transportation in the Russian Arctic with a Western Fleet," *ibid.*; and, A. Backlund, "Seaborne Oil and Gas Transportation from Northern Russia," *ibid.*

Finnish presentation confirmed that the initial expenditures were significantly in excess of profits and that any investment would have to be long term. It was also emphasized that marine insurance costs were very high and that Russian cooperation was variable. For example, Russian icebreaker assistance involved extremely high charges, far beyond cost recovery.

On the other hand, the Tokyo meeting heard relatively little on the expected use of the Northern Sea Route by commercial operators. The importance of the Route was re-emphasized by Dr. Willy Østreng, the then Director of the Oslo-based FNI, one of the principal INSROP organizations, in his opening address.[12] He reiterated that the attraction of the Northern Sea Route had two aspects: firstly, the fact that 80 per cent of global industrial production takes place north of 30 degrees North latitude, and 70 per cent of all major global metropolitan areas are located north of the Tropic of Cancer; and secondly, that there is an obvious distance advantage in the use of the NSR between Pacific and Atlantic ports, when compared to the Suez and Panama Canals. On the latter point, the Northern Sea Route distance from Yokohama to Hamburg is about 6,600 nautical miles, as against 11,400 nautical miles via Suez. This could lead to a freight rate saving of some 42 per cent. Several other examples of NSR use illustrated its potential as a viable alternative which, in the summer season, could reduce the Japan to Europe passage by 10 to 15 days. Yet,

> ...shipping companies the world over have shown scant interest in the distance savings offered by the NSR. More important to them is whether reduced distances can be translated into reduced carrying times, thereby lowering costs on a year-round basis. What Soviet and Russian vessels have demonstrated to date is that reduced distances can be translated into reduced carrying time during certain periods of the year. What remains to be shown is whether it is possible to achieve reduced freight times all year round at costs lower than those involved in using the existing sea routes.[13]

At the Tokyo Symposium only one paper attempted to address this issue and then only in a rather theoretical fashion.[14] Furthermore, the Head of Research of the Norwegian Shipowners' Association, one of the few shipping industry representatives present at the Symposium, stated that the shipping industry was not yet convinced that the NSR was viable. He

---

[12] W. Østreng, "Introduction to INSROP," *ibid.*

[13] *Ibid.*

[14] J. Schwarz, "Can the Northern Sea Route be Profitable?" *ibid.*

pointed out that a traditionally conservative industry would not abandon its well-established routeing practices unless there were very clearly measurable advantages that would reduce cost as well as economic and other risks. The same address also suggested that if the NSR was only viable for a certain part of the year, ship operators would be faced with having to reposition their vessels at least twice a year. Although this could be done, it was not the normal practice and would have to be economically feasible.[15]

In the Tokyo Symposium closing session some other problems facing potential shipping operations were pointed out. These included:

- The uncertainty of Russian icebreaker charges
- The lack of reliable radio and other communications systems
- The lack of reliable Russian weather and ice reports
- The uncertainty of Russian legislation relating to the use of the route
- The ageing Russian Arctic fleet available to carry out some of the transportation requirements

Further INSROP Research

Several later INSROP working papers addressed some of the commercial and economic information necessary to form the basis of any NSR viability. One study that examined the potential cargoes from and to Europe via the NSR concluded that most shippers are totally unaware of the NSR alternative and base much of their scepticism on lack of information.[16] Another research paper actually made some economic forecasts of the potential of the NSR for regular cargo service, but concluded that only the right types of cargoes and suitable vessels could achieve economic returns equal to what may be possible on other routes during the season when minimal icebreaker support was required. However, the paper also concluded that the NSR could be a viable alternative to the Suez Canal for commercial trades where cost is more critical than fixed schedule.[17]

---

[15] N. Telle, "Discussion Period," *ibid.*

[16] Ø. Andersen, T. J. Heggeli and T. Wergeland, *Assessment of Potential Cargo from and to Europe via the NSR*. INSROP Working Paper No. 11–1995, III.10.1 (Oslo: FNI/INSROP, 1996).

[17] B. Buchan, *The Potential of the Northern Sea Route for a Regular Cargo Service*. INSROP Working Paper No. 15–1995, III.07.2 (Oslo: FNI/INSROP, 1996).

INSROP's NSR risk assessment research, especially in the London insurance market, as well as in Russia, indicated an increased awareness that full disclosure of relevant Russian NSR information would be absolutely essential.[18] In response, Russian researchers indicated that experience and statistics in this area were being compiled and updated constantly.[19] In addition, Russian shipping has a lengthy relationship with London and Norwegian marine underwriters as well as P&I insurers. Furthermore, it was already apparent during the INSROP project that the rapidly expanding Russian private commercial and shipping sector was not only developing increasingly "Western" commercial approaches but, at the same time, inducing the former State shipping and insurance companies to join in this process through healthy competition. For example, the Murmansk Shipping Company has 55 years of experience in the Arctic and operates a 50-vessel fleet, including several nuclear-powered and numerous standard icebreakers.

In its final stages INSROP further expanded its information base. After almost six years of concentrated, interdisciplinary research, INSROP had established the most advanced Northern Sea Route knowledge base ever assembled. Although very little data was available when INSROP commenced in 1993, this had significantly changed in all four major research areas: Natural Conditions and Ice Navigation; Environmental Factors; Trade and Commercial Shipping Aspects; and, Political and Legal Aspects, when the work ceased late in 1999. In addition to the interdisciplinary nature of the research carried out, a considerable amount of important data was assembled through INSROP's multinational approach. Russian data, previously unknown, unobtainable or restricted, had generally become available, and Russian researchers and experts were extensively involved in all aspects of the project. Russian input and interest was, of course,

---

18 S. Loset, and S. Vefsnomo, *Content of Database, Planning and Risk Assessment.* INSROP Working Paper No. 5–1994, I.5.1 (Oslo: FNI/INSROP, 1995); E. Gold, J. A. Cantello and P. L. Wright, *Marine Insurance for the Northern Sea Route: Towards a New Risk Regime.* INSROP Working Paper No.46–1994, IV.3.3 (Oslo: FNI/INSROP, 1995); E. Gold, J. A. Cantello, and P. L. Wright, *Marine Insurance for the Northern Sea Route: The Feasibility of a New Risk Regime.* INSROP Working Paper No. 85–1997, IV.3.3 (Oslo: FNI/INSROP, 1998); E. Gold, J. A. Cantello, and P. L. Wright, *Shipping and Marine Insurance on the Northern Sea Route: Conclusions 1993-1998.* INSROP Working Paper No. 124–1998, IV.3.3 (Oslo: FNI/INSROP, 1999).

19 M. Gavrilo and B. Sirenko, *Initial Survey of Russian Data Sources.* INSROP Working Paper No. 9–1995, II.1 (Oslo: FNI/INSROP, 1996).

not surprising as Russia has the most direct interest in the success and viability of the NSR. Not only would Russian vessels, icebreakers, crews and pilots be centrally involved, but Russian resources, ports, services and infrastructure specifically, and the Russian Arctic region generally, would be direct beneficiaries.

INSROP Conclusions and Results

It could be expected that in a complex research project of this magnitude there would be a number of problems regarding data development. The four research areas developed their data acquisition somewhat unevenly, reflecting their different disciplinary cultures. For example, a significant part of INSROP's research data has been assembled on important aspects related to ship design for ice navigation, ice-breaking criteria and ice conditions. Much of this type of data was acquired through traditional, systematic scientific and technical research. It was concluded that in more severe ice conditions, vessels navigating the NSR will continue to require icebreaker assistance. Furthermore, even in the summer season, vessels without ice-strengthening would be at greatest risk. In general, the data supports the construction of vessels specially built or strengthened for ice navigation. This has obvious implications for potential NSR users.[20]

INSROP's environmental research provided a variety of data further confirming that the Arctic's natural environment is significantly more vulnerable to oil and chemical spills than most other marine areas.[21] As a result, it was shown that environmental factors, conceptual design and environmental impact assessment must be a prime consideration for navigating the NSR. Useful data on particularly vulnerable areas, fauna and flora, and other aspects was also assembled. It was concluded that, despite the very severe land-based and inshore pollution in the Russian Arctic, the ecological safety of the region was being given a very high

---

[20] L. M. Brigham, V. D. Grishenko and K. Kamizaki, "The Natural Environment, Ice Navigation and Ship Technology," in Østreng, *supra* note 3, at Ch. 2.

[21] D. Vidas, ed., *Protecting the Polar Marine Environment* (Cambridge: Cambridge University Press, 2000); D. Brubaker, *Environmental Protection of Arctic Waters – Specific Focus: The Russian Northern Sea Route,* University of Stockholm, Unpublished Doctoral Dissertation, 2002; K. A. Moe and G. N. Semanov, "Environmental Assessments," in Østreng, *supra* note 3, at Ch. 3.

priority by the Russian central and regional governments.[22] Research also concluded that navigation in the region will face environmental regulations as strict, or even stricter, than anywhere in the world. At the same time, INSROP's environmental data provides a very valuable basis for risk coverage, marine spill control, the identification of vulnerable areas, and environmental administration for the future. Undoubtedly, continuing and additional scientific research on the subject in Russia and elsewhere will further strengthen this knowledge base.

As already indicated, INSROP's research on commercial aspects of NSR utilization was somewhat fragmented. There were three reasons for this. First, the collapse of the Soviet Union's central planning system had a particularly disastrous effect on the Russian Arctic regions that lost their guaranteed State support and investment. Continuing regional Russian economic difficulties further exacerbated this decline as the free-market system was only slowly implemented. As a result, commercial development in the Arctic regions was almost at a standstill and the formidable Russian NSR fleet of cargo ships, icebreakers and other support vessels was virtually idle. Ports and other ancillary services were also steadily deteriorating. Second, although Russia openly welcomes and encourages foreign investment, joint ventures and other business and industrial activities for the Russian Arctic, these are developing slowly. As already referred to above, two Finnish projects, involving oil and gas exports from the region, have been in operation for some years. However, even these projects had difficulties in becoming profitable. Contacts with the Japanese energy industry revealed very little appetite for development and investment in the region at this stage. Finally, as indicated above already, during INSROP's work period there was very little interest in utilizing the NSR by international shipping and cargo interests. Although INSROP data reveals the commercial viability and other advantages of the NSR, this information has not, at least at this stage, captured the full interest of the shipping industry.[23]

---

[22] See, for example, Y. M. Ivanov, A. P. Ushakov and A. N. Yakovlev, *Russian Administration of the Northern Sea Route – Central or Regional?* INSROP Working Paper No. 106–1998, IV.2.5 (Oslo: FNI/INSROP, 1998).

[23] M. Tamvakis, A. G. Granberg and E. Gold, "Economy and Commercial Viability," in Østreng, *supra* note 3, at Ch. 4.

INSROP's fourth major research area on political and legal aspects has provided a rich variety of data ranging from geo-political to sociological aspects.[24] It was generally concluded that some confusion still exists in terms of administrative responsibilities between central and regional governments that could cause difficulties for those using the Route.[25] It was also shown that, in terms of international law, some problems still persist regarding maritime boundaries[26] and international straits.[27] A significant part of the data revealed concerns for the protection of the rights of native and indigenous peoples in the Russian Arctic region.[28] Much of this work broke new ground and revealed an aspect that was not considered under Russia's former regime. It was shown that, although the development of the NSR will be accompanied by much needed regional economic development,[29] the social, cultural and environmental impact on the indigenous peoples must be fully taken into account.[30]

In November 1999, INSROP presented its findings to a Northern Sea Route User Conference in Oslo, Norway.[31] As the title of the Conference

[24] W. Østreng, et al., "Military, Political, Legal and Human Affairs," in Østreng, *supra* note 3, at Ch. 5.

[25] A. L. Kolodkin, O. V. Kulustikova and E. M. Mokhova, *Matters of Responsibility for Marine Pollution under the Legislation of the Russian Federation.* INSROP Working Paper No. 88–1997, IV.3.1 (Oslo: FNI/INSROP, 1997).

[26] R. D. Brubaker, *The Legal Status of Russian Baselines in the Arctic.* INSROP Working Paper No. 37–1996, IV.3.1 (Oslo: FNI/INSROP, 1997).

[27] R. D. Brubaker, *Jurisdiction Governing the Straits in Russian Arctic Waters.* INSROP Working Paper No. 52–1996, IV.3.1 (Oslo: FNI/INSROP, 1996); R. D. Brubaker, *The Legal Status of Straits in the Russian Arctic.* INSROP Working Paper No. 57–1997, IV.3.1 (Oslo: FNI/INSROP, 1997).

[28] G. Osherenko, et al., *The Northern Sea Route and Native Peoples: Lessons from the 20th for the 21st Century.* INSROP Working Paper No. 93–1997, IV.4.1 (Oslo: FNI/INSROP, 1998).

[29] A. G. Granberg, *The Significance of the NSR for Regional Development in Arctic Areas of Russia.* INSROP Working Paper No. 19–1995, IV.4.1 (Oslo: FNI/INSROP, 1996).

[30] D. G. Anderson, *Northern Sea Route Social Impact Assessment: Indigenous Peoples and Development in the Lower Yenisei Valley.* INSROP Working Paper No. 18–1995, IV.4.1 (Oslo: FNI/INSROP, 1996); S. I. Boyakova, et al., *Influence of the Northern Sea Route on Social and Cultural Development on Indigenous Peoples of the Arctic Zone of the Sakha Republic (Yakutia).* INSROP Working Paper No. 49–1996, IV.4.1 (Oslo: FNI/INSROP, 1997).

[31] C. L. Ragner, ed., *The 21st Century – Turning Point for the Northern Sea Route? Proceedings of the Northern Sea Route User Conference, Oslo, 18–20 November 1999* (Dordrecht/Boston/London: Kluwer, 2000).

implies, this meeting provided an opportunity to present INSROP's work to those who actually may be interested in the development of the NSR and the ancillary regions. In addition to presentations from senior representatives of the Russian, Norwegian and Japanese governments, as well as the three sponsoring research institutions, there was strong attendance from and participation by almost all sectors that may have an interest in the Northern Sea Route. During the Conference's sessions it became quite clear that the INSROP project had demonstrated that the NSR was open for transit and for business.

During INSROP's final stages, the collected project data was consolidated in a geographical information system (GIS). The GIS consists of an extensive data base for the whole NSR area providing data for analytical, statistical and presentational purposes. It contains data from all INSROP sub-programmes, including all available information on climate, ice conditions, shipping lanes, indigenous peoples, wildlife, ports, economy and infrastructure, commercial aspects, and legal, administrative and jurisdictional aspects. In addition, the data collected separately by the Norwegian, Russian and other international partners has been merged into a single data base that is available on-line for external users.[32]

The Arctic Operational Platform Project

Background

In the Tokyo Symposium keynote address cited above,[33] FINI director and INSROP coordinator, Dr. Østreng, referred to the NSR's access to the rich resources of northern Russia. He emphasized that this access would not only provide a new route to world markets for these products but, at the same time, develop an important Russian region. An INSROP research project also examined the current and potential oil product export from

---

32 Using the "ArcView for Windows" software. Users are able to obtain documentation, charts and maps, query and select data in an interactive manner, analyze data, and print the results. See S. M. Løvås and O. W. Brude, "INSROP – GIS," in Ragner, *ibid.*, p. 261.

33 Østreng, *supra* note 12.

northwest Russia and came to some cautiously optimistic conclusions.[34] Although the importance of access to Russian resources was a consideration throughout the INSROP project, its overall potential as a basis for further NSR development would only become clear with the advent of the ARCOP project in 2003.

ARCOP was a three-year (2003–2006) research project that concentrated specifically on accessing the oil and gas resources of the Arctic regions of Russia, which are considered to hold the largest energy reserves outside the OPEC countries.[35] ARCOP was principally funded through the European Union's (E.U.) Directorate-General of Transport and Energy.[36] The project was coordinated by Aker Finnyards Inc. in Helsinki, Finland, in cooperation with the Ministry of Trade and Industry, Finland, with participation of 21 organizations from five E.U. Member States (Finland, Germany, the Netherlands, Italy, and the United Kingdom), as well researchers and research institutions in Norway and Russia.

ARCOP was designated and supported as a research and development project under the E.U.'s "Competitive and Sustainable Development" programme. It was designed to examine the development of the transport of the natural resources, particularly oil and gas, of the Russian Arctic regions and consisted of the following six work programmes (WP):

- Development of collection methods for ice information and ice forecasts in view of choosing transport routes (WP1)
- Assessment of the rules and regulations on transport by sea and of insurance and payment systems (WP2)
- Development of a united transport system for Arctic oil and gas transport (WP3)
- Development of an environmental impact assessment method and environmental hazard management system (WP4)
- Practical trial of solutions and recommendations developed during an actual transport assignment (WP5)

---

34 T. R. Ramsland, *Oil Product Export from North West Russia.* INSROP Working Paper No. 8-1995, III.01.3 (Oslo: FNI/INSROP, 1996).

35 "Background," ARCOP Website, <www.arcop.fi>. It should be noted that the Website indicated that its contents would only remain available until the end of 2007.

36 ARCOP's total budget was EUR 5.23 million, with EUR 3.02 million contributed by the European Union.

- Organization of expert meetings between industry, government authorities and technological specialists in order to direct the project, assess its results and provide final recommendations (WP6)

As already indicated above, there were significant differences between INSROP and ARCOP. The latter specifically concentrated on the practical and commercial viability of Russian Arctic resource access, development and transportation with four sharply focussed objectives:

1. As European energy supplies cannot be built on a transportation system that might lead to environmental hazards, the critical assessment of environmental safety in shipping must be considered to be an ARCOP priority.
2. As the economies of transport dictate the viability of the transport solution that has to be developed, the most viable technology-based cost efficiency for the entire transportation system has to be identified.
3. As the legal and regulatory base of sea transportation is of prime importance in terms of environmental safety and commercial efficacy, the most appropriate legislative system for Arctic marine transportation has to be identified.
4. Informed discussion, based on acceptable research data shared between the various stakeholders and other interested parties, is the best method of achieving consensus for developing Arctic marine transportation.

As already indicated, there were also some important similarities between the two projects that permitted ARCOP research to build on what had already been achieved by INSROP. This was especially apparent in the area of risk assessment, considered to be the key to NSR development generally and access to Russian resources specifically.

The ARCOP/INSROP Risk Assessment Interface

The conclusions of the INSROP marine insurance project focussed on responding to a series of specific questions in order to determine if risk coverage for NSR navigation was feasible.[37] These responses provided the

[37] D. Brubaker, "Marine Insurance Related to the Russian Northern Sea Route and Russian Barents Sea," Paper presented to the ARCOP Workshop, Helsinki, 25–27 March 2003;

interface between the two projects that existed in this area at the commencement of ARCOP.

Shipowners were reluctant to risk high-value vessels, such as large, new-generation container vessels on the Northern Sea Route. This was due to the fact that such vessels were not generally constructed for navigation in ice. In addition, the shipping industry still needed to carry out a substantive economic analysis of the overall NSR advantages.[38]

If year-round navigation on the NSR was not feasible, vessels would have to be re-positioned at least once or twice per year. This has significant cost implications even if feeder services were established. On the other hand, it was found that limited low-value bulk cargo operations might take place for most of the year, depending on ice conditions.

If extensive NSR navigation became feasible, ice-strengthened and/or purpose-built vessels would be required by marine underwriters. However, standard, non-strengthened vessels might be able to navigate in the ice-free season. Of particular interest to marine insurers were the additional navigational skills required for vessels operating in ice and also in close proximity to icebreakers, especially nuclear-powered vessels.

Although there appear to be certain time- and cost-saving advantages in using the NSR over the Panama/Suez Canal routes, the exact economic advantages were not known.[39] This was due to the lack of detailed knowledge related to costs for icebreaker assistance, ice-pilots, possible delays and deviation, cargo damage due to temperature variations, possible ice damage, and higher marine insurance costs.

In the short term, the existing Russian ice-strengthened merchant fleet was probably capable of meeting the initial international charter requirements.[40] However, this had to be seen against the background of the general economic difficulties that exist in Russia. There was little indication of fleet renewal in Russia.

see also, Gold, Cantello and Wright, *Shipping and Marine Insurance on the Northern Sea Route: Conclusions 1993–1998*, *supra* note 18.

38 R. Bergman, "What does the Shipping Industry Expect as to Administration of the Northern Sea Route?" in Ragner, *supra* note 31, at pp. 193–194.

39 E. Gold, "Summary: Transit Traffic on the Northern Sea Route: Cargo Potential and Commercial viability vs. Suez, Trans-Siberian Railway and Other Alternatives," in Ragner, *ibid.*, pp. 215–216.

40 A. P. Ushakov, "The Real Face of Northern Sea Route Shipping," in Ragner, *ibid.*, pp. 153–156.

Although single-hull tankers and bulk carriers were in operation to and from Russian Arctic region ports, it was determined that such operations would not last very long, given the increasing ship construction requirements of the IMO as well as tightening IMO, European Union and Russian environmental regulations.[41]

Questions also remained about the capability and cost of Russia's icebreaker fleet. Russia owns and operates one of the largest and most advanced ice-breaking fleets in the world, and it appears that this fleet would be capable of providing services to any foreseeable NSR navigation. On the other hand, if this fleet seeks full cost-recovery for such services, the economic advantage of the NSR may well be lost. In addition, there were some concerns about the age, maintenance standards, and safety of Russian nuclear-powered icebreakers.[42]

There appeared to be insufficient interest by cargo exporters and importers in Japan and Europe in the NSR. It appeared that other trade routes continued to provide cost-effective and reliable alternatives. There was only very limited interest by European and Japanese energy and other mineral resource interests in having access to Russian resource areas.[43] Only Finland, which had strong investments in certain Russian Arctic resource regions, appeared to have a commitment.[44]

There was only a limited commitment by the various levels of Russian government in providing the necessary, reliable services required for operating a regular shipping service to and from the Russian Arctic/Barents Sea region. This was due to the general disarray that pervaded government operations in Russia, including overlapping or inadequate jurisdictions, ineffective legislation, and a general shortage of funds.[45]

---

[41] M. Valenzuela, "Enforcing Rules against Vessel-Source Degradation of the Marine Environment: Coastal, Flag and Port State Jurisdiction," in D. Vidas and W. Østreng, eds., *Order for the Oceans at the Turn of the Century* (The Hague/London/Boston: Kluwer Law International, 1999), Ch. 35; E. Gold, *Gard Handbook on the Protection of the Marine Environment*, 3rd edition (Arendal: Gard AS, 2006), Ch. 6–8.

[42] V. Ruksha, "Russian Experience in Ensuring Regular Navigation along the Northern Sea Route," in Ragner, *supra* note 31, at pp. 185–187.

[43] J. L. Anselmo, "Marine Transport of Oil and Gas from Northwest Russia to the European Union," in Ragner, *ibid.*, pp. 133–135.

[44] J. Säävälä, "Suggestions for Future Northern Sea Route Shipping Based on a Northern Sea Route Experience," in Ragner, *ibid.*, pp. 157–161.

[45] V. Smirnov, et al., "Russian Ice Information Services in the Future," in Ragner, *ibid.*, pp. 169–176.

It was concluded that the international marine insurance market would be prepared to underwrite NSR navigational and related risks. In other words, marine underwriters voiced the firm opinion that if shipowners wished to use the NSR, marine insurers would provide the necessary risk coverage.[46]

Changes Affecting NSR Development 1998–2003

As already indicated, the findings on NSR navigation and its marine insurance coverage, when INSROP ended, provided a very helpful starting point for ARCOP. As a result, one of the principal purposes of ARCOP's research was to determine what had changed in the intervening three-year period. In order to illustrate the changed scenarios, it is first necessary to point to the major global aspects that have occurred since 1998 and that impact on the development of NSR navigation generally and its directly linked risk coverage area specifically.

A growing world economy requires ever more natural resources, including energy resources. Global crude oil prices have frequently reached more than USD 100/barrel, i.e., the highest level in history. The world's premier energy-producing region, the Gulf area, is still beset by periodic political instability. A number of other oil producing regions have or are reaching depletion. China is very rapidly becoming one of the world's largest economies and will continue to strengthen its role as a major import and export market with commensurate natural resource demands. Russian oil production has been significantly increasing, i.e., from 303 million tons in 1996 to 421 million tons in 2003, with exports increasing from 145 million tons to 230 million tons during the same period.[47]

Post-9/11 global security concerns have also focussed on the safety of energy supplies including its transportation aspects. As well, increasing global, regional and national marine environmental concerns have focussed on the shipping industry and its regulatory regimes and this will result in the phasing out of older vessels, including single-hull tankers. This is

---

[46] E. Gold, "Transiting the Northern Sea Route: Shipping and Marine Insurance Interests," in Ragner, *ibid.*, pp. 113–122.

[47] A. Moe, *Oil Transportation through the Barents Sea*. Working Paper 11, Barents Russia 2015, Project, ECON Analysis AS.

directly linked to strengthened liability and compensation regimes for marine pollution.[48]

The world shipping industry, including its ship-construction sector, has been experiencing considerable growth and significantly increased profitability to unprecedented levels. The global marine insurance market, despite regular, serious losses, continues to be strengthened by new investment interests, the broadening of its base and ever greater competitiveness. The willingness by the market to accept new risks was fully confirmed by ARCOP research.[49]

Although still beset by a number of internal, administrative and economic difficulties, the Russian Federation has, nevertheless, demonstrated a willingness and ability to become an integral part of the international commercial system through the stabilization of its legal, banking, regional and national administrative systems, as well as its insurance and shipping sectors.

ARCOP Conclusions

There were also a number of changes, more specifically related to NSR development that had become clear at the conclusion of ARCOP. There was still no indication that shipowners were willing to operate high-value vessels, such as new-generation container vessels on the NSR. Instead, Russian vessels were still chartered for specific, break-bulk or bulk cargo operations. Only certain Finnish shipowners had commenced specific NSR operations that involve the export of Russian energy and other resources from the Russian Arctic/Barents Sea region. The vessels involved were insured in the international insurance market. New vessels were in the building or design stage. However, increased purpose-built vessel construction continued to be dependent upon the development of shore infrastructure, i.e., ports and terminals. This was lagging behind due to lack of international investment.

Although there continued to be some year-round NSR operations, mainly involving Russian and Finnish vessels with Russian icebreaker assistance, most operations are seasonal, i.e., weather and ice condition

48 Gold, *supra* note 41, at Pt. IV.

49 E. Gold and P. L. Wright, *Marine Insurance Coverage for the Sea Carriage of Oil and Other Energy Materials on the Northern Sea Route: Moving from Theory to Reality*. ARCOP Project Paper WP2.4-2006 (Helsinki: ARCOP, 2006).

dependent. Furthermore, the advantage of the NSR passage over the Suez/ Panama Canal routes had become less relevant for ARCOP, as resources from the Russian Arctic/Barents Sea region will not involve NSR transit passage. On the other hand, if Japan and, possibly, the United States and China become involved, further substantive economic feasibility studies will have to be undertaken.

Russia's present Arctic cargo fleet was ageing and would soon no longer be operative. There has been relatively little new construction, although vessels that have been built have been constructed to the highest international standards. Most of the older vessels would soon not be able to meet international standards under new IMO regulations.

Although single-hull vessels are not yet precluded from operating on the NSR (or anywhere), their operational viability is strictly time limited. New IMO regulations (strongly supported by the European Union) will phase out most single-hull tankers during the period 2007–2009.[50]

The current status of the Russian icebreaker fleet was uncertain. Although several new vessels have been constructed, many others are laid up, including most of the larger nuclear-powered vessels. There appears to be insufficient funding for proper maintenance. Furthermore, it continues to be questionable whether ice-breaking should operate on a full cost recovery basis. Some ice-breaking services operate on a specifically negotiated private fee basis.[51]

There is no new evidence that training in ice navigation is offered outside Russia, Finland and, to a limited extent, Canada.[52]

At the conclusion of ARCOP, there appeared to be little new evidence that cargo exporters and importers have expressed specific interest in the use of the NSR. However, it was also obvious that should Middle East instability affect the Suez Canal route and/or if continued drought conditions or delays in upgrading affect the Panama Canal, this situation may change rapidly. On the other hand, as there was less concern with

[50] Gold, *supra* note 41, at Ch. 6.

[51] V. I. Peresypkin, "Further Development of the NSR in the Coming Decades," in Ship and Ocean Foundation, *New Era in Far East Russia & Asia* (Tokyo: SOF, 2006), pp. 29–44.

[52] V. Santos-Pedro, "Canadian Experience in Arctic Development and Regulating Shipping in Ice," in SOF, *ibid.*, pp. 231–242.

actual transit passage, this aspect may not be considered essential for the time being.

There was, however, clear evidence of renewed interest by European and North American energy and other natural resource importers in the development of Russian Arctic/Barents Sea resources. Although Finland has been the leader in this thrust, there is also investment interest from other States. Given the general, global energy situation, the major focus is on crude oil and natural gas. However, other mineral resources, such as ores and coal,[53] as well as forestry products, are also of increasing interest.[54] This has and continues to result in limited infrastructure investment in order to access these resources. It was not clear if Japan, which during the INSROP research period expressed little interest in Russian resources, has had a change in policy.[55]

Russian Arctic/Barents Sea administrative policies have been further developed. Much of this is due to a certain amount of decentralization as well as privatization in areas such as exporting companies, ports, terminals, etc. On the other hand, it was less clear if services to shipping, i.e., charting, pilotage, towage, salvage, repair facilities, electronic and other communications systems, were sufficiently developed for increasing NSR navigation. There still appeared to be severe budgetary difficulties combined with policies that seek total cost recovery for all services. This will continue to cause concerns for ship operations and, obviously, marine underwriters.

The ARCOP project not only updated and revised a considerable amount of INSROP information, but also brought NSR research into the present. It has shown that NSR navigation has become a reality, at

---

[53] G. Y. Serebrjansky and T. P. Patrakova, "Transporation of Coal from the Russian Far East via Land, Rovers, the Sea of Okhotsk and the Northern Sea Route," in SOF, *ibid.*, pp. 109–122.

[54] Finland has also developed container ships for the NSR. In December 2005, the Finnish shipbuilder Aker Finnyards launched and named the first ice-class container ship, the *Norilsk Nickel.* This vessel, equipped with diesel and electric-powered propulsion machinery, was built for MMC Norilsk Nickel for its transportation needs on the NSR. The vessel is unique in design and is capable of moving stern first in heavy ice conditions. It is also designed to operate for most of the year without icebreaker assistance. The 14,500 DWT vessel was delivered in February 2006 and underwent ice trials in real winter conditions in March 2006. The company intended to build five similar vessels by mid-2008 at a cost of EUR 70 million each. See Press Report, "Unique Ice-Class Boxship Launched," *Psi Daily Maritime Clippings*, 2005:263, 12 December 2005.

[55] This policy may have changed. See the section on JANSROP II below.

least in terms of access to resources from the Russian Arctic. Nevertheless, questions still remain whether the NSR as an important international "transit" route will gain wider acceptance. It is still too early to forecast if the NSR will become a viable alternative to the Suez and Panama Canals options, or if increasing "global warming," and the commensurate reduction of Arctic ice, will add viability to the NSR. In any case, NSR and other related Arctic research has continued elsewhere, although perhaps not quite at the INSROP/ARCOP levels.

Other Multinational and National Developments

International Maritime Organization

One of the first, new indications that NSR navigation was moving from theory to reality was the publication of *Guidelines for Ships Operating in Arctic Ice-covered Waters* by the IMO at the end of 2002.[56] These Guidelines were developed by the IMO's Maritime Safety and Marine Environment Protection Committees in recognition of the need for recommendatory provisions applicable to ships operating in Arctic ice-covered waters. The Guidelines were developed in addition to the mandatory and recommendatory provisions already contained in existing IMO instruments,[57] as well as other approved industry guidelines.[58] The IMO Arctic Guidelines are designed to set out the technical requirements for ships operating under Arctic conditions in a systematic way. The four parts of the guidelines cover:

- Part A: Construction Provisions: Structures; Subdivision and Stability; Accommodation and Escape Measures; Directional Control Systems;

[56] IMO Doc. MSC/Circ.1056-MEPC/Circ.399 of 23 December 2002.

[57] Such as the SOLAS (Safety of Life at Sea), MARPOL (Marine Pollution from Ships), STCW (Standards of Training, Certification and Watchkeeping), and LL (Load Line) Conventions, as well as the ISM (International Safety Management) Code. See Institute of Maritime Law, *Ratification of Maritime Conventions*. 4 Vol. up-dated, loose-leaf service (London: Lloyd's Press, 2001/2007).

[58] Such as the IACS (International Association of Classification Societies) *Unified Requirements for Polar Class Ships*.

Anchoring and Towing Arrangements; Main Machinery; Auxiliary Machinery Systems; Electrical Installations
- Part B: Equipment: Fire Safety; Life-Saving Appliances and Survival Arrangements; Navigational Equipment
- Part C: Operational: Operational Guidelines; Crewing; Emergency Equipment
- Part D: Environmental Protection and Damage Control

If the Guidelines are implemented by the ship construction industry, as well as ship operations, and there is already evidence that they are, they will also greatly assist in risk assessment decision-making.

Japan Northern Sea Route Project Phase II

As indicated above, the Ship and Ocean Foundation (SOF) of Japan had been one of the major sponsors[59] of and participants in INSROP. Initially Japan had been mainly interested in the NSR as a speedier navigational track from Japanese ports to Western Europe. However, with the transformation of Russia after the collapse of the Soviet Union, there was a discernible, growing interest on the part of Japan in Russian Arctic resources. As a result, the SOF completed a project entitled "Development and Operation Programme for Environmental Sustainability in East Eurasia" (JANSROP II) during the period 2003–2006.

JANSROP II concentrated its work on the eastern part of the NSR, including parts of Siberia, Far East Russia and the Sea of Okhotsk. The project's primary objective was to stimulate interest in the NSR by Asian States. This was done through the presentation of up-to-date information on regional natural resources as well as the developmental and transportation scenarios. The project was able in 2005 to create JANSROP-GIS, the world's first geographical information system containing data on the natural resources and natural environment of Far East Russia. The core concept of JANSROP II was sustainable development in harmony with the preservation of the environment in the regions adjoining the NSR. A pilot study that centred on the Sea of Okhotsk, involving Arctic and environmental specialist researchers from Japan, Russia, Norway and Canada, was also completed. The results of this and other JANSROP

[59] Through funding from the Nippon Foundation.

II work was presented at an international symposium and subsequently published in 2006.[60]

The Arctic Council

Since its establishment in 1996, the Arctic Council, which is a high-level forum for cooperation, coordination and interaction between Arctic States,[61] indigenous communities and other Arctic residents, has also become interested in Arctic navigational issues. However, this interest concentrates on almost all Arctic activities, development and related navigational aspects rather than the more specific NSR navigation that were examined by INSROP and ARCOP. In general terms, the Arctic Council was created in response to concerns about the region's vulnerability to the impacts of projected climate change as well as the increase in accessibility and marine activity due to technological development. However, it was also realized that a balance between increased environmental risk and new opportunities for social and economic development, through new investment and infrastructure that would provide improved access to goods, services and supplies to a once isolated region, would have to be achieved.

In 2004 the Arctic Council approved a major study on "Arctic Climate Impact Assessment" (ACIA) that is being conducted during the period 2005–2008 with the main task of documenting substantial observational evidence that Arctic sea ice has been undergoing significant changes in recent years. These changes would have a profound effect on Arctic navigation as they involve a steady decrease in ice cover, with larger areas of open water during the summer season, a decrease in coverage of multi-year ice in the Central Arctic Ocean and the thinning of sea ice throughout the Arctic Ocean.[62]

As a response to these environmental changes and their expected impact on the Arctic marine regions, the Arctic Council has approved the Arctic Marine Strategic Plan developed by working group on Protection of the

---

[60] Ship and Ocean Foundation, *New Era in Far East Russia & Asia* (Tokyo: SOF, 2006).

[61] The Member States of the Arctic Council are Canada, Denmark, Finland, Iceland, Norway, Russia, Sweden, and the United States.

[62] L. W. Brigham, "Future Marine Access in the Arctic Ocean," in SOF, *supra* note 51, at pp. 21–28.

Arctic Marine Environment (PAME). The Arctic Council also requested PAME to conduct a comprehensive Arctic Marine Shipping Assessment (AMSA) during the period 2005–2008, with Canada, Finland and the United States as lead States, but in collaboration with other Arctic Council working groups[63] and Member States.[64] AMSA is a circumpolar survey that attempts to capture all shipping activities, ranging from small fishing vessels to cruise ships that utilize Arctic waters. Once concluded, it will provide one of the first, comprehensive data systems on actual Arctic navigation.

Canadian Aspects

Canada is a major Arctic State with one of the largest polar coastlines in the world. As a result, successive Canadian federal governments have developed limited national Northern policies from time to time and even established ministerial responsibilities for Canadian Arctic regions and its indigenous and other inhabitants. Since the late 1960s, Canada has been at the forefront of global concerns to protect the marine environment from ship-source marine pollution.[65] Further concern for Canada in this area appeared with the successful Arctic transit of the U.S. tanker *Manhattan* in 1969. The potential of oil and gas exploration, exploitation and transportation in Canadian Arctic areas resulted in the passage of strict

---

63 Such as the Emergency, Prevention, Preparedness and Response (EPPR) Working Group and the Arctic Oil and Gas Assessment Working Group.

64 Brigham, *supra* note 62. Dr. Lawson Brigham is one of the world's leading Polar researchers and is a Director of the U.S. Arctic Research Commission, as well as Vice-Chairman of PAME, and Chairman of AMSA. Dr. Brigham also supplied up-to-date information on PAME and AMSA activities and supplied the Arctic navigational routes map included in this contribution. This assistance is most gratefully acknowledged.

65 E. Gold, "Pollution of the Sea and International Law: A Canadian Perspective," *Journal of Maritime Law and Commerce* 3 (1971): 13–44; R. M. M'Gonigle and M. W. Zacher, *Canadian Foreign Policy and the Law of the Sea* (Vancouver: University of British Columbia Press, 1977); D. VanderZwaag, "Regionalism and Arctic Marine Environmental Protection: Drifting between Blurry Boundaries and Hazy Horizons," in Vidas and Østreng, *supra* note 41, at pp. 231–248; E. Gold, A. Chircop and H. M. Kindred, *Maritime Law* (Toronto: Irwin Law, 2003), Ch. 17.

federal legislation.[66] This legislation established a pollution prevention area off all Canadian coastlines above 60° North latitude.[67]

Nevertheless, despite fairly significant federal budgetary expenditures on various aspects of Canadian Arctic regions, national interest in these regions has been quite spasmodic. Interest has been revived from time to time when oil and gas exploration and development in the region were in the news. However, even resource interest has been irregular. At other times, periodic U.S.-Canadian irritations over Arctic jurisdiction have surfaced and provided media and political interest. The Canadian scientific community, a number of federal government agencies, as well as the resource industry have kept up a certain amount of research in the region. As a result, the Canadian knowledge base on the area has steadily expanded. However, for quite some time, there has been very little overall Canadian federal government policy that has focussed sharply either on Canadian Arctic interests generally or on Arctic navigational issues specifically.[68] However, the present Canadian government has finally expressed significant interest in developing a more focussed Canadian "Northern" policy, and has moved relatively quickly to rectify previous disinterest not only by reasserting Canadian sovereignty over its Arctic regions, but implementing new "Northern" policies backed with the necessary funding.[69]

As indicated in the Introduction, it is regrettable that although a number of very competent, credible Canadian researchers were directly involved in the INSROP and ARCOP projects, no direct, official Canadian involvement took place. Several Canadian specialists involved in both projects were, in fact, members of Canadian government departments, but participated in the work in their private or unofficial capacities. Although Canada will be a direct beneficiary of the INSROP and ARCOP knowledge, it

---

[66] *Arctic Waters Pollution Prevention Act* (AWPPA), R.S.C. 1985, c. A-12. Although passed in 1970, the legislation, which has been amended a number of times, was not proclaimed until 1972.

[67] See Gold, Chircop and Kindred, *supra* note 65, at p. 681.

[68] There are a number of exceptions. See F. Griffiths, "The Northwest Passage in Transit," in Vidas and Østreng, *supra* note 41, at pp. 249–260. But see also, D. M. McRae, "Arctic Sovereignty: Loss by Dereliction," *Northern Perspectives* 22(4) (1994–1995).

[69] It is interesting to note that in his final years, Douglas Johnston had been advising the Government of British Columbia, which had become frustrated with the lack of federal government action, on "Northern" development. It would not be surprising if these efforts also assisted the latest policy developments.

would have been significantly more beneficial if Canada had been formally or officially involved. However, as already indicated above, Canadian Arctic policy priorities did not allow this to happen.

On the other hand, Canada is a Member State of the Arctic Council, as well as one of the lead States in PAME and AMSA. Although this a commendable development, there is still need for much more focussed Canadian Arctic research interest that must be linked to significant budgetary contributions. The Arctic Council itself is a rather "soft" organization attempting to achieve much with relatively little.[70] Yet this may provide an opportunity for Canada to catch up on at least a lost decade of NSR research.

Conclusion

Regular shipping in Arctic regions and use of the Russian NSR, as well as the Canadian Northwest Passage (the Canadian NSR), is inevitable. The projects outlined in this essay have created much of the knowledge base required for the commercial opening-up of one of the last regions to global development. Although the environmental vulnerability of the region is abundantly clear, modern maritime technology and a well-established international regulatory system are available to ensure that the region can be adequately protected. The maritime industry and its risk underwriting sector are prepared to initially accept shipping access to Arctic region resources and subsequently for wider NSR transit operations. The Russian political, regulatory and commercial system is on the edge of being sufficiently stable to permit access to important, globally scarce resources. Canadian Arctic resources have already been identified and are ready to be exploited. One of the positive aspects of global warming will be the reduction of ice infestation in the Arctic. When combined, all of these factors will ensure that the Arctic – the last frontier – will soon be open for regular, safe and environmentally protective navigation.

---

[70] Communication from Dr. Lawson Brigham, *supra* note 64.

New Standards for International Shipping

Consequences of Securing Merchant Shipping for Contractual Relations in the Carriage of Seaborne Trade

Hugh M. Kindred and Mary R. Brooks*

Introduction

Fear of terrorism has induced a whole new maritime security regime for the 21st century. Steps were taken to criminalize hijacking and other forms of attacks on shipping after the *Achille Lauro* affair in 1985.[1] The Convention for the Suppression of Unlawful Acts Against the Safety of Maritime Navigation (SUA Convention) was concluded in 1988.[2] Yet the promotion of safety rather than security was the paramount concern of the shipping community towards the end of the 20th century. However, the attacks upon the United States on 11 September 2001 changed that focus in the shipping world, as in so many other areas of global affairs. In rapid response to the perceived threats to maritime communications,

---

* Respectively Professor of Law and William A. Black Chair of Commerce, both of Dalhousie University, Halifax, Canada. As former colleagues of Douglas Johnston and long time admirers of his creative scholarship, intellectual leadership and generous hearted friendship, our contribution to this volume in his honour is made in grateful and happy memory of him. We also thank our former students, Sonja Mills, LL.B. 2007, and Victoria Ford, M.B.A./LL.B. 2007, for their research assistance along with the Dalhousie Law School for financial support.

1 The *Achille Lauro* was an Italian cruise ship that was hijacked by an armed group who demanded the release of a number of Palestinians imprisoned in Israel.

2 Done 10 March 1988, in force 1 March 1992. At the same time a *Protocol for the Suppression of Unlawful Acts Against the Safety of Fixed Platforms Located on the Continental Shelf* was also concluded. Both treaties were published in *I.L.M.* 27 (1988) 672 and are referenced at <www.imo.org/home.asp?topic_id=910>. Piracy has been an offence of narrow scope at customary international law for a very long time and was reiterated in the United Nations Convention on the Law of the Sea in 1982, but no other treaty addressed any other criminal activity on and against ships until the SUA Convention.

elaborate new, worldwide measures of protection for ships and ports were quickly devised and instituted.

The central piece of the modern maritime security regime is the International Ship and Port Facility Security Code (ISPS Code) adopted by the International Maritime Organization (IMO) on 12 December 2002.[3] The ISPS Code imposes measures on ships and ports that are intended to provide much greater security in maritime transportation. Its international impact is enhanced by new measures for trans-border movements instituted by other international organizations, like the World Customs Organization (WCO), the International Labour Organization (ILO) and the International Standards Organization (ISO), as well as a number of significant national initiatives, in particular the U.S. *Maritime Transportation Security Act 2002*[4] (MTSA) and the U.S. *Security and Accountability for Every Port Act 2006*[5] (SAFE Port Act).

The significance of the complex new maritime security regime for world trade is demonstrated by the proportionate importance of marine transportation as the means of delivering that trade. Although the volume of trade moved by water relative to other modes is not known, it is generally argued to be the majority in tonnage terms. As Figure 1 displays, world seaborne trade, measured in ton-miles, is dominated by the movement of crude oil, oil products and dry bulk. Merchandise trade is moved in containers and is included in the growing category of "all other," but not reported separately.[6]

---

3 IMO, *Amendments to the Annex to the International Convention for the Safety of Life at Sea, 1974* [SOLAS] made by the Conference of Contracting Governments in Resolution 1 which inserted a new Chapter XI-2 on Special Measures to Enhance Maritime Security [SOLAS Annex XI-2], which, in turn, imported the ISPS Code adopted by Resolution 2, London, 12 December 2002 (in force 1 July 2004).

4 *Maritime Transportation Security Act of 2002*, Pub. L. No. 107–295, 116 Stat. 2064 (codified at 46 USC § 2101).

5 *Security and Accountability for Every Port Act of 2006*, Pub. L. No. 109–347, 120 Stat. 1884 (codified at 6 USC § 901).

6 What is also not disaggregated is how much of this ton-mile demand is accrued to the repositioning of empty containers, which demand ton-miles to carry air. If the number of import boxes to the United States exceeds the number of export boxes by two times, as reported by U.S. Department of Transportation Maritime Administration, then the number is substantial. Anecdotal evidence suggests that the number is more than 20 per cent of all container moves.

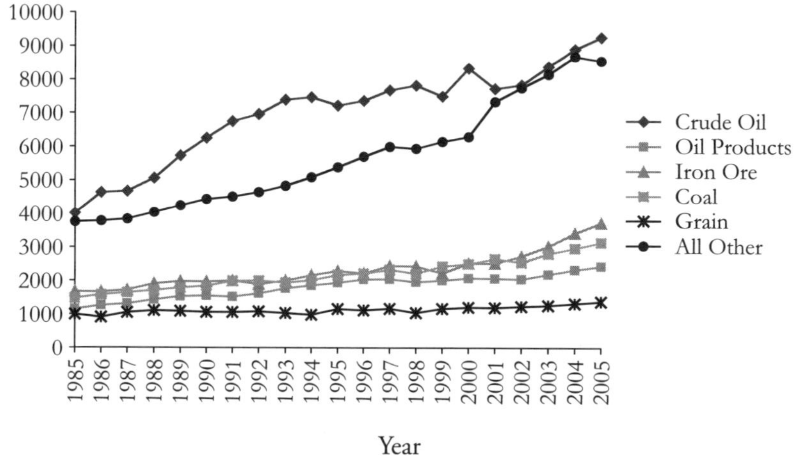

Source: Created from Table 5 of the annual UNCTAD *Review of Maritime Transport,* (Geneva: United Nations Conference on Trade and Development).

Figure 1. Global seaborne trade

The purpose of this essay is to assess the impact of the new maritime security regime on cargoes carried by sea in the global supply chains of international trade. Through a legal and commercial analysis of the regulatory provisions for maritime security and corporate responses to them, this contribution will investigate the consequences for the contractual relationships involved in the transport of seaborne trade. Although the maritime security regime is principally addressed to carriers and terminals, since the sole purpose of merchant shipping is to move the world's trade, the new regulations inevitably also affect cargo owners. Any delay or congestion arising from the application of the security regulations invariably has a financial as well as a physical impact on cargo owners, particularly those whose commercial operations are dependent upon just-in-time delivery.

Every contract between shipper and shipowner will be subject to the international and national security responsibilities laid on the carrier. Hence it is necessary to analyze their impacts at the level of individual trade transactions to determine the consequences for typical contracts for the carriage of goods by sea. In other words, this analysis will investigate the consequences of the security regime for the terms of seaborne trade (in the sense of the standard conditions of sea carriage between individual shippers and shipowners). By comparison, merchants' decisions about moving goods in trade depend upon the efficiency and reliability of the overall supply chains, of which sea carriage is but a part. Hence, it is also necessary to review the effects of the new maritime security regime on the global supply chains in order to assess the qualitative changes that have occurred in the way that seaborne trade is processed.

This essay is organized in four sections. The first section provides a very brief overview of the new maritime security regime being implemented internationally. The following two sections engage in the micro and macro level analyses of the impacts of the new regime respectively. The concluding section integrates the results of these analyses in order to expose and evaluate the overall consequences of the maritime security regime for the contractual relations between carrier and cargo interests in seaborne trade.

Overview of the New Maritime Security Regime

Internationally, the new maritime security regime is built around the ISPS Code promulgated by the IMO. It has been adopted around the world,[7] including by Canada,[8] and has been widely disseminated and discussed[9]

---

[7] Pursuant to the tacit amendment mechanism of SOLAS, *supra* note 3, Article VIII, for all State parties.

[8] SOR/2004-144 and SOR/2006-269 made pursuant to the *Marine Transportation Security Act*, S.C. 1994, c. 40 as amended.

[9] A cross section of opinion from both legal and commercial perspectives includes: R. Balkin, "The International Maritime Organization and Maritime Security," *Tulane Maritime Law Journal* 30 (2006): 1; P. Barnes and R. Oloruntoba, "Assurance of Security in Maritime Supply Chains: Conceptual Issues of Vulnerability and Crisis Management," *Journal of International Management* 11 (2005): 519; A. Goulielmos and A. Anastasakos, "Worldwide Security Measures for Shipping, Seafarers and Ports: An Impact Assessment of ISPS Code," *Disaster Prevention and Management* 14(4) (2005): 462; T. A. Mensah, "The Place of the ISPS Code in the Legal International Regime for the Security of International Shipping," *World Maritime University Journal of Maritime Affairs* 3(1) (2004): 17; A. Rodriguez, "When Your Ship Is in the Bull's Eye: The Maritime Transportation Security Act and Potential Vessel Owner Liability to Third Parties Resulting from a Terrorist Attack," *U.S.F. Maritime Law Journal* 17 (2004–2005): 241; T. Schoenbaum and J. Langstom, "An All Hands Evolution: Port Security in the Wake of September 11th," *Tulane Law Review* 77 (2002–2003): 1333; B. Soyer and R. Williams, "Potential Legal Ramifications of the International Ship and Port Facility Security (ISPS) Code

so only the briefest outline is presented here. The Code focuses on the security of ships and port facilities, imposing standards on governments, port authorities and shipping companies as well as terminal operators, through an agreed risk management framework. Three levels of security are demanded of both ships and ports. Security level 1 is the ordinary level, which must be maintained at all times. Level 2 requires additional protective measures in the face of a heightened risk of a security incident. Level 3 involves special measures of protection for a limited time when a security incident is believed to be probable and imminent. When a ship calls at a port or interfaces with another vessel, the security levels of each party are ordinarily expected to match. To achieve these security levels, all ships and ports must first make a security assessment and establish a security plan. This will involve the appointment of port, ship and company security officers, as the case may be, who bear numerous responsibilities in connection with the execution of their particular security plan, its maintenance and verification, and the security equipment it calls for as well as the training of personnel in its operation. The port or flag State, respectively, is responsible for setting security standards, approving security plans and ensuring compliance with the ISPS Code, which in the case of ships includes issuing an International Ship Security Certificate for carriage on board and production as required at foreign ports of call.

The impact of the ISPS Code is enhanced by broader criminal prescriptions that have been added to the SUA Convention.[10] It is also furthered in its application and operation by international steps towards standardization in the processing of world trade. In particular, the WCO[11] has taken several initiatives to facilitate both the efficiency and security of moving trade shipments across international borders. These measures include security standards for international supply chain management, an unique consignment referencing system and a process for granting authorized status to "economic operators" such as shippers and carriers and their agents in the international supply chain who achieve minimum security compliance

on Maritime Law," *Lloyd's Maritime and Commercial Law Quarterly* 4 (2005): 515; Paul Todd, "ISPS Clauses in Charterparties," *Journal of Business Law* (2005): 372.

10 2005 Protocol to the SUA Convention, *supra* note 2 (not yet in force). A similar 2005 protocol was added to the 1988 Protocol concerning attacks against fixed platforms: *ibid.*

11 Officially known as the Customs Co-operation Council.

standards.[12] Other significant contributions to the worldwide maritime security regime include action by the ISO to introduce standards for high security mechanical seals along with steps towards the future application of electronic seals for containers.[13] In addition, the ILO has revised and modernized the convention on seafarers' identification documents,[14] by which it established a standardized form of seafarer's credentials, including biometric data.[15]

None of these initiatives would have been undertaken without the attacks on New York and Washington on 11 September 2001, and the pressure the United States placed on its allies and trading partners. In the United States, the predecessor department to Customs and Border Protection moved to put Customs agents into three Canadian ports (Vancouver, Montreal and Halifax) even before 2001 ended, while Congress passed the MTSA[16] authorizing U.S. maritime security measures to

---

[12] WCO, Customs Co-Operation Council High Level Guidelines for Co-Operative Arrangements Between Members and Private Industry to Increase Supply Chain Security and Facilitate the Flow of International Trade, Annex VII to Doc. SP0122E1, Doc. No. TF0004E3 (2003); Customs Guidelines on Integrated Supply Chain Management; Framework of Standards to Secure and Facilitate Global Trade, (2005); Resolution of the Customs Co-Operation Council on Security and Facilitation of the International Trade Supply Chain, (2002); SAFE Framework of Standards: Authorized Economic Operator Guidelines, (2006); and Unique Consignment Reference (UCR). These guidelines were promulgated by the WCO pursuant to *The International Convention on the Simplification and Harmonization of Customs Procedures* (*as amended by Protocol*), (*Revised Kyoto Convention*), Brussels, 26 June 1999 (in force 3 February 2006), <www.wcoomd.org/kybodycontent.htm>.

[13] See ISO/PAS 17712:2006 and ISO 18185-1-5:2007 in ISO, *ISO Standards Handbook: Freight Containers*, 4th ed (2006) and <www.iso.org/iso/en/CatalogueDetailPage.CatalogueDetail?CSNUMBER=42791&ICS1=55&ICS2=180&ICS3=10>.

[14] *C185 Seafarers' Identity Documents Convention* (*Revised*), Geneva, 19 June 2003 (in force 9 February 2005), ILOLEX, <www.ilo.org/ilolex/cgi-lex/convde.pl?C185>.

[15] Two fingerprints are used to create a biometric template which is then employed to load a chip in the Seafarer's Identity Document that may be read as an international barcode: see DDCOM, "Seafarers Identity Becomes Clearer: New International Labour Convention for Seafarers' ID Documents Comes into Force," *World of Work Magazine* 53 (2005): 35 and J. Erdmann, "Biometrics Identification Rides Wave of the Future," IEC E-Tech Online News, <www.iec.ch/online_news/etech/arch_2005/etech_0205/focus.htm>.

[16] *Supra* note 4.

be taken by the newly established Department of Homeland Security, founded in June 2002.

The Department of Homeland Security quickly introduced a series of new, voluntary programmes, including the Container Security Initiative (CSI) at selected ports worldwide and the Customs-Trade Partnership Against Terrorism (C-TPAT) for U.S. cargo interests.[17] These programmes for U.S. trade have since been endorsed and enhanced by the SAFE Port Act.[18] The legislation establishes the framework for these initiatives and gives the Department of Homeland Security authority to fashion and execute them. It has chosen to exercise its powers in partnership with the business community. Since these border control programmes of the United States have international as well as national impact, they influence decisions throughout the global supply chains of U.S. trade interests, and add a significant extra dimension to the internationally formulated maritime security regime. More details on these U.S. measures may be found in Appendix 1.

Impacts on Individual Sea Carriage Contracts

International trade consists of millions of individual commercial transactions for the sale and delivery of goods. Each transaction has to be separately processed and performed, and thus is facilitated by a globally integrated supply chain. A central link in this supply chain for the largest portion of international trade is the sea transport of goods between ports in one country and another. Hence ocean carriers play a crucial

---

[17] For commentary on the U.S. container security measures see A. Roach, "Container and Port Security: A Bilateral Perspective," *International Journal of Marine & Coastal Law* 18 (2003): 341; J. Romero, "Prevention of Maritime Terrorism: The Container Security Initiative," *Chicago Journal of International Law* 4 (2003): 597. For an assessment of the execution of the U.S. initiatives see U.S. Government Accountability Office, "Maritime Security – One Year Later: A Progress Report on the SAFE Port Act," GAO-08-171T (16 October 2007). For discussion of container trade security issues generally see R. Banomyong, "The Impact of Port and Trade Security Initiatives on Maritime Supply-Chain Management," *Maritime Policy and Management* 32(1) (2005): 3; S. Noda, "Container Shipping and Security Issues: The Carrier's Responsibility in the Fight Against Terrorism," *Maritime Economics & Logistics* 6 (2004): 157.

[18] *Supra* note 5, at s. 205 & s. 211 (6 USC § 945 & § 961) respectively.

role in the execution of a trade transaction. The security and efficiency with which they achieve delivery of a shipment will affect performance of the trade transaction.

The discussion of impacts of maritime security begins with the division of interests between buyer and seller in the international contract of sale. If the delivery fails to be executed as expected by the shipper or consignee, the delay suffered will likely cause commercial loss, especially where just-in-time inventory management is practised. The consequences of that delay will depend upon the terms of the international sale contract. If the party affected by the delay is the one who chose the terms of sale (and hence allocated the risk inherent in the choice of the carrier), the impact will be less than if the party most affected is the one who did not choose the terms of trade. Naturally, the party choosing the carrier has an expectation of reliability in delivery. For example, suppose the buyer of the cargo is able to exercise its choice of the terms of trade and selects FCA (free carrier) so that it has the freedom to choose the carrier. Then, if delivery is delayed, the buyer will have to absorb the consequential loss, perhaps through a risk management plan that incorporates buffer stock relative to the expected reliability. If a serious consequential loss results, the buyer may, in future, choose another carrier in an effort to minimize this risk. On the other hand, supposing the seller of the cargo has the commercial power to choose the terms of trade and selects DDU (delivered duty unpaid), it will have control over the selection of the carrier. Even so, if delivery is delayed, the consequential loss arising from the delay will still be borne by the buyer, who may decide, in future, to choose a different cargo supplier (seller) to minimize the risk in the transaction. This becomes particularly important when delivery windows are specified in the contract of sale. The consequential loss in the case of the latter sale arrangement is greater than in the case of the former, and arises from the loss of future business as well as the breach of contract in the current transaction.

Although the relative bargaining power of the seller/exporter and the buyer/importer in agreeing to the sale contract may play into the consequences of delay in delivery, the terms of the delivery contract subsequently made with the chosen carrier will also be significant for the commercial risks in the transaction. These contracts for the carriage of goods by sea, and the laws that govern them, are of several general types depending on the kinds of goods and materials being sold. Immensely varied as trade consignments may be, they can be roughly divided into bulk and non-bulk or unitized cargoes. The division springs from the fundamentally different ways ships are employed and the trade that they carry is processed. Bulk cargoes generally consist of low value, unprocessed products like raw materials, be they liquid, such as oils and gases under pressure, or dry, such as mineral ores, grains and timber. Non-bulk cargoes are made up of the myriad different manufactured products and hi-tech goods, some of which are of very high value and low density. They are typically packaged goods that are further containerized or unitized (such as grouped on pallets or in crates).

Bulk cargoes, as the term suggests, are loaded directly into the ship's hold; the vessel then sails to the delivery port selected by the shipper. Thus, bulk carriers tramp around the world from port to port at the behest of their charterers, who are the cargo owners. Traditionally the contracts of carriage for these tramp ships are inscribed in charter parties that are negotiated between the shipper and the shipowner without regulation by any international rules of sea carriage.

In contrast, the modern method of handling finished products and other merchandise goods is by 20 and 40 foot standardized containers.[19] Indeed, global supply chains have been built upon the ubiquitous use of containers. This kind of traffic is carried under bills of lading or sea waybills on liner ships that operate to schedules within a global network of hub and feeder ports. The routes, ports of call and schedules are set by the shipowners, and their ships load cargoes delivered for carriage by thousands of different shippers. The largest liner ships today can carry as many as 11,000 containers, which may contain the shipments of many more merchants.

Such traffic has been regulated for a great many years largely by an internationally uniform set of rules. Most of the world's major trading nations adopted the Hague Rules,[20] formulated in 1924, and many, including Canada,[21] have since moved to the application of their amended version,

---

[19] A TEU is a 20-foot Equivalent Unit and consists of an ISO standard container measuring 8 feet × 8 feet × 20 feet. A 40-foot container is two TEU.

[20] *International Convention for the Unification of Certain Rules of Law Relating to Bills of Lading* (Hague Rules), Brussels, 25 August 1924, 120 *L.N.T.S.* 155. The United States continues to apply the Hague Rules: see the *Carriage of Goods by Sea Act, 1936*, 49 Stat. 1207 (codified at 46 U.S.C. §1300).

[21] See *Marine Liability Act*, S.C. 2001, c. 6, s. 43 & Schedule 3.

known as the Hague-Visby Rules,[22] or some national variant thereof. The central function of the Hague-Visby Rules is to allocate the risks and responsibilities of sea carriage between the two transacting parties, the carrier and the cargo owner, in an internationally agreed uniform manner. These rules govern the liner trades, but not tramp shipping, because they attach to bills of lading and similar documents of title but not to charter parties.

Obviously, the imposition of new security requirements on carriers will coincidentally affect their contractual relations with cargo owners, so it is necessary to investigate how those relations will be altered. This section will analyze the specific consequences that are likely to be incurred by the parties, first to an individual consignment on a liner ship under the Hague-Visby Rules, and then to a bulk cargo on a tramp ship under a voyage charter party. From this review it will be possible to assess how the new maritime security regime may have altered the standard trading conditions for the movement of seaborne trade.

The principal obligations of the parties to a carriage contract are six in number. The carrier is expected 1) to provide a seaworthy ship and 2) to care for the cargo put on board. It is also required to deliver the cargo 3) without deviation and 4) without delay. In return, the cargo owner 5) must not load dangerous goods without due notice and proper precautions, and 6) must pay the freight, i.e., the cost of carriage. In bulk trades, the scope of these duties are negotiated by the parties, but in liner trades they are all, except for freight, internationally regulated. The Hague-Visby Rules establish mandatory minimum standards for the conduct of the carrier and the cargo owner. In particular, the carrier must exercise due diligence to make the ship seaworthy and must properly and carefully look after the cargo.[23] In addition, the carrier may not deviate in transit or delay in delivery in an unreasonable manner.[24] In return for

[22] *Protocol to Amend the International Convention* (Visby Rules), *supra* note 20, Brussels, 23 February 1968, *U.K.T.S.* 1977 No. 83, together with the *SDR Protocol to the International Convention*, *supra* note 20, Brussels, 21 December 1979, *U.K.T.S.* 1984 No. 28. Other competing sets of rules include the *United Nations Convention on the Carriage of Goods by Sea* (Hamburg Rules), Hamburg, 31 March 1978, 17 *I.L.M.* (1978) 608 and UNCITRAL, Draft Convention on the Carriage of Goods [Wholly or Partly] [by Sea], A/CN.9/621 (17 May 2007).

[23] Hague-Visby Rules, *ibid.*, Arts. II & III.

[24] *Ibid.*, Arts. IV(4) & III(2).

bearing these responsibilities, the carrier will be excused for cargo loss and damage in a large number of defined circumstances and may limit the compensation paid in the event of liability for a casualty in accordance with a fixed rate per package or per kilo of cargo.[25] The cargo owner's obligations with respect to dangerous goods are also regulated in a general way by the Hague-Visby Rules,[26] but they are additionally governed in a much more detailed manner by the International Maritime Dangerous Goods Code (IMDG Code) promulgated by the IMO.[27]

It is not difficult to imagine ways in which a carrier might fail to fulfil a requirement of the ISPS Code, and thus cause loss to the other contracting party. The carrier bears considerable risk since it is responsible for both the seaworthiness of the ship and the care of the cargo. Hence, in preparing for a voyage, the carrier must take due account of the requirements of the new security regime in loading, sailing and delivering. For instance, the carrier is likely to be judged as failing to exercise due diligence to provide a seaworthy ship if the ISPS Code is not complied with, and thus the vessel is detained in sailing from the loading terminal or denied entry to the destination port. Any number of errors or omissions by the carrier might result in liability to the cargo owner. If the carrier has not made a ship security assessment,[28] has not provided on board a ship security plan,[29] has not adequately trained the crew in security procedures,[30] has not kept records of the security training given[31] or has failed on any other grounds to acquire an International Ship Security Certificate from its State of registry,[32] the vessel may be subjected to constraints in its trading.

Even if the carrier, the ship and the crew are in compliance, the vessel must still meet the safety requirements of the ports it visits. The ISPS Code imposes three levels of security in ascending order of severity on

---

25 *Ibid.*, Arts. IV(2) & (5).

26 *Ibid.*, Art. IV(6).

27 Applied in Canada by the *Transportation of Dangerous Goods Act*, S.C. 1992, c. 34 and the regulations in SOR/85-77. See also the *Canada Shipping Act, 2001*, S.C. 2001, c. 26, s. 274 and the dangerous goods regulations in SOR/81-951, SOR/87-24 & SOR/93-4.

28 ISPS Code, *supra* note 3, at Part A, s. 8.

29 *Ibid.*, s. 9.

30 *Ibid.*, s. 13.

31 *Ibid.*, s. 10.

32 *Ibid.*, s. 19.

both ships and ports.[33] The level of security of the ship must be the same as that of the port. If a ship is cleared to sail from the loading port at level 1 it may be prevented from entering and delivering its cargo in a destination port operating at security level 2. If the carrier has prepared the ship properly under the ISPS Code, it should be able to upgrade its security level en route so that it is not prevented from entering the destination port. If the ship is delayed or denied entry on account of its lower security level, such inadequacy may be viewed as a failure to exercise due diligence in preparing the ship for a readily foreseeable contingency in the voyage and so render the carrier liable to the cargo owner. Even if the failing is not a breach of the carrier's duty of seaworthiness, it may yet result in an unjustified deviation and an unreasonable delay.

Furthermore, the carrier takes a risk if the ship calls at a port that does not comply with the ISPS Code. A subsequent destination port may demand information about the ship's security compliant interface with its previous ten ports of call,[34] which it would obviously have difficulty in demonstrating and so could well result in its investigation, delay and even detention. Thus the carrier's act of contractual delivery to the first cargo owner may result in a breach of contract towards other cargo owners if, as is likely, the carrier's lack of foresight about the consequences under the ISPS Code of entering a non-compliant port are viewed as a failure in diligence towards the ship or care towards the cargo. The reverse course of action is no less unsatisfactory. If the carrier refuses to deliver at the first port because it is not ISPS compliant, while it would protect its trade to subsequent ports, it would risk breaching its contract for carriage with the first cargo owner. In liner trades, this scenario is not to be expected as the carrier controls the schedule of ports and sailings and therefore has only itself to blame if it elected to use a non-compliant port. The difficulty is more likely to arise in tramp shipping when the cargo owner or charterer may select the ports of call.[35]

---

33 *Ibid.*, ss. 2.9–2.11, 7 & 14.

34 SOLAS Annex XI-2, *supra* note 3, at Reg. 9.2. See also, C. Strickland, "International Ship and Port Security (ISPS) Code: Potential Commercial Disputes," presented to the National Judicial Institute's Federal Court and Federal Court of Appeal Education Seminar: Maritime Law, Ottawa, Canada, 5 November 2004, pp. 4–5.

35 However, the carrier, being free of the Hague-Visby Rules in bulk shipping, may protect itself by inclusion of a suitable term in the charter party akin to the standard unsafe ports clause.

The likelihood of a ship being delayed in entering, or even refused entry to, a destination port is not great if the owner has made a genuine effort to comply with the ISPS Code. Port authorities must have "clear grounds for believing that the ship is not in compliance" before they may interdict it[36] and they further need to have "clear grounds to believe that the ship poses an immediate threat to the security or safety of persons, or of ships or other property and there are no other appropriate means for removing the threat" before they may deny it entry into port.[37] However, what constitutes "clear grounds" is itself far from clear. Port authorities in different jurisdictions may view an approaching ship's responses to identical enquiries as more or less clearly raising suspicion of non-conformity with the ISPS Code. In the absence of a definition of this phrase or much known practice around it, there is room for enough difference of administrative policy and professional judgment amongst security officers in different countries to pose a distinct risk for the shipowner. If the vessel is stopped and loss results for the cargo owner, a claim of unseaworthiness or delay will surely be forthcoming. However, the shipowner may respond that, having made best efforts to comply with the ISPS Code, the interdiction of the ship was not reasonably foreseeable. If this assertion is true, the shipowner may rely on one of the numerous exceptions in the Hague-Visby Rules. The most obvious one favouring the shipowner is the exemption from liability in the event of "restraint of princes,"[38] which, in more contemporary language, would include the interdiction by port authorities of a foreign State. However, the shipowner has first to establish that the preparation of the ship for the voyage was diligent and the voyage itself was conducted reasonably, so there is no advance assurance of escape from liability.

Even when the carrier complies fully with the ISPS Code, it also needs to take account of the requirements of national laws of the countries at which it calls. The United States imposes especially demanding requirements that reach far beyond its ordinary territorial jurisdiction. The U.S. security legislation expresses a policy of protecting American borders

[36] SOLAS Annex Part XI-2, *supra* note 3, at Reg. 9.1.1.

[37] *Ibid.,* Reg. 9.3.3.

[38] Hague-Visby Rules, *supra* note 22, at Art. IV(2)(g). The exemption in Article IV(2)(q) for "any other cause arising without the actual fault and privity of the carrier…" might also be applicable.

from possible terrorist incursions by establishing a series of ring-fences or layers through regulations applicable to ships and cargoes long before they reach the United States. One of these is a requirement to give the U.S. Customs and Border Protection 96 hours advance warning, with prescribed security information, of a ship's intended arrival in American waters.[39] Failure to do so may delay the ship while it is investigated or cause it to oversail its destination and deliver at another place. Both results would constitute breaches of the carrier's contract, either by delay or by deviation respectively, which could hardly be justified and hence would incur liability towards the cargo owner.

An additional U.S. regulation requires the carrier to submit a manifest of all cargo destined to pass through or be delivered at an American port to the U.S. Customs and Border Protection at least 24 hours in advance of loading.[40] This requirement affords U.S. officials the opportunity to screen the cargo documentation in its Automated Targeting System[41] and to decide whether to inspect any particular cargo either in the foreign port before loading or on arrival in the United States. As a result, carriers have become more concerned about the cargoes they are asked to carry and the completeness and accuracy of the information provided by the cargo owner. If the carrier is suspicious of the cargo, or anticipates that U.S. Customs and Border Protection will be, the practical course of action for the carrier, which has a schedule to keep and other cargo interests to satisfy, is to leave the dubious cargo behind. However, in doing so, the carrier would breach the agreed contract of carriage with that shipper, unless it has a term in its standard trading conditions that gives it the liberty to move the goods by a different ship or at a later time, or to refuse to carry a cargo, or that disclaims all liability for the consequences of failing to load and carry such cargo. Such a clause would be a valid protection for the carrier and would not contravene its obligations under the Hague-Visby Rules since they apply only "tackle to tackle" and place

[39] 33 C.F.R. § 160 212 (2003).

[40] 19 C.F.R. § 4.7 (2003). Canada imposes a similar requirement of 96 hours notice of arrival in a Canadian port together with a long list of detailed information and documentation about the ship, its cargo, its past trading activity, its SOLAS certificates and its security experience at its last ten ports of call: see SOR/2004-144, s. 221(1).

[41] See the description under CSI in Appendix 1.

no restraints on the contractual terms by which the parties may fix their respective responsibilities for the cargo prior to loading.[42]

In effect, this U.S. regulation gives carriers a degree of influence over cargo owners that they did not have before: they are now able to demand precise and complete descriptions of the goods to be carried. Bills of lading or other transport documents that express the cargo as so many containers "said to contain..." are no longer acceptable. Shippers are now bound to declare the contents of containers they have stuffed and will not only be held by the Hague-Visby Rules to have warranted the accuracy of their declarations to the carrier[43] but will also be subject to the full force of the requirements of the Department of Homeland Security. Hence responsibility for carrying mis-described cargo that turns out to be terrorist materiel will fall on the shipper, not the carrier.

This requirement upon the shipper to declare fully and to describe accurately the nature of the goods to be carried does not, in fact, shift the risks between the two contracting parties. The Hague-Visby Rules have always imposed on the shipper full responsibility for its declarations about the cargo. In the past, usually the carrier simply adopted the shipper's statement of the cargo's weight or quantity, and description on the transport document, typically including the protective phrase "condition unknown," in full knowledge that the Hague-Visby Rules would allow it to pass any liability incurred towards the consignee or receiver of the goods on account of inaccuracy by the shipper onto the shipper. Thus, the new stringency of the U.S. law regarding cargo notifications and declarations does not alter the allocation of risks between the carrier and the shipper but does increase the latter's responsibility. It also affords the carrier sufficient clout to extract concessions from shippers when markets are tight.

This catalogue of instances are but examples of how the new maritime security regime meshes with the operation of the Hague-Visby Rules, yet it shows clearly that liner carriers have incurred considerably more responsibility in the running of their ships that they cannot readily transfer to cargo owners. Only the new imposition on shippers to provide a complete and accurate description of the cargo to the carrier so it can

[42] Hague-Visby Rules, *supra* note 22, at Arts. I(e) & VII.

[43] *Ibid.*, Art. III(5).

submit a cargo manifest at least 24 hours before loading for American destinations offers any relief to carriers. All other infractions of both the ISPS Code and national security rules are the responsibility of the carriers, who, if their ships are detained, may in consequence become liable to the cargo owners for one or more breaches of their obligations under the Hague-Visby Rules. Since the security requirements are readily foreseeable, errant carriers may be found wanting in due diligence to provide a suitably seaworthy ship, or in exercising proper care for the cargo, or by unreasonably deviating on the voyage or delaying in delivery of the cargo, or a combination of these defaults. In sum, the new maritime security regime has added another layer of risk to the liner carrier's business for the sake of public security, which the international rules of sea carriage largely inhibit it from allocating to, or at least sharing with, the cargo owner.

In comparison the tramp shipowner would appear to be in a more favourable position than the liner carrier. The difference lies in the tramp shipowner's freedom from international regulation. Owners and charterers are constrained by few rules even at the national level, where freedom of contract is generally the applicable legal principle. In practice, large international shipowners' associations, like the Baltic and International Maritime Council (BIMCO), issue a variety of charter party forms for different trades that contain standardized terms and conditions. In other words, the shipowners, through their organizations, try to set the terms of trade for the movement of bulk cargoes, to which the merchants, as shippers and charterers, have to negotiate changes as best they can according to the relative bargaining strengths of the parties.[44]

In the tramp operator's business environment, the imposition of the new maritime security regime has different consequences from the liner trades. Although all shipowners bear the same obligations under the security regime, tramp shipowners, unlike liner shipowners, have some leeway to shift the risks it presents to them. The principle of freedom of contract means that, though they have security responsibilities, they do not have to bear the liabilities consequent on error, omission or default if by agreement they allocate these risks to the charterers. For example,

[44] In practice, shipowners tend to have the greater bargaining power in agreeing the terms of carriage of dry bulk cargoes, while charterers have the edge in fixing charter parties for tankers: see Todd, *supra* note 9, at p. 383.

as discussed in regard to liner ships, any number of failures to comply with the ISPS Code may render the ship unseaworthy or delay delivery of the cargo. While the shipowner cannot escape the consequences of the mandatory security requirements – the public law obligations – it may avoid the private law consequences of breach towards the cargo owner if their charter party contains clauses exempting or disclaiming liability.

It should not be surprising to learn that shipowners' associations that provide standard charter party terms of trade have been active in preparing suitable new clauses to cope with the shipping community's obligations under the new maritime security regime. One might expect these contractual clauses at least to dilute, if not to exclude, shipowners' liabilities towards cargoes and cargo owners. In fact, they appear only to have taken modest advantage of this opportunity.[45] BIMCO has drafted a number of model clauses that take account of the ISPS Code's demands and of the U.S. legal requirements for both voyage and time charter parties. For comparison with liner shipping, voyage charter parties are the better choice of these two main types of tramp shipping contracts as shipowners have nearly as much control over the arrangements for carriage as they do under bills of lading. Time charters inevitably provide the charterer with much greater freedom to determine voyages, ports of call and transit times within the period of hire and consequently with more responsibility for employment of the ship in a lawful manner. Voyage charter parties, like bills of lading, basically anticipate the shipowner will be responsible for all decisions and operations of running the carrying ship once the parties have agreed on the scope of the voyage, the cargo and the freight.[46]

---

[45] However, existing disclaimer clauses in charter parties may also cover liability incurred by carriers. For instance the "owners responsibility clause" in BIMCO's GENCON charter party states that the shipowners "are to be responsible for loss of or damage to the goods or for delay in delivery of the goods only in case the loss, damage or delay has been caused by the personal want of due diligence on the part of the Owners or their Manager to make the vessel in all respects seaworthy and to secure that she is properly manned, equipped and supplied, or by the personal act or default of the Owners or their Manager. And the Owners are not responsible for loss, damage or delay arising from any other cause whatsoever, even from the neglect or default of the Master or crew. . . ."

[46] For analyses of the impacts of the ISPS Code on both voyage and time charter parties see Todd, *supra* note 9, and L. C. Bulow, "Charter Party Consequences of Maritime

BIMCO offers an ISPS/MTSA Clause for Voyage Charter Parties revised in 2005 (BIMCO ISPS/MTSA Clause),[47] along with several specific clauses to address U.S. Customs and Border Protection regulations and the C-TPAT programme. The ISPS/MTSA Clause requires the shipowner to comply with both the ISPS Code and the U.S. MTSA on pain of compensating the charterer/shipper for all direct losses resulting from any default. For its part, the charterer is obliged to supply the shipowner with all the details concerning itself and its business as are required to comply with the Code and the MTSA, bearing any loss consequent on breach and any delay as laytime used. To this extent, the parties carry the duties they are uniquely qualified to fulfil. However, if the shipowner is in compliance and the ship has arrived in port, notice of readiness to load or unload may be given even though the vessel is not cleared due to security measures imposed by the port. As a result, unless the security measures are taken pursuant to the negligence of the shipowner, the master or crew or the previous trading of the ship, the laytime of the charterer for loading or unloading will begin. In other words, the risk of delay in port on account of security checks on the ship, its contents and documentation, provided it is not due to the shipowner's negligence, is allocated to the cargo owner. Obviously, one of the two parties must bear this risk,[48] which is not controllable by either of them, but the charterer might reasonably expect the shipowner to absorb it as part of its exclusive responsibility for operating its ship. To this extent, the ISPS/MTSA Clause shifts the legitimate responsibility of the shipowner for the incidence of port security inspired delays to the cargo interests. The Clause also adds that any costs associated with the imposition of these security measures are also for the cargo owner's account.

BIMCO's clauses referring to U.S. legal requirements also impose additional responsibilities on the charterer. The U.S. Trade-Unique Bill

---

Security Initiatives: Potential Disputes and Responsive Clauses," *Journal of Maritime Law and Commerce* 37(1) (2006): 79.

[47] BIMCO, <www.bimco.org/Corporate%20Area/Documents/Clauses/ISPS-MTSA_Clause_for_Voyage_Charter_Parties_2005.aspx>.

[48] This risk is commercially significant to both parties since delay to a ship may cause costly losses in trading while delay in delivery of the cargo probably reduces the shipper's or consignee's margin of profit on their trade transaction.

of Lading Identifier Clause[49] demands that the charterer warrants that U.S. bound cargo bears a unique bill of lading identifier in compliance with U.S. Customs and Border Protection regulations. The U.S. Customs Advance Notification /AMS Clause for voyage charter parties[50] requires the charterer to provide the information necessary for the shipowner to make a timely and accurate cargo declaration to the U.S. Customs and Border Protection. These obligations are appropriately allocated to the charterer as it is uniquely in possession or control of the required information. Likewise, the AMS Clause obliges the shipowner to put in place a standard carrier alpha code and an international carrier bond and to submit an automated manifest (all as required by U.S. Customs and Border Protection regulations for advance screening of cargo information within its Automated Targeting System) as only it is in a position to do. Additionally, if the charterer is a trader who has entered the C-TPAT programme with U.S. Customs and Border Protection, the shipowner undertakes by the BIMCO C-TPAT Clause[51] to make reasonable efforts to assist the charterer to comply with its C-TPAT agreement, but without any liability for any failure to do so. In other words, BIMCO's U.S. directed clauses distribute the risks and responsibilities of compliance with U.S. laws as the trading community might expect, that is to each party according as it is uniquely able to fulfil the U.S. Customs and Border Protection requirements.

It must always be remembered that these clauses of BIMCO are models for voluntary adoption in the chartering trades. Unlike the liner trades, no international or national law compels these or any other arrangements. BIMCO's clauses have no effect unless they are incorporated by agreement in a charter party, and the parties may also alter the terms and so stiffen the obligations of one of them as they may negotiate. Yet, as model clauses presented by a leading shipowner association, they offer a division of responsibilities for maritime security measures that do not much advantage shipowners over cargo owners. Apart from the risk of

49 BIMCO, <www.bimco.org/Corporate%20Area/Documents/Clauses/US_Trade_Unique_Bill_Of_Lading_Identifier_Clause.aspx>.

50 BIMCO, <www.bimco.org/Corporate%20Area/Documents/Clauses/US_Customs_Advance_Notification_AMS_Clause_for_Voyage_Charter_Parties.aspx>.

51 BIMCO, <www.bimco.org/Corporate%20Area/Documents/Clauses/US_Customs_Trade_Partnership_Against_Terrorism_C-TPAT_Clause.aspx>.

delay in port for security checks being imposed on the charterer as part of its laytime, the obligations for ships and cargoes fall on the side that holds the information or means to fulfil them.

In summation, it matters little whether the goods are moved on liner ships under bills of lading and similar documents of title or on tramp ships under voyage charter parties. The BIMCO model clauses for chartered ships suggest a similar allocation of responsibilities for the new maritime security regime as the international Hague-Visby Rules impose on liner trades. By and large, only when the information or ability to meet the security requirements are within the control of cargo owners are they expected to take responsibility for compliance. In general, the shipowners must absorb the legal responsibility for fulfilment of the security requirements of each trade transaction, but maybe this affords them an opportunity to extract greater returns for their services. Just how the shipping industry as a whole is coping with these new responsibilities and costs is the subject of the next section.

Impacts on Sea Links in the Global Supply Chains

Over the course of the 1990s, as global supply chains developed and logistics management grew ever more sophisticated, cargo interests streamlined their choices of carriers, transportation supplier reduction strategies were implemented, and cargo interests concentrated their contracts of sale on fewer trade terms. A preference for FCA and CIF or DDU terms developed as part of managing the cargo risk in container shipments. Many cargo interests contracted with carriers that offered larger geographic scope in an effort to gain greater control over their business processes and to manage the complexities of multiple suppliers or to minimize supply chain costs. Supply chains also grew more dependent on confidential contracts for liner shipping, as these became possible under the U.S. *Ocean Shipping Reform Act of 1998.*[52] By the turn of the century, the majority of container shipments involved large multinational supply chains, were transactions in which buyer and seller had contracted for long-term supply arrangements, or moved between related parties. The old norm of a single transaction-based relationship accounts for fewer and fewer transactions.

---

[52] Pub. L. No. 105–258,112 Stat. 1902 (codified at 46 U.S.C. App. § 1701 (1998)).

In some cases, individual cargo interests have banded together in buying groups to tender their transport contracts, as is the case in Canada with the Canadian Retail Shippers' Association.[53]

Prior to 2001, cargo interests had long been concerned about the security of the supply chain in order to minimize loss due to theft. However, loss and damage experience was not a key factor, and definitely not a determinant one, in ocean container carrier selection by exporters.[54] Accidents will happen at sea. For this reason, cargo interests have used insurance markets to ensure that loss through weather, accident or theft did not put their business interests at risk.

Since 2001, and particularly after the advent of the new security measures, the largest cargo interests have viewed physical security of their premises and membership in C-TPAT as the required extra layer of diligence that will ensure continuing business interests are protected. In a study of the top 100 U.S. importers and exporters of containerized cargo and their responses to the new security measures, Brooks and Button found:

> For many of the manufacturers and retailers, cargo theft had already driven logistics processes in a direction that the new security requirements after 9–11 extended. Therefore, it is not surprising to find that, when companies were asked whether they had switched to outsourcing distribution/logistics or changed terms of sale, nine out of the 10 answered neither and only one responded that a change in terms of sale occurred... Some of those who had not changed terms of sale indicated that they had preferred to control the transport decisions prior to 2001 for risk, cost control, or other reasons, and so no change was necessary. The one large retailer that changed its terms of sale did so "to ensure that origin trucking is controlled by our company or a trusted 3PL [third party logistics supplier], as an increased security measure."[55]

---

[53] B. Gerrior, Presentation to the Atlantic Provinces Transportation Forum Access North America Conference, Halifax, Canada, 12 October 2007.

[54] M. R. Brooks, "An Alternative Theoretical Approach to the Evaluation of Liner Shipping, Part II: Choice Criteria," *Maritime Policy and Management* 12(2) (1985): 145; M. R. Brooks, "Ocean Carrier Selection Criteria in a New Environment," *The Logistics and Transportation Review* 26(4) (1990): 339.

[55] M. R. Brooks and K. J. Button, "Maritime Container Security: A Cargo Interest Perspective," in K. Bichou, M. Bell and A. Evans, eds, *Port, Maritime and Supply Chain Security: Frameworks, Models and Applications* (London: Informa, 2007): 221 at p. 228.

Brooks and Button learned that membership in C-TPAT and adding requirements to transport contracts were the two most common responses of U.S. cargo interests to new security requirements; for some, greater security has become a part of the company's corporate social responsibility mandate.[56] (In fact, corporate social responsibility impacts have been found to be of greater concern to cargo interests than cargo delay.)[57] Finally, most importers are highly compliant with security regimes; U.S. Customs and Border Protection has reported a compliance rate of 97 per cent, and notes that the remaining three per cent tend to be repeat offenders.[58] However, it is the repeat offenders that create the uncertainty for the rest.

More than anything else, what the commercial parties do not like is uncertainty – uncertainty in regulations and legislation, and uncertainty that cargo delay brings to the just-in-time system inherent in global supply chain development. The unevenness associated with the U.S. regime, and in particular the relentless march of legislative change,[59] has been clearly identified as the strongest issue for cargo interests in the new security environment. Since these findings were published in 2006, the United States has introduced the SAFE Port Act[60] and Congress has set 1 July 2012 as the date for implementing the scanning of 100 per cent of the goods bound for the United States prior to their departure for the United States,[61] further exacerbating the commercial uncertainty. No amount of focus on terms of trade of liner shipments can ameliorate these concerns about scheduled services.

From the perspective of the shipowner, managing the interchange points in a transport chain is critical to the success of a security programme. Cargo is most vulnerable to hijacking, invasion or attack when not moving,

---

56 *Ibid.*

57 J. Goepfert and C. Boone, "Safe Commerce: Securing Goods In and Out of Ports," IDC, May 2004, <www.unisys.com/commercial/insights/white__papers/index.htm>.

58 B. Mongelluzzo, "Customs finds 97% Compliance with Import Regulations," *Journal of Commerce Online* (20 April 2007).

59 M. Thibault, M. R. Brooks and K. J. Button, "The Response of the U.S. Maritime Industry to the New Container Security Initiatives," *Transportation Journal* 45 (2006): 5.

60 *Supra* note 5.

61 R. G. Edmonson, "A 100 Percent Mess," *Journal of Commerce Online* (30 July 2007).

Table 1. U.S. waterborne foreign container trade by U.S. customs port (000 TEUs)

Port	1997	2000	2003	2005	2005 Imports	% Import
Los Angeles, CA	2,090	3,228	4,664	4,864	3,821.3	78.6
Long Beach, CA	2,723	3,204	3,091	4,378	3,354.7	76.6
New York, NY	1,739	2,200	2,803	3,387	2,415.2	71.3
Charleston, SC	956	1,246	1,250	1,509	893.5	59.2
Savannah, GA	530	720	1,124	1,469	799.7	54.4
Oakland, CA	844	989	1,064	1,374	762.7	55.5
Seattle, WA	954	960	815	1,339	875.4	65.4
Norfolk, VA	771	850	1,093	1,319	779.0	59.1
Houston, TX	612	733	933	1,222	622.9	51.0
Tacoma, WA	551	647	931	1,155	792.5	68.6
Miami, FL	623	684	764	772	447.8	58.0
Port Everglades, FL	455	439	423	578	276.4	47.8
Baltimore, MD	261	276	307	382	244.5	64.1
San Juan, PR	143	152	185	213	165.1	77.4
Gulfport, MS	122	156	204	182	108.6	59.7
New Orleans, LA	231	229	237	174	73.2	42.1
Wilmington, DE	104	123	195	162	120.5	74.5
West Palm Beach, FL	113	130	140	159	38.5	24.2
Philadelphia, PA	91	83	103	158	138.7	87.6
Jacksonville, FL	199	110	113	144	45.0	31.2
Boston, MA	62	74	93	130	73.9	56.7
Total top 20	14,173	17,232	20,533	25,071	16,848.9	
Share of Total	95.37%	96.07%	96.45%	96.92%		
Total	14,860	17,938	21,289	25,868	17,290.4	66.8%

Notes: The statistics exclude postal and military shipments. A TEU is defined in footnote 19 of this essay.

Source: U.S. Department of Transportation Maritime Administration, *Vessel Calls at U.S. & World Ports 2005*, Office of Statistical and Economic Analysis, Washington, D.C., April 2006, <www.marad.dot.gov/marad_statistics>, 26 February 2007.

Table 2. U.S. waterborne foreign container trade by trading partner in TEUs

Trading Partner	1997	2000	2003	2005
China	1,835,555	3,414,946	5,655,895	9,041,159
Japan	1,759,910	1,746,935	1,602,625	1,660,834
Hong Kong	1,119,923	1,315,236	1,618,808	1,156,227
South Korea	689,875	863,531	898,154	1,007,059
Taiwan	940,753	919,128	946,202	981,319
Germany	469,637	624,025	649,788	726,412
Brazil	417,797	426,376	532,787	658,125
Italy	412,511	552,746	601,928	593,107
Thailand	326,963	455,452	496,056	552,003
India	191,338	249,452	388,845	507,163
Top 10 total	8,164,261	10,567,827	13,391,089	16,883,409
Other	6,696,107	7,369,843	7,897,456	8,984,749
Share of Top 10	54.9%	58.9%	62.9%	65.3%
Total	14,860,367.5	17,937,669.6	21,288,545.1	25,868,158.3

Source: U.S. Department of Transportation Maritime Administration, Waterborne Databanks, <www.marad.dot.gov/marad_statistics>, 26 February 2007.

such as in port.[62] U.S. bound cargo is divided into two types, tanker and liner, in U.S. maritime trade statistics. Of the liner cargo, 20 ports in the United States account for 96.9 per cent of all container trade (Table 1). This concentration of activity enables U.S. authorities to focus their efforts on securing a known land area. Once these port facilities are secured through meeting the physical requirements of the ISPS Code,[63] the weak link is the feeder traffic that serves them from other domestic ports, and the land traffic that arrives by truck or rail.

As Table 2 shows, approximately 55 per cent of the boxes used to come from the U.S.' top 10 container trading partners in the 1990s but, by 2005, almost two-thirds of the container traffic came from them, with the majority of the growth attributable to the burgeoning China trade. Indeed, China's trade has grown to 35 per cent of the total maritime container trade passing through U.S. ports. It is also interesting to note that, while 44 ports participated in the CSI in 2005, rising to 58 in 2007, and the

62 Brooks and Button, *supra* note 55, at p. 225.

63 As well as the personnel management requirements of the U.S. Transport Worker Identification Credential Program.

majority of the biggest and busiest 50 container ports worldwide were members, of the eight ports in China in the top 50, only Hong Kong and Shanghai belonged to the CSI as of 30 July 2007.[64]

In tonnage terms, however, the container trade is the smallest component of U.S. seaborne trade. It is quite clear that the majority of vessels calling at U.S. ports, in terms of tonnage, fall into the tanker and non-liner categories (Table 3). In the most recent year for which data are available (2003), scheduled vessel service (liner) only accounted for 13.1 per cent of the total traffic. This proportion of U.S. seaborne trade has been consistent for the past 15 years, well before the ISPS Code changed shipowner, operator and port responsibilities. Yet, the majority of U.S. security efforts have focused on the maritime container trades, i.e., those falling into the liner category. The top 10 source countries of U.S. imports overall (Table 4) reflect the dependence of the United States on imported energy; some of these countries might be of greater security concern than those from which container imports are received.

Table 3. U.S. waterborne foreign trade by type of service (2003–1956) (million metric tons)

Year	Total	Liner	% of total	Tanker	% of total	Non-Liner	% of total
*2003	1,220	160	13.1	660	54.1	400	32.8
2002	1,168	168	14.4	621	53.2	379	32.4
2001	1,161	150	12.9	628	54.1	384	33.0
2000	1,158	149	12.8	601	51.9	408	35.2
1995	980	136	13.9	425	43.4	419	42.7
1990	877	100	11.4	387	44.1	390	44.5

Note: * Preliminary Data
Source: U.S. Department of Transportation Maritime Administration, *Vessel Calls at U.S. & World Ports 2005*, Office of Statistical and Economic Analysis, Washington, D.C., <www.marad.dot.gov/marad_statistics>, 26 February 2007.

[64] Anonymous, "The JoC Top 50 World Container Ports," *The Journal of Commerce* (30 July 2007): 46–54.

Table 4. U.S. waterborne foreign trade by trading partner (in metric tons)

Trading Partners	1997	2000	2003	2005
Canada	67,227,122	77,317,868	89,662,372	121,813,261
Mexico	105,317,646	119,970,327	121,813,787	117,268,846
China	29,655,805	45,824,767	65,241,944	88,612,025
Venezuela	96,027,736	92,593,331	78,740,251	83,231,220
Japan	79,041,930	70,803,676	56,496,746	55,665,547
Gulf of Mexico	101,461,644	119,642,222	99,824,870	43,369,340
Brazil	29,105,904	34,261,847	36,102,171	41,156,844
Saudi Arabia	8,483,476	20,572,029	24,662,255	38,505,193
Nigeria	15,697,410	18,723,769	23,088,590	35,247,616
Colombia	24,548,721	32,824,855	32,176,266	34,952,256
South Korea	25,634,425	29,610,844	25,436,255	29,328,880
Top 10 Trading Partners	582,201,819	662,145,535	653,245,507	689,151,027
Share with Top 10	54.8%	55.9%	56.3%	54.8%
Total	1,062,250,848	1,184,161,712	1,161,269,976	1,258,240,424

Source: Port Import Export Reporting Service (PIERS), collected from Vessel Manifests and Bills of Lading, as reported by U.S. Department of Transportation Maritime Administration, Waterborne Databanks.

Research by Thibault, Brooks and Button[65] found that, since 2001, the maritime container shipping industry has made major efforts to improve the security of the sea links in the global supply chains. This research provided a detailed assessment of what liner carriers and ports thought of the security requirements and how they responded. For the most part, carriers viewed the requirements as an additional layer of activity that would be added to existing efforts to reduce smuggling, cargo theft, and the problems associated with stowaways. All liner operators that responded to enquiries had re-examined established procedures and business processes, and had made adjustments to better manage operations from a security perspective, but with varying levels of intensity. All parties indicated that they had re-evaluated their physical security and operational procedures, making changes where and when required. The carriers were not asked if they altered their documentary requirements, but they were asked if they opted to participate in C-TPAT, and most did. The researchers also

[65] *Supra* note 59.

contacted maritime security consultants and found that there was very little evidence of switching of transportation suppliers for security reasons.

In conclusion, from a commercial perspective, it appears that the tramp operator is better off than the liner operator. The largest share of international cargo security oversight by the United States has been visited on the liner industry and maritime containers. Very little has been written on the commercial response of tramp ship operators to the advent of the ISPS Code. The fear of "a bomb in a box" may have focused more attention on the marine container industry than is considered warranted in proportion to its percentage share of U.S. seaborne trade and to the nature and scale of maritime security risks. It would appear on the surface that there is as much to fear from the threat of the hijacking of a heavily laden tanker or an LNG carrier, or the disruption of a port by the blockage of a main channel with a submerged dry bulk vessel, as there is from the potential impact from terrorist activities of a non-maritime nature.

Conclusion: Consequences for Carrier-Cargo Contractual Relations in Seaborne Trade

The new maritime security regime, while being implemented on a multilateral scale by the International Maritime Organization, the International Labour Organization, and the World Customs Organization, is largely driven by U.S. legislation and U.S. desires for greater security of the nation. As one of the largest economies in the world, the U.S. market is highly attractive to foreign traders who are quite prepared to adjust to the requirements imposed on U.S. transactions in return for access to American consumers. The new security prescriptions have added requirements to overall trade documentation processes but have not necessarily forced changes of trading partners, although delay arising from the intervention of security officers might lead a buyer to reconsider a source of supply if security-induced delay causes significant consequential loss.

Overall, contractual arrangements for the movement of trade by sea have not changed as a result of the new maritime security requirements, although some adjustments in the form of security clauses and liberty clauses may have been added to contracts of sale and/or contracts of carriage. Trade has become more costly as both carriers and shippers have tightened business processes and added to the activities undertaken as part of executing a transaction.

Although global supply chain formation was well-advanced before the imposition of the new maritime security regime, power has been further concentrated in the hands of the largest multinational cargo interests and a smaller number of global liner operators. According to Brooks and Button, multinational cargo interests have found that they must ensure their own security by joining C-TPAT (or their national equivalent) and assisting their developing country suppliers to meet the new security requirements. This trend has induced a chilling effect on trade opportunities for the smallest and weakest, just when they thought that globalization offered them a new hope of growing their economies.

From a carrier perspective, the new maritime security regime has forced them to streamline their trading processes. Operating as well as capital costs have increased, but benefits have also been realized. For example, cargo documentation supplied by shippers is more likely to be both complete and accurate, as advance notification rules encourage this and U.S. Customs and Border Protection will no longer accept cargo reports using the phrase "said to contain." The security requirements place more leverage in the hands of the carrier in dealing with the once-careless cargo owner or with those seeking to ship a suspicious or dubious box.[66]

The new and constantly changing maritime regime may shore up security on the sea leg of a trade movement, but the weakest link in the international supply chains has always been and remains located elsewhere: generally it's a land-based "first-" or "last-mile" challenge.

Appendix 1. U.S. Security Programmes

Customs-Trade Partnership Against Terrorism (C-TPAT): This U.S. voluntary programme is designed to develop a cooperative relationship between U.S. companies with global supply chains and the U.S. authorities responsible for security. Companies that wish to join C-TPAT conduct a

[66] It is not known if liner carriers have seen any insurance benefits arising from these tightened business processes in the absence of research on this aspect. Conversely, there is the risk that a non-compliant carrier will lose its insurance cover because its acts or omissions that may render it liable to cargo owners for failure to exercise due diligence to provide a seaworthy ship may also invalidate its insurance contract. See *Marine Insurance Act*, S.C. 1993, c. 22, s. 37 and Strickland, *supra* note 34, at p. 27.

comprehensive self-assessment of their security practices using guidelines developed by U.S. Customs and Border Protection (CBP) in consultation with private industry. After developing a security enhancement plan that incorporates C-TPAT guidelines, participants then submit a supply chain security profile to CBP. It is expected that C-TPAT participants will build the guidelines into their relationships with other firms in their global supply chains. In this way, U.S. standards are absorbed along the supply chain. C-TPAT certification was originally only available to U.S. companies and some Mexican *maquiladoras*, forcing countries like Canada to develop their own C-TPAT-like programs. (The Canadian programme is called Partners in Protection, or PIP.) The U.S. Government Accountability Office has attacked the programme for its difficulty in handling applications in a timely manner, its inadequate auditing of participants, and other problems arising from insufficient funding of the programme.[67] Policy commentators have been vocal about the inability of the United States to deliver a system of security because of funding issues,[68] and industry has also weighed in with its concerns about timeliness and the failure to deliver on the benefits promised.[69]

Container Security Initiative (CSI): Based on a risk management approach that sees security as a layered process, CBP places its staff at foreign ports to work with foreign counterparts to inspect high-risk cargo *before* it is shipped to the United States. These placements are governed by bilateral agreements between the United States and key trading partners, and are reciprocal in that inspectors have the authority to inspect containers being shipped by sea to their respective countries. Using the documents filed under the advanced notification rules (the bill of lading or sea waybill data from the ship's manifest), the Automated Targeting System (ATS) analyzes the information for each shipment and ranks it in

---

67 U.S. Government Accountability Office (GAO), "Cargo Security: Partnership Program Grants Importers Reduced Scrutiny with Limited Assurance of Improved Security" (D-05-404) (Washington, D.C.: U.S. Government Accountability Office, 2005) and the 2006 GAO report, *supra* note 17.

68 R. A. Clarke, G. P. Aga, R. W. Cressey, S. E. Flynn, B. W. Mobley, E. Rosenbach, S. Simon, W. F. Wechsler and L. S. Wolosky, *Defeating the Jihadists: A Blueprint for Action* (New York: Century Foundation Press, 2004).

69 M. Ojah, "Securing and Facilitating U.S. Land Border Trade: A Critical Analysis of the C-TPAT and FAST Programs," *Transportation Research Record* 1938 (2005): 30.

order of risk (its ATS score). CBP officers located at the ports use ATS scores to help them make decisions on which shipments to inspect. As of 28 September 2007, 58 ports participated in CSI, accounting for more than 85 per cent of U.S. container trade.[70]

[70] U.S. Customs and Border Protection, "U.S. Customs and Border Protection Achieves Container Security Initiative Milestone with 58 Operational Ports," <www.cbp.gov/xp/cgov/newsroom/news_releases/09282007_5.xml>.

"Making Labour History"[1] and the Maritime Labour Convention, 2006: Implications for International Law-Making (and Responses to the Dynamics of Globalization)

Moira L. McConnell*

Introduction

On February 23, 2006, the Maritime Labour Convention, 2006[2] (MLC, 2006), comprising more than 100 pages of text, was adopted by the 94th (Maritime) Session of the International Labour Conference[3] (ILC) of the

---

* Professor of Law and Associate, Marine & Environmental Law Institute, Dalhousie University, Halifax, Canada. Since 2003 Dr. McConnell has also been a Special Advisor to the International Labour Office of the ILO, assisting with the development of the Maritime Labour Convention, 2006 and with research regarding the economic dynamics of International Labour Standards (ILS). The author wishes to express her gratitude to the Dr. Cleopatra Doumbia-Henry, Director, International Labour Standards Department, ILO and Mr. Dominick Devlin, former Legal Advisor to the ILO and Special Advisor to the ILO on this Convention, for their helpful comments and discussion throughout the last few years. The views expressed by the author in this chapter remain her sole responsibility.

1 Dr. Juan Somavia, Director-General of the International Labour Office, International Labour Organization (ILO), speech after the vote to adopt the Convention by the International Labour Conference, 94th (Maritime) Session, Ninth Sitting, *Provisional Record* No. 17 (Geneva: ILC, 23 February 2006), p. 9.

> ...because the instrument is historic. Because history is being written, Article by Article, within what you have approved and what you have decided today.

2 *Maritime Labour Convention*, Geneva, 7 February 2006, <www.ilo.org/global/What_we_do/InternationalLabourStandards/MaritimeLabourConvention/lang--en/index.htm>. It is available online in eight languages.

3 7–23 February 2006, Geneva. This was also the 10th Maritime Session of the ILO since 1920. For the complete Conference documents see <www.ilo.org/global/What_we_do/Officialmeetings/ilc/ILCSessions/94thSession/lang--en/index.htm>.

International Labour Organization (ILO) by a record vote[4] of 314 in favour, none against, and two[5] abstentions for reasons unrelated to the substance of the Convention. This show of tripartite support and the relative dearth of amendments[6] submitted to the final text proposed by the International Labour Office[7] for such a comprehensive and complex convention was almost unprecedented in ILO history. This situation is even more remarkable when one considers the fact that the MLC, 2006:

- adopts an entirely new format for ILO conventions;
- brings in a new approach to the updating/amendment of ILO conventions;[8]

[4] Under the *Constitution of the International Labour Organisation* the "State" (the "Member") is conceived of as a tripartite entity comprising workers, employers and government. Although only governments can ratify conventions (and be held accountable in international law), all three vote on the adoption of international legal instruments using a weighted voting formula. The 314 vote in favour reflects the fact that 106 countries that are Members of the ILO attended the three-week Conference and that tripartite support was obtained in 104 Members. The ILO Constitution is available at <www.ilo.org/ilolex/english/constq.htm>.

[5] The government representatives of two ILO Members, the Bolivarian Republic of Venezuela and Lebanon (4 votes) abstained for reasons unrelated to the substance of the Convention. The government representatives of Venezuela abstained for reasons relating to its views on the reference in the Preamble of the MLC, 2006 to the 1982 United Nations Convention on the Law of the Sea but placed on record its support for the substance of the Convention: see *Provisional Record* No. 17, *supra* note 1, at pp. 1–2. The representatives of Lebanon also abstained and placed on record its support for the Convention but felt it could not vote in favour because of its lack financial capacity to implement Conventions: see *Provisional Record* No. 17, *supra* note 1, at p. 2.

[6] In total, 115 amendments, of which 57 were adopted either as proposed or subamended, International Labour Conference, 94th (Maritime) Session, Eighth Sitting, *Provisional Record* No. 16 (Geneva: ILC, 22 February 2006), p. 3. Of these, a number related to areas, such as the formula for coming into force and amendment of the MLC, 2006 and its Code, that had been specifically left to be determined at the Conference. See *Tripartite Intersessional Meeting on the Follow-up to the Preparatory Technical Maritime Conference*, Geneva, 21–27 April 2005, Report PTMC2005/23, paras 48–58.

[7] International Labour Conference, 94th (Maritime) Session, 2006, *Proposed consolidated maritime labour Convention*, Report I(1B) (Geneva: International Labour Office, 2005).

[8] P. Bolle, "The ILO's new Convention on maritime labour: An innovative instrument," *International Labour Review* 145(1–2) (2006): 135–142.

- covers close to the full gamut of socio-economic issues in the maritime sector, including the very controversial question of social security protection, shipowner liability and repatriation; and
- also, sets out a new comprehensive enforcement and compliance system based on flag State inspection and certification of the requirements of the Convention complemented by port State control inspections and on board and onshore complaint handling systems.

At the time of its adoption, the Director-General of the ILO was moved to describe this Convention, and the process by which it was developed, as "historic" and a model for a way forward to achieving "fair globalization".[9] The Convention is also variously described as the "super convention,"[10] the "seafarers' bill of rights," and a "charter of rights." It has also been described by many, including the Secretary-General of the International Maritime Organization (IMO), as "the fourth pillar" of the international maritime regulatory regime.[11]

---

[9] *Supra* note 1. The term "fair globalization" stems from the work and the report of the World Commission on the Social Dimension of Globalization, *A Fair Globalization: Creating Opportunities for All* (Geneva: ILO, 2004) at p. xi, which argues the problems including social conflict and polarization of views are

> ...not due to globalization as such but to deficiencies in it governance. Global markets have grown rapidly without the parallel development of economic and social institutions necessary for their smooth and equitable functioning. At the same time, there is concern, about the unfairness of key gobal rules on trade and finance and their asymmetric effects on rich and poor countries...There is a serious democratic deficit at the heart of the system....workers and the poor have little or no voice in this governance process.

On 10 June 2008, the 97th ILC adopted an important new document reflecting their ideas, the "Declaration on Social Justice for a Fair Globalization."

[10] See, for example, Speech to the ILC by Mr. Mitropoulos, Secretary-General, International Maritime Organization, in International Labour Conference, 94th (Maritime) Session, Fourth Sitting, *Provisional Record* No. 10 (Geneva: ILC, 20 February 2006), p. 2.

[11] *Ibid.* The IMO, a specialized agency of the United Nations, has adopted numerous conventions, however, three are usually considered key or "cornerstones": the International Convention for the Safety of Life at Sea, 1974, as amended (SOLAS); the International Convention on Standards of Training, Certification and Watchkeeping, 1978, as amended (STCW); and the International Convention for the Prevention of Pollution from Ships, 73/78 (MARPOL).

Whether it fulfils or will fulfil all of these aspirations is a question for the future.[12] What is certain, however, is that this Convention, and the process by which it was adopted, signals a revitalization of interest in, and a recommitment to, the importance of international law and the development of the global regulatory regime. Whilst this is an era that is experiencing a crisis of faith in international law and institutions and, perhaps, even the rule of law generally, paradoxically there is also an unrelenting demand for, and production of, international law: the plethora of international law-making and law implementing institutions testifies to this phenomenon. In fact, the excess of international obligations, and the related lack of capacity on the part of many countries to keep up with implementation of their obligations, has led to a growing concern about the problem of "effectiveness" of international law as well as an increased focus on dispute resolution and the expanding number of international tribunals.[13]

The MLC, 2006, aside from being of obvious sectoral interest and impact, should be understood as a first[14] but important step – an experiment if you will – by the ILO in specifically addressing concerns raised by a number of commentators regarding imbalances in the global trade

---

[12] I am mindful of the analysis presented by Ed Morgan, *The Aesthetics of International Law* (Toronto: University of Toronto Press, 2007) at pp. 43–44, with respect to a tendency in public international law, in the LOS Convention related doctrinal pronouncements, in particular, to lay claim to represent the progressive development of the law and much more besides. He argues that:

> By so situating the text historically between an instant of closure and a wide open process, the reader becomes aware of the momentum generated by international legal discourse, and it is from the dust this non-static phenomenon that the sovereign states emerge as 'parties' to the law of the of the sea system.... Modern doctrinal pronouncements seem to be positioning themselves as the energy or momentum coming between stasis and motion – that is, as both the potential and fulfilment of the law's promise.

[13] See, for example, United Nations Environment Programme, *Advancing the Millennium Development Goals through the Rule of Law*, (draft) Background Paper, UNEP Round-Table Dialogue, Nairobi, 16–17 February 2005, see <www.unep.org/gc/gc23/documents/AgendaMDGsRoundTable18-02-05.pdf>.

[14] A related convention that adopts similar ideas *mutatis mutandis* for the fishing sector was adopted by the ILO in June 2007 at the 96th ILC: see *Work in Fishing Convention, 2007* (No. 188), Geneva, 14 June 2007, <www.ilo.org/ilolex/english/convdisp1.htm> and the associated Recommendation (R 199), <www.ilo.org/ilolex/english/recdisp1.htm>.

and economic law regime and its failure to truly address the relationship between the costs of labour and trade.[15] It is of particular interest in this respect that the MLC, 2006 is explicitly based on the twin goals of helping to secure decent work[16] for workers (seafarers) *and* helping to achieve fair competition. The latter goal is sometimes described as creating

[15] Clearly the cost of transporting goods is, in fact, an important aspect of the market competitiveness of many of the world's goods, aside from revenue related to ships' operations. Governance scholars, Errol Mendes and Ozay Mehmet, *Global Governance, Economy and Law, Waiting for Justice* (New York, London: Routledge, 2003) have provided an extensive critique of the failure to integrate the issue of labour with trade in the "strong" World Trade Organization (WTO) process and the failure of the present system in the Trade Policy Review Mechanism (TPRM) of the WTO or the supervisory system of the ILO to require that States provide a full picture of the relationship between labour costs and trade competitiveness, see esp. Chapter 2 at p. 79ff and Chapter 4 at pp. 153–177. A recent joint study by the ILO and WTO, M. Jensen and E. Lee, *Trade and Employment: Challenges for Policy Research* (Geneva: ILO and WTO, 2007), examined the interaction of trade and employment policy. In addition, the ILO has begun to carry out research with respect to the economic dynamics of international labour standards: see the Report to the Governing Body of the ILO, *Project on economic dynamics of international labour standards*, ILO Doc No GB.300/LILS/10, 300th Session, Governing Body, Geneva, November 2007: <www.ilo.org/global/What_we_do/Officialmeetings/gb/GBSessions/GB300/lang--en/commId--LILS/WCMS_083598/index.htm>. The MLC, 2006 may provide a response to critiques such as Philip Alston, "'Core Labour Standards' and the Transformation of the International Labour Rights Regime," 15 *European Journal of International Law* 15(3) (2004): 457–521, which raised concerns (see page 519) about the "eschewing of legalism" and, in his view, the related "weakening" of the labour standards regime, especially with the adoption in 1998 of the "Declaration on Fundamental Principles and Rights at Work" and a principles-based approach, relative to the stronger or "hard" trade law system.

[16] The term "decent work" encapsulates the contemporary agenda of the ILO:

> The Decent Work concept was formulated by the ILO's constituents – governments and employers' and workers' organizations – as a means to identify the Organization's major priorities and to modernize its approach for the twenty-first century. It is based on the understanding that work is a source of personal dignity, family stability, peace in the community, democracies that deliver for people, and economic growth that expands opportunities for productive jobs and enterprise development. In a relatively short time this concept has led to an international consensus among governments, employers, unions and civil society organizations that productive employment and Decent Work are key elements to achieving a fair globalization and the reduction of poverty."

See ILO, "Facts on Decent Work," (2006), <www.ilo.org/global/About_the_ILO/Media_and_public_information/Factsheets/lang--en/docName--WCMS_082654/index.htm>.

a "level playing field"[17] among employers (shipowners) by addressing the unfair competition resulting from shipowners and flag States condoning or failing to regulate poor or substandard (below international norms) working conditions.

The maritime sector, oft described as one of the first "internationalized" or "globalized"[18] sectors, presents particular challenges to the international legal system.[19] It is perhaps one of the most concrete examples of the "cutting edge" or the future of many of the underlying issues relating to globalization, with workers drawn from many countries working in workplaces that move between jurisdictions and involving employer/ownership arrangements that are similarly fluid. This is the future face of many sectors in connection with the impact of electronic communication, and the equally fluid "e-highway," on the location of workplaces and workers, and relates to diverse issues including, *inter alia*, extra-territoriality, the tension between the "North" and the "South," and the challenges to principles of State sovereignty as manifest in the concept of flag State control.[20]

---

[17] This concept is, of course, one that can be debated especially in the context of the developing economies.

[18] B. Wu and H. Sampson, "Reconsidering the Cargo Sector's Seafarer Labor Market: A 21st Century Profile of Global Seafarers," *Ocean Yearbook* 19 (2005): 357–380.

[19] From one perspective it may be a relative easy or vulnerable target for experimenting with change. The high degree of internationalization and the extensive international regulatory regime already in place under the IMO and the ILO means that, depending on the extent of economic reliance on the sector, some governments are more easily able to withstand domestic lobbying to resist international regulation and standards.

[20] See, for example, the recently resurrected call for a "genuine link": Resolutions 58/240 (at para. 42) and 58/14 adopted by the General Assembly of the United Nations at its 58th session invited the IMO and other relevant agencies to study, examine and clarify the role of the "genuine link" in relation to the duty of flag States to exercise effective control over ships flying their flag, including fishing vessels. Resolutions 59/24 (para. 41) and 59/25 (para. 30) also requested the Secretary-General to report to the General Assembly at its 61st session on the study undertaken by IMO in cooperation with other competent international organizations on the role of the genuine link and the potential consequences of non-compliance with duties and obligations of flag States described in relevant international instruments. The IMO reported 23 June 2006. The lengthy document, comprising briefs from the various organizations concerned, was reported to the General Assembly of the United Nations, Doc. A61/160, 17 July 2006, Item 69(a) of the provisional agenda in connection with oceans and the law of the sea. But see recent views of the ITF that suggests that there may be a shift in its strategy for

Although considered in this book specifically under the topic New Standards for International Shipping, this contribution raises broader issues that were also preoccupations of Douglas Johnston, namely, development of international law as a system, ocean regime-building, and the revitalization of the law of the sea including revising the 1982 United Nations Convention on the Law of the Sea (LOS Convention), or "RUNCLOS" as he termed it. Accordingly, a discussion of the MLC, 2006 is of interest to maritime lawyers and law of the sea scholars in that it constitutes an elaboration of issues referenced only *en passant* in the LOS Convention.

Under Article 94(1) of the LOS Convention, a State is obliged to "effectively exercise its jurisdiction and control in administrative, technical, and social matters over ships flying its flag". The MLC, 2006 is an important step to develop this obligation into a more effective regime[21] complementing, as suggested earlier, the regulatory regime in connection with ship-source marine pollution, seafarer training and competence, and ship safety and security. The LOS Convention addresses many aspects of ocean use but, aside from the indirect impact on workers as a result of the resource-based allocation of jurisdiction, it does not deal with the use of the ocean, particularly the high seas, as a workplace or a site for human rights.[22] These matters are simply left to the "default" jurisdiction of the relevant flag State of the ship.[23] The MLC, 2006 is a significant step forward in that it develops, in detail and at the level of a multilateral convention, explicit responsibilities for both flag States, in connection with the inspection and certification of labour conditions on ships, and

dealing with flags of convenience (FOC): "ITF Softens FOC Stance," *Fairplay Daily News*, 24 July 2007.

> SINGAPORE 24 July – The ITF has apparently softened its stand against the flags of convenience system. "Just because the vessel is registered in Panama or Liberia does not mean that it is a substandard vessel," the federation's maritime coordinator Stephen Cotton told the Singapore Organisation of Seamen yesterday…

21 See generally papers in, for example, O. Young, ed., *The Effectiveness of International Environmental Regimes: Causal Connections and Behavioural Mechanisms* (Cambridge, MA: MIT Press, 1999).

22 There are, however, some human rights provisions relating to slavery and genocide, although human security rights are also encompassed in connection with, for example, the provisions governing the protection of the marine environment or living resources.

23 *United Nations Convention on the Law of the Sea*, Geneva, 10 December 1982, 1833 *U.N.T.S.* 396, Art. 92ff.

port States that choose to inspect foreign ships. It also introduces a new player in the flag State, coastal/port State jurisdictional web – States with labour-supply responsibilities that arise independently of those of the flag State.

This contribution also argues that the MLC, 2006 is relevant to broader questions regarding "governance," the emergence of international organizations as a regulatory bureaucracy, and the development of the compliance or supervisory systems of these organizations.[24] Specifically it provides an opportunity for further reflection on the problem of fragmentation and the need for greater integration of activities between United Nations (U.N.) specialized agencies.[25]

From a treaty drafting perspective (regulatory design), the MLC, 2006 provides an interesting example of the potential cross-fertilization of approaches and format and concepts across and within regimes.[26] The MLC, 2006 reflects a conscious and explicit development of approaches adopted in previous international labour instruments. It adopts and builds upon a format and approach taken from conventions developed under the IMO in connection with a range of structural approaches to improving compliance (e.g., certification of ships) and an approach that allows for more rapid updating of technical parts of the Convention through a

24 *Supra* note 19. For a discussion of the potential arrangement including delegation to international organizations, see M. Koenig-Archibugi, "Mapping Global Governance," in D. Held and A. McGrew, eds, *Governing Globalization: Power, Authority and Global Governance* (Cambridge: Polity Press, 2002), p. 46.

25 See generally, M. L. McConnell, "Inter-Agency Collaboration or Inter-Agency Competition – A Challenge for the UN System," in A. Kirchner, ed., *International Marine Environmental Law* (The Hague, New York, London: Kluwer Law International, 2003).

26 Various scholars in the field of regime theory have elaborated frameworks for understanding the relationship between regimes: see, for example, V. K. Aggarwal, "Reconciling Multiple Institutions: Bargaining, Linkages and Nesting," in V. Aggarwal, ed., *Institutional Design for a Complex World: Bargaining, Linkages and Nesting* (Ithaca: Cornell University Press, 1998); O. R. Young, *Governance in World Affairs* (Ithaca: Cornell University Press, 1999). For a good overview of the vocabulary and a discussion also about national level implementation see K. Abbot and D. Snidal, "Nesting, Overlap and Parallelism: Governance Schemes for International Production Standards," paper prepared for the Alter-Meunier Princeton Nesting Conference, February 2006, <www.princeton.edu/~smeunier/Abbot%20Snidal%20memo.pdf>. As the word itself suggests, the concept of "nesting" refers to the idea of a narrower or more issue specific regime (institution and rules) operating within a broader regime.

combination of format and tacit acceptance techniques. It has also been designed to operate, as seamlessly as possible, within the regional port State memorandum of agreement (MOU) system[27] and related enforcement mechanisms. Importantly, the MLC, 2006 addresses an issue that hitherto existed only at the level of guidance in the IMO in relation to the explicit recognition of the fact that many States delegate the operational implementation of their international responsibilities to private sector organizations (recognized organizations, or ROs).[28] Finally, with respect to the need to better ensure effectiveness and a global impact, the MLC, 2006 introduces a "no more favourable treatment"[29] clause that will apply to ships of countries that do ratify the MLC, 2006 and will help to ensure a level playing field for shipowners. In addition, the entry into force requirements are unusually demanding for an ILO convention.[30]

This contribution first provides a brief overview of maritime labour in the context of the maritime sector and the international institutional

---

[27] See, for example, the Paris Memorandum of Understanding (MOU) on Port State Control (PSC), <www.parismou.org> or the Tokyo MOU on PSC, <tokyo-mou.org>.

[28] The issues, and the potential conflict of interest, that may arise for such organizations are of particular interest. The extent to which the international maritime regulatory edifice rests on the lynchpin of these private sector relatively unregulated organizations is a point of some interest.

[29] MLC, 2006, *supra* note 2, at Art. V, para. 7:

> Each Member shall implement its responsibilities under this Convention in such a way as to ensure that the ships that fly the flag of any State that has not ratified this Convention do not receive more favourable treatment than the ships that fly the flag of any State that has ratified it.

[30] *Ibid.*, Art. VIII:

> 2. This Convention shall come into force 12 months after the date on which there have been registered ratifications by at least 30 Members with a total share in the world gross tonnage of ships of at least 33 per cent.

The MLC, 2006 also draws up the approach used in an earlier effort at consolidation in 1976, the *Convention concerning Minimum Standards in Merchant Ships* usually known as *The Merchant Shipping (Minimum Standards) Convention, 1976* (No. 147), Geneva, 29 October 1976, <www.ilo.org/ilolex/cgi-lex/convde.pl?C147>, Art. 6:

> 2. It shall come into force twelve months after the date on which there have been registered ratifications by at least ten Members with a total share in world shipping gross tonnage of 25 per cent.

framework. It then provides a detailed overview of the MLC, 2006 and highlights several provisions that relate to the points mentioned above before concluding with a consideration of the potential implications of the Convention from a broader systemic perspective. It argues that the MLC, 2006 provides, as suggested by the Director-General of the ILO, a way forward for the ILO to ensure more effective labour standards and reflects an analysis underway in the ILO with respect to the need to design conventions to take account of the economic dynamics of international labour standards.[31] It is also clear, however, that existing ILO principles and approaches regarding national flexibility must be explicitly considered in connection with other principles such as transparency and accountability. The process by which the MLC, 2006 was developed is also instructive with respect to the need to ensure the commitment of the two sectors that will affect governmental law and policy-making: the industry – the workers and the employers.

Context

The Maritime Sector

The terms "maritime sector" and "shipping" are often loosely used to capture a range of activities spanning shore-based activities, such as arranging the carriage of goods by sea, and ship-board operations. Shipping is inseparable from world trade and economic development: access to inexpensive and reliable transportation impacts directly on the competitiveness of exports and imports which, in turn, affects a wide range of domestic economic activities. Like other economic sectors, the cost of operating ships involves both the cost of the plant itself and the labour costs. As noted in the introduction, it is often described as one of the first globalized sectors or multinational sectors in that, although there is domestic shipping in most countries, trade in goods is largely between countries and involves ships registered in many different countries.

---

[31] *Supra* note 15, and see also, M. L. McConnell, "Economic Dynamics of International Labour Standards: Implications for International Labour Law and Practice," Special Presentation to the Workshop on the Economic Dynamics of International Labour Standards, ILO, Geneva, 21–22 June 2007.

Historically, in many countries, this sector – the merchant marine – was heavily protected from international competition and ownership, as an aspect of national public policy. Similarly, many countries had restrictions with respect to which ships or flags could trade to their ports and on what terms. This is no longer the case with the emergence in the 1950s of what are sometimes called flag of convenience (FOC) or more usually now "open" international registries.[32] These registries allow shipowners to register ships and operate under the jurisdiction of countries in which the owner and the workers on board the ship are not necessarily resident or nationals. Ships operating under such a system can transfer easily between flags and can be owned by a company in one State, managed by a company based in another country, with seafarers drawn from multiple jurisdictions. This is still the case today and, in fact, many countries that formerly had restrictions with respect to registration have now established "international registers" and adopted other incentives to attract ships to their flags. Much of the movement between registers initially related to increased regulatory activity with respect to safety, labour, and environmental standards. However, with the rise of port State control, decisions as to where to flag are now more related to corporate regulation and taxation and, in the case of fishing vessels, conservation measures.

The Law of the Sea and the International Maritime Regulatory Regime

Conventions directed to conditions both of ships and on board ships predate the LOS Convention and, in some cases, even the earlier attempts to codify the law of the sea in the 1958 Geneva Conventions[33] by many years. This is particularly the case with conventions addressing working and living conditions on ships. For example, among the earliest of the ILO conventions are the *Minimum Age (Sea) Convention, 1920* (C.7), the *Placing of Seamen Convention, 1920* (C.9), and the *National Seamen's*

[32] See M. L. McConnell, "Business as Usual: An Evaluation of the 1986 United Nations Convention on Conditions for Registration of Ships," *Journal of Maritime Law and Commerce* 18 (1987): 435–449; M. L. McConnell, "Darkening Confusion Mounted Upon Darkening Confusion: The Search for the Elusive Genuine Link," *Journal of Maritime Law and Commerce* 6 (1985): 365–396.

[33] See, for example, the 1958 *Convention on the High Seas*, Geneva, 29 April 1958, 6465 *U.N.T.S.* 450, pp. 82–103.

Code Recommendation, 1920 (R.9). Similarly, the IMO also developed conventions on marine protection in the 1950s and early 1970s.

The LOS Convention was an effort to develop a comprehensive and over-arching international legal regime for the law of the sea. In addition to issues with respect to the jurisdictional and resource related rights of coastal States, it also dealt with the issue of (flag State) responsibility for ships in waters outside territorial seas (high seas) and related inspection rights and obligations vis-à-vis foreign ships in ports.

In this respect the LOS Convention attempted to capture and codify the extant elements and principles in the ILO and IMO regimes. Interestingly, despite the much more developed international maritime labour regime that existed at the time,[34] marine environmental protection from ship-source pollution received significant attention in the LOS Convention (Part XII), while the issue of working and living conditions on ships is sparsely addressed. In fact, one of the earlier conventions of the ILO, the *Merchant Shipping (Minimum Standards) Convention, 1976*, set in place the concept of port State inspections, and also required ratification of several IMO ship safety conventions[35] as a minimum condition for ratification of the Convention. (This may be one of the earliest attempts to expressly integrate the regimes and coordinate activities among U.N. organizations.)

The key provisions of the LOS Convention are (emphasis added):

---

[34] By 1982 the ILO had developed an extensive body of conventions and recommendations.

[35] *Supra* note 30, at Art. 5:

> 1. This Convention is open to the ratification of Members which –
>   (a) are parties to the International Convention for the Safety of Life at Sea, 1960, or the International Convention for the Safety of Life at Sea, 1974, or any Convention subsequently revising these Conventions; and
>   (b) are parties to the International Convention on Load Lines, 1966, or any Convention subsequently revising that Convention; and
>   (c) are parties to, or have implemented the provisions of, the Regulations for Preventing Collisions at Sea of 1960, or the Convention on the International Regulations for Preventing Collisions at Sea, 1972, or any Convention subsequently revising these international instruments.

Article 92
Status of ships

1. Ships shall sail under the flag of one State only and, save in exceptional cases expressly provided for in international treaties or in this Convention, shall be subject to its exclusive jurisdiction on the high seas....

Article 94
Duties of the flag State

1. *Every State shall effectively exercise its jurisdiction and control in administrative, technical and social matters over ships flying its flag.*
2. In particular every State shall:

    (a) maintain a register of ships containing the names and particulars of ships flying its flag, except those which are excluded from generally accepted international regulations on account of their small size; and
    (b) *assume jurisdiction under its internal law over each ship flying its flag and its master, officers and crew in respect of administrative, technical and social matters concerning the ship.*

3. Every State shall take such measures for ships flying its flag as are necessary to ensure safety at sea with regard, inter alia, to:

    (a) the construction, equipment and seaworthiness of ships;
    (b) *the manning of ships, labour conditions and the training of crews, taking into account the applicable international instruments*;
    [...]

5. In taking the measures called for in paragraphs 3 and 4 each State *is required to conform to generally accepted international regulations, procedures and practices* and to take any steps which may be necessary to secure their observance.
6. *A* State which has clear grounds to believe that *proper jurisdiction and control with respect to a ship have not been exercised may report the facts to the flag State.* Upon receiving such a report, the flag State shall investigate the matter and, if appropriate, take any action necessary to remedy the situation.

Although directed to responsibility on the high seas these provisions, *de facto,* frame the overall regime for the regulation in this sector. There are additional provisions in the LOS Convention addressing responsibilities and rights with respect to ship-source marine pollution and other matters, but not with respect to working and other conditions on ships. Instead, as suggested by Article 94, this is left as a matter for the domestic law (internal law) of the flag State, which is obliged to effectively exercise its jurisdiction and control. Paragraph 3(b) refers to labour conditions and only requires States to "take account" of international instruments. Paragraph 5 does, however, require *conformity* with "generally accepted" international regulations. Paragraph 6 allows for what could be seen as some limited form of port State intervention but, unlike the provisions found elsewhere in the LOS Convention regarding marine pollution, it does not go beyond reporting to the flag State on these matters. In principle, however, it can be seen that the 157 State parties[36] to the LOS Convention have an obligation to address labour and social conditions on ships at least to the extent just indicated.

It is also useful to briefly consider the key elements of the maritime regulatory regime that has been elaborated under the IMO. The IMO conventions[37] all generally follow a similar formula to addressing ship-related concerns. This approach is of interest as it expressly responds to the transnational character of the sector primarily regulated by IMO,[38] focusing on effectiveness whilst remaining firmly founded on traditional international law principles of State sovereignty. IMO conventions usually[39] speak to flag State administrations requiring that the ships flying the flag of that State, which are covered by the convention, operate according to the international regulations set out in the convention (usually in an annex). Ships are required to be inspected and certified by the flag State or by an organization acting on its behalf in carrying out the technical inspection/survey (the RO). The certificate is carried on the ship and subject to inspection in other (port) States under a voluntary regime.

36 As of 25 September 2008, see <www.un.org/depts/los>.

37 See <www.imo.org>.

38 Generally but not solely concerned with ships 500 GT and above that are engaged in voyages outside the flag State.

39 More recent conventions reflect landward "creep" and address coastal State and even traditionally land-based activities such as the environmental and safety standards in the operation of ship recycling facilities.

Port States essentially carry out inspections of the documentation required by the conventions and, if there are grounds to suggest a problem, a further inspection is carried out and, if necessary, the ship detained. Reports are filed with flag States regarding the problem and under the current regional arrangements (PSC MOUs or Port State Agreements (PSA)) the information is also put into an electronic database. This system could be seen as an important form of international cooperation in that the port State, while acting in its own interest in seeking "safe and secure ships and cleaner seas" to avoid harm to its waters and coastline, is also supporting and extending the regulatory control exercised by the flag State. In ensuring continual or on-going compliance with international standards, this system can also be seen as a way of assisting weaker flag State administrations that must, irrespective of their lack of capacity, ensure that their ships are operated in accordance with international standards. Obviously the system also operates indirectly as a check on the State's implementation of its international obligations[40] and provides an incentive to shipowners to operate under flag States that have a record of reliable flag State certification.

A further aspect of the IMO conventions is the inclusion of a "no more favourable treatment clause" under which State parties agree that ships of non-party States will not be treated more favourably with respect to the matters under the convention. This helps to establish a "level playing field" in that shipowners that register in States that have not ratified the conventions will still be subject to inspection for the same matters in foreign ports but will not have the benefit of a convention certificate. The entry into force formula of the IMO conventions combines a number of States with a percentage of the world fleet (by gross tonnage) in order to ensure that flag States with a significant interest in and responsibility in the sector are party.

Although the IMO has a large number of conventions, three (STCW, MARPOL and SOLAS)[41] can be considered core and, as mentioned earlier, are sometimes called the three pillars of the regime dealing with ship safety and security, ship-source marine pollution, and seafarer training and

---

40 Appendix "Examination and clarification of the role of the "genuine link" in relation to the duty of the States to exercise effective control over ships flying their flags, including fishing vessels," submitted by the International Labour Office, *supra* note 20, at para. 19.

41 *Supra* note 11.

competence. Together these conventions provide the framework for the IMO mandate to promote safe and secure ships and cleaner seas.

In terms of structure, IMO conventions generally follow a standard design. They comprise fairly brief articles spelling out general State obligations regarding implementation, and often, inspection, certification and port State control, scope and coverage, entry into force, amendments, and include a requirement to comply with the convention and its annex. The more detailed mandatory "technical" provisions, the "regulations," are attached in one or more annexes. Further instruments such as codes (sometimes with both mandatory and non-mandatory provisions) and voluntary guidelines are also often developed to complement the convention and deal in greater detail with specific elements of the regulations in the annex. The different parts of the conventions often have differing amendment provisions, with the tacit acceptance procedure used to allow for more rapid updating of the technical provisions in the annexes.

Overview of the International Labour Organization

Aside from being one of the earliest of the international organizations, predating the United Nations,[42] the ILO differs in several important respects from other U.N. organizations, the most predominant of which is its tripartite nature. As described above, States are represented in the ILO by national delegations comprising representatives from the government, employers, and workers organizations, all of whom have a vote to adopt instruments. Since 1920, the ILO has adopted 188 conventions,[43] 5 protocols and 199 recommendations. Until recently, ILO conventions tended to be either issue and/or sector specific, were usually fairly brief, and followed a standard format of specific requirements on the particular topic, with no preamble and all the mandatory ("shall") provisions set out in articles. Normally, relatively few ratifications are required to bring the conventions into force. Recommendations are often adopted along with conventions and contain additional technical provisions framed as "should." A convention, once adopted by a two-thirds majority of the annual ILC, does not move to signature then ratification. Rather, there

[42] The ILO was established in 1919 as an agency of the predecessor to the United Nations, the League of Nations.

[43] Also sometimes known as international labour standards (ILS).

is an immediate obligation under the ILO's *Constitution* on each Member to place it before the "competent authority" for ratification within 12 or 18 months[44] and, if not ratified, to report on national coverage on the matter and efforts to move to ratification. There is a supervisory system[45] under the ILO's *Constitution* with mandatory reporting on implementation measures, reviews, as well as provision for complaints by employer or workers' organizations or other ILO Members.[46] Flexibility with respect to national implementation and concerns with respect to the stage of economic development are usually accommodated through declarations but without reservations.[47] Under the *Constitution*, a new ILO convention cannot operate to reduce existing rights at the national level.[48]

In the last decade, the ILO has increasingly focussed it efforts on achieving change by working with governments and relevant national organization to secure decent working conditions, including social protection, largely by adopting a rights-based approach. This has also involved attention to ensuring effective implementation – that is, moving beyond *de jure* to *de facto* implementation of legal obligations. Related to this has been a concern that, while there are many conventions in force, key interested States have not ratified them for reasons such as technical aspects that are out-dated or otherwise impede implementation, even in cases where there is agreement with the principle of the convention. Not surprisingly, the interaction of the economic system and the international trade regime and labour standards, particularly in the context of sustainable economic development is a central concern. The ILO has not taken a position against globalization per se but rather it has long argued for "fair globalization," that is the need to ensure that exploitation of workers does not become a competitive advantage.[49] All of these factors provide the backdrop against which the Maritime Labour Convention, 2006 was developed.

[44] *Supra* note 4, at Art. 19.

[45] *Ibid.*, Art. 34.

[46] There is also the possibility of a reference to the International Court of Justice after a Commission of Inquiry (Art. 29) or for disputes regarding interpretation (Art. 37).

[47] *Supra* note 4, at Art. 19.

[48] *Ibid.*, Art. 19.

[49] World Commission on the Social Dimension of Globalization, *A Fair Globalization: Creating Opportunities for All* (Geneva: ILO, 2004).

The Maritime Labour Convention, 2006

Introduction

One of the first legal instruments that the ILO adopted was the *National Seamen's Code Recommendation, 1920* (R.9), a year after the ILO was created. It called for the establishment of an international seafarers' code that would clearly set out the rights and obligations relevant to this sector. Between 1920 and 2006 more than 40 maritime labour conventions were adopted.[50] As suggested by the volume of legal activity, seafarers, as internationalized workers, have occupied a special place in the ILO, both in its legal regime and in its institutional arrangements. There are dedicated maritime sessions of the ILC and a special entity established by the ILO's Governing Body, the Joint Maritime Commission.[51] Shipowners, represented by the International Shipowners' Federation (ISF), and workers, represented by the International Transport Workers Federation (ITF), have worked together to negotiate working conditions, including for example, minimum wages, on an international basis. In fact, these representatives of the maritime industry initiated the move to develop the MLC. The foundation for the MLC, 2006 and many of its provisions was an industry agreement known as the "Geneva Accord" of 2001. The Geneva Accord set out an agreement between the international representatives of shipowners and seafarers on the need for a new convention

---

[50] There were also a number of conventions in connection with fishing and dock/port workers.

[51] See <www.ilo.org/public/english/dialogue/sector/techmeet/jmc/index.htm>. The Joint Maritime Commission (JMC) is

> ...a bipartite standing body that provides advice to the Governing Body on maritime questions including standard setting for the shipping industry. It is composed as follows: Chairman of the Governing Body; two Governing Body members (one Worker, one Employer); twenty regular shipowner members; twenty regular seafarer members; four deputy shipowner members; and four deputy seafarer members. The Standing Orders of the JMC provide guidance on how to deal with the various procedural questions that may arise in the course of the Commission's work.

For the formal rules adopted by the ILO Governing Body in 1948 see <www.ilo.org/public/english/bureau/leg/publ/marit/index.htm>.

and a new approach focussed on effectiveness and ensuring widespread ratification.[52]

From a labour perspective, the problems identified in the maritime sector included:

- poor working conditions in the industry
- changes in ownership, financing and the rise of ship management companies resulting in significant shifts in the labour market for seafarers
- development of consciously composed mixed nationality crews in a highly organized global network linking shipowners, ship managers, crew managers, and labour supplying agencies
- increased internationalization of ship registries and "flags of convenience"
- a need to provide a "level playing field" and avoid exploitation of workers
- increased stress and complexity in the maritime work place that has an impact on the health and social security of workers
- the high level of detail combined with the large number of ILO conventions led to problems for compliance and enforcement and a relatively low ratification level for some key conventions

Put simply the goal of the MLC, 2006 was to provide comprehensive and effective protection of a seafarer's right to decent work, and to secure a "level playing field" for governments and shipowners, with flexibility as to the means of delivering this protection and accommodating diversity. As noted by Doumbia-Henry et al.,

> The basic purpose of the *Maritime Labour Convention, 2006* was not... simply to consolidate the existing *corpus* of international maritime labour standards and to bring clarity and coherence into the various texts (of Conventions and Recommendations) adopted since 1920. The new Convention was principally designed to meet the serious concerns expressed by both the seafarer and shipowner communities, and later endorsed by Governments, that the existing instruments were losing their relevance. Their substance was of good quality, but they were unevenly applied and enforced and often poorly ratified. Moreover, the existing procedures for the revision of international labour Conventions have not allowed the technical aspects to be rapidly updated so as meet modern conditions in the industry. From

[52] As noted in the Office Commentary prepared for the International Labour Conference, 94th (Maritime) Session, 2006, *Adoption of an instrument to consolidate maritime labour standards*, Report I(1A) (Geneva: International Labour Office, 2005), p. 1.

> a broader systemic perspective the underlying concern was that, with the increasing level of regulation in other aspects of shipping operations and increased international competition, labour would be the one remaining area for cutting costs. Paralleling these sectoral concerns, since the mid-1990s the ILO had also been undergoing a review of its Conventions and standard-setting activities with a view to making them more effective and to increasing and broadening ratifications.[53]

The process by which the MLC, 2006 text was negotiated is a story that is of interest[54] but is outside the scope of this contribution. It should, however, be noted that between the Geneva Accord in 2001 and the final adoption of the MLC, 2006 in February 2006 at the 10th Maritime Session, 94th Session of the ILC, there were six high level tripartite consultative meetings,[55] a preparatory technical conference, and an additional intersessional meeting. In addition, there were several officers' meetings to resolve particular issues in the text, as well as strategic development of special procedures within the ILO framework to facilitate consideration of such a comprehensive text.

The role of the industry or "civil society," as it could also be described, was a very important aspect of this lengthy negotiating process that cul-

---

[53] C. Doumbia-Henry, D. Devlin and M. L. McConnell, "The *Maritime Labour Convention, 2006* Consolidates Seafarers' Labour Instruments," *ASIL Insight* 10(23) (13 September 2006), <www.asil.org/insights/2006/09/insights060913.html>.

[54] As noted by Doumbia-Henry, et al., *ibid.:*

> ...the process of development of the text, over a period of nearly five years, by the International Labour Office under the guidance of the tripartite constituents of the Organization, reflects a shift in ILO practice in the drafting of texts for submission to the International Labour Conference. This process of more extensive consultation and international "social dialogue" led to the development of concrete tripartite solutions to address highly sensitive issues such as the provision of social security protection, employer liability and worker complaint mechanisms, before the Convention was considered for adoption. In addition, strategic approaches needed to be devised to facilitate discussion of such a comprehensive and lengthy instrument within the ILO's constitutional framework. The fact that very few amendments were submitted to the proposed text considered by the Conference and, of these, very few were controversial, testifies to the effectiveness of the tripartite process and of international social dialogue.

See also, J. I. Black, Jr., "Reflections on the Negotiation of the Maritime Labour Convention 2006 at the International Labor [*sic*] Organization," *Tulane Maritime Law Journal* 31 (2006): 35–56.

[55] Two of which were subgroup meetings. See Report 1(1A), *supra* note 52, at pp. 1–4.

minated in the MLC, 2006. Essentially the international representative organizations of shipowners and seafarers negotiated solutions to many specific issues and helped broker compromise texts in areas where some governments were having difficulty as a result of national interests. Despite some obvious areas of difference in perspective there was, in fact, a very high level of cohesion as to what constitutes decent work in the context of modern "quality shipping." The most prescriptive areas of the MLC, 2006, for example, the accommodation standards on ships, essentially reflects standards for berths, sleeping room sizes, etc., that were agreed in discussion between the shipowners' and seafarers' organizations based on industry norms. Government representatives were often primarily concerned about issues such as the impact of these standards on smaller ships and ships that do not go on international voyages, the need to respect cultural and other differences in some aspects of onboard conditions, the need to account for differing approaches to social protection, as well as the potential administrative costs that might result from implementation of the Convention.

The representatives of these shipowner and seafarer organizations also attended and closely followed the IMO meetings and processes, and had an interest in ensuring that labour standards were given the same profile as environmental and safety and security issues. As much as possible, efforts were made to ensure that the two regimes were consistent in approach and avoided uncertainty resulting from conflicting or differing requirements, especially in overlapping areas.[56] In terms of integrating forces in the international system, it was really the pressure from these civil society representatives, far more than governments, perhaps because of the differing composition of delegations to the IMO and the ILO and the related national departmental assignments, that helped to ensure the cross-fertilization of ideas and cooperation to create complementary regimes.

The MLC, 2006 is, as noted earlier, over 100 pages in length and equally authoritative in English and French. The MLC, 2006 consolidates the substantive content of 37 existing conventions[57] since 1920, including

[56] For example, seafarer training and certification, hours of work and rest, medical examinations, onboard medical personnel requirements, aspects of liability, and human factors in safety system management.

[57] Four conventions are not revised: Convention Nos. 108, 185 relating to seafarers identity documents, 71 (pensions), and 15 (an outdated convention relating to stokers). The

areas on which many countries had not agreed previously such as social security protection.[58] The text also includes a new accelerated procedure to update (by amendment) the Convention. The MLC, 2006 was adopted by a record vote of 314 in favour and none against and involved 106 ILO Member States.[59]

The new strategic approach adopted in the Convention, aimed largely at achieving nearly universal ratification, has been described as a "blend of firmness on rights and flexibility with respects to approaches to implementation of the more technical requirements and because of the advantages it gives to ship of countries that ratify it."[60] In order to achieve this flexibility without decreasing existing levels of protection to seafarers, the Convention contains several innovative approaches engaging the interaction of both its structure and content. The next section briefly outlines the Convention's structure and substantive content. The innovations,[61] in part, reflect the development of existing ILO approaches as well as its focus on securing effective implementation and improving enforcement and encouraging social dialogue at a national level.

Structure and Content

From a structural perspective the MLC adopts an approach similar to the IMO's STCW Convention. It has three different[62] but related parts:

---

related recommendations, as well as the listed conventions that have not yet entered into force, can be withdrawn by a decision of the ILC in accordance with Article 45bis of its Standing Orders.

58 Once the MLC, 2006 enters into force, the 37 conventions will be closed to further ratification and *ipso facto* denounced for States party to the MLC, 2006. They will then be phased out.

59 See comments *supra* notes 4, 5 and 6.

60 ILO, *Maritime Labour Convention, 2006, Frequently asked questions* (Geneva: ILO, 2006), 9, <www.ilo.org/global/What_we_do/InternationalLabourStandards/MaritimeLabourConvention/FAQs/lang--en/index.htm#P39_7456>.

61 See also, *supra* note 8.

62 It should be noted that even the adoption of this terminology was a significant change for an ILO convention, which normally only has articles. "Recommendations" refer to a separate instrument. The terms "guidelines" and "codes" are adopted in some sectors, such as occupational safety and health, to refer to nonbinding advisory publications. In addition, in the ILO, the term "standards" is used interchangeably with conventions, e.g., international labour standards (ILS). Thus the use of the term standard as distinct category within a convention/ILS is also a change.

articles, regulations and a two-part Code (Part A – mandatory Standards, Part B – non-mandatory Guidelines). There is an important "Explanatory Note" found after the articles which, although not officially part of the Convention text, is intended to provide further information and assistance, especially to governments about the relationship between the parts of the Code. In addition, the MLC, 2006, again unusually for an ILO convention, contains a lengthy preamble referring to the LOS Convention and to the IMO conventions, as well as to several core ILO conventions.[63]

The MLC, 2006 is organized into 16 articles (Articles I-XVI) and five titles:

Title 1: Minimum requirements for seafarers to work on a ship
Title 2: Conditions of employment
Title 3: Accommodation, recreational facilities, food and catering
Title 4: Health protection, medical care, welfare and social protection
Title 5: Compliance and enforcement

These five titles comprise the substance of the existing 37 conventions with some modernization and updating. The MLC, 2006 also

---

[63] Desiring to create a single, coherent instrument embodying as far as possible all up-to-date standards of existing international maritime labour conventions and recommendations, as well as the fundamental principles to be found in other international labour conventions, in particular:

- the Forced Labour Convention, 1930 (No. 29);
- the Freedom of Association and Protection of the Right to Organise Convention, 1948 (No. 87);
- the Right to Organise and Collective Bargaining Convention, 1949 (No. 98);
- the Equal Remuneration Convention, 1951 (No. 100);
- the Abolition of Forced Labour Convention, 1957 (No. 105);
- the Discrimination (Employment and Occupation) Convention, 1958 (No. 111);
- the Minimum Age Convention, 1973 (No. 138);
- the Worst Forms of Child Labour Convention, 1999 (No. 182); and
- Mindful of the core mandate of the Organization, which is to promote decent conditions of work, and Recalling the ILO Declaration on Fundamental Principles and Rights at Work, 1998, and Mindful also that seafarers are covered by the provisions of other ILO instruments and have other rights which are established as fundamental rights and freedoms applicable to all persons,...

> ...occasionally contains new subjects, particularly in the area of occupational safety and health to meet current health concerns, such as the effects of noise and vibration on workers or other workplace risks.
>
> The provisions relating to flag State inspections, the use of "recognized organizations" and the potential for inspections in foreign ports (port State control) in Title 5 are based on existing maritime labour Conventions; however, the new Convention builds upon them to develop a more effective approach to these important issues, consistent with other international maritime Conventions that establish standards for quality shipping with respect to issues such as ship safety and security and protection of the marine environment.[64]

Each of the five titles comprises regulations, standards, and guidelines related to various topics[65] in what is described as a "vertically integrated"

---

[64] *Supra* note 60.

[65] E.g., **Title 1. Minimum requirements for seafarers to work on a ship**
Regulation 1.1 – Minimum age
Regulation 1.2 – Medical certificate
Regulation 1.3 – Training and qualifications
Regulation 1.4 – Recruitment and placement
Title 2. Conditions of employment
Regulation 2.1 – Seafarers' employment agreements
Regulation 2.2 – Wages
Regulation 2.3 – Hours of work and hours of rest
Regulation 2.4 – Entitlement to leave
Regulation 2.5 – Repatriation
Regulation 2.6 – Seafarers' compensation for the ship's loss or foundering
Regulation 2.7 – Manning levels
Regulation 2.8 – Career and skill development and employment opportunities for seafarers
Title 3. Accommodation, recreational facilities, food and catering
Regulation 3.1 – Accommodation and recreational facilities
Regulation 3.2 – Food and Catering
Title 4. Health protection, medical care and social security protection
Regulation 4.1 – Medical care on board ship and ashore
Regulation 4.2 – Shipowners' liability
Regulation 4.3 – Health and safety protection and accident prevention
Regulation 4.4 – Access to shore-based welfare facilities
Regulation 4.5 – Social security
Title 5. Compliance and Enforcement
Introductory paragraphs
Regulation 5.1 – Flag state responsibilities

presentation. The regulations and Parts A and B of the Code have a numbering system that links to the related regulations, standards, and guidelines.

Each regulation also has a "plain language" purpose clause.[66] Title 5, Part A of the Code has three appendices and Part B has one appendix.

Appendix A5-1 – List of matters for flag State inspection
Appendix A5-III – List of areas subject to detailed inspection in a port State
Appendix A5-II – "Model" documents relating to the inspection and certification system established in Title 5:
- a Maritime Labour Certificate
- a Declaration of Maritime Labour Compliance

Appendix B5-I – An example, to provide guidance as to the way a declaration might be filled out

These appendices set out the list of 14 areas of the Convention[67] that are to be verified by the flag State and are the subject of detailed port State control inspections under the new certification system. They also contain the model documents that are mandatory on ships 500 GT and above on international voyages (as defined in the Convention)[68] and voyages from or between ports in another country.

Regulation 5.2 – Port state responsibilities
Regulation 5.3 – Labour supplying responsibilities

[66] This reflects the view that the Convention's coverage should be clearly stated and easily accessible by the workers it was intended to benefit.

[67] Minimum age; medical certification; qualifications of seafarers; seafarers' employment agreements; use of any licensed or certified or regulated private recruitment and placement service; hours of work or rest; manning levels for the ship; accommodation; on-board recreational facilities; food and catering; health and safety and accident prevention; on-board medical care; on-board complaint procedures; payment of wages.

[68] *Regulation 5.1.3 – Maritime labour certificate and declaration of maritime labour compliance*

1. This Regulation applies to ships of:
   (a) 500 gross tonnage or over, engaged in international voyages; and
   (b) 500 gross tonnage or over, flying the flag of a Member and operating from a port, or between ports, in another country.

The articles set out the overarching obligations and principles of the MLC, 2006. The structure of the MLC, 2006 can be understood as a "cascade" of increasing specificity and flexibility to deal with the rights and obligations (under each title) with the most technical matters set out in the non-binding Guidelines in Part B of the Code.

In addition,

> ...[o]ne of the main purposes of the division between the Regulations and the Code concerns the accelerated ("tacit consent") amendment procedure;..., this can only be used for the Code provisions and any amendment must be consistent with the general provisions in the Articles and Regulations. The purpose of the Guidelines is to provide clear guidance as to the way in which each Standard should be implemented, while allowing flexibility for ratifying countries to provide for national implementation in a different way.[69]

The basic obligations of the State party are presented with respect to three roles: flag State, port State and a new area, a State with labour supplying responsibilities. Article V sets out these responsibilities.

Article V: Implementation and enforcement responsibilities

1. Each Member shall implement and enforce laws or regulations or other measures that it has adopted to fulfil its commitments under this Convention with respect to ships and seafarers under its jurisdiction.
2. Each Member shall effectively exercise its jurisdiction and control over ships that fly its flag by establishing a system for ensuring compliance with the requirements of this Convention[[70]], including regular inspections, reporting, monitoring, and legal proceedings under the applicable laws.
3. Each Member shall ensure that ships that fly its flag carry a maritime labour certificate and a declaration of maritime labour compliance as required by this Convention.

For the purpose of this Regulation, "international voyage" means a voyage from a country to a port outside such a country.

[69] Doumbia-Henry, et al., *supra* note 53.

[70] A term that is defined in the MLC, 2006, *supra* note 2, at Art. II, para. 1(e), as "*requirements of this Convention* refers to the requirements in these Articles and in the Regulations and Part A of the Code of this Convention."

4. A ship to which this Convention applies may, in accordance with international law, be inspected by a Member other than the flag State, when the ship is in one of its ports, to determine whether the ship is in compliance with the requirements of this Convention.
5. Each Member shall effectively exercise its jurisdiction and control over seafarer recruitment and placement services, if these are established in its territory.
6. Each Member shall prohibit violations of the requirements of this Convention and shall, in accordance with international law, establish sanctions or require the adoption of corrective measures under its laws which are adequate to discourage such violations.
7. Each Member shall implement its responsibilities under this Convention in such a way as to ensure that the ships that fly the flag of any State that has not ratified this Convention do not receive more favourable treatment than the ships that fly the flag of any State that has ratified it.

Articles III and IV, in particular, are seen as declaring the "seafarers' bill of rights" that are articulated in the balance of the Convention.

Article III: Fundamental rights and principles

Each Member shall satisfy itself that the provisions of its law and regulations respect, in the context of this Convention, the fundamental rights to:

(a) freedom of association and the effective recognition of the right to collective bargaining;
(b) the elimination of all forms of forced or compulsory labour;
(c) the effective abolition of child labour; and
(d) the elimination of discrimination in respect of employment and occupation.

Article IV: Seafarers' employment and social rights

1. Every seafarer has the right to a safe and secure workplace that complies with safety standards.
2. Every seafarer has a right to fair terms of employment.

3. Every seafarer has a right to decent working and living conditions on board ship.
4. Every seafarer has a right to health protection, medical care, welfare measures and other forms of social protection.
5. Each Member shall ensure, within the limits of its jurisdiction, that the seafarers' employment and social rights set out in the preceding paragraphs of this Article are fully implemented in accordance with the requirements of this Convention. Unless specified otherwise in the Convention, such implementation may be achieved through national laws or regulations, through applicable collective bargaining agreements or through other measures or in practice.

Paragraph 5 of this article is especially important in that contains one of the main areas of flexibility at a national level in connection with the precise mechanisms adopted for implementation.

Article VI clarifies the relationship between the various parts of the Convention. It also includes another important area of national flexibility – substantial equivalence.[71]

Article VI: Regulations and Parts A and B of the Code

1. The Regulations and the provisions of Part A of the Code are mandatory. The provisions of Part B of the Code are not mandatory.
2. Each Member undertakes to respect the rights and principles set out in the Regulations and to implement each Regulation in the manner set out in the corresponding provisions of Part A of the Code. In addition, the Member shall give due consideration to implementing its responsibilities in the manner provided for in Part B of the Code.
3. A Member which is not in a position to implement the rights and principles in the manner set out in Part A of the Code may, unless expressly provided otherwise in this Convention, implement Part A through provisions in its laws and regulations or other measures which are substantially equivalent to the provisions of Part A.
4. For the sole purpose of paragraph 3 of this Article, any law, regulation, collective agreement or other implementing measure shall

[71] This does not apply to Title 5, MLC, 2006, *supra* note 2, Introductory para. 3.

be considered to be substantially equivalent, in the context of this Convention, if the Member satisfies itself that:

a) it is conducive to the full achievement of the general object and purpose of the provision or provisions of Part A of the Code concerned; and
b) it gives effect to the provision or provisions of Part A of the Code concerned.

The interaction of these elements, in addition to some further national level flexibility under Article II with respect to smaller ships (200 GT and below that do not go international voyages) and the coverage of seafarers and ships has been explained[72] as follows:

> Flexibility is provided through the distinction drawn between Standards and Guidelines. In addition, the Convention has also been guided (especially in the case of social security protection) by the programmatic approach in the *1966 International Covenant on Economic, Social and Cultural Rights.*[73] In addition, it includes some provisions to provide national flexibility in implementation based on principles of transparency and accountability. For example, the definition and scope provisions for the Convention, set out in

[72] Doumbia-Henry, et al., *supra* note 53.

[73] *Ibid.*, at footnote 6 states:

> UNGA Resolution 2200A (XXI) of 16 December 1966, 993 UNTS 3 (in force 3 January 1976). See, in particular, the approach suggested in Article 2 of the Covenant and related commentary (see: CESCR General Comment 3, 1990 on Art. 2, para. 1), which explicitly recognizes the concept of "taking steps" and "achieving progressively the full realization" of rights. This approach is consistent with the ILO Constitution (Article 19, *Constitution of the International Labour Organisation*) which expressly recognizes the importance of recognizing differing climatic, industrial or special circumstances of countries that may affect their ability to implement Convention related obligations. The objective is to encourage initial commitment and implementation with a view to securing increased and, finally, full implementation as national conditions allow (see, e.g., MLC, 2006, Standard A4.5, para. 3, "Each Member shall take steps according to its national circumstances to provide the complementary social security…).

> Article II, relating to the terms "ship"[74], "seafarer"[75] and "shipowner,"[76] are intentionally broad and inclusive. There is national flexibility explicitly provided to address the problems some States will encounter in applying some or all of the Convention's requirements to particular categories of workers or ships or to smaller ships engaged only in domestic voyages. Any exemptions or exclusions made by national authorities require consultation with shipowners and seafarers organizations and reporting of the exemption or exclusion to the ILO. The ship certification system, consistently with other maritime Conventions, is mandatory only for ships of 500 gross tonnage and above engaged in international voyages [or other specific voyages defined the Convention]. The Convention also includes provision for recognition of national standards or practices which are "substantially equivalent" to the Convention requirements. Finally, in order to encourage fair competition, the Convention requires port States to ensure that ships of non-ratifying States do not receive more favourable treatment during port State inspections than ships of ratifying States receive.

The MLC, 2006 also formally recognizes the "on the ground" reality that most maritime administrations delegate the execution of the flag State ship verification inspections and certification to ROs (usually ship classification

---

[74] Article II(i) provides that "*ship* means a ship other than one which navigates exclusively in inland waters or waters within, or closely adjacent to, sheltered waters or areas where port regulations apply."

This definition is similar to that found in STCW. The Convention applies to all ships that fall under this definition, irrespective of tonnage or voyage, however some ships are excluded under Article II, para. 4:

> Except as expressly provided otherwise, this Convention applies to all ships, whether publicly or privately owned, ordinarily engaged in commercial activities, other than ships engaged in fishing or in similar pursuits and ships of traditional build such as dhows and junks. This Convention does not apply to warships or naval auxiliaries.

[75] Art. II, para. 1(f) provides that "*seafarer* means any person who is employed or engaged or works in any capacity on a ship to which this Convention applies" and para. 2 provides that "Except as expressly provided otherwise, this Convention applies to all seafarers." This definition reflects the greater awareness of the full range of people employed at sea (especially workers on passenger ships). There is some national flexibility provided in the "event of doubt as to categories," which would relate to exclusion from the entire Convention. A resolution was adopted by the Conference to provide guidance to national authorities on this matter.

[76] This definition is generally consistent with the definition of "Company" in para. 2 of SOLAS, Chapter IX adopted in 1994 in connection with the International Safety Management Code (ISM Code).

societies).[77] It also contains provisions regarding seafarer complaints to port State officers as well as various requirements for reports in connection with Title 5 to the International Labour Office. The latter may in future be related to the development of the existing supervisory system.[78]

Finally, as noted as in the Introduction to this contribution, the entry into force formula for the MLC, 2006 is much more stringent than previous ILO conventions and is specifically aimed at ensuring key actors in the sector are party.[79]

The foregoing has briefly summarized the MLC, 2006 and highlighted some of its key elements. Obviously much more could be said on the details of the Convention. However, this contribution has focussed on the elements of interest relating to the further elaboration of the international maritime regulatory regime under the LOS Convention as it pertains to labour and social conditions on ships on the high seas and an approach to responding effectively to a globalized sector and better ensuring implementation and enforcement. The move to certification of working and living conditions – essentially a rights-based regime that uses the economic realities in the

---

77 Despite prior work in connection with the IMO's ISM Code, this is a relatively new area for the ship classification societies. See also, comment, *supra* note 28.

78 Appendix, *supra* note 40, at para. 19.

79 Interestingly, the recently adopted Work in Fishing Convention, 2007 (*supra* note 14) requires ratification by coastal States:

> Article 48
>
> 1. This Convention shall be binding only upon those Members of the International Labour Organization whose ratifications have been registered with the Director General of the International Labour Office.
> 2. It shall come into force 12 months after the date on which the ratifications of ten Members, eight of which are coastal States, have been registered with the Director General.

However, coastal States would not necessarily have flag State responsibility for distant water fleets.

The Recommendation (R 199) adopted by the Conference with the Convention provides in para. 55:

> 55. A Member, in its capacity as a coastal State, when granting licences for fishing in its exclusive economic zone, may require that fishing vessels comply with the requirements of the Convention. If such licences are issued by coastal States, these States should take into account certificates or other valid documents stating that the vessel concerned has been inspected by the competent authority or on its behalf and has been found to be in compliance with the provisions of the Convention.

sector as leverage – is an especially important step by the ILO that may be of much wider significance. In terms of increased integration of U.N. organizations and international law-making, the high level of cross-fertilization between the IMO and ILO is clearly evidenced. The MLC, 2006 also contains several specific structural and substantive innovations related to important ILO values such as tripartism and encouraging social dialogue at the national levels as a mechanism for resolving specific national implementation issues. This contribution concludes with a reflection on the MLC, 2006 and the broader questions of international law-making.

Reflections and Implications

The introduction to this contribution argued that the MLC, 2006 is of interest not only to international labour law scholars and those concerned with the maritime sector but also to international law theorists and scholars, such as Douglas Johnston, with an interest in wider systemic questions. The MLC, 2006 is of particular interest because of the "regulatory design" approach that was adopted. As pointed out at a recent ILO international standards workshop: "design matters."[80] It is also of interest because of the extent of conscious effort in one U.N. organization to experiment with a new strategic approach to squarely tackle the problem of ways to secure universal ratification and more effective implementation of a sector closely tied to international economic and trade matters. The combination of structure and content both play into this approach. As pointed out by Doumbia-Henry et al.,

> The principal challenge – and thus one of the reasons why innovation was essential – was to endow the new Convention with a far higher prospect for widespread ratification than had been achieved in the case of more traditional international labour Conventions. Much of the answer lay in allowing sufficient flexibility so as to accommodate national circumstances and economic diversity; but this flexibility had to be provided without prejudicing the strength of protection to be given to seafarers. The inno-

[80] Presentation by Raymond Torres, Head of the Employment Analysis and Policy Division, OECD, ILO Workshop on the Economic Dynamics of International Labour Standards, Geneva, 21–22 June 2007. See also, R. B. Mitchell, "Regime Design Matters: Intentional Oil Pollution and Treaty Compliance," *International Organization* 48(3) (Summer 1994): 425–458.

> vations relate not so much to the solutions adopted, but rather to their development in the Convention
>
> [...]
>
> These innovations in substance and in process reflect a renewed vitality and commitment in the ILO which will be of broader interest to the international community. They signal a vigor and interest in seeking new ways to encourage the use of international law and international standards to balance the broader economic and social changes affecting most sectors in an era of increased globalization.[81]

However it is also reflects a "lessons learned" approach or, as it has been described, "...a marriage of the traditional and the new"[82] and a cross-fertilization with approaches that have worked elsewhere. The approach of IMO in this sector has been markedly successful in its ability to deal effectively with the international law concerns about State sovereignty. At the same time, through promotion of international cooperation in the form of voluntary regional arrangements such as the port State MOUs or Agreements related inspections of foreign ships (port State control),[83] IMO has successfully tackled the problem of failures of flag State responsibility, a problem largely encapsulated by concern for the existence of a "genuine link." This finding is suggested in the 2006 report of the IMO and other

---

[81] Doumbia-Henry, et al., *supra* note 53.

[82] C. Doumbia-Henry, "The Consolidated Maritime Labour Convention: A Marriage of the Traditional and the New," in ILO, *Les normes internationals du travail: un patrimoine pour l'avenir. Melanges en l'honneur de Nicolas Valticos* (Geneva: Bureau International du Travail, 2004), at p. 319. See also, Doumbia-Henry, et al., *supra* note 53,

> Important concepts in the new Convention can already be seen in the ILO's Merchant Shipping (Minimum Standards) Convention, 1976 (No. 147). The format of the new Convention and its terminology build upon and further develop the well-established format of IMO Conventions, but with adjustments to meet ILO values and approaches. Article XV, relating to a new "accelerated amendment" procedure (to allow for rapid updating of more technical detailed provisions in the Code which is part of the Convention), is a good example of how an IMO procedure has been adapted to a tripartite environment and to the specificity of international labour Conventions. In essence, the procedure gives individual States parties to the Convention an opportunity to opt out of amendments to the Code approved by the tripartite General Conference of the ILO, which would otherwise apply to them, by tacit consent, if they do not opt out within a stated time.

[83] Also reflected in the LOS Convention in the context of protection of the marine environment. PSC inspections are usually understood as a "may" obligation. However, recent discussion in the European Union suggests that, at least on an EU-wide regional basis, PSC may no longer be voluntary.

concerned agencies to the U.N. General Assembly on the "genuine link,"[84] despite some resurgence of interest in the meaning of the genuine link in cases before the International Tribunal for the Law of the Sea.

> 28. Participants in the Meeting took the view that the exclusivity attached by the United Nations Convention on the Law of the Sea to the right of States to fix conditions for the grant of nationality, as reaffirmed by the authoritative interpretations of the International Tribunal for the Law of the Sea in the M/V *Saiga* (No. 2) and subsequent cases, as well the other agreements referred to in section 2 above, indicated that the questions relating to the precise criteria or conditions adopted by a State with respect to the grant of its nationality to a ship were a matter beyond the purview of the organizations participating in the Meeting. However, participants in the Meeting also considered that issues relating to securing the objective and purpose of the "genuine link" requirement, that is, assuring the ability of the flag State to effectively exercise its jurisdiction over ships flying its flag, were matters of central concern to all of the organizations and formed a substantial part of their programmes of regulatory initiatives and technical cooperation activities in the shipping and fishing sectors.

However, the test of the many innovations in the MLC, 2006, which were based on the principle of "firm on rights: flexible on implementation" and intended to attract widespread ratification and more effective implementation, will be whether it is indeed ratifiable[85] and ratified and, ultimately, whether the Convention's goals are achieved.

Will it be able achieve these objectives through its mix of structure (Code, Parts A and B) and the flexibility/scope clauses and substantial equivalence and the certification system all of which are linked to ideas of tripartism, transparency, and accountability?

Is this Convention a good example of a way that international conventional law can be designed to be (more) effective and, in particular, can it be designed in such a way as to affect the behaviour of non-State actors[86] directly while at the same time operating primarily at the level of State responsibility?

---

[84] *Supra* note 20.

[85] The primary areas of difficulty encountered by governments relates to the application to smaller domestic trade ships and the obligations with respect to social protection for seafarers. The IMO conventions often do not apply the international requirements on smaller domestic trade ships. The MLC, 2006, as noted earlier does apply but with some flexibility.

[86] For example, through detention of ships or no more favourable treatment of ships of non-ratifying States.

The answer to these questions may not be known for several years. However, these changes in approach and the ambitious goals of the MLC, 2006 also mandate a strategic approach to its promotion and securing implementation. In September 2006, the International Labour Office adopted a five-year *Action Plan to achieve rapid and widespread ratification and effective implementation of the Maritime Labour Convention, 2006*. The Action Plan explicitly targets promotional efforts at three levels: international, regional, and national. It also reflects an evolution for the ILO in terms of adopting a more active tripartite-based approach to securing ratification, including promoting government-to-government dialogue. The Action Plan seeks to achieve sufficient ratification to allow entry into force by 2011. As of the date of writing this appears possible. Three key flag States have ratified the MILC, 2006, Liberia, Bahamas, and the Marshall Islands, the second, third and fourth largest flag States (e.g., nearly 20% of the world fleet based on percentage of gross tonnage).[87] In addition, the European Union has taken a decision to ratify by 2010.[88] Countries such as Canada, China, Egypt, Germany, Japan, Panama, the Philippines, and the Russian Federation have been taking specific steps to enable ratification in the next few years. In addition, in late 2008 the ILO adopted international guidelines for flag State inspections and port State control officers[89] and to develop training programmes to assist with national level capacity-building for inspections. The private sector, in particular several major ROs, has also taking steps to prepare their staff

[87] As at 28 April 2008.

[88] Council of the European Union, Transport, Telecommunications and Energy, 2805th Council Meeting, Luxembourg, 6–8 June 2007, as stated in Press Release 10456/07 (Presse 133):

> The Council adopted a Decision authorising Member States to ratify, in the interest of the European Community, the 2006 Maritime Labour Convention of the International Labour Organisation (ILO). The 2006 Maritime Labour Convention was adopted in February 2006 by the maritime session of the International Labour Conference. The Convention will bring a major input at international level in the shipping sector by promoting decent living and working conditions for seafarers and fairer competition conditions for operators and ship owners. Member States are invited to ratify the Convention before 31 December 2010.

[89] Called for in two of the Resolutions (Resolution IV and XIII) adopted by the 94th ILC when it adopted the MLC, 2006. See, "Guidelines for Flag State Inspection Under the Maritime Labour Convention, 2006" and "Guidelines Port State Control Officers Carrying Out Inspections Under the Maritime Labour Convention, 2006," at <www.ilo.org>.

to carry out labour inspections. In fact, in early 2008, one RO has issued a certificate confirming MLC, 2006 compliance by a seafarer recruitment and placement service.[90]

A broader question in the context of globalization is whether some aspects of this approach are transferable to other sectors and issues. The process by which the MLC, 2006 was developed may be a way to help governments to address the policy dilemma posed by the need to both protect workers and to improve their economic and social conditions and, at the same time, to remain competitive in the global marketplace.

The Convention could also be seen as relevant to questions regarding the role of international law and global standards – are these elements relevant in the face of economic diversity and a trend to regionalism? Again the answer, so far, appears to be yes.

The MLC, 2006 offers an example of one approach addressing the current problems in sustainable economic development and the relative lack of effective governance with respect to transnational corporations. This Convention, particularly its inclusion of social protection and other elements such as occupational safety and health standards, also has implications for transnational or globalized workers and workplaces.

From an ocean regime and a "systems" perspective, the MLC, 2006, as well as the recently adopted Work in Fishing Convention, 2007,[91] are finally tackling the idea of the ocean space, especially outside national waters, as a workplace and dealing with the problems of extra-territoriality. Finally, the Convention offers an important example of an effort by a U.N. organization, albeit in response to strong encouragement from its civil society (employer and worker representatives) constituents, to integrate its approach as a strategy to securing better implementation. In principle, efforts at the international level should also serve to encourage better national level integration and regulatory design.[92]

---

90 Det Norsk Veritas, "OSM: The First Ever to be Certified per the ILO Maritime Labour Convention!" 1 April 2008, <www.osm.no/view_casearticle.asp?id=388&news=news>. "OSM is now the first company in the shipping world to be awarded the much-coveted certification after a series of audits from Det Norske Veritas (DNV)."

91 *Supra* note 14.

92 For example, implementing the MLC, 2006 will typically engage maritime administrations, labour departments, and agencies or departments involved in social security.

The Contribution of the 2006 ILO Maritime Labour Convention to Global Governance

Peter B. Payoyo*

Introduction

International law at the service of human welfare and human dignity. This is the *élan* of international law that I was extraordinarily privileged to be immersed in as a student of Professor Douglas M. Johnston. What a calling! It was more than a principled vision, more than the conviction on the *telos* of a world ordered by law. It was a dedication to a way of thinking and living, a way of breathing the work that one does in society. Most fundamental in this engagement is a disciplined technique to comprehend normative facts, processes and discourses, a feat that Professor Johnston demonstrates with such mastery and luminosity in his last essay, *World Constitutionalism in the Theory of International Law.*[1]

"Sustain an observational standpoint" was his memorable advice to me as I struggled to finish my dissertation.[2] It was not at all an easy prospect. The observational standpoint requires a panoramic frame of mind to canvass facts and analyze doctrine, a longer reading of history and philosophy, a deeper commitment to interdisciplinarity, and a willingness to constantly and rigorously observe one's standpoint in relation to others'. The beauty and, if I may add, the fun of such an approach to legal scholarship can not be more rewarding: in the end, the cerebrally-intense observational

* Programme Manager, Philippine Seafarers Assistance Programme, Rotterdam, and Vice-Chairman, International Committee on Seafarers Welfare, United Kingdom.

1 D. M. Johnston, "World Constitutionalism in the Theory of International Law," in Ronald St. John Macdonald and Douglas M. Johnston, eds, *Towards World Constitutionalism: Issues in the Legal Ordering of the World Community* (The Hague: Martinus Nijhoff, 2005), pp. 3–29.

2 P. B. Payoyo, *Cries of the Sea: World Inequality, Sustainable Development and the Common Heritage of Humanity* (The Hague: Martinus Nijhoff, 1997).

standpoint in international law leads to an authentically ethical standpoint. International law cares!

The development of the international normative system as seen in the evolution of law and policy under the United Nations Convention on the Law of the Sea was a domain close to the heart of Professor Johnston. This was also the field of study that I chose to devote myself to at Dalhousie University, a truly vibrant centre for maritime affairs studies. In my own little quest for the true, the beautiful and the just in the law of the sea, I count myself immeasurably privileged to have been mentored by Professor Johnston. He, along with two other towering amici of a kinder and gentler world, Professor Elisabeth Mann Borgese and Professor Ronald St. J. MacDonald, showed me that the love, awareness and wisdom encapsulated in law can and will change the world – planet, seas, people and all. I have no doubt that the legacy of caring passed on by Professor Johnston, which shone brilliantly in his works and in his life, points to the sunnier way, the worthier goal, in international law.

The purpose of this essay is to briefly explore current aspects of the international law of human welfare and human dignity in the realm of the oceans. It will describe the emerging regime to protect the rights of seafarers, embodied in the 2006 International Labour Organization (ILO) Maritime Labour Convention, and examine the contribution of this regime to the transformation of international law in the era of globalization.

The 2006 Maritime Labour Convention: An Overview

The 2006 Maritime Labour Convention, or MLC, was adopted unanimously on 23 February 2006 by the International Labour Conference (Maritime).[3] It was negotiated over a period of five years by the tripartite constituents of the ILO. Its provisions are spread to over a hundred pages

[3] When the vote on the MLC was taken, the government representatives of Venezuela and Lebanon indicated their support for the MLC regime per se but cast abstention votes for reasons unrelated to the merits of the MLC. International Labour Conference, 94th (Maritime) Session, Ninth Sitting, *Provisional Record* No. 17, 23 February 2006. A full range of ILO documents on the MLC, including records of its negotiation history, are available online at <www.ilo.org/global/What_we_do/InternationalLabourStandards/MaritimeLabourConvention/lang--en/index.htm>.

of text.[4] It is also called the "consolidated convention" because it consolidates, codifies, and updates 68 international maritime labour instruments adopted by the ILO since 1920.[5] The MLC is the very first comprehensive consolidation of international labour standards.

The MLC is also considered as innovative in other ways.[6] The structure of the MLC is unique and departs from the usual approach of allocating agreed-upon provisions under either a "convention" or a "recommendation" track. Instead, the MLC is one integral package consisting of "articles," "regulations," and "the Code," the articles and regulations setting out the core rights and principles and basic obligations of ratifying States, and the Code containing the details of the implementation of the regulations spelled out as "mandatory standards" as well as "non-mandatory guidelines."[7] The MLC, structured in this manner, is said to serve a two-fold purpose: to lay down a firm set of rights and principles, and to allow considerable flexibility in the implementation of these rights and principles. Thus, the motto on the central virtue of the MLC: "Firmness with respect to rights, Flexibility with respect to implementation."

The regulatory scope of the MLC covers the full range of subject areas on the living and working conditions of seafarers and deals with numerous aspects of a seafarer's "right to decent employment": conditions of employment, recruitment, hours of work and rest, accommodation, recreational

[4] Mr. Bell of the Bahamas government, speaking as a Reporter for the Committee of the Whole during the ILC refers to the MLC as "the largest Convention that the ILO has ever produced." International Labour Conference, 94th (Maritime) Session, Seventh Sitting, *Provisional Record* No. 16, 22 February 2006, p. 4.

[5] Article X of the MLC lists 37 conventions revised by the MLC. Excluded from the consolidation exercise were the following ILO Conventions: the 2003 ILO Convention on Seafarers' Identity Documents (No. 185) and its predecessor, ILO Convention No. 108; ILO Convention No. 71, on seafarers' pensions; and ILO Convention No. 15, referring to minimum age for trimmers and stokers.

[6] For authoritative accounts of the MLC, see C. Doumbia-Henry, "The Consolidated Maritime Labour Convention: A Marriage of the Traditional and the New," in *Les normes internationales du travail: un patrimoine pour l'avenir: mélanges en l'honneur de Nicolas Valticos* (Geneva: ILO, 2004), pp. 319–334; C. Doumbia-Henry, D. Devlin and M. McConnell, "The Maritime Labour Convention, 2006 Consolidates Seafarers' Labour Instruments," *ASIL Insight* 10(23) (13 September 2006), <www.asil.org/insights/2006/09/insights060913.html>.

[7] See *Explanatory Note to the Regulations and the Code of the Maritime Labour Convention*, which is appended but "does not form part" of the MLC.

facilities, food and catering, health protection, medical care, welfare and social security protection. Every regulation in the MLC is introduced by a curt statement of "purpose," which gives a short-hand statement about the intent behind any particular regulation, standard or guideline. Another important regulatory area in the MLC is on "compliance and enforcement" which specifies the details of flag State responsibilities, port State responsibilities, and labour-supplying responsibilities.

When the MLC was adopted, tripartite representatives at the conference converged in the conclusion that history was being made, taking turns to praise the "historic achievement" that was the MLC. The MLC was a turning point in history of the maritime industry, the world's most globalized sector. The MLC had set a precedent in international standard setting. ILO Director-General Juan Somavia underlined the epoch-making character of the MLC as follows:

> The Maritime Labour Convention 2006 covers a sector that has become a driving force of globalization, increasing productivity and demand over the past decades and moving 90 per cent of world trade. And by setting solid and uniform rules for the workers, employers and governments involved in commerce at sea, it provides a model for tackling the most pressing globalization challenges of our time.... This exceptional vision and capacity for social dialogue among seafarer and shipowner organizations has thus helped build the foundation for an innovative approach to social policy that represents a pioneering contribution to making globalization fair.[8]

The MLC is a veritable accomplishment in international regulatory policy for several reasons. First, in its form, the MLC combines both "hard law" and "soft law" formulations and approaches in a single instrument. Secondly, from the perspective of treaty-making, the MLC is an exceptional international agreement because it codifies labour standards for an entire global industry, received unanimous endorsement not only from States but also from the social partners of government in this industry, is designed to be universally applied and effectively enforced, and provides a mechanism for the continuous review of the MLC, including the introduction of rapid amendments thereto. Thirdly, the MLC *ratione materiae* is indeed a "bill of rights" because the language and substance of rights pervade this

[8] J. Somavia, "A New 'Bill of Rights' for the Maritime Sector: A Model for Fair Globalization," Statement from the ILO Office of the Director General, April 2006, <www.ilo.org/public/english/bureau/dgo/speeches/somavia/2006/maritime.pdf>.

instrument. And lastly, as a normative tool in governance, the MLC is regarded as a "best practice" response to the challenge of globalization.

These closely-interwoven qualities and characterizations of the MLC invite reflection on the why's and wherefore's of the MLC.

Why International Maritime Labour Standards?

The significance of the MLC as a body of norms inquires into the rationale for treaty-based labour regulation. The original motivation for labour standards applicable internationally can be traced back to the mid-19th century in Europe, when it was felt that national labour legislation put a country in a competitive disadvantage vis-à-vis another country or countries which had inferior labour conditions. As a strong argument against those opposed to the adoption of national labour policies, international labour standards were then advocated in order to realize fair competition or, in current terminology, to remove labour conditions as a factor in international competition, or in order to avoid social dumping and a race to the bottom. When the ILO was established in 1919, this "fair international economic competition" justification for international labour standards was upheld and found its way in the celebrated clause in the Preamble of Part XIII of the Treaty of Versailles. The clause asserts that "the failure of any nation to adopt humane conditions of labour is an obstacle in the way of other nations which desire to improve the conditions in their own countries."[9]

Commentators agree that the "fair international competition" justification for international labour standards was not borne by the facts of international economic trade, and its appeal had steadily lost steam.[10] With the founding of the ILO, the pursuit of "social justice" and the promotion of "peace" were advanced as the more primordial and persuasive objectives of international social standard setting. After the Second World War, the 1944 Declaration of Philadelphia, which amended the Charter

---

[9] Third para., Preamble, Part XIII "Labour," Treaty of Versailles, 28 June 1919. Text of the Treaty, <history.sandiego.edu/gen/text/versaillestreaty/vercontents.html>.

[10] J.-M. Servais, *International Labour Law* (The Hague: Kluwer Law International, 2005), pp. 26–35; N. Valticos, *International Labour Law* (Deventer: Kluwer, 1979), pp. 21–22.

of the ILO, placed the expansive social justice objective as the true end and aim of international labour standards. The meaning of social justice was captured in the memorable statement that "all human beings, irrespective of race, creed or sex have the right to pursue both their material well being and their spiritual development in conditions of freedom and dignity, of economic security and equal opportunity."[11]

The watershed Declaration on Fundamental Rights and Principles at Work, adopted by the ILO in 1998, reiterated the argument that social justice, which was now linked to the economic growth objective, provides the governing framework for international labour standards. Equally significant, the 1998 ILO Declaration points out a *negative* justification for international labour standards such that "labour standards should not be used for protectionist trade purposes, and that the comparative advantage of any country should in no way be called into question" in the application of international labour legislation.[12] This stance on international labour standards reflected the prevailing consensus about the inadvisability of utilizing "social clauses" in international trade agreements. These social clauses would have linked the trade policy of a country, usually in the North, to the observance of international labour standards in another country, usually in the low-wage South – an approach to the application of international standards that had met with general disfavour at the WTO Ministerial Meeting in Singapore in 1996.[13]

Throughout the 20th century, the "social justice" rationale prevailed as the sole ideological backdrop for the adoption of international labour

[11] *Declaration concerning the aims and purpose of the International Labour Organization*, 10 May 1944. The text of this Declaration is annexed to the ILO Constitution, <www-old.itcilo.org/actrav/actrav-english/telearn/global/ilo/law/phila.htm>.

[12] *ILO Declaration on Fundamental Principles and Rights at Work*, 86th Session, Geneva, June 1998, <www.ilo.org/dyn/declaris>.

[13] World Trade Organization, "Singapore Ministerial Declaration," WT/MIN(96)/DEC, 18 December 1996, <www.wto.org/english/theWTO_e/minist_e/min96_e/wtodec_e.htm>. For the debate on using social clauses in international trade agreements, see, e.g., E. Potter, "A New Paradigm for International Labor Standards," in A. Morris and S. Estreicher, eds, *Cross-border Human Resources, Labor and Employment Issues* (The Hague: Kluwer Law International, 2005), pp. 727–752; L. Rosenberry, "Labor Versus the World Trade Organization: The Clash Between Economic, International Law, and Labour Law Theories," in R. Blanpain and B. Flodgren, eds, *Corporate and Employment Perspectives in a Global Business Environment* (The Hague: Kluwer Law International, 2006), pp. 165–184.

standards applicable to seafarers and for evolution of social policy in the maritime sector. Standard setting for seafarers to meet social justice objectives basically meant giving special attention to seafarers considered as a separate category of workers needing specialized protection on account of the special conditions of their work at sea. The special treatment was institutionalized early on in the history of the ILO with the convening of separate maritime sessions of the International Labour Conference for the framing of employment standards at sea. This procedure has been faithfully followed to the present day.

A remarkable characteristic of the MLC is its utilization or resuscitation of the "fair economic competition" justification for international labour regulation. "Fair competition" is a categorical justification for the MLC and is to be assumed as its underlying motivation. The genesis of the ILO project to realize a framework labour convention is usually traced to the so-called Geneva Accord of January 2001. This was a joint resolution of shipowner organizations and seafarers' trade unions to develop "an instrument which brings together into a consolidated text as much of the existing body of ILO instruments as it proves possible to achieve" as a matter of priority "in order to improve the relevance of those standards to the needs of all the stakeholders of the maritime sector."[14] It was in fact the shipowners group which put the seminal idea of a consolidated convention on the table, arguing that they "wanted sensible and impartially applied [international] labour standards, so that a level playing field could be created, where standards of service instead of poor labour conditions would dictate customer preference." Furthermore, the shipowners, more than the trade unions, gave urgency to the case for consolidating maritime labour standards, stating that "if the ILO could not satisfy the needs of the sector another forum would have to be found."[15]

[14] *Resolution concerning the review of relevant ILO maritime instruments*, Appendix 2 of the *Final Report*, 29th Session of the Joint Maritime Commission, Geneva, 22–26 January 2001, JMC/29/2001/14, <www.ilo.org/public/english/dialogue/sector/techmeet/jmc01/jmcfr.pdf>.

[15] *Final Report*, Joint Maritime Commission (29th Session), Geneva, 22–26 January 2001, MC/29/2001/14, p. 4 and Annex 15. See also, J. I. Blanck, Jr., "Reflections on the Negotiation of the Maritime Labour Convention 2006 at the International Labour Organization," *Tulane Maritime Law Journal* 31 (2006), at p. 39.

The MLC is thus illustrative of a novel phenomenon in global governance: the direct regulation of an entire industry by an ILO convention.[16] As it now stands, the MLC is the only regime which acknowledges and legitimizes the fair economic competition argument for the adoption of international labour regulations. While "levelling the playing field" among firms or "ensuring quality shipping everywhere" is not expressly indicated anywhere in the text of the MLC, the principle of fair competition in the shipping industry was taken for granted throughout the negotiation of the MLC and can be conclusively glossed as a principal objective of the new regime.[17] As Mr. Bruce Cariton of the United States, who chaired the Committee of the Whole, stated, "What is fundamentally different about this Convention is that it is about quality shipping.... Beyond improving the working conditions of the seafarers, it is also about further marginalising the bad shipowners who end up costing the entire industry. This is a very sound economic benefit for the entire industry."[18] To paraphrase a resurrected clause of the ILO *Constitution*: "the failure of any shipowner to adopt humane conditions of labour is an obstacle in the way of other shipowners which desire to improve the conditions in their own ships."[19]

The guarantee of fair competition among shipowners and shipping operators on the basis of the impartial and universal application of maritime labour standards is given teeth through the enforcement and compliance provisions of the MLC. Under the MLC, flag State implementation is accorded the leading role to accomplish this objective consistent with the

---

16 B. A. Langille, "The ILO is Not a State, Its Members are Not Firms," in G. Politakis, ed., *Protecting Labour Rights as Human Rights: Present and Future of International Supervision* (Proceedings of the International Colloquium on the 80th Anniversary of the ILO Committee of Experts on the Application of Conventions and Recommendations, Geneva, 24–25 November 2006), p. 267, <www.ilo.org/global/What_we_do/InternationalLabourStandards/InformationResources/Publications/lang--en/docName--WCMS_087817/index.htm>.

17 See, e.g., "Advantages of the Maritime Labour Convention and FAQs," No. 1, <www.ilo.org/global/What_we_do/InternationalLabourStandards/MaritimeLabourConvention/Advantages/lang--en/index.htm>.

18 Quoted in "The New ILO Consolidated Maritime Labour Convention – Charting Labour Standards for the Future," *P&I Club Steamship Mutual* (January 2007), <www.simsl.com/publications-articles.html>.

19 See Preamble to the *International Labour Organization Constitution*, <www-old.itcilo.org/actrav/actrav-english/telearn/global/ilo/law/phila.htm>.

customary law principle in Article 94 of the UN Convention on the Law of the Sea that "Every State shall effectively exercise its jurisdiction and control in administrative, technical and social matters over ships flying its flag."[20] However, it is most likely that in practice the institution of port State control will provide the linchpin of the MLC's enforcement and compliance infrastructure to achieve the goal of fair competition among shipping enterprises. Port State control, as a modality of implementing labour, maritime safety and marine pollution prevention standards under IMO conventions, delivers immediate and direct economic consequences that impact on the shipowners whose vessels are adjudged "sub-standard." As a sanctioning mechanism, port State control has the virtue of impartial application to all ships on account of the "no more favourable treatment" principle in the inspections of ships in port. Existing regional and inter-regional port State control arrangements worldwide have established a fairly successful record of policing and targeting sub-standard ships.[21] It is to be expected that port State control interventions will be further expanded and intensified with the entry into force of the MLC.[22] Fair competition among shipping industry participants brought about essentially by port State control is the hallmark of the MLC as a regulatory norm directly applicable to a global industry. This certainly makes the MLC an exemplary governance tool in the age of globalization.

If inducing fair competition among firms is a legitimate objective of the MLC, can it also be said that enforcing fair economic competition among States is part of its mission? The question has reference to the reality of an internationalized ship registration market which allows shipowners to pick and choose the country where they flag their ships.

---

[20] *United Nations Convention on the Law of the Sea*, Montego Bay, 10 December 1982, UN Doc. A/Conf.62/122 (1982), 21 *I.L.M.* 1261 (1982).

[21] See the annual reports published by the secretariats of some PSC regions. The IMO publishes flag State comments on port State control, which also contains details on the vessel detained and the classification society involved, reinforcing the "name and shame" tool that contributes to the effectiveness of PSC. See, e.g., Doc. FSI.2/CIRC.6, 4–1–2007, "List of flag State comments on detentions for the years 2003, 2004, 2005 and 2006," <www.imo.org/includes/blastDataOnly.asp/data_id%3D16735/6.pdf>. The IMO has established a list of contact points nominated by flag States exclusively for the notification of port State control detentions.

[22] The ILO will consider more specific guidelines on flag State implementation as well as port State implementation of the MLC in the latter part of 2008.

The question also has reference to countries involved in the labour supply chain, providing alternative sources of seafaring labour for the shipping industry. Once again, there is no explicit preambular or normative statement in the MLC supporting the goal of "levelling the playing field" with respect to the competition of flag States among themselves for tonnage or competition among crewing States for the supply of shipboard labour. Does the MLC have a role in regulating these two markets such that fair competition among flag States and among labour-supply States *inter se* is likewise to be included in the remit of the MLC's objectives? The answer is in the affirmative.

The intention to have a universally applied MLC is backed up by provisions in the MLC to penalize flag States or labour-supplying States that choose to stay out of the regulatory principles and standards of the MLC.[23] For example, ships that are registered with non-ratifying flag States, as well as countries whose nationals are not trained and recruited in accordance with MLC standards, will risk being subjected to frequent and costly port State control inspections, and being "named and shamed" in the maritime industry. Supervisory procedures in the ILO can also lead to embarrassing conclusions against States deciding to stay out of the MLC regime, inevitably affecting those States' market reputation and standing. Other States will then be justified in taking protective measures against non-ratifying States, or States whose national standards are lower than MLC standards, even as these measures enhance their own competitive advantage in the maritime field vis-à-vis non-conforming, and therefore "pariah," States.

In this sense, the MLC is a form of a social clause in international trade relations which can be invoked by a ratifying State to justify actions or policies which protect that State's maritime interests or which seek to remove the competitive advantage of sub-standard flag States and labour-supplying States.[24] Levelling the playing field in all arenas of economic

[23] The preamble of the MLC refers to the determination to have a regime "designed to secure the widest possible acceptability among governments, shipowners and seafarers committed to the principles of decent work." *Maritime Labour Convention*, Geneva, 7 February 2006, <www.ilo.org/global/What_we_do/InternationalLabourStandards/MaritimeLabourConvention/lang--en/index.htm>.

[24] *Cf.* The 1998 ILO Declaration on Fundamental Rights and Principles at Work, *supra* note 12, which commits to the position that international labour standards should

competition in the industry – among shipowners, among States of registry, and among labour-supply States – is the grand ambition of the MLC.[25]

Social Justice and International Labour Standards

The shipping industry is blighted by its rather notorious historical record in the maltreatment of seafarers. Throughout history, the romantic profession of seafaring has been incessantly littered with stories of cruel and tyrannical ship masters, long hours of debilitating work, ill-nourishment, psychological torment, neglect and abandonment, and the continuous exposure to hazards to life and limb. Alongside the progress in international standard setting for sea workers in the 20th century, and the technological and communications revolution in the industry and the explosion of trade in the period of globalization, the narrative of exploitation and victimization at sea continues. Dehumanization of sea workers has long been outlawed by international policy, but recurring and even systemic episodes of oppression put the issue high on the agenda of international standard setting.[26]

---

not be used for protectionist purposes or to remove the competitive advantage of countries.

[25] In the aftermath of the Irish Ferries controversy in 2005 (i.e., Irish Ferries had sought to outsource jobs to foreign workers) some European Union officials have expressed the view that the European Union's endorsement of the MLC is a first step to combating "social dumping" as well as establishing fair wages and revitalizing career options for E.U. seafarers. See Website of MEP Mary Lou McDonald, <www.sfguengl.com/news/entry/507>. Seafarers' trade unions in Panama, the world's largest register, also see the MLC as an opportunity to increase the number of Panamanian seafarers on board Panamanian flagged vessels. The head of the ITF Seafarers Section is reported to have counselled authorities in the Philippines, the world's largest crew-supply country, that "once the Convention is in force, shipowners will have to ensure that the seafarers on their ships have been recruited in accordance with the Convention" and that "they may well get to the stage where they feel it is simply easier to go to another country which has ratified, where mechanisms are in place to help ensure procedures are carried out in compliance with the MLC, and then the Philippines could find it is starting to lose its position." In "Delivering Global Rights," *Transport International Online*, Issue 28, July 2007, <www.itfglobal.org/transport-international/ti28.cfm>.

[26] P. Chapman, *Trouble on Board: The Plight of International Seafarers* (New York: ILR Press, 1992); A. D. Couper, *Voyages of Abuse* (London: Pluto Press, 1999); *Ships, Slaves and Competition*, Report of the International Commission on Shipping (2001), includ-

One of the classic first accounts of brutalization at sea in the modern era is the autobiographical *Two Years Before the Mast* (1840) by Richard Dana. This perceptive seaman had wondered whether all that suffering and dehumanization were really an inevitable part of his vocation. At the time, the depth of human pain experienced by seafarers became the rallying cry of pioneer slave trade abolitionist the deacon Thomas Clarkson (1790–1846), who drew public attention to the fact that the crews of slave ships were as abused and equally savaged as the slaves themselves.[27]

While the abolition of slave trade had been decisively won, slavery at sea in various shapes and hues has persisted. The unenviable situation of a seafarer is partly explained by his problematic legal status. To illustrate, the movement that began in America to emancipate the seafarer by giving him the guaranteed right to quit his work when his vessel is in safe harbour, did not catch on internationally. The seafarer, who was deemed bonded to the vessel where he lived and worked, had obligations that were considered more compelling than his alleged freedom from involuntary servitude.[28] The legal status of the seafarer was given poignant expression in the "Appeal to the World" put forward by the International Seamen's Union of America to support the right of a seafarer to quit his employment in mid-contract:

> We stand in the same relation to vessel as the serf did to the estate, as the slave to his master.... At sea, the law of common hazard remains. There must be discipline and self-sacrifice, but in any harbor, the vessel and you are safe, and we beseech you to give us that freedom which you claim for yourself and which you have bestowed on others to the end that we may be relieved of that bitterness of soul that is the heavy burden of him who knows and feels that his body is not his own.[29]

The legal status of seafarers in admiralty and *lex mercatoria* did not also presuppose their full rights as free and autonomous individuals. U.S. Justice

---

ing its *Final Report* (2005), <www.itfglobal.org/seafarers/icons-site/news.html>. See also the "sweatships campaign" against abuse in the cruise industry, <www.waronwant.org/Dream%20versus%20reality%20+2860.twl>.

[27] R. Kverndal, *Seamen's Missions: Their Origin and Early Growth* (USA: William Carey Library, 1986) pp. 49–51.

[28] C. Klee, "Desertion and the Freedom of the Seaman," *International Labour Review* 13 (1926): 649–672, 808–849.

[29] Cited in P. Bolle, "The ILO's New Convention on Maritime Labour: An Innovative Instrument," *International Labour Review* 145 (2006): 135–137.

Story famously described this *sui generis* legal status of seafarers: "They are emphatically the wards of the admiralty; and though not technically incapable of entering into a valid contract, they are treated in the same manner as courts of equity are accustomed to treat young heirs, dealing with their expectancies, wards with their guardians, and *cestuis que trustent* with their trustees."[30]

Progressive improvements in the working and living conditions of seafarers were thus brought about by admiralty courts acting as wards of seafarers and through positive regulation and national legislation. This situation holds true today.[31]

With the onset of standard-setting at the ILO, the rights of seafarers were codified and given shape in instruments recognizing universal rights for seafarers and specifying standards to govern their working and living conditions onboard. Sixty-five conventions and recommendations had been adopted by the ILO prior to the MLC, defining and updating minimum rights and labour standards and addressing a wide range of social issues that were considered to need international regulatory attention. Other general instruments applicable to all workers, such as the Maternity Protection Convention (No. 103, 1952) or the Freedom of Association Convention (No. 87, 1948) were also extended to seafarers.[32] The International Seafarers Code, as these instruments came to be known

[30] *Harden v. Gordon*, Fed. Cas. No. 6047, p. 480, 485 (1823) cited in *Garrett v. Moore-McCormack Co.*, 317 U.S. 239 (1942). In 1825, the leading British Admiralty judge, Lord Stowell, gave a less flattering account of this status, referring to seafarers as "a set of men, generally ignorant and illiterate, notoriously and verbally reckless and improvident, ill provided with the means of obtaining useful information and almost ready to sign any instruments that may be proposed to them; and on all accounts requiring protection, even against themselves." For the early history and evolution of seafarers legal rights, see D. Fitzpatrick and A. Anderson, eds, *Seafarers' Rights* (UK: Oxford University Press, 2005), Chapter 1.

[31] In the United States, the recent judicial policy to deny international seafarers their right to access U.S. courts for the vindication of their rights has been criticized. M. Nickson, "Closing U.S. Courts to Foreign Seamen: The Judicial Excision of the FAA Seamen's Arbitration Exemption from the New York Convention Act," *Texas International Law Journal* 41 (2006): 103–144; L. Sechelski, "Forum Selection Clauses in Seamen's Contracts: Are We Protecting Commercial Progress by Denying Seamen their Rightful Day in Court?" *Houston Journal of International Law* 26 (2003): 203–243.

[32] Valticos, *supra* note 10, at pp. 192–193.

collectively, supplies the normative framework for achieving social justice in the maritime sector.

The International Seafarers Code was indubitably built on the foundation of national sovereignty, or the exclusive prerogative of the flag State to translate and transpose these international labour standards to national standards applicable to its merchant fleets. This at once became the Achilles heel of the impressive normative structure of ILO labour standards. The agency of the flag State is the *sine qua non* to the concrete enjoyment of labour standards by seafarers. And since there is no rule or principle in international law that would compel a State to ratify a maritime labour treaty, or any treaty for that matter, much of the rights and standards remained fictional. In the decentralized international society of sovereign flag States, the regime of seafarer's rights was in reality a confusing gamut of flag State jurisdictions displaying varying commitments or non-commitments to ILO labour standards. The generally uneven and low rate of ratifications of maritime labour conventions by flag States testifies to the basic difficulty of achieving globally effective minimum standards. As a consequence, some seafarers working under some flags were more endowed in rights and entitlements than other seafarers working under other flags.[33]

After the Second World War, the flagging out of ships to non-traditional maritime powers and the proliferation of flags of convenience put up even bigger obstacles to the progressive realization of minimum labour standards. Likewise, the globalization of the international labour market for seafarers and the liberalization of ship registration did not at all prove helpful in creating the environment that would make flag States and shipowners more conscious of maritime labour standards. Shipowners were naturally more inclined to register their vessels in jurisdictions where compliance or non-compliance with labour standards did not present too much cost.[34] Notwithstanding calls and arguments to establish a "genuine link" between the flag State and the ship, and in spite of campaigns to

---

[33] The ILO website gives information on the ratification record of all maritime labour conventions at <www.ilo.org/public/english/dialogue/sector/sectors/mariti/standards.htm> and <www.ilo.org/ilolex/english/convdisp1.htm>.

[34] ILO, *The Impact on Seafarers' Living and Working Conditions of Changes in the Structure of the Shipping Industry*, Report for discussion at the 29th Session of the Joint Maritime Commission, JMC/29/2001/3 (Geneva: ILO, 2001), <www.ilo.org/public/english/dialogue/sector/techmeet/jmc01/jmc-r3.pdf>.

eliminate flags of convenience, the power of free market forces held sway, affecting the global level of qualitative and quantitative compliance with international labour standards. The principle of flag State jurisdiction has proven dysfunctional in so far as flag States were failing to adhere to an increasing number of maritime labour standards produced by the ILO. To the extent that flag States were not ratifying ILO conventions, much of the international seafarers' rights and maritime social standards were mere norms on paper. The moral: exclusive reliance on the principle of flag State sovereignty to deliver social justice for seafarers globally has been rather misplaced in a historical context where a free-for-all, race-to-the-bottom competition among industry players and maritime nations was unfolding.

Poor labour conditions as a result of the dysfunctional operation of the flag State sovereignty also entailed costs that proved too much too bear for a significant part of the industry. It was already mentioned that shipowners lamented the confused and confusing regulatory environment that was distorting customer preference in favour of sub-standard operators. The chaos in labour regulations and standards was seen as bad for business and "the mere existence of widely varied national standards hampered the smooth flow of shipping operations."[35] Thoroughgoing reform was in order.[36]

The MLC regime is a historic intervention by the joint will of the international community and shipping industry because it proposes to restore sanity and order in maritime labour regulations, consistent with the goal of social justice. The underlying strategy to accomplish this purpose is straightforward: put integrity in the exercise of flag State jurisdiction and install checks on the exercise of flag State sovereignty. The test to gauge the success of this strategy is also unambiguous: whether the exercise of flag State jurisdiction and the checks made on it are effective from the point of view of the recipients or beneficiaries of MLC standards.[37]

---

[35] "Statement from the Representative of the ISF," *Final Report*, Joint Maritime Commission (29th Session), Geneva, 22–26 January 2001, JMC/29/2001/14, p. 4.

[36] *Cf.* the approach to reform suggested in T. Mensah, "Seafarers in the Modern Maritime Industry: The Need for Reform," *Lloyd's Maritime and Commercial Law Quarterly* 2 (2006): 172–180.

[37] Upholding the effectiveness test, see T. Novitz, "On Social Participation, Public Awareness and Social Capacity," in Politakis, *supra* note 16, at pp. 105–108, and the comment of E. Riedel, *ibid.*, at pp. 125–126.

Realizing Seafarers' Rights and Labour Standards

Integrity in the Exercise of Flag State Sovereignty

It is often observed that flag States do not ratify labour conventions, or do not implement conventions they have ratified, because of lack of will, lack of capacity, or both. The ILO deals with these issues through its regular supervisory and promotional/technical cooperation activities. There is every reason to believe that, given its established record on these matters,[38] the ILO will be able to assist member States to overcome their lack of will and/or capacity in ratifying the MLC and implementing it.[39]

The obligation of a ratifying flag State to "fully implement" the MLC and give "complete effect"[40] to its provisions means, above all, that the numerous labour standards indicated under the five titles of the convention are incorporated in a national legal framework. The MLC gives States the flexibility and wide leeway on how this is to be done: through national laws or regulations, through applicable collective bargaining agreements, or through other measures or in practice.

On the matter of compliance and enforcement, Regulation 5.1 of the MLC states the purpose of the regulation, which is "to ensure that each Member implements its responsibilities under this Convention with respect to ships that fly its flag." The inspection and certification obligations of flag State authorities in fulfilment of this purpose are drawn in great detail. With respect to the obligation to inspect its vessels, the flag State is duty-bound to deploy qualified inspectors and is urged to recruit the assistance or input from seafarers themselves, by way of a complaint or otherwise, in the implementation of MLC requirements and seafarers' rights.[41] With respect to its certification duties, the flag State is directed to comply with the requirements on the issuance of the "maritime labour certificate" and

---

[38] Servais, *supra* note 10, at pp. 289–331; K. Tapiola, "The ILO System of Regular Supervision of the Application of Conventions and Recommendations: A Lasting Paradigm," in Politakis, *supra* note 16, at pp. 29–36.

[39] See the ILO's *Maritime Labour Convention 2006: Action Plan 2006–2011*, <www.ilo.org/wcmsp5/groups/public/---ed_norm/---normes/documents/publication/wcms_088034.pdf>.

[40] MLC, *supra* note 23, at Arts. I(1) and IV(5).

[41] *Ibid.*, Regulation 5.1.4.

the "declaration of maritime labour compliance."[42] These documents are subject to review during port State control inspections.[43]

Port State Control

Two historical factors have contributed to the erosion of the absolute discretion of the flag State to engage or not engage in the realization of international labour standards. These factors are recognized in the MLC.

The first is the trend in international maritime law and policy to give port States an increased authority and legitimacy to monitor and control international labour standards and IMO-promulgated safety and technical standards on board ship. This puts pressure on flag States to abide by internationally-established minimum standards. In the field of labour standards, the Merchant Shipping (Minimum Standards) Convention of 1976, ILO No. 147, played a central role in initiating this trend. ILO Convention 147 also pioneered a model or a methodology for consolidating different maritime conventions, thereby representing "a compact statement of what basic standards should be guaranteed in relation to the work and employment of seafarers."[44] A ship would be considered "sub-standard" if it failed to pass the requirements under the 15 agreements whose provisions have been incorporated in ILO Convention 147. Port States would be able to intervene to ensure observance of the relevant standards.

To the credit of 14 European States, the basic authorizations for port State control that were found in disparate ILO and IMO instruments were consolidated and regionally coordinated through a creative administrative arrangement called the Paris Memorandum of Understanding on Port State Control. The significance of this regional experiment to the international normative process cannot be overstated.[45] This regional approach to port

---

[42] *Ibid.*, Regulation 5.1.3.

[43] *Ibid.*, Regulation 5.2.1, para. 2.

[44] "Introduction," *Labour Standards on Merchant Ships* (Geneva: ILO, 1990), <training.itcilo.it/ils/CD_Use_Int_Law_web/Additional/Library/English/ILO_S_B/general_gurveys/25902.htm>.

[45] For an evaluation of the first ten years of the Paris MOU, see P. Payoyo, "Implementation of International Conventions Through Port State Control: An Assessment," *Marine Policy* 18 (1993): 379–392.

State control gave practical definition and yielded valuable information on "substandard flags," "substandard ships," and "substandard conditions onboard." The Paris MOU also became the basis for a reinvigorated regional campaign to reduce substandard shipping in European Union waters.[46] It also served as the blueprint for the creation of similar regional arrangements worldwide, in the Asia-Pacific rim, Latin America, the Mediterranean, and the Caribbean.

The MLC is said to be the fourth pillar of the international regulatory regime for quality shipping, the other three being the International Convention for the Safety of Life at Sea (SOLAS), 1974, the International Convention for the Prevention of Pollution from Ships, 1973, as modified by the Protocol of 1978 relating thereto (MARPOL 73/78), and the International Convention on Standards of Training, Certification and Watchkeeping for Seafarers (STCW), 1978. All four pillars are reinforced in an integral manner by the institution of port State control, a tried and tested platform to enforce international standards. The rise of the port State in international law is one of the salient aspects of the normative quest to have more uniformity and predictability in the enforcement of international standards.[47]

The introduction of on-shore seafarer complaint handling procedures under the MLC surely raises the profile of both port States and seafarers in building and defending the compliance culture for maritime standards that underlies the new regime. For this purpose, the MLC introduces an innovative "right to report" for seafarers, whereby complaints related to MLC requirements and seafarers' rights are given effect. To protect a seafarer in his exercise of this right, the MLC directs port States to safeguard the confidentiality of complaints made by seafarers.[48] The invitation to seafarers to participate in the process of port State control is a key element of the "right to decent work" under the MLC.

---

[46] See EU Council Directive 95/21/EC of 19 June 1995 *Concerning the enforcement, in respect of shipping using Community ports and sailing in the waters under the jurisdiction of the Member States, of international standards for ship safety, pollution prevention and shipboard living and working conditions (port State control)*, <eur-lex.europa.eu/LexUriServ/LexUriServ.do?uri=CELEX:31995L0021:EN:HTML>.

[47] The "rise of the port State in international law" is described in P. Payoyo, *Port-State Control in the Asia-Pacific: An International Legal Study of Port State Jurisdiction* (Manila: University of the Philippines Press, 1993).

[48] MLC, *supra* note 23, Regulation 5.2.2.

A Rights-based Approach to Social Justice

The second factor contributing to the erosion of exclusive flag State discretion in the implementation of labour rights and standards relates to historical ascent of human rights principles and approaches justifying the wider realization of labour standards. As early as 1960, Wilfred Jenks underscored the complementarity and mutual enrichment of human rights norms, on the one hand, and labour standards, on the other hand, in a developmental process which he believed would lead to "a new birth of freedom."[49] The utilization of the human rights argument to justify the universal and automatic application of certain core labour standards was the achievement of the 1998 ILO Declaration on Fundamental Principles and Rights at Work. The Declaration posits certain labour rights or standards as "core" and partaking of fundamental human rights. And because they are fundamental, States are bound to respect them even if they have not formally ratified any of the instruments containing these fundamental rights.[50] The Declaration buttresses the Decent Work agenda of the ILO.[51]

The MLC, as a bill of rights, enumerates the rights pertaining to all seafarers in two categories, namely the "fundamental rights and principles" under Article III and the "seafarers' employment and social rights" listed under Article IV:

> **Fundamental Rights and Principles**
> Article III
> Each Member shall satisfy itself that the provisions of its law and regulations respect, in the context of this Convention, the fundamental rights to:

---

[49] C. W. Jenks, *Human Rights and International Labour Standards* (London: Stevens & Sons Ltd, 1960).

[50] The 1988 Declaration, *supra* note 12, provides "...all [ILO] Members, even if they have not ratified the Conventions in question, have an obligation arising from the very fact of membership in the Organization to respect, to promote and to realize, in good faith and in accordance with the Constitution, the principles concerning the fundamental rights which are the subject of those Conventions...."

[51] *Report of the Director-General: Decent Work*, June 1999, <www.ilo.org/public/english/standards/relm/ilc/ilc87/rep-i.htm#Human%20rights%20and%20work>.

(a) freedom of association and the effective recognition of the right to collective bargaining;
(b) the elimination of all forms of forced or compulsory labour;
(c) the effective abolition of child labour; and
(d) the elimination of discrimination in respect of employment and occupation.

Seafarers' Employment and Social Rights
Article IV
1. Every seafarer has the right to a safe and secure workplace that complies with safety standards.
2. Every seafarer has a right to fair terms of employment.
3. Every seafarer has a right to decent working and living conditions onboard ship.
4. Every seafarer has a right to health protection, medical care, welfare measures and other forms of social protection....

The fundamental rights enumerated in Article III refer to the eight ILO Conventions identified in the Preamble.[52] It is to be noted that ratification of the MLC by a State does not imply an obligation to ratify these eight Conventions as Article III only requires that State to "satisfy itself" that its laws and regulations respect the four fundamental rights listed. This contrasts with the obligation of a ratifying State to ensure that the rights under Article IV are "fully implemented in accordance with the requirements" of the MLC. These rights under Article IV summarize the range of consolidated conventions, all 37 of them and listed under Article X.

It appears that the two categories of rights in the MLC – "fundamental rights" and "seafarers' rights" – contemplate distinct implementation regimes, with seafarers' rights being given priority and dominance in respect of their compliance and enforcement. In comparison, fundamental rights are not further elaborated or made more specific in the MLC. There is

[52] The eight conventions are the Forced Labour Convention, 1930 (No. 29); the Freedom of Association and Protection of the Right to Organize Convention, 1948 (No. 87); the Right to Organize and Collective Bargaining Convention, 1949 (No. 98); the Equal Remuneration Convention, 1951 (No. 100); the Abolition of Forced Labour Convention, 1957 (No. 105); the Discrimination (Employment and Occupation) Convention, 1958 (No. 111); the Minimum Age Convention, 1973 (No. 138); and the Worst Forms of Child Labour Convention, 1999 (No. 182).

also no provision in the MLC that defines the limitations, qualifications or checks regarding the basic obligation of a State "to satisfy itself" that its laws respect these fundamental rights. This can only mean that a violation of this obligation by a State, or complaints about its breach, is to be redressed by resort to the general supervisory procedures of the ILO.[53] A seafarer, for instance, whose right to association is denied by the flag State will not be able find remedial prescriptions in the MLC or in the compliance and enforcement provisions under Title V of the MLC. His remedy will have to be found in more general procedures, like resorting to the judicial fora in flag States, port States, or labour supplying States, if these are feasible and available, or, at the international level, by invoking the supervisory procedures established under the ILO Constitution, *viz.*, a "representation" or a "complaint" filed on his behalf.

This situation contrasts with the regime established by ILO Convention 147, which recognized the legal position that labour standards embodied in the Freedom of Association Convention, 1948 (No. 87) and the Right to Organize and Collective Bargaining, 1949 (No. 98) are appropriate subjects for international regulation through port State control. While these "fundamental" Conventions have been excluded, in practice, in the inspection procedures of regional port State control authorities, it could be observed that the enforcement possibilities for these fundamental Conventions (Nos. 87 and 98) are more expansive under ILO Convention 147 than under the MLC.[54] Under the MLC, Conventions 87 and 98 have been removed from the oversight and control authority of port States.

---

[53] These supervisory mechanisms apply to a State even if it has not ratified these fundamental conventions by virtue of their membership in the ILO. Servais, *supra* note 10, at pp. 289 *et seq*. See also, 1998 ILO Declaration on Fundamental Principles and Rights at Work, *supra* note 12.

[54] The 1982 Paris MOU and EU Council Directive 95/21/EC of 19 June 1995 do not refer to these fundamental conventions. Assuming that a port State control inspection discovers serious violations of these conventions in relation to the situation of a crew, such is not considered a "clear ground" to justify further inspection of a ship that may result in its detention. The most that can be done is to file a report on the violation or notify the flag State about it. See generally, I. Christodoulou-Varotski, "Port State Control of Labour and Social Conditions," *Annuaire de droit maritime et oceanique* 21 (2003): 251–285. Under port State control procedures in the United States, deficiencies concerning freedom of association, the protection of the right to organize, and collective bargaining "are forwarded to the U.S. Department of Labor for appropriate action." See "Commandant Instruction 16711.12A: The Merchant Shipping (Minimum

On the other hand, the specific content of "seafarers' employment and social rights" is given particularity in the MLC and is transposed as the regulations, standards and guidelines in the MLC. The seafarer's right to a safe and secure workplace is elaborated in Title 1, the right to fair terms of employment is expounded in Title 2, the right to decent working and living conditions is explained in Title 3, and the right to health protection, medical care, welfare and social protection is dealt with by Title 4. In sum, the requirements of these rights have been fully elaborated as the labour standards in Titles 4 to 5. These labour standards are in turn justified on the basis of the existence of, and commitment to, seafarers' rights defined as the employment and social rights listed in Article IV.

The redress procedures for violations of these seafarers' rights are spelled out under the compliance and enforcement provisions in Title 5 which refer specifically to the standards under Titles 1 to 4 of the MLC. In particular, redress procedures are directly accessible to seafarers for violations of their seafarers' rights and consist of provisions for the filing of complaints on-board, at port, and in the administrative and judicial bodies of flag States and labour supplying States. With respect to their enforcement, therefore, seafarers' rights, although not denominated as "fundamental," are in effect accorded more importance in the practical implementation of the "Bill of Rights for Seafarers." These rights look forward to a robust "continuing compliance culture" that is hoped to be engendered by the universal application of the MLC.

A final note on the relationship between "seafarers rights" under the MLC and "human rights" in general: in the context of "human rights" defined under U.N. human rights conventions, such rights are legally invoked by individuals against their own government. Constitutionally recognized entitlements under the "bill of rights" in the municipal law context are also invoked by individuals against their government by direct recourse to judicial remedies. Under the MLC, the nature of "seafarers' rights" does not partake of such rights which an individual seafarer can claim against his own government. The "seafarers' employment and social rights" are not individual rights granted to or endowed on seafarers *qua* individuals which would give them a basis to legally take action against their government for violation of these rights.

Standards) Convention, 1976 (ILO 147) and Port State Control (PSC)," <www.uscg.mil/hq/g-m/nmc/compl/labor.pdf>.

Instead, seafarers' rights are to be "fully implemented" by States in accordance with the requirements of the MLC, and a large modicum of flexibility is thereby given to States on how to implement or realize such seafarers rights.[55] Unlike human rights instruments, the MLC does not give individual seafarers the direct warrant to claim these rights, in either a judicial or administrative sense, against State agencies. These rights are to be implemented in such a way that living and working conditions of seafarers complies with MLC regulations and standards, and on this score it is the shipowners who are primarily addressed as the bearers of the obligation to respect these rights.

Under the MLC then, individual seafarers are not to be regarded as holders of inherent rights that they can independently and juridically assert against a State. And if these rights are violated or not implemented, the compliance and enforcement mechanisms in the MLC come into play which, as mentioned above, involves the supervisory activities of the ILO as well as the countervailing actions by flag States, port States, and labour supply countries who are called upon to ensure the fulfilment of these rights on behalf of the seafarer, or through redress actions by the individual seafarer as authorized under subsidiary legislation or implementing regulations adopted by States. Additionally, violations of these rights may be vindicated in admiralty, which accords a *sui generis* status to seafarers, whereby seafarers are treated not as autonomous individuals but as wards of admiralty courts.[56]

Conclusion

Social justice for seafarers has proven largely elusive, if not illusory, in modern history. The MLC, as an international standard, promises to deliver the goal of social justice for seafarers as workers and human beings. This large ambition is as well founded as it is well grounded.

First, the MLC confronts head-on the challenge of globalization by levelling the playing field among competing economic units in the maritime industry. The direct regulation of the international shipping industry

---

55 The scope of this flexibility is described in Doumbia-Henry, *supra* note 6.

56 The compliance and enforcement provisions of the MLC do not determine legal jurisdiction or venue. MLC, *supra* note 23, at Title 5, para. 4.

is the pre-eminent contribution of the MLC to the globalization debate. The international community is indeed better off with this paradigm of governance offered by the MLC.

Secondly, the MLC distils the experience of standard-setting in the past century and utilizes the attendant lessons learned to ensure that the normative structure of maritime labour standards is established on the principles of integrity and effectiveness. The principles of international cooperation and checks and balances are also given life in the MLC. An important cornerstone of the new regime is, without doubt, the greater scope of action of port States to secure compliance with minimum standards.

Finally, the rights-based approach which underlies the architecture of norms and standards in the MLC puts social justice and human dignity at the core of the new regime. The MLC regime is a voice that participates in the on-going debate about the proper place of human rights in the work and in the actual lives of people around the world. To the extent that the MLC advocates a standpoint on global governance, the MLC argues that rights are, and must remain, a practical matter and the true test of their relevance and value is the measure by which they enrich the lives and foster the human values of *everyone* involved in the enterprise of community.

> Being thus arrived in a good harbor and brought safe to land, they fell upon their knees and blessed the God of heaven, who had brought them over the vast and furious ocean, and delivered them from all the periles and miseries thereof, againe to set their feet on the firm and stable earth, their proper elemente.[57]

[57] *The Voyage of the Mayflower* by William Bradford (1590–1657), in *Classic Sea Stories* (London: Bracken Books, 1994), p. 224.

Fisheries Conservation and Human Rights Protection

The Management of Internationally Shared Fish Stocks: A Law and Economics Approach

Gordon R. Munro*

Introduction

The Food and Agriculture Organization of the United Nations (FAO) has declared that the management of marine shared capture fishery resources stands as one of the great challenges that must be met, if the goal of achieving long term sustainable marine fisheries is to be realized.[1] There can be no question about the importance of these resources, given that they account for up to one-third of total world marine capture fishery harvests.[2] There can also be no question that the management of these fishery resources present difficulties and problems – legal, economic, and biological – that are not to be found in the management of fishery resources confined to strictly national waters.

The economics of the management of internationally-shared fish stocks can be analyzed effectively only with explicit reference to international law. The new international law of the sea, specifically the 1982 United Nations Convention on the Law of the Sea (LOS Convention)[3] and the

---

* Professor Emeritus, Department of Economics and Fisheries Centre, University of British Columbia, Vancouver, Canada, and Visiting Professor, Centre for the Economics and Management of Aquatic Resources, University of Portsmouth, England.

1 Food and Agriculture Organization of the United Nations (FAO), *Report of the Norway-FAO Expert Consultation on the Management of Shared Fish Stocks, Bergen, Norway, 7–10 October, 2002, FAO Fisheries Report* No. 695 (Rome: FAO, 2002).

2 G. Munro, A. Van Houtte and R. Willmann, "The Conservation and Management of Shared Fish Stocks: Legal and Economic Aspects," *FAO Fisheries Technical Paper* 465 (Rome: FAO, 2004).

3 *United Nations Convention on the Law of the Sea*, Montego Bay, 10 December 1982 UN Doc. A/Conf. 62/122 (7 October 1982), 21 *I.L.M.* 1261 (1982) [hereinafter LOS Convention].

1995 United Nations Fish Stocks Agreement,[4] provides the framework within which the economics of shared fish stock management has to be examined. Specify the legal framework incorrectly, and the relevant resource management economics is certain to be worthless, at best. The problem is to be properly seen as that of constructing a regime for the management of these resources that is sound, both legally and in terms of the underlying economics.

A rapidly growing field of study, almost unknown forty years ago, is that of Law and Economics. Courses on the subject are to be found both within faculties/schools of law and departments of economics in many leading universities, certainly within North America.[5] The economics of the management of internationally shared fish stocks is a topic that falls well within the purview of Law and Economics.

To begin, it is necessary to review the definitions employed by FAO of the several classes or categories of internationally shared fish stocks.

Categories of Internationally Shared Ocean Fishery Resources and Cooperative Resource Management

The FAO categorizes internationally shared fishery resources as follows:

1. Transboundary fish stocks – Fishery resources that cross the exclusive economic zone (EEZ) boundary of one coastal State into the EEZ(s) of one, or more, neighbouring coastal States.
2. Highly migratory fish stocks – Highly migratory species, as set forth in Annex 1 of the 1982 LOS Convention, consist primarily of the major tuna species. In light of their highly migratory nature, they are to be found both within the coastal State EEZ and the adjacent high

[4] United Nations Conference on Straddling and Highly Migratory Fish Stocks, *Agreement for the Implementation of the United Nations Convention on the Law of the Sea of 10 December 1982 Relating to the Conservation and Management of Straddling Fish Stocks and Highly Migratory Fish Stocks*, New York, 4 December 1995, UN Doc. A/Conf. 164/37 (4 August 1995) [hereinafter United Nations Fish Stock Agreement].

[5] See, for example, R. Cooter and T. Ulen, *Law and Economics*, 4th edition (Boston: Pearson Addison Wesley, 1994). While both authors hold PhDs in Economics, Cooter is a faculty member of the School of Law, Stanford University, and T. Ulen is a faculty member of the College of Law, University of Illinois.

seas. In the adjacent high seas, the stocks are subject to exploitation by distant water fishing States (DWFSs).
3. Straddling fish stocks – All other fish stocks, with the exception of anadromous/catadromous stocks, that are to be found both within the coastal State EEZ and in the adjacent high seas, where they are subject to exploitation by DWFSs.
4. Discrete high seas fish stocks – Fish stocks to be found wholly within the high seas.[6]

The four categories are, of course, not mutually exclusive (with the exception of category 4). Many transboundary stocks are also either highly migratory or straddling stocks.

While both the 1982 LOS Convention and the 1995 Fish Stocks Agreement make a clear distinction between category 2 and category 3 stocks,[7] this distinction reflects neither biological nor economic reality.[8] In terms of the economics of fishery resource management, there is no meaningful distinction between category 2 and 3 stocks. It is, therefore, helpful to merge categories 2 and 3 into what shall henceforth be termed straddling fish stocks – broadly defined. Thus, there are the following categories:

A. transboundary fish stocks
B. straddling fish stocks – broadly defined
C. discrete high seas fish stocks

In this essay, the focus of attention is on categories A and B.[9]

With respect to management, it will be concluded that, with a few exceptions, all categories of shared fish stocks require that the States/entities engaged in exploiting the resources manage the resources cooperatively, if the resources, and the fisheries upon which they are based, are to be sustainable through time. The late John Gulland of the FAO, over a quarter of a century ago, pointed out that there are two levels of such

---

[6] Munro, et al., *supra* note 2, at p. 3.

[7] See in particular, LOS Convention, *supra* note 3, at Arts. 63(2) and 64.

[8] See W. T. Burke, *The New International Law of the Sea: UNCLOS 1982 and Beyond* (Oxford: Clarendon Press, 1994); R. Hilborn and J. Sibert, "Is International Management of Tuna Necessary?" *Marine Policy* 12 (1988): 31–39.

[9] Length constraints prevent us from giving a thorough review of the complex issue of the management of discrete high seas stocks.

cooperation. The first is the primary level, which consists of scientific cooperation alone, without reference to coordinated management programmes.[10] The secondary level, involving "active management," almost by definition requires the establishment of coordinated joint management programmes. If the secondary level of cooperation is to succeed, then, according to Gulland, the cooperating States/entities must deal effectively with the following:

a) allocation of harvest shares among the participating States (or entities)
b) determination of an optimal management strategy through time, including *inter alia*, the determination of optimal global harvests over time
c) implementation and enforcement of coordinated management arrangements[11]

Obviously, achieving effective cooperation at the secondary level can be a formidable undertaking.

Transboundary Fish Stocks

Category A fish stocks, transboundary stocks, are those crossing the boundary of the coastal State zone into those of neighbouring coastal States. The Pacific salmon resources, shared by Canada and the United States, provide an example that will be explored in detail at a later point.

The reasons for commencing with this category are two in number. First, the problems to be faced in achieving effective resource management are less formidable than they are in category B stocks. Secondly, transboundary fish stocks were recognized as a major resource management issue well before the second category was so recognized. As a consequence, the management of transboundary fish stocks has been extensively studied by economists, legal experts, as well as others. With respect to the economic aspects of management, the economics of the management of transbound-

---

[10] J. A. Gulland, "Some Problems in the Management of Shared Fish Stocks," *FAO Fisheries Technical Paper* No. 206 (Rome: FAO, 1980).

[11] Cited in Munro, et al., *supra* note 2, at p. 5.

ary stocks has come to serve as the foundation for the economics of the management of straddling fish stocks – broadly defined.

The first question to be raised is: What, in fact, is to be allocated between, or among, the coastal States sharing the resource? Is it the agreed upon total allowable catch (TAC), or its equivalent, that is to be allocated among the separate coastal States' fleets, or is it the net economic benefits that arise from the total harvests (resource rent in economic jargon) that is to be allocated among the coastal States? The two are not necessarily the same.

Next, a comment on the underlying legal framework surrounding the management of these resources, to be found in the 1982 LOS Convention, is in order. The LOS Convention contains but one article pertaining to the management of these resources, namely article 63(1). The Article imposes a duty on relevant coastal States to negotiate cooperative arrangements for the management of these resources. What the article does not do is impose a duty on the coastal States to reach an agreement. If the States negotiate in good faith but are unable to reach an agreement, then each State is to manage its share of the resource (i.e., that part occurring within its EEZ), in accordance with the relevant rights and duties laid down by the LOS Convention.[12] This can be referred to as the default option.

With the default option in mind, economists find that they have before them two issues, which they must attempt to analyze:

1. The consequences, if any, of the relevant coastal States adopting the default option and not cooperating in the management of the resource, at least not beyond the primary level of coordinating scientific research.
2. The conditions that must prevail if a cooperative resource management regime at the secondary level is to be stable over the long run.

If, in investigating the first issue, it is found that non-cooperative management does not carry with it significant negative consequences, then, of course, the second issue ceases to be of interest and can be safely ignored.

[12] A. Van Houtte, "Legal Aspects in the Management of Shared Fish Stocks: A Review," in FAO, *Papers Presented at the Norway-FAO Expert Consultation on the Management of Shared Fish Stocks, Bergen, Norway, 7–10 October 2002, FAO Fisheries Report* No. 695 Supplement (Rome: FAO, 2003), pp. 30–42.

Examples of shared fish stocks can be found in which such a conclusion can be reached with respect to the first issue.

If it is discovered that non-cooperation does, in fact, yield results much inferior to cooperation, then the second issue must be addressed head-on. The investigation of the second issue – the conditions that must be met if cooperative resource management regimes are to be stable through time – will have to take place in the knowledge that any cooperative resource management arrangement must be self-enforcing, as Scott Barrett emphasizes in his 2003 book, *Environment and Statecraft*.[13]

By the term self-enforcing, Barrett means that no participant in an arrangement or agreement can turn to a third party to enforce the arrangement/agreement. Those entering into a domestic contract can turn to the courts, local or national, to enforce the contract, should a dispute arise. The International Court of Justice notwithstanding, States entering into an international treaty arrangement, or some other form of contractual international agreement, have no such recourse. Hence, the arrangement/agreement must be self-enforcing and so it is with cooperative fisheries management arrangements.[14]

Consider now the simplest case of a transboundary resource, one that is shared by two, and two only, neighbouring coastal States. An example is the rich scallop resource off the Atlantic coast of North America, shared by the United States and Canada. With the advent of extended fisheries jurisdiction, and the 200 nautical mile EEZ regime, and the resolution of the maritime boundary delimitation dispute between the two coastal States in the Gulf of Maine, it was found, for various biological, seabed terrain and technological reasons, that the harvesting of scallops in the Canadian zone had no significant impact upon American harvesting opportunities and vice versa. The default option provided by the LOS Convention, while perhaps not perfect, was deemed to be adequate. There was no strong case to be made for secondary level cooperation.[15]

The case of the northwest Atlantic scallop fishery resource shared by the United States and Canada is the exception, not the rule. The normal

---

[13] S. Barrett, *Environment and Statecraft: The Strategy of Environmental Treaty-making* (Oxford: Oxford University Press, 2003).

[14] *Ibid.*

[15] G. Munro, "The Management of Shared Fishery Resources under Extended Jurisdiction," *Marine Resource Economics* 3 (1987): 271–196.

situation is one in which the harvesting activity of one coastal State, sharing a fishery resource, will have a significant impact upon the harvesting opportunities of its neighbours. Thus, a strategic interaction will inevitably arise between the two States. The resource management decisions made by the first coastal State will influence the resource management decisions of the second coastal State. The reverse will be equally true. Once again, the Pacific salmon resources shared by the United States and Canada come to mind.

If, in what has been termed the standard, or normal, transboundary fish stock case, there will inevitably be strategic interaction between or among the coastal States sharing the resources, then economists have no choice but to incorporate such strategic interaction explicitly into their analysis. As a consequence, the economics of the management of transboundary fish stocks is a blend of the standard fisheries economics applied to single State fisheries and the theory of strategic interaction (or interactive decision theory), more commonly known as the theory of games. Economists studying other shared resources, e.g., water resources, the atmosphere and global warming, also find themselves compelled to incorporate game theory into their analysis.

Game theory is becoming increasingly widely used in many different branches of economics[16] and is applied in numerous other fields. One field where game theory is gaining increasing prominence is legal studies.[17]

For the purposes of this contribution, all that is necessary is a broad overview of game theory – theory of strategic interaction – and the insights

---

[16] So widespread has the use of game theory in economics become that the Nobel Prize in Economic Sciences has twice been awarded to game theorists in the last decade and a half, with the second time being in 2005. The press release announcing the awarding of the Prize for 2005 to Laureates Thomas Schelling and Robert Auman reads as follows:

> Why do some groups of individuals, organizations and countries succeed in promoting cooperation while others suffer from conflict? The work of Robert Aumann and Thomas Schelling has established game theory – or interactive decision theory – as the dominant approach to this age-old question. <nobelprize.org/economics/2005/press.html>

This age-old question is precisely the one that we confront in a fisheries context, when addressing the issue of the management of shared fish stocks.

[17] See, for example, D. Baird, R. Gertner and R. Picker, *Game Theory and the Law* (Cambridge: Harvard University Press, 1994).

which the theory can offer when dealing with the two central issues.[18] Some basic terminology is required.

To begin, those engaging in the strategic interaction, e.g., coastal States sharing a transboundary fishery resource, are referred to as players. The players are assumed to be rational and to have various courses of action open to them, which are referred to as strategies. The expected return to a player in following a particular strategy is referred to as a payoff. The size of the expected return or payoff will, needless to say, be dependent upon the known or expected reactions of other players. The interaction between or among the "players" as they execute their strategies is the game. The stable outcome of a game, if it exists, is termed the solution. Finally, the game may be a once only affair or it may be repeated.

There are two broad categories of games – competitive or non-cooperative, and cooperative. In a cooperative game, the players are assumed to be motivated entirely by self-interest but have some incentive to attempt to cooperate. Of critical importance is the fact that the players are able to communicate with one another effectively. In competitive, non-cooperative games, the lines of communication between and among the players are more often than not faulty or simply non-existent.

Having said all of this, however, it must be emphasized in passing that open lines of communication between and among players do not, in themselves, guarantee a stable solution to a cooperative game. Communication among players is a necessary, but not sufficient, condition for a stable outcome to the cooperative game.

In exploring the first issue, the consequences of non-cooperative resource management, not surprisingly, it is the theory of non-cooperative games that is relevant. The key conclusion arising from non-cooperative game theory is that the players will be driven inexorably to adopt strategies that they know perfectly well will produce decidedly undesirable results. This outcome is referred to as a Prisoner's Dilemma outcome after a famous non-cooperative game developed to illustrate the point.[19]

---

[18] See as well, *Recommended Best Practices for Regional Fisheries Management Organizations: Report of An Independent Panel to Develop a Model for Improved Governance by Regional Fisheries Management Organizations* (London: Chatham House, 2007), Chapter 2.

[19] A. Tucker, "A Two-Person Dilemma," unpublished paper, Stanford University, 1950. See as well, Baird, et al., *supra* note 17, and Munro, et al., *supra* note 2, at Appendix.

The basic nature of the Prisoner's Dilemma outcome, in a fisheries context, can be illustrated as follows. Consider a transboundary fishery resource shared by two coastal States A and B. A's harvesting activities will have an impact upon B and vice versa. Suppose further that there is no significant resource management cooperation between the two. A and B adopt the default option and manage their respective segments of the resource on their own.

If A undertakes to restrict harvests in order to invest in the resource, the benefits from this action will not be enjoyed by A alone but will be shared with B. What assurance does A have that B will also undertake to conserve the resource? Since there is no cooperation, the answer is none. It is only too possible that B would be content to free ride off of A's resource investment efforts. In these circumstances, it is likely that A will conclude that the return on its resource investment would be less than the cost and that its best course of action is to do nothing. B could be expected to come to the same conclusion.

Worse, A has to allow for the possibility that B might deliberately deplete the resource. If A seriously believes this, then it could decide that its best strategy is to strike first. Once again, B could follow the same line of reasoning.

The Pacific salmon shared between Canada and the United States provides a real world example. The cooperative fishery management arrangement between the two States is arguably one of the most complex in the world. Historically, Pacific salmon was the single most important resource for the fishing industries of the U.S. states of Washington, Oregon and Alaska, and the Canadian province of British Columbia. The resource is shared because U.S. fishermen inevitably intercept (i.e., harvest) salmon produced in Canadian rivers and streams, while Canadian fishermen inevitably intercept U.S. produced salmon. The United States and Canada, two developed coastal States with extensive fisheries management capabilities and experience, came together in the late 1960s to cooperate in the management of all Pacific salmon fishery resources from northern California to the Gulf of Alaska. The negotiations were long, arduous and difficult. A successful conclusion was not reached until 1985. What drove the negotiators on was the manifestation of the Prisoners' Dilemma.

It was believed at the time that salmon production could be increased substantially through enhancement projects (e.g., fish ladders) on major salmon rivers, such as the Fraser River, which empties south of Vancouver, and the Columbia River, which marks the boundary between Washington and Oregon. If Canada and the United States both engaged in such projects, the mutual benefits could have been impressive. Each country held back from enhancement projects for fear that the other would free ride on its efforts.[20]

In addition, there were outbreaks between the two States of Pacific salmon fish wars, which the U.S. legal expert, Thomas Jensen, defines as deliberate overexploitation of the fishery resource for the purpose of denying harvest opportunity to the other party or parties.[21] When negotiations finally reached a successful conclusion in 1985, in the form of the Canada-United States Pacific Salmon Treaty,[22] Jensen commented that the Treaty could best be described as a "peace treaty memorializing the end of the Pacific salmon war."[23] In the early 1990s, the Treaty seized up for a period of several years. During this period of paralysis, the Prisoners' Dilemma re-emerged with a vengeance to the detriment of the resources.[24]

The conclusion is that, with few exceptions, cooperation beyond the primary level does matter in the management of transboundary stocks. Hence, the second issue, the conditions that must be met if the cooperative resource management arrangement is to prove to be stable through time given that the arrangement must be self-enforcing, must be explored.

In examining this issue, the second branch of game theory, the theory of cooperative games, which is essentially a theory of bargaining, is drawn upon. From the theory arises the first, and one might say, critical condition for stability, a condition, which on the face of it, is stunningly obvious. It goes under the heading of the individual rationality condition. No cooperative arrangement will be stable unless, at every point in time,

---

[20] G. Munro, T. McDorman and R. McKelvey, *Transboundary Fishery Resources and the Canada-United States Pacific Salmon Treaty*, Occasional Papers: Canadian-American Public Policy No. 33 (Orono: Canadian-American Center, University of Maine, 1998).

[21] T. Jensen, "The United States-Canada Pacific Salmon Interception Treaty: An Historical and Legal Overview," *Environmental Law* 16(3) (1986): 363–422, at p. 371.

[22] *Treaty Between the Government of the United States of America and the Government of Canada Concerning Pacific Salmon*, Ottawa, 18 March 1995, TIAS 11091.

[23] Jensen, *supra* note 21, at p. 372.

[24] K. Miller, G. Munro, T. McDorman, R. McKelvey and P. Tydemers, *The 1999 Salmon Agreement: A Sustainable Solution?*, Occasional Papers: Canadian-American Public Policy No. 47 (Orono: Canadian-American Center, University of Maine, 2001).

each and every player is assured a return (payoff) from the cooperative arrangement at least as great as that which it would enjoy by refusing to cooperate. The Report of the 2002 Norway-FAO Expert Consultation on the Management of Shared Fish Stocks notes that, while this proposition should be obvious, it is often ignored in practice.[25] An accompanying, one might say overarching, condition for stability is that of perceived fairness and equity. If only one player believes that its share of the overall economic returns from the cooperative arrangement is inequitable, the arrangement will be inherently unstable.

The problem is to devise a cooperative arrangement, which is deemed to be fair and which will deter defection. A cooperation arrangement in which no participant has an incentive to defect is, by definition, self-enforcing. Obviously, in order to achieve this goal, the allocation of the economic returns from the cooperatively managed fishery among the participants must be seen as equitable, but there is more to it than this. Return to the Gulland list of the three problems that a cooperative regime must address. In addition to allocations, he includes implementation and enforcement, and determining the optimal resource management programme through time.

Consider first implementation and enforcement. If weak monitoring promises that cheating will go largely undetected, then even players with a strong moral sense may calculate that cheating by other players will reduce their expected payoffs from cooperation below what they could expect to receive under cooperation regardless of how fair their promised allocations may have been. They will refuse to cooperate – the individual rationality condition once again.

With regards to determining the optimal resource management programme through time, cooperative game theory uses the term cooperative surplus. This refers to the difference between the sum of payoffs to the players under cooperation and the sum of the payoffs to the players under non-cooperation. Achieving optimal resource management through time will maximize the cooperative surplus. The larger the cooperative surplus, the more the players have to gain through effective cooperation, and the easier it is to ensure that the individual rationality constraint is met for each player.

25 FAO, *supra* note 1.

The Barents Sea and a cooperative fisheries management arrangement involving Norway and the Soviet Union/Russia is an example. The cooperative resource management arrangement was established in the mid-1970s when the Cold War was in full force. The arrangement, involving three groundfish species, has been successful in spite of complications arising from the fact that some of its stocks are straddling, as well as transboundary, in nature. The arrangement has also survived the political transformation in the former Soviet Union. A key factor in the success of the arrangement was, and is, the magnitude of the cooperative surplus.

Two Norwegian economists, Claire Armstrong and Ola Fläten, undertook in 1999 an empirical analysis of the Barents Sea cooperative fisheries management arrangement, focussing on the most important of the three resources, cod.[26] They argue that the cooperative arrangement was far from perfect in terms of maximizing the global economic returns from the fisheries. Nonetheless, they estimate that the combined economic returns from the cod fishery to the two coastal States under cooperation is more than 50 times greater than what it would be under non-cooperation. While the study was completed over a decade and a half ago, there is no reason to believe that the returns from cooperation have diminished significantly.[27] Thus, both players have a great deal to gain from the cooperative management arrangement and much to lose from its collapse. The arrangement is indeed self-enforcing.

Having said all of this, there is one important caveat. The discussion in the previous paragraphs holds true given that the monitoring of the arrangement is effective. If the monitoring is weak, large gains from cooperation will increase the temptation to cheat. With weak monitoring, one can encounter the paradoxical situation in which large potential gains from

[26] C. Armstrong and O. Fläten, "The Optimal Management of a Transboundary Fishery Resource: The Arcto-Norwegian Cod Stock," in R. Arnason and T. Bjørndal, eds, *Essays on the Economics of Migratory Fish Stocks* (Berlin: Springer-Verlag, 1991), pp. 137–152.

[27] O. Stokke, "Management of Shared Fish Stocks in the Barents Sea," in FAO, *Papers Presented at the Norway-FAO Expert Consultation on the Management of Shared Fish Stocks Bergen, Norway, 7–10 October 2002, FAO Fisheries Report* No. 695 Supplement (Rome: FAO, 2003), pp. 180–191.

cooperation will actually undermine the prospects for long-term stability of the cooperative arrangement.[28]

If the theory of cooperative games is essentially a theory of bargaining, this suggests that an obvious step towards ensuring a stable outcome to the bargaining is that of broadening the scope for bargaining as much as possible. This involves the willingness to use transfers, or the equivalent, which in game theory parlance are referred to as side payments. The terminology is unfortunate, because it carries with it the connotation of bribes and other unsavoury practices. The term, in fact, means nothing of the kind. With respect to cooperative fisheries management, the willingness to use such transfers means that the focus should be on the dividing of the net economic returns from the fishery over time, rather than dividing the total allowable catch between or among national fleets. A classic example of the use of side payments is provided by the case of the North Pacific fur seal fishery.

From the late 19th century onwards, the North Pacific fur seal fishery was shared by four States: Canada, Japan, Russia and the United States. When the fishery became significant in the late 19th century, there was no cooperative management. The Prisoner's Dilemma played itself out and the resource was subject to severe overexploitation. Fearing the outright collapse of the resource, the four States came together and transformed the non-cooperative game into a cooperative one, which took the form of the 1911 Convention for the Preservation and Protection of Fur Seals,[29] which was to last, with one lengthy hiatus, until 1984.[30]

The four players were not identical. Two, Russia and the United States, were low cost harvesters, harvesting the seals on land; the other two, Canada and Japan, were high cost harvesters, harvesting the seals at sea. Moreover, Russia and the United States received higher prices for their seal skins than did the other two States. From an economic standpoint, the optimal cooperative management programme would have been one in which all of the harvesting was undertaken by Russia and the United States. Under the terms of the 1911 Convention, this is exactly what

[28] S. Barrett, "Self-enforcing International Environmental Agreements," *Oxford Economic Papers* 46 (1994): 878–894.

[29] *Convention between the United States, Great Britain, Russia and Japan for the Preservation and Protection of Fur Seals*, Washington, D.C., 7 July 1911, 104 *B.F.S.P.* 175.

[30] Barrett, *supra* note 13.

happened. The season-by-season allocations to the Canadian and Japanese fleets were set equal to zero. All harvesting was done by the United States and Russia. The Canadians and Japanese were compensated by being given fixed percentages of the total allowable harvest (TAC) in the form of seal skins each and every season. This pure side payments cooperative arrangement proved to be profitable for all four players. Moreover, it also had beneficial conservation consequences. It was estimated by the FAO that, between 1911 and 1941, when the hiatus in the Convention, referred to earlier, commenced, the seal herds had increased eighteen fold.[31]

Even if enforcement is strong and a good attempt to maximize the cooperative surplus has been made, there remains an additional condition that must be met if the cooperative resource management regime is to be stable through time. This can be referred to simply as resilience. It can be anticipated that the cooperative resource management regime will be subject to unpredictable shocks through time, due to environmental, economic or political factors. The consequence can be that what was initially seen as a stable cooperative management regime will cease to be so unless the cooperative resource management regime has sufficient flexibility to adjust to the shocks.

For an example, we turn again to the case of the Canada-U.S. Pacific Salmon Treaty. In the cooperative fisheries game that was the Treaty, it was reasonable to regard Canada as a single player because, within Canada, jurisdiction over marine fisheries rests wholly with the federal government. In the United States, on the other hand, significant power respecting salmon rests with the individual states. The United States was and is, therefore, not a single player. The United States can be seen as a four player coalition, consisting of Washington/Oregon, Alaska, the U.S. federal government, and because of key U.S. federal court decisions regarding the allocation of salmon harvests off of Washington/Oregon, the 24 Treaty Native American tribes of Washington, Oregon and Idaho.[32] The cooperative game is a two-stage one in which the players in the U.S. coalition bargain among themselves and, upon achieving a consensus, proceed to bargain with Canada. The Alaskans have always had the least

---

[31] FAO, "Marine Fisheries and the Law of the Sea: A Decade of Change, *FAO Fisheries Circular* No. 853 (Rome: FAO, 2002).

[32] The U.S. federal government plays a relatively minor role in the U.S. coalition. See Miller, et al., *supra* note 24.

to gain from a treaty, and indeed the treaty negotiations ground to a halt in the early 1980s because of Alaskan dissatisfaction. Through a complex bargaining process, the problem was resolved and the Pacific Salmon Treaty came in to place in 1985.

At the time of the signing of the Treaty, there was a rough balance between Canadian interception of American produced salmon and the American interception of Canadian produced salmon. It was recognized by all that the cooperative surplus was substantial and it appeared that the allocations of the economic benefits from the fishery were more or less fair.[33] What was not recognized at the time was that there was a climate regime shift underway. The regime shift was to have a markedly negative impact upon salmon stocks off Washington, Oregon and southern British Columbia, and a decidedly positive impact upon salmon stocks off of Alaska. The equitable division of benefits was upset and the Treaty regime was thrown into disarray, with Alaska effectively being pitted against Canada, Washington/Oregon and the American Treaty tribes. To all intents and purposes, the individual rationality condition with respect to Alaska was not being met.[34]

The cooperative resource management arrangement proved to lack the resilience to withstand the major environmental shock in the form of the climate regime shift. One weakness of the Treaty was that bargaining between Canada and the U.S. coalition was constrained by the fact there was no allowance whatsoever for side payments.[35]

After almost six years of paralysis, Canada and the United States signed the Pacific Salmon Agreement in 1999[36] designed to patch-up the Treaty. Interestingly, the 1999 Agreement contained modest provisions for side payments, although they were certainly not labelled as such.[37] At the time of writing, the Treaty, as modified by the Agreement, has worked reasonably. Whether the Treaty proves to be sustainable over the long run remains to be seen.

---

[33] Munro, et al., *supra* note 20.

[34] Miller, et al., *supra* note 24; K. Miller and G. Munro, "Climate and Cooperation: A New Perspective on the Management of Shared Fish Stocks," *Marine Resource Economics* 19 (2004): 367–393.

[35] Miller, et al., *supra* note 24.

[36] United States Department of State, Diplomatic Note No. 0225 from Canada to the United States; reply: attached Agreement, 30 June 1999, <www.www.state.gov>.

[37] Miller, et al., *supra* note 24.

Straddling Fish Stocks Broadly Defined

In this contribution, straddling fish stocks are being defined broadly to include all fish stocks found within the coastal State EEZ and the adjacent high seas. To repeat an earlier comment, in terms of the economics of resource management, there is no meaningful difference between straddling stocks, as defined by the FAO, and highly migratory stocks.[38]

The legal framework surrounding the management of these resources is found in the 1982 LOS Convention, Part V, Articles 63(2) and 64, and Part VII, Articles 87 and 116–120. Articles 63(2) and 64 admonish relevant States to cooperate in the conservation and management of straddling stocks (narrowly defined) and highly migratory stocks respectively. Article 87 sets forth the freedom to fish in the high seas, but this right is heavily circumscribed by Articles 116–120.[39]

Be that as it may, it has been argued that under the aforementioned set of articles the rights, duties and obligations of coastal States to the high seas portions of straddling fish stocks (broadly defined), as opposed to those of distant water fishing States (DWFSs), were, prior to 1995, opaque. This, in turn, made it very difficult to establish effective cooperative resource management arrangements. As will be noted, this led in turn to Prisoner's Dilemma type of situations, resulting in extensive overexploitation, and caused the United Nations to mount the Fish Stocks Conference in 1993. The Conference brought forth the 1995 Fish Stocks Agreement,[40] which was designed not to replace any part of the LOS Convention, but rather to buttress the Convention.[41]

Two observations are in order. First, under the terms of the 1995 Fish Stocks Agreement, the resources are to be managed through regional fisheries management organizations (RFMOs), as exemplified by the Western and Central Pacific Fisheries Commission, the Northwest Atlantic Fisheries Organization (NAFO) and the Northeast Atlantic Fisheries Commission (NEAFC). Pursuant to Article 8(4) of the 1995 Agreement:

---

[38] See note 8.

[39] Munro, et al., *supra* note 2.

[40] See *Recommended Best Practices for Regional Fisheries Management Organizations, supra* note 18, at Chapter 1.

[41] Munro, et al., *supra* note 2.

> only those States which are members of such an organization or participants in such an arrangement, or which agree to apply the conservation and management measures established by such organization or arrangement, shall have access to the fishery resources to which these measures apply.

The second observation is that the fleets of non-participating States, which engage in exploitation of the high seas portion of stocks subject to RFMO management in a manner contrary to the management provisions of the RFMO, because the fleets' home States are not parties to the 1995 Fish Stocks Agreement, are engaging in unregulated as opposed to illegal fishing.[42]

Economists, in exploring the economics of the management of straddling stocks, commence with the economics of the management of transboundary stocks and then ask themselves what additions, or modifications, to the analysis, if any, are required. With respect to non-cooperative management of straddling stocks, the answer is simple. The economics of non-cooperative management of transboundary stocks applies without modification. Non-cooperative management of straddling type of stocks leads directly to Prisoners' Dilemma type of outcomes.

Munro, Van Houtte and Willmann give support to the argument that weaknesses in the LOS Convention pertaining to high seas fishery resources, particularly Part VII, ensured that these resources would be managed non-cooperatively prior to 1995. The result was discord and overexploitation, as exemplified by the pollock resources of the Bering Sea "Doughnut Hole" and groundfish resources on the Nose and Tail of the Grand Banks of Newfoundland. The three then go on to state that

> ...the overexploitation of straddling stocks [broadly defined] worldwide, which provided the rationale for the UN Fish Stocks Conference, bears powerful testimony to the predictive power of the economic analysis of the non-cooperative management of such resources.[43]

---

42 For the distinction between illegal and unregulated fishing see FAO, *International Plan of Action to Prevent, Deter and Eliminate Illegal, Unreported and Unregulated Fishing* (Rome: FAO, 2001). The nebulous nature of unregulated fishing with respect to RFMOs reflects, in part the fact that the provisions of the 1995 Fish Stocks Agreement are not fully recognized as a part of customary international law. See A. Serdy, "Trading of Fishery Commission Quota in International Law," in A. Chircop, S. Coffen-Smout and M. McConnell, eds, *Ocean Yearbook 21* (Leiden: Martinus Nijhoff, 2007), pp. 265–288.

43 Munro, et al., *supra* note 2, at p. 45.

It is in cooperative resource management that significant differences appear between transboundary and straddling stock management. The first difference is one of degree rather than kind. In cases of cooperative management of transboundary stocks, the number of players, with few exceptions, is small. In the cooperative management of straddling type stocks, a large number of players is commonplace. As a general rule of thumb, the larger the number of players, the more difficult is it to achieve a stable cooperative arrangement. The two following differences are differences in kind.

The second difference is in terms of the nature and number of participants through time. In the case of transboundary stocks, the identity of those who should be involved in the cooperative management of a resource, or set of resources, is, with few exceptions, obvious. Moreover, the number of participants through time will be constant, again with few exceptions. As far as straddling type stocks are concerned even the identity of the initial, or charter, members of a given RFMO may not be entirely clear. Article 8(3) of the 1995 Fish Stocks Agreement maintains that "…States having a real interest in the fisheries concerned may become members of such organizations," i.e., RFMOs. Does this imply that the charter members of a RFMO should include, for example, DWFSs, which had hitherto never been involved in the relevant fisheries, but now would like to become so involved, and express a real interest to this effect?[44]

With regards to the number and nature of participants of a RFMO over time, this may be anything but constant since some of the participants are DWFSs, whose fleets are nothing if not mobile. Possibly some initial charter members of the RFMO may withdraw and abandon the relevant fishery(ies). More importantly, new members/participants may appear and demand to be allowed entrance. Articles 8, 10 and 11 of the 1995 Fish Stocks Agreement make it apparent that charter members of a RFMO cannot outright bar prospective new members who are prepared to adhere to the RFMO management regime.[45] The question is under

[44] For sharply divergent views on this issue see E. Molenaar, "The Concept of 'Real Interest' and Other Aspects of Cooperation Through Regional Fisheries Management Mechanisms," *The International Journal of Marine and Coastal Law* 15 (2000): 475–531 and F. Órrego Vicuña, *The Changing International Law of High Seas Fisheries* (Cambridge: Cambridge University Press, 1999).

[45] See Munro, et al., *supra* note 2.

what terms are prospective new members to be promoted to enter, e.g., what allocations are to be made to new members?

The third major difference between cooperative management of transboundary stocks and that of straddling stocks revolves around the threat of free riding. Munro, Van Houtte and Willmann make a distinction between non-compliance and free riding, while conceding that the boundary between the two may be blurred. In using the term non-compliance, they essentially mean cheating by participants in a cooperative arrangement. By free riding, on the other hand, they refer to enjoyment of the benefits of, or returns from, a cooperative arrangement by non-participants.[46] Non-compliance is obviously an issue in the management of transboundary stocks. While the authors agree that free riding is conceivable in the case of transboundary stock management, they are hard pressed to come up with any real world examples.[47]

By way of contrast, free riding is very much an issue in the cooperative management of straddling stocks given the possibilities open for unregulated fishing in the high seas adjacent to the EEZs. The consequences of uncontrolled free riding are straightforward enough. With the prospect of much of the economic return from cooperative management being bled off by free riders, charter members of a RFMO may calculate that their expected payoff from participating in, or remaining in, the cooperative arrangement would fall below their payoffs from non-cooperation – the individual rationality condition yet again. The stability of the RFMO could collapse. In addition to free riding through unregulated fishing, the new member and real interest issues carry with them a more subtle variant of the free rider problem, quite separate from unregulated fishing. It arises in the following manner.

McDorman, Sigurjonsson and Örebech, in their 1998 article, maintain that, under the 1995 Fish Stocks Agreement, new members must be allocated just and reasonable shares of the TAC(s) available under the RFMO management plan.[48] In their 1997 article, Kaitala and Munro demonstrate

46 *Ibid.*

47 *Ibid.*

48 P. Örrego, K. Sigurjonsson and T. L. McDorman, "The 1995 United Nations Straddling and Highly Migratory Fish Stocks Agreement: Management, Enforcement and Dispute Settlement," *The International Journal of Marine and Coastal Law* 15 (1998): 361–378.

the following. If just and reasonable implies that new members/participants upon joining a RFMO should be allocated, at no further cost as it were, shares of the TAC or the equivalent on a pro-rata basis, then, when planning is undertaken for the establishment of a RFMO, prospective charter members could well calculate that their expected payoffs from cooperation would fall below their respective non-cooperation payoffs. Hence, the RFMO would be stillborn, in essence because of potential new member free riding.[49]

The Kaitala-Munro argument can be explained in terms of the following example. Suppose that a hitherto overexploited straddling type of stock comes under the management of a RFMO consisting of coastal State V, and three DWFSs, W, X and Y, all of which had a history of involvement in the fishery. The four charter members undertake the cost and sacrifice of rebuilding the resource over a seven year period. In the eighth year, the four are in a position to enjoy a return on their resource investment through harvesting. At the beginning of the eighth year, a prospective new member, DWFS Z, appears. It demands access to the RFMO, agrees to abide by the resource management rules, but demands, "free of charge," a pro-rata share of the harvest, and by implication, a pro-rata share of the net economic returns from the fishery. If DWFS Z's demands were acceded to, Z would effectively be a free rider. Having incurred none of the costs and sacrifices of investment in the resource, it will enjoy, at no cost, a pro-rata share of the return on the investment. A straightforward application of game theory demonstrates that the impact of this new form of free riding is no different from the impact of the free riding associated with unregulated fishing.[50]

The real interest issue raises a similar free rider problem. Munro, Van Houtte and Willmann (1998) argue that, if real interest as expressed in Article 8 of the Fish Stocks Agreement is interpreted to mean that States not currently engaged in exploiting resources to come under the management of a RFMO must be invited to become charter members of the RFMO, then the same sort of free rider problem threatened by the new member issue can readily arise.

---

[49] V. Kaitala and G. Munro, "The Conservation and Management of High Seas Fishery Resources Under the New Law of the Sea," *Natural Resource Modeling* 10 (1997): 87–108.

[50] *Ibid.*; Munro, et al., *supra* note 2.

Suppose, as before, that States V, W, X, and Y come together to establish a RFMO to oversee the management of a straddling or highly migratory stock that had been overexploited. Suppose, also as before, that the four had been actively involved in the fishery prior to any thought being given to establishing a RFMO. The four plan to rebuild the resource over a seven year period. Suppose that DWFS Z is a State, which had never participated in the exploitation of the resource, but which has developed a "real interest" in the resource now that it may come under effective management. Rather than wait to come in later as a new member, Z demands full and undiluted charter membership. The four charter members feel compelled to accede to Z's demand. Z incurs no real sacrifice in the rebuilding of the resource because it had not hitherto been engaged in harvesting the resource. Z will simply bide its time over the seven-year period and, when the eighth year arrives, will come to enjoy an allocated share of the return on the resource investment as the free rider that it most certainly is. Once again, the possibility of such free riding could undermine the viability of the RFMO.

Willock and Lack, in their 2006 study of RFMO practices, maintain that there have been two broad approaches to addressing the new member problem. The first approach, for which NAFO and NEAFC provide examples, involves informing prospective new members that the RFMO fisheries are fully subscribed and that they can expect allocations from the "Others" category and new fisheries only. Willock and Lack aptly describe this approach as "effectively closing the door on new members."[51]

The second approach, as exemplified by the International Commission for the Conservation of Atlantic Tunas (ICCAT) and the Commission for the Conservation of Southern Bluefin Tuna (CCSBT), is to grant prospective new members allocations at the expense of charter members. Some RFMOs attempt to mask the pain to charter members by adding the allocations to new members to the existing TACs. Rational charter members will, however, soon strip the mask away. South Africa, a charter member of ICCAT, referred to this practice, then being carried out by

[51] A. Willock and M. Lack, *Follow the Leader: Learning from the Experience and Best Practices of Regional Fisheries Management Organizations* (Sydney: WWF International and TRAFFIC International, 2006), p. 27.

ICCAT, as "nothing less than ICCAT-sanctioned overfishing in complete violation of our convention."[52]

The two approaches combined pose a dilemma. If allocations offered to prospective new members, or hitherto non-participants in the fishery(ies) now claiming a real interest, are too generous, then the RFMO may be undermined for the reasons discussed. If, however, States/entities found in these two groups deem the offered allocations to be insufficient, they may refuse to join the RFMO and turn to unregulated fishing in the adjacent high seas, agreement or no agreement. How then is the dilemma to be resolved?

A group of European fisheries economists at the cutting edge of the application of game theory to the management of shared fish stocks have addressed this very problem.[53] Their conclusion is that, if restrictions on unregulated fishing are weak, there will be instances in which no resolution of the dilemma is possible regardless of how ingenious the allocation schemes might be.[54]

There remains a third approach, which is coming up for increasing discussion. This is to allow, to enable, prospective new members to buy or lease quota from existing RFMO members. This is similar to prospective new entrants to a domestic individual transferable quota (ITQ) fishery attempting to buy quota from existing ITQ holders. The alternative was discussed at the 2002 Norway-FAO Bergen Expert Consultation. The Report states:

---

52 *Ibid.*, p. 26.

53 See in particular, P. Pintassilgo, "A Coalition Approach to the Management of High Seas Fisheries in the Presence of Externalities," *Natural Resource Modeling* 16 (2003): 175–197; P. Pintassilgo and M. Lindroos, "Coalition Formation in Straddling Stock Fisheries: A Partition Function Approach," *International Game Theory Review* (forthcoming).

54 The key issue, to which the theory points, is the impact of the establishment of an RFMO upon those who remain outside. If the impact is positive, then it may opt to remain outside the RFMO and free ride. The cooperative management arrangement that is the RFMO then becomes inherently unstable. If the impact on outsiders is negative, then the cooperative management is likely to be stable – it pays to join the club. If there are no effective checks on unregulated fishing, then the aforementioned impact on outsiders will be positive, if the RFMO succeeds in improving the management of the relevant resources. The outsiders will reap the benefits of the improved management at no cost to themselves. The trick, obviously, is to construct a scheme such that the impact upon outsiders of establishing an RFMO will be negative.

> If... it were possible for prospective New Members to purchase quotas from existing members of RFMOs, this would serve to ease the problem of quota allocation to New Members.[55]

It was recognized that, if this approach were to be adopted, by implication, the existing members of the RFMO would be granted *de facto* collective property rights to the fishery resources encompassed by the RFMO.[56]

The approach is of academic interest only, of course, if it is found to be in violation of international law. In a recently published article, Andrew Serdy argues that the proposal is fully compatible with international law.[57] He goes further. As noted, whether the provisions of the 1995 Fish Stocks Agreement constitute customary international law remains very much in doubt, a fact that does, it can be claimed exacerbate the unregulated fishing problem. Serdy argues that allowing for the transferability (i.e., sale or lease) of national quota between existing RFMO members and prospective new members

> ...will tend to hasten the parallel crystallisation of the customary rule of cooperation in international fisheries law into a requirement that non-members abide by the RFMOs rules in order to fish, as long as these are non-discriminatory. This test should not be hard to satisfy, since a would-be new entrant can, at any time, by becoming a member of the RFMO... make itself eligible to buy quota from an existing member – and refusal of an offer is non discriminatory.[58]

The implications of the preceding discussion on new members and unregulated fishing are clear enough. In his keynote address to the 2004 International Institute of Fisheries Economics and Trade (IIFET) Conference, FAO Assistant Director General Ichiro Nomura stated: "...if Regional Fisheries Bodies [including RFMOs] are not able to fulfil their mandates because of IUU fishing, the outlook for the sustainable utilization of many of the world's commercially important fish stocks is bleak."[59] If the newly emerging RFMO regime is to prosper, it is of utmost importance that

---

[55] FAO, *supra* note 1, at para. 63.

[56] Munro, et al., *supra* note 2.

[57] Serdy, *supra* note 42.

[58] *Ibid.*, p. 281.

[59] I. Nomura, "State of World Fisheries and Future Sustainability Issues," *Proceedings of the Twelfth Biennial Conference of the International Institute of Fisheries Economics and Trade: What Are Responsible Fisheries?* (Corvallis: Department of Agricultural and Resource Economics, Oregon State University, 2004).

unregulated fishing be eliminated, with such fishing becoming *de facto*, if not *de jure*, illegal fishing.[60]

Conclusion

The FAO has declared that the management of internationally shared fish stocks stands as one of the great challenges in the attempt to achieve the goal of long-term sustainability of world capture fisheries. This essay focuses on the management of the two most important categories of shared fish stocks, namely transboundary stocks and straddling stocks, broadly defined. The analysis of the management of these resources is seen as an exercise in Law and Economics, where international treaty law can be viewed as providing the framework within which the economic management of the resources is to take place.

The examination of the economic aspects of the management of shared fish stocks commenced with two fundamental propositions. The first is that, with few exceptions, there will be a strategic interaction between and among the States/entities sharing the fishery resource. The second is that cooperative arrangements for the management of such resources must be self-enforcing if they are to be sustainable over the long run.

Given the first proposition, economists analyzing the management of internationally shared fishery resources have been compelled to view the problem through the lens of the theory of strategic interaction, or game theory, a mode of analysis now used widely by both economists and legal scholars.[61] In so doing, the first conclusion is that, if the cooperative management arrangement breaks down, the consequences can be severe with there being little left to share over the long term.

Self-enforcing cooperative resource management regimes are those in which no participant (player), now or in the future, has or will have an incentive to defect and go it on its own. To achieve this goal, cooperative

---

[60] The recently released Chatham House report on Recommended Best Practices for RFMOs, *supra* note 18, at p. 18, states:

> ...society has learned painfully over the past several decades [that] the freedom to fish on the high seas is now incompatible with the goals of conservation, sustainable use and optimum utilization of the world's capture fishery resources.

[61] See note 17.

management regimes for both transboundary and straddling stocks must be resilient in that they can withstand unpredictable shocks. Straddling stock cooperative management regimes face an additional threat in the form of free riding. It is concluded that, unless the problem of unregulated fishing can be dealt with effectively, the threat posed by free riding will be chronic and could undermine the emerging RFMO regime.

One approach for addressing the problem of unregulated fishing confronting RFMOs is to establish a scheme that would enable prospective new members to buy or lease quota from existing members of the RFMO. Designing a workable scheme for implementing this approach would, as Serdy emphasizes, be necessarily an interdisciplinary undertaking involving natural scientists, economists and legal experts.[62]

[62] Serdy, *supra* note 42, at p. 288.

To Catch a Thief: Canadian Law and Practice in the Northwest Atlantic, 1992–1994

Wendell Sanford*

Author's Note

One thing I learned from Douglas Johnston was that, notwithstanding his prodigious theoretical intellect, he was a man looking for practical solutions to the real problems of the day. The following is an account of how the Government of Canada used the international and domestic law available to it to address the problem of non-contracting party vessels fishing within the Northwest Atlantic Fisheries Organization (NAFO) area while, at the same time, developing new legal and operational tools to achieve more lasting solutions.[1]

Background

Freedom of the high seas and control of vessels on the high seas solely by their flag States is a customary international law concept dating from the 16th century.[2] With the completion of the United Nations Convention on the Law of the Sea (LOS Convention) in 1982 a new series of

---

* Director, Oceans and Environmental Law Division, Department of Foreign Affairs and International Trade, Ottawa, Canada.

[1] The views contained in this essay are those of the author alone and do not represent the policies or positions of Foreign Affairs and International Trade Canada (DFAIT). The author wishes to thank Amos Donohue, Senior Counsel, Justice Canada, for his comments which have aided in the drafting of this essay.

[2] D. P. O'Connell, *The International Law of the Sea*, Vol. II, Shearer, ed. (Oxford: Clarendon Press, 1984), pp. 792–801.

international legal norms began to come into play.[3] The relevant article of the LOS Convention states:

Article 116
Right to fish on the high seas

All States have the right for their nationals to engage in fishing on the high seas subject to:

(a) their treaty obligations;
(b) the rights and duties as well as the interests of coastal States provided for, *inter alia* in article 63, paragraph 2, and articles 64 to 67; and
(c) the provisions of this section.

While recognizing the basic right of freedom to fish on the high seas, the reference to Article 63, paragraph 2 indicates that the right is not an unfettered one.

> Where the same stock or stocks of associated species occur both within the exclusive economic zone and in an area beyond and adjacent to the zone, the coastal State and the States fishing for such stocks in the adjacent area shall seek, either directly, or through appropriate sub-regional or regional organizations to agree upon the measures necessary for the conservation of these stocks in the adjacent areas.[4]

Also, Article 117 states that "All States have the duty to take, or to co-operate with other States in taking, such measures for their respective nationals as may be necessary for the conservation of the living resources of the high seas." The duty to cooperate with Regional Fisheries Management Organizations (RFMOs) was made clear as well.

> States shall co-operate with each other in the conservation and management of living resources in the areas of the high seas. States whose nationals exploit identical living resources, or different living resources in the same area, shall

[3] *United Nations Convention on the Law of the Sea*, Montego Bay, 10 December 1982, UN Doc. A/Conf.62/122 (1982); 21 *I.L.M.* 1261 (1982) [hereinafter LOS Convention]. While it is understood that the LOS Convention did not come into force until 1994 and a number of countries did not ratify until later dates (including Canada in 2003), for the purposes of this essay it is asserted that by 1992 the fishing provisions in the LOS Convention had been accepted as reflecting customary international law.

[4] *Ibid.*, Art. 63(2).

> enter into negotiations with a view to taking the measures necessary for the conservation of the living resources concerned. They shall, as appropriate, cooperate to establish subregional and regional fisheries organizations to this end.[5]

In the northwest Atlantic these provisions coincided with the establishment of the Northwest Atlantic Fisheries Organization (NAFO) in 1979.[6] The principal purpose of NAFO, which was developed approximately midway through the LOS Convention negotiating process, was to take into account the growth of 12 nautical mile territorial seas and the proclamation of 200 nautical mile exclusive economic or fishing zones by members of the earlier International Commission for Northwest Atlantic Fisheries organization (ICNAF).[7] While NAFO came into force before the negotiations on the new law of the sea were completed, nevertheless NAFO contained the key concepts within the LOS Convention.

In September 1992 there were 14 parties to NAFO[8] whose responsibility it was to "establish and maintain an international organization whose object shall be to contribute through consultation and cooperation to the optimum utilization, rational management and conservation of fisheries resources of the Convention Area."[9] The Convention Area was essentially south from Cape Farewell, Greenland, to 35 degrees north and then westward to the American coastline.[10] The Regulatory Area within which NAFO had jurisdiction was "that part of the Convention Area which lies beyond the areas in which Coastal States exercise fisheries jurisdiction."[11]

While it was clear that many of the vessels flying the flags of the Member States of NAFO were exceeding their quotas and using illegal

---

[5] *Ibid.*, Art. 118.

[6] The *Convention on Future Multilateral Cooperation in the Northwest Atlantic Fisheries*, Ottawa, 24 October, 1978, *U.N.T.S.* 1135, 369 [hereinafter NAFO] was the successor to the International Commission for Northwest Atlantic Fisheries (ICNAF), established in 1949.

[7] *International Convention for Northwest Atlantic Fisheries Organization*, 157 *U.N.T.S.* 157 (1949).

[8] Bulgaria, Canada, Cuba, Denmark in respect of the Faroe Islands and Greenland, Estonia, the European Economic Community, Iceland, Japan, Latvia, Lithuania, Norway, Poland, Romania, and the Russian Federation.

[9] NAFO, *supra* note 6, at Art. II.

[10] *Ibid.*, Art. 1, para. 1.

[11] *Ibid.*, Art. I, para. 2.

equipment and fishing practices they were nonetheless vessels subject to regulation by NAFO.

Fishing Outside the NAFO Regime

From 1985 to 1992, an average of 200 vessels flagged to NAFO Member States per year were fishing within the NAFO Regulatory Area (NRA) under NAFO rules. A further yearly average of 33 non-contracting party (NCP) vessels (vessels flagged in States that were not a member of NAFO) were observed on an annual basis. These totals varied from a low of 29 in 1987 to a high of 47 in 1989. In 1992, the departure point for this analysis, there were 35 NCP vessels fishing within NAFO waters.

Vessels from Europe (Spain and Portugal) were flagged in various Caribbean "open registry" or flag of convenience States.[12] Panama had by far the largest fleet, growing from four European-owned vessels in 1985 to 25 in 1992. The distribution amongst other States was widely spread but small in numbers: Cayman Islands–1, Honduras–2, Morocco–1, Sierra Leone–1, St. Vincent and the Grenadines–1, and Venezuela–2. Among the States with vessels fishing in the NAFO area, Korea, a non-member of NAFO, had the distinction of having both its own flagged NCP vessels, varying in number from two to six, plus a further two to five fishing under the Panamanian flag.[13] The United States, at the time a non-member of NAFO, had vessels fishing in the NRA as well but, given their coastal State status, this essay does not consider the U.S. vessels as being of interest.

The situation of the NCP vessels operating in the NRA was typical of how illegal, unregulated, and unreported (IUU) fishing practices were conducted. The vessels were distributed among a range of flags to make it more difficult to address. The vessels moved from flag to flag when pressure

[12] States which offer their flag to vessels with little or no substantial connection to their country call themselves "open registry" States. The more commonly used term "flag of convenience," while reflective of reality, is not appreciated by these countries.

[13] All figures for this section are taken from the report for the 28–30 April 1993 Meeting of the NAFO Standing Committee on Fishing Activities of Non-Contracting Party in the NAFO Regulatory Area (STACFAC). NAFO/GC 93/2, Serial No. N.2234, Dartmouth, Canada.

was applied.[14] Small States appreciated the income stream, albeit a small amount per vessel as registry fees are based on the tonnage of vessels and fishing vessels are almost all less than 1,000 gross registered tonnes.

Normally the NCP vessel activities act as a low grade infection within an RFMO. The vessel owners cut corners and operate outside the rules and by doing so make money, but do not cause sufficient concern to make it worth an RFMO's while to track them down and attempt to curtail their activities.

However, the situation in the late 1980s and early 1990s in the northwest Atlantic was vastly different from the norm. The fisheries situation on Canada's east coast was dire. Canadian federal Fisheries Minister John Crosbie, a political powerhouse from Newfoundland, faced a monumental predicament:

> In all my years of political life this was the most serious political and economic crisis I faced. The first indication of any real problem with the northern-cod stocks had been delivered to the Department of Fisheries and Oceans by our scientists in 1989. Only three years later it appeared that the very survival of the species was at risk.[15]

In July 1992, Crosbie closed the northern cod fishery for two years to allow recovery of the stocks. The social, political and economic effects of the closure within Newfoundland and Labrador have been analyzed and discussed by others.[16] Sixteen years later the fishery still remains closed.

In this devastating context within Canada, NCP activity in the NRA grew from 840 days in 1984 to 4,400 days in 1990. Even with the collapse of the cod stocks in 1992, NCP vessels fished 3,775 days on the Grand Banks. Catches also grew dramatically from 12,000 metric tonnes (mt) in 1984 to a peak of 47,600 mt (a four-fold increase) in 1991 and remained at 42,600 mt in 1992.[17] The total of this fishing activity between 1985 and 1991 was 215,750 mt of all species. Most dramatically, this increase

---

[14] R. G. Rayfuse, *Non-Flag State Enforcement in High Seas Fisheries* (Boston: Martinus Nijhoff Publishers, 2004), p. 35.

[15] John C. Crosbie with Geoffrey Stevens, *No Holds Barred – My Life in Politics* (Toronto: McClelland and Stewart, 1997), p. 38.

[16] See, for example, the Special Issue, "After the Collapse," *Dalhousie Journal* 18(1) (1995): 5–167.

[17] All catch figures are taken from NAFO/GC 93/2, *supra* note 13. These are, at best, rough estimates and almost certainly underestimates as NCP vessels are only "hailed" or permit "courtesy boardings" to provide voluntary information. They cannot be boarded

occurred at the same time that the legitimate fishery within the NRA was in jeopardy due primarily to the actions of the European Economic Community, which, on behalf of Spain and Portugal (who had become members of the European Economic Community in 1986), used the NAFO objection procedure from 1986 until 1992 to increase their total allowable catch (TAC) for cod beyond the TAC agreed upon under the NAFO rules by 155,400 mt.[18]

The Canadian Reaction

International Law

In light of the NCP vessel activities in the NRA, the use of the objection procedure by the European Community to increase its catch in the NRA and the warnings of stock problems within Canada's 200 nautical mile zone, Canada began a full-scale defence of its interests. In 1989, Alan Beesley, a Canadian Ambassador who had chaired the Drafting Committee of the law of the sea negotiations in the 1970–1980s, was appointed as Ambassador for Fisheries Conservation. Changing attitudes and international law with respect to the heretofore sacred concept of freedom of the high seas was to be a major initiative for the Ambassador. This was not going to be easy since the major proponents of freedom of the high seas included the United States, the European Economic Community/European Union,[19] and the Soviet Union/Russian Federation who were averse to any significant changes in the great power system of the oceans of which they were principal beneficiaries. Nonetheless action had to be taken and, it was decided in Canada that such action had to be on all fronts.

In 1992, a series of multilateral initiatives commenced a process of change. In May, the International Conference on Responsible Fishing was held in Mexico, which resulted in the Cancun Declaration calling for the

for a thorough and compulsory inspection. In addition, their flag States rarely, if ever, reported their activity to NAFO.

[18] NAFO SCS DOC 91/19.

[19] During the timeframe covered by this essay, the European Economic Community (EEC) adopted the nomenclature European Union (EU). Its executive arm remained the European Commission (EC). These terms have been used interchangeably, as seems most appropriate for each date in question.

Food and Agriculture Organization of the United Nations (FAO) to draft an International Code of Conduct on Responsible Fishing.[20] The United Nations Conference on Environment and Development (UNCED) followed in June of 1992. The so-called "Earth Summit" adopted Agenda 21, chapter 17 which, in Programme Area C, dealt with "Sustainable Use and Conservation of Marine Living Resources of the High Seas."[21] Among the elements in this Programme Area was effective monitoring and enforcement with respect to fisheries activities and effective action to deter reflagging of vessels.

From this foundation work began within the U.N. system. First, a Technical Consultation on High Seas Fishing was held in September 1992 at the FAO headquarters in Rome.[22] This led to a further series of negotiations at the FAO, the outcomes of which were the FAO Compliance Agreement,[23] completed in 1993, and the Code of Conduct for Responsible Fisheries,[24] finalized in 1995. The key provision in the Compliance Agreement[25] is Article III, "Flag State Responsibility," which reads: "Each Party shall take such measures as may be necessary to ensure that fishing vessels entitled to fly its flag do not engage in any activity that undermines the effectiveness of international conservation and management measures."[26] This is the linkage which Canada sought for flag States to take more effective control of their vessels. However, as is the norm in treaty development, it took until 2003 to obtain the 25 instruments of acceptance necessary to bring the Compliance Agreement into force.

---

[20] FAO, *Code of Conduct for Responsible Fisheries*, UN Doc. A/CONF. 151/15 (Rome: FAO, 1995), Annex.

[21] Chapter 17 is reproduced in Netherlands International Law of the Sea Institute, ed., *International Organizations and Law of the Sea Documentary Yearbook 1992*, Vol. 8 (The Hague: Kluwer Law International, 1992), pp. 400–432.

[22] FAO, *Report of the Technical Consultation on High Seas Fishing*, FAO Fisheries Report No. 484 (Rome: FAO, 1992).

[23] *Agreement to Promote Compliance with International Conservation and Management Measures by Fishing Vessels on the High Seas*, Rome, 24 November 1993, (1994) 33 *I.L.M.* 968.

[24] FAO Doc. 95/20/rev. 1(1995) Reproduced in Netherlands International Law of the Sea Institute, ed., *International Organizations and the Law of the Sea. Documentary Yearbook 1995*, Vol. 11 (The Hague: Kluwer Law International, 1995), pp. 700–734.

[25] It is worth noting that the initial discussions were with respect to a "Re-Flagging Agreement" but this proved impossible to achieve.

[26] Code of Conduct, *supra* note 20.

Canada ratified on 20 May 1994. Among the NAFO NCP States, none had ratified as of 2000.[27] In addition, the Compliance Agreement, at the insistence of flag States, had no enforcement mechanism. The Code of Conduct meanwhile, as its title implies, is a non-binding agreement. With respect to flag State control, the Code states that flag States

> should ensure that the activities of such vessels (those fishing and fishing support vessels entitled to fly their flag) do not undermine the effectiveness of conservation and management measures taken in accordance with international law and adopted at the national, subregional, regional or global levels.[28]

However, notwithstanding these efforts at the FAO, Churchill and Lowe are correct in their analysis that these measures "have had little impact in practice on redressing the fundamental problems of high-seas fisheries."[29]

A more broadly-based initiative was undertaken at the U.N. headquarters in New York. Canada utilized the agreement reached at UNCED as the basis for including in a 1992 U.N. General Assembly Resolution[30] a call for a conference to deal with the issue of "straddling stocks," being those stocks which exist both within a coastal State's exclusive economic zone (EEZ) and the adjacent high seas, and "highly migratory fish stocks," which essentially means the tunas that exist across vast areas of the world's oceans. A small number of States had similar interests and concerns to Canada, including Iceland, Norway, Chile, Peru, and Argentina with respect to straddling stocks, but these were not enough to move the United Nations to action. Fortunately, the search for allies did not have to stretch far. New Zealand was a leading member of the Forum Fisheries Agency (FFA), a 16-member organization which included 14 Pacific Island Countries (PICs), New Zealand and Australia. Through the waters of the FFA Member States pass over 40 percent of the world's tuna stocks, known in LOS Convention parlance as highly migratory

---

27 Internet Guide to International Fisheries Law, <www.oceanlaw.net/texts/summaries/fascomp.htm>.

28 Code of Conduct, *supra* note 20, at Art. 6.11.

29 R. R. Churchill and A. V. Lowe, *The Law of the Sea*, 3rd ed. (Yonkers, N.Y.: Juris Publishing, 1999), p. 304.

30 United Nations, General Assembly, "United Nations conference on straddling fish stocks and highly migratory fish stocks," 93rd Plenary Meeting, 22 December 1992, A/RES/47/192.

species.[31] New Zealand and its fellow FFA Member States proved to be natural allies with Canada and the straddling stock States as they brought another key issue to the table, added significantly to the number of States interested in instigating a process, and balanced the focus by including 14 less developed nations. Thus, a series of negotiations was begun in 1993 and completed in 1995 with the conclusion of the United Nations Fish Stocks Agreement (UNFSA).[32]

While UNFSA dealt with a broad array of issues, the provision of greatest interest with respect to fishing by NCP vessels was Article 17. It holds the NCP directly accountable for fishing activities in waters under the regulatory control of an RFMO, stating that:

> Such States (NCPs) shall not authorize vessels flying its flag to engage in fishing operations for the straddling fish stocks or highly migratory fish stocks which are subject to the conservation and management measures established by such organization or arrangement.[33]

Furthermore, in Articles 18 and 19, responsibilities for all flag States are established, including NCPs. Article 18, paragraph one, sets out the duties of flag States in great detail, declaring that:

> A State whose vessels fish on the high seas shall take such measures as may be necessary to ensure that vessels flying its flag comply with subregional and regional conservation and management measures and that such vessels do not engage in any activity which undermines the effectiveness of such measures.[34]

Article 18, paragraph two, further elaborates:

> Where there is a subregionally, regionally or globally agreed system of monitoring, control and surveillance in effect, States shall ensure that the measures that they impose on vessels flying their flag are compatible with that system.[35]

---

[31] LOS Convention, *supra* note 3, at Art. 64.

[32] *Agreement for the Implementation of the Provisions of the United Nations Convention on the Law of the Sea of 10 December 1982 Relating to the Conservation and Management of Straddling Fish Stocks and Highly Migratory Fish Stocks*, New York, 4 December 1995, (1995) 34 *I.L.M.* 1542 [hereinafter UNFSA].

[33] *Ibid.*, Art. 17, para. 2.

[34] *Ibid.*, Art. 18, para. 1. This provision supplants the Compliance Agreement.

[35] *Ibid.*, Art. 18, para. 4.

Article 19 sets out a detailed series of enforcement measures while reiterating that "A State shall ensure compliance by vessels flying its flag with subregional and regional conservation and management measures for straddling fish stocks and highly migratory fish stocks."[36] These measures with UNFSA, which Canada ratified in 1999 and which came into force in 2003, have the necessary strength to address the NCP issue forcefully.

From an international law perspective, UNFSA is seen as having made great progress in dealing with fishing activities on the high seas. As of 1 February 2008, when the Republic of Korea ratified the Agreement, UNFSA has 68 parties representing over 73.6 percent of world fish imports and 53.6 percent of world fish exports.[37] The challenge remains to have the open registry States accept UNFSA. It is of particular note, however, that Belize, an open registry State with a long history of NCP activity in the NRA and elsewhere, ratified UNFSA on 14 July 2005. This contributed significantly to Belize being accepted as a cooperating non-member by the Western and Central Pacific Fisheries Commission (WCPFC) in December 2007.[38]

Diplomatic Offensive

While the foregoing initiatives were working their way forward, Canada, with Atlantic coast stocks plummeting and unable to await the outcome of the international legal process, began its diplomatic offensive. Canada's relations with the European Economic Commission (EEC) were heavily strained by the issue of the objections to NAFO quotas and by its Member States, Spain and Portugal, although beneficial owners of NCP vessels, expressing their inability to control the vessels which they did not flag. Accordingly, Canada set about engaging all NCP flag of convenience States at every opportunity as well as Korea, which had both Korean flagged and

36 *Ibid.*, Art. 19, para. 1.

37 Comparisons of the number of UNFSA versus LOS Convention ratifications are not valid as there are provisions within the LOS Convention of national interest to both landlocked and shelflocked States (an estimated 48 States). These provisions would encourage these States to ratify the LOS Convention but have no interest in UNFSA.

38 The author Chairs the Technical and Compliance Committee of the WCPFC and attended the 2007 WCPFC Annual Meeting as Alternate Head of the Canadian delegation.

flag of convenience vessels in the NRA and, at the same time, wished to become a member of NAFO.

The Honourable John Crosbie, Canada's Minister of Fisheries, began the offensive by meeting with his Panamanian counterpart in June 1992 and obtaining an agreement that Panama would apply sanctions to its vessels that fished contrary to NAFO measures and that no new vessels would be taken on its registry unless they agreed to comply with NAFO rules.[39] This was, at best, a first step. Fines were minimal – US$2,500 for a first offence and US$5,000 for a second offence, representing no more than a small cost for doing business to the fishing companies.

During the fall of 1992 Canada kept up diplomatic representations which resulted in Venezuela (European vessels, *Pescagel* and *Bacanova*), Morocco (Korean vessel, *Ain Chanech*) and Vanuatu (Korean vessel, *Marsopla*) withdrawing vessels from their respective registries. The *Marsopla* represented a typical case in that it had been a Panamanian flagged vessel before it took the Vanuatu flag.[40]

The most important fact which these activities revealed was Canada's need for a detailed profile of each vessel, both to determine beneficial ownership and to find ways of levering vessels whose stock in trade was to stay one step ahead of the law. Fortunately the Department of Fisheries and Oceans (DFO) efficient offshore patrol and boarding operation had an excellent database of photos and information derived from daily (weather permitting) flights over the NRA, plus hailings[41] and the occasional courtesy boarding. With this information it was possible to follow the fishing patterns of the NCP vessels; their changes of name and flag; and, through the Department of Foreign Affairs and International Trade (DFAIT) extensive system of embassies and high commissions, to make representations in the capitals of the flag States involved.

However, this approach to dealing with NCP flag States was only able to operate on a reactive basis to the moves of the fishing vessels and their owners. To become proactive required a significant upgrading

[39] L. S. Parsons, *Management of Marine Fisheries in Canada* (Ottawa: National Research Council of Canada, 1993), p. 282.

[40] STAFAC, *supra* note 13.

[41] "Hailing" is the practice of a fisheries patrol vessel asking NCP vessels it encounters to indicate their catch by a marine radio request. As no physical inspection occurs, figures are highly unreliable.

in Canada's effort. It was decided by DFAIT that the best approach was to enhance the operations of the Office of the Ambassador for Fisheries Conservation. A lawyer with admiralty law experience was added to the staff whose full time role was to seek all possible information available anywhere in the world on these NCP vessels. Lloyd's Registry and other databases were searched to find information. Canadian diplomatic posts were authorized to hire admiralty lawyers in the port of registry of all 35 vessels involved. They were requested to make complete searches of the vessels and their history. In the best tradition of Watergate's "Deep Throat," Canadian officials followed the money trail. Maritime registries in one country often led to company registries in another. These were searched as well. Who owned the vessel? Who owned the company that owned the vessel? Was there a mortgage? How much? Who held the mortgage? Who insured the vessel? For how much was the vessel insured? What type of vessel and fishing license did it hold and from what country or countries? When were these due to expire? With this information in hand Canada was able to gradually build a picture of each vessel's corporate structure and, most importantly, to identify the vulnerabilities. If the vessel had a mortgage from a government institution, then the next Canadian démarche would ask the country to call the loan. If the vessel had a Panamanian flag, Spanish name and captain, but one Korean company owned it and another held the mortgage, then the request to withdraw the vessel would be sent to Korea as well as the flag State.

Korea

Addressing the Republic of Korea proved to be an interesting problem as there were three levels of approach required. Firstly, Korea was seeking membership in NAFO and to receive quotas as a fishing State. Secondly, Korea was taking advantage of the loophole within the NAFO Convention involving NCP States to have two of its fishing vessels regularly in the NAFO area fishing without quota, inspection, or scientific reporting. Finally, and simultaneously, there were four Korean vessels flagged in Panama and fishing as NCP vessels under a flag of convenience. Canadian officials estimated that Korea alone was responsible for half the catch by NCP vessels in the NRA.

With respect to NAFO, Korea had been attending the annual meetings as an observer for a number of years. In September 1993 at the annual meeting held in Dartmouth, Nova Scotia, the author met with the Korean delegation for a Canada–Korea bilateral. As was our custom, Canada asked Korea to withdraw its flag vessels from the NRA pointing out that the Korean vessels were NCP vessels fishing without a quota and they were undermining endangered stocks. The Koreans replied that their fishery was a "historic" fishery which had been in progress since 1985 and that they wished to join NAFO and receive quotas based on their historic share. The author, on behalf of Canada, replied that although there were provisions within the NAFO Convention to "take into account the interests of Commission members who have traditionally fished within the Area,"[42] the word "traditionally" within the NAFO Convention was to be read in the context of a six centuries old fishery in the northwest Atlantic[43] not in decades – or in Korea's case seven years. It was stressed that an Asian fishing nation should understand this concept of time as a matter of centuries and respect it. Nothing more was heard of the argument.

The next round of approaches was made in Seoul by the Canadian Embassy, which provided detailed information to the Korean Foreign and Fisheries ministries regarding the actions of specific Korean-flagged vessels. In addition, every high level Canadian delegation that met with Koreans ministers worldwide raised the matter. For each such meeting the Canadian ministers were armed with "talking points" to register the Canadian dissatisfaction with the Korean fishing practices in the northwest Atlantic. The purpose was to show to the Korean government that these fishing practices would have an impact on the entire bilateral relationship.

Next came the question of the Korean-owned vessels not flagged in Korea. The Koreans were reluctant to acknowledge that these vessels were under their control and stated that their hands were tied with respect to domestic law. Nonetheless, Canada kept up the pressure. In October 1993 the author, in a better effort to understand the problem, was on one of the regular patrol flights out of St. John's, Newfoundland and Labrador, when the vessels *Danica* and *Pilgrim* were overflown. The electronic systems operator noted that these vessels always "chatted" to one another when the Canadian patrol aircraft was in their area. He invited me to listen in. Notwithstanding the Panamanian flags and Hispanic names of the vessels,

---

[42] NAFO, *supra* note 7, Article XI, paragraph 4.

[43] H. A. Innis, *The Cod Fisheries, The History of an International Economy*, revised edition (Toronto: University of Toronto Press, 1954), Chapters I–IV.

it was clear that the conversation was in an Asian tongue. The operator was pleased to make a tape and DFAIT translators provided a text which made it clear that these vessels were owned and operated by Koreans. They were accordingly added to our list to raise at the first opportunity.

Following up on their lack of success at the NAFO annual meetings, the Koreans decided that they needed to send a delegation to Ottawa in November 1993 to lobby the Canadian government directly. Fisheries officials from Seoul and senior representatives of their industry association made arrangements to meet with the Ambassador for Fisheries Conservation, by then Randolph Gherson, and Fisheries Minister Crosbie. In the Ambassador's board room they were met by a presentation which identified, with photographs and tables, the activity of all of the Korean-flagged and non-flagged vessels in the NRA, including the ones under Panamanian flag which they had been denying existed. Several of the vessels were even named as being owned by some of the Korean fishing interests in the room. This was followed by a meeting with the Fisheries Minister, where they were accompanied by the Economic Counsellor of the Korean Embassy in Ottawa. Minister Crosbie made it clear in no uncertain terms that there would be economic repercussions if the Korean vessels were not withdrawn immediately. This latter point was not lost on the senior Korean Embassy official although it no doubt meant little to the Korean fishing companies.

The outcome was surprisingly prompt. Korea submitted its application for membership to NAFO in December 1993 without a claim to an "historic share" and was accepted as a member. All six Korean-flagged or interest vessels were withdrawn within the first four months of 1994. Since that time, Korea has been a member of NAFO in good standing receiving quotas in 2008 totalling 1,185 mt, a figure, it is worth noting, less than five percent of its former catch as a NCP to NAFO.[44]

The Open Registry States

With an average of 33 NCP vessels registered in the Caribbean and elsewhere at the beginning of 1992 it was apparent that a concerted effort

---

44 NAFO Annual Quota Table 2008, NAFO Conservation and Management Measures 2008, Annex I. A, NAFO/FC Doc. 08/01.

was required with respect to these States. As already noted, Panama was by far the most important of the group with 27 vessels (25 of which were western European and two Korean). Other States involved were Honduras, Venezuela, the Cayman Islands, St. Vincent and the Grenadines, and Belize. The NCP vessels in total caught an estimated 42,600 mt of groundfish, including 8,600 mt of cod which was under a total moratorium both within the NRA and in Canadian waters.[45]

Following up on the ministerial visit of June 1992 to Panama noted above, Canadian officials sent two further delegations to Panama in December 1992 and November 1993. In each instance individual cases of NCP vessels were brought to the attention of the Panamanian officials. The action sought by Canada was for Panama not to issue licenses to fish in the NRA, to fine those who breached their license, and to de-register repeat offenders. The process had limited success. As already noted, the fines were insignificant, but they were within the limits allowed under domestic Panamanian law. The question of de-registering vessels presented a real challenge as NCP fishing did not constitute an offence under Panamanian domestic legislation punishable by de-registration. Additionally, Panama did not wish their registry to get a bad name in the shipping industry as being seen to bend to the interests of specific States, but they also did not wish to be used by other countries to export their problems. The Panamanian officials understood the magnitude of the problem in the northwest Atlantic and the very limited income which the small tonnage vessels brought to their treasury. Accordingly, they took a pragmatic approach and remained open to Canadian initiatives.

An opening presented itself during the second round of discussions when Panama allowed that, although it would find it difficult to remove a vessel from its registry, they would give careful consideration to letting a registration lapse. This proved an important development as will be apparent later in this narrative.

Meetings in Honduras and Belize proved difficult as these States were highly resistant to outside pressure and particularly protective of their perceived sovereign interest. In St. Vincent and the Grenadines, the Canadian representatives received a warm reception from the foreign minister who was very favourably disposed towards Canada as he had family in Montreal.

[45] STAFAC, *supra* note 13, Data on Non-Contracting Parties Activities in the NAFO Regulatory Area, Table 3.

Nonetheless, he candidly admitted that the registry was controlled by a lawyer in Geneva who "sold" registrations at US$500, without any prior advance notice or consideration by any ministry of his government.

The Problem with Canadian Domestic Law

Together with the process of pressuring flag and beneficial ownership States to withdraw their flags or their vessels from the NRA there was a second process that was taking place within Canada. This consisted of a concerted effort to assess what could be done with vessels flying NCP flags or found flagless on the high seas in the NRA. An episode that took place in October 1993 proved instructive. The FV *Gadus* was struck from the Honduran flag after concerted diplomatic efforts by Canada. A Canadian naval destroyer HMCS *Iroquois* was passing through the area in which the vessel was known to be fishing. While Canada had the authority under international law to visit and inspect a vessel believed to be flagless,[46] there existed no provision under Canadian domestic legislation that would permit a Canadian government vessel or warship to proceed against such a vessel on the high seas. The *Canada Shipping Act* section 7(4), for example, permitted the escorting of a vessel to port and the detention of the vessel.[47] However, a legal analysis of this provision took the view that it was inapplicable on the high seas.

An interdepartmental committee began the complicated task of identifying legislative changes to the *Coastal Fisheries Protection Act*[48] and its associated regulations to deal both with stateless fishing vessels and NCP vessels. By the spring of 1994, the legislation was ready for presentation to Parliament. To the surprise of all in attendance, the new legislation passed[49] all three readings in the House of Commons in one afternoon by unanimous consent, even garnering the support of the separatist Bloc Québécois based on a briefing which their leader, Lucien Bouchard, received from Justice Canada.

[46] LOS Convention, *supra* note 3, at Art. 110, para. 1(d).

[47] *Canada Shipping Act*, R.S.C. 1985, c. S-9.

[48] *Coastal Fisheries Protection Act*, S.C. 1994, c. C-12.

[49] Amendments to the *Coastal Fisheries Protection Act*, known a Bill C-29.

With the new legislation, regulations had also been prepared and adopted which included a Schedule naming the NCP countries whose vessels were to be targeted.[50] The States identified on the Schedule were informed by Canadian ambassadors of the situation. While all of the identified States were concerned with "naming and shaming," none presented diplomatic protests and all grudgingly agreed that they would be unlikely to react should one of the listed vessels flying their flag be seized by Canada in the NRA. Additionally, all the relevant vessels operating in the NRA were notified by Canadian marine radio in Spanish and Portuguese and their respective agents in St. John's were informed of the new legislation.

As Canada was testing the limits of international law by proclaiming the right to arrest foreign vessels on the high seas (the NRA), Canada took the precautionary step of amending its declaration of acceptance of the compulsory jurisdiction of the International Court of Justice to prevent other States from bringing an action against Canada as a consequence of the legislation.[51]

Canada "Catches a Break"

As 1994 began progress in dealing with NCP vessels in the NRA was being made, but it remained at a slow pace. Twenty-nine NCP vessels still fished within the NRA at some point in 1993, catching 27,000 mt of fish, an estimated 20,000 mt of which was the high value and vulnerable cod, flounder, and redfish stocks.[52] As part of Canada's efforts to force NCP vessels out of the NRA, officials in the Office of the Ambassador for Fisheries Conservation noted that a vessel with what appeared to be Portuguese beneficial ownership was coming up for renewal of

---

50 Belize, Cayman Islands, Honduras, Panama, Saint-Vincent and the Grenadines, and Sierra Leone. Table III, Part B, *Coastal Fisheries Protection Regulations* (C.R.C., c. 413).

51 This action proved beneficial when the Spain took Canada before the ICJ following the arrest of the Spanish fishing vessel *Estai* on the high seas in the NRA in 1995. The ICJ declined jurisdiction due to the Canadian reservation. *Fisheries Jurisdiction Case (Spain v. Canada)*, 5 *I.C.J. Rep.* 363 [hereinafter *Fisheries Jurisdiction Case*].

52 NAFO, STACFAC Working Paper 94/1 NAFO, 25–26 April 1994, Dartmouth, Canada.

its Panamanian registry on 29 March 1994. The vessel was the *Kristina Logos* with a Portuguese company address. Canada's lawyer in Panama was placed on standby to provide a certified true copy of the vessel's lack of Panamanian registry at the opening of the office in Panama City on 30 March 1994. With action in the offing a more focused examination of the records resulted in an amazing turn of events.

A detailed search of the Portuguese companies registry determined that the owner of the Portuguese company was in fact a Canadian with a firm in Nova Scotia which had recently been struck from the Nova Scotia registry of companies. In addition, there was an outstanding mortgage on the vessel in Canada, so that the transfer to the Panamanian registry itself was not valid under Canadian admiralty law. Adding into the mix was the fact that it was the week before Easter and there was little activity in Panama City so that the vessel's owner was not well placed either to know what was happening or react. Fortunately, the Panamanian shipping registry was open 30 March 1994 (a Wednesday) and the certified copy of the required document was obtained by the lawyer retained by Canada in Panama City and forwarded to St. John's. It was faxed to the Canadian Fisheries protection vessel *Leonard J. Cowley*, which had been waiting in the vicinity, and she proceeded to the *Kristina Logos* location. On approaching the fishing vessel on a Saturday morning of the Easter long weekend (2 April 1994) the *Cowley* hailed the *Logos* enquiring what flag she held. The reply received was Panama. The *Cowley* responded that she did not believe this to be a valid flag and requested permission to board and verify. The *Logos* granted permission and the Fisheries Protection Officer from the boarding party provided the *Logos* skipper with the documentation and indicated that the vessel was either Canadian flagged by virtue of the outstanding mortgage keeping it on the Canadian registry or that it was a vessel without nationality and under the international law of the sea assumed the nationality of the government vessel from which it had been boarded.[53] The *Kristina Logos* was arrested for violations of Canadian fisheries regulations and escorted to St. John's.[54] Court proceedings followed. The *Kristina Logos* was found to be a Canadian vessel fishing

[53] H. Meyers, *The Nationality of Ships* (The Hague: Martinus Nijhoff, 1967), pp. 318–321.

[54] "Portuguese Captain Declares He's Innocent," *The Globe and Mail*, Toronto, 5 April, 1994, p. A4.

in the NRA without the benefit of license or quota.[55] A CDN$30,000 fine was levied and the vessel (CDN$750,000 in value) and its cargo (CDN$150,000 in value) seized and sold.

The Outcome

Canada expended considerable diplomatic effort and political will in order to prepare itself to act against NCP vessels which were fishing on the high seas within the NRA and which it had managed to force into a flagless situation. As luck would have it, the first vessel found in this situation proved to be both Canadian-owned and under Canadian law still a Canadian-flagged vessel, thus minimizing the strain on international relations. Nonetheless, Canada was able to demonstrate its determination to act. It was made clear that Canada would no longer countenance NCP vessel activity on the vulnerable cod stock breeding grounds of the Nose and Tail of the Grand Banks. As a consequence, the NCP vessels began to disappear and by the end of June 1994, after Bill C-29 entered into force, none remained. In the process, Canada put its domestic legislative house in order, deposited its amended acceptance at the International Court of Justice, and put the rest of the international community on notice regarding both its willingness and ability to act. The following year when the European Community again objected to the Greenland halibut quota which it received at the NAFO annual meeting and proceeded to adopt a higher quota for the stocks, Spain and Portugal were added to the Schedule in the revised *Coastal Fisheries Protection Act*, and the Spanish vessel *Estai* was seized. The rest is history.[56]

NCP Redux

While the rest of the 1990s remained quiet on the NCP vessel front in the NRA, early in the new century a group of ships appeared in the NRA

---

[55] There were a number of trials and appeals by the captain and company. The first appeal by the captain sets out the facts and decisions reached. *R. v. Traveres*, 131 *Nfld and P.E.I.R.* 271.

[56] See Fisheries Jurisdiction Case, *supra* note 51.

without the benefit of NAFO flags and fished as NCP vessels. Each of the vessels name began with an "O" and became known as the "O boats."[57] The first flag they appeared under was that of Dominica, with registry searches pointing towards Russian ownership. DFAIT resurrected its old methods. A NAFO resolution was passed and a joint NAFO démarche made without effect in the fall of 2004. Representations were made with the Caribbean community in early 2005. Finally, Dominica de-registered the vessels in late 2005 after a meeting between the Canadian High Commissioner and the Permanent Secretary of the Ministry of Foreign Affairs of Dominica.

By this time the vessels were in Rostock, Germany, where the German government and the European Community found that they had no authority, either within European Union or German law, to act against these flagless vessels found in a NAFO State's port. Additionally, Germany could not prevent the vessels from registering under the flag of Georgia, where they obtained new names and became the so-called "Spanish dancing girls."[58] The vessels fished under these names and flag in the NRA in 2006. Diplomatic representations were made in Tbilisi in 2006 by the Canadian Embassy and an offer made to transport Georgian fisheries officers to the Grand Banks to enforce their national fisheries law on the vessels. The vessels wintered in Kanilgrad in 2006–2007, where once again pressure was applied, this time on the Russian government, another Member State of NAFO, by both Canada and the European Community as the vessels were also fishing without license in the area under the authority of the Northeast Atlantic Fisheries Commission. This time it worked. The vessels were removed from service and broken up before the 2007 fishing season.

Lessons Learned

To catch a thief involved in the high seas fishery, a State must use the "Three Ps": (1) *persistence*; (2) *patience*; and (3) a willingness and ability to *pounce* when the occasion presents itself.

---

[57] *Okhotino, Olcan, Ostroe, Ostrovets, Oyra, Ozherelye, Lisa,* and *Pavlovsk.*
[58] *Carmen, Eva, Isabel, Juanita, Rosita,* and *Ulla.*

(1) Persistence is required to determine the nature of the enterprise, to apply pressure at all possible points of access, and to locate weaknesses.
(2) Patience is essential because the fishing companies are used to operating on the edge of the law and have made a good living at it for many years. There are many loopholes in international and domestic law and practice, which they know and exploit. Only by continuing to follow their every move and endeavouring to counter each one will the chink in their armour be found or an inevitable mistake made which can be exploited.
(3) Finally, States must have their domestic law and their operational procedures in place and be prepared to exercise the requisite political nerve to pounce on an NCP vessel as and when it is vulnerable.

Human Rights Aspects of Legislation in the Fisheries Sector

William R. Edeson*

Introduction

The purpose of this essay is to explore how human rights issues relate to national legislation in the fisheries sector. There are two principal aspects. First, we need to identify those human rights issues which you would expect to find covered in fisheries legislation, usually because they are required by international law, in particular, the 1982 United Nations Convention on the Law of the Sea (LOS Convention),[1] or the 1995 United Nations Fish Stocks Agreement,[2] or the FAO Compliance Agreement.[3] Second, we need to indicate what other human rights issues underlie fisheries legislation. In some countries, these are specifically dealt with in the fisheries law, in others, they are not. There is no dominating reason why these might or might not be included in the fisheries law, in fact, the answer will depend on the particular legal system.

A number of related questions will also be explored. For example, what do we mean by human rights; which human rights impact on the fisheries

* Professorial Fellow, Australian National Centre for Ocean Resources and Security, University of Wollongong, Australia; Professorial Visiting Fellow, Law Faculty, University of New South Wales. This contribution is based on work undertaken by NORAD in connection with the human rights dimension of fisheries, and is intended to complement their important work in bringing a human rights perspective to aspects of development assistance. Their help, in particular that of Ms. K. Bjoru, is gratefully acknowledged.

1 *United Nations Convention on the Law of the Sea*, Montego Bay, 10 December 1982, UN Doc.A/Conf. 62/122; 21 *I.L.M.* 1261 (1982) [hereinafter LOS Convention].

2 *Agreement for the Implementation of the Provisions of the United Nations Convention on the Law of the Sea of 10 December 1982 Relating to the Conservation and Management of Straddling Fish Stocks and Highly Migratory Fish Stocks*, New York, 4 December 1995, 2167 *U.N.T.S.* 88; (1995) 34 *I.L.M.* 1542 [hereinafter U.N. Fish Stocks Agreement].

3 Its full title is the *Agreement to Promote Compliance with International Conservation and Management Measures by Fishing Vessels on the High Seas*, 1993, approved by the FAO Conference at its twenty seventh session in November 1993.

sector; what is the role of ethical considerations in considering the subject; and what is the role of administrative law in protecting human rights. While the subject of human rights is widely explored in both national law and international law, there has been little attention paid to its specific application to fisheries. This contribution seeks to fill this gap in the area of legislation in the fisheries sector.

What Do We Mean by "Human Rights"?

An underlying question is: what do we mean by human rights? In this contribution it will be taken to refer to rights found in such standard international instruments as the International Covenant on Civil and Political Rights,[4] the International Covenant on Economic, Social and Cultural Rights[5] and the Universal Declaration on Human Rights.[6] There are also certain International Labour Organization (ILO) conventions that apply to the fisheries sector, either specifically or as part of a more general application, which can be broadly described as involving human rights. Likewise, certain aspects of the work of the International Maritime Organization (IMO) relating to rights of seafarers can be characterized as having a human rights dimension.

The discussion of what are human rights has another dimension which cuts across much of which has already been said, namely, the ethical dimension to fisheries. This is emerging as an important point of discussion principally in the FAO context and will be dealt with here, at least in an introductory way. Therefore this essay adopts a practical or functional approach to what is meant by human rights as opposed to a theoretical or conceptual definition.

---

4 *International Covenant on Civil and Political Rights*, 16 December 1966, G.A. res. 2200A (XXI), 21 U.N. GAOR Supp. (No. 16) at 52, U.N. Doc. A/6316 (1966), 99 *U.N.T.S.* 171.

5 *International Covenant on Economic, Social and Cultural Rights*, 16 December 1966, G.A. res. 2200A (XXI), 21 U.N. GAOR Supp. (No. 16) at 49, U.N. Doc. A/6316 (1966), 993 *U.N.T.S.* 3.

6 *Universal Declaration on Human Rights*, 10 December 1948, United Nations General Assembly A/RES/217 A (III).

Why the Fisheries Sector?

Why address the legal aspects of fisheries from a human rights perspective? It could be argued that all economic sectors warrant such a study and that fisheries is not sufficiently different to warrant its own study. It might also be argued that human rights are of such basic importance that focusing on one sector could lead to distortions of analysis. One answer to this is that, in general terms, while much has been written on the subject of human rights, little work has been done on the application of human rights to specific sectors. Indeed, there are some factors specific to fisheries which warrant particular attention from a human rights perspective. For example, the fisheries sector raises a very unusual mix of national and international legal issues, which have so far received only scant attention from a human rights perspective.

If we look at fisheries conducted at the national or local level, there are numerous opportunities for the application of a human rights based approach. In some countries, this might involve the application of well-established, possibly constitutionally-protected, rights to fish. On the other hand, it may be that, in some societies or in some regions, the human rights perspective has not been given any consideration. The fishers themselves may not have the resources to know what their rights are. The needs might, indeed probably will, vary within the sector. The wealthier members, such as the owners of fishing fleets, may well have adequate access to legal remedies to protect their rights, whatever they are. However, the crews and the poorer members of the sector might have few if any resources to access their legal rights. It can be added that, even in countries with a well-developed legal and administrative system, there is often insufficient information available to participants in the fisheries sector to be able to protect their rights.

There may also exist within some societies certain groups entitled, largely for historical reasons, to special treatment, which might prevail over the immediate application of human rights. For example, the Maoris in New Zealand and the First Nations Peoples in Canada have special fishing rights. These rights may sit uncomfortably with the general human rights available to the community at large, sometimes giving rise to issues of discrimination under human rights provisions. Special treatment could also raise questions of who are members of the group entitled to preference or special protection. In addition, there may be poor coastal communities heavily dependent on access to fisheries resources for their livelihood that merit special consideration.

Where foreign fishers are involved, in addition to the national regime which would apply to them in the exclusive economic zone (EEZ), the territorial sea and internal waters, there would be applicable to them a number of rights found in the principal legal instruments governing these areas which could be described as human rights. These rights are better known. However, depending on the country where their application arises, they may be not fully understood by the fishers, by the government, or the judiciary applying them. The rights in this situation are often not viewed as human rights in the typical sense.

For fishers on the high seas, who for the most part are likely to be large-scale fishers, the legal regime is relatively well known to many of the actors. However, there are still uncertainties in aspects of its application. For example, cooperation between States in at-sea enforcement relying on the hot pursuit doctrine, which was settled for many decades, is now being reassessed by several States in order to find a more effective means of controlling illegal fishing on the high seas. Further, when such foreign fishing vessels and crew are brought into local ports for possible judicial or administrative proceedings, there will almost always be human rights type issues involved.

What Human Rights are Covered in International Instruments?

Before examining some specific law of the sea and fisheries related human rights, it is important to consider what are the most important basic human rights recognized by the international community.

The NORAD *Handbook in Human Rights Assessment*[7] provides a very useful list of the principal economic and social rights as well as the civil and political rights found in the two principal U.N. covenants, namely the 1966 International Covenant on Economic, Social and Cultural Rights and the 1966 International Covenant on Civil and Political Rights. This list has been relied upon below, except that certain of the rights which have special legal significance have been included or elaborated on.

[7] Published by the Norwegian Agency for Development Cooperation, Oslo, February 2001.

The principal rights found in the Economic, Social and Cultural Rights Covenant are:

- the right to work
- the right to enjoyment of just and favourable conditions of work
- the right to form and join trade unions
- the right to social security, including social insurance
- the right to protection of the family
- the right to an adequate standard of living
- the right to the enjoyment of the highest attainable standard of physical and mental health
- the right of everyone to education
- the right to take part in cultural life

Further, the Covenant in Article 2 states:

> Each state party to the present Covenant undertakes to take steps, individually and through international assistance and cooperation, especially economic and technical, to the maximum of its available resources, with a view to achieving progressively the full realization of the rights recognized in the present Covenant by all appropriate means, including particularly the adoption of legislative measures.

This article is important because it both underlines the importance of development assistance in the achievement of these rights and the fact that the economic, social and cultural rights referred to in the Covenant are goals to be worked towards. The provision is also important because it acknowledges the fundamental importance of legislation in protecting these rights.

The principal rights found in the Civil and Political Rights Covenant are:

- the right to life
- freedom from torture or cruel, inhuman or degrading treatment or punishment
- freedom from slavery
- freedom from arbitrary arrest and detention
- the right to humane and respectable treatment of persons lawfully deprived of their liberty
- freedom of movement and liberty to choose one's place of residence for everyone lawfully within the territory of a State

- freedom to leave any country and to enter one's own country
- equality before courts and tribunals
- equal protection before the law
- the right to be recognized as a person before the law
- freedom from arbitrary and unlawful interference into one's privacy, family, home, or correspondence or unlawful attacks on one's honour or reputation
- freedom of thought, conscience and religion
- the right to hold opinions without interference
- freedom of expression, including the freedom to seek, receive and impart information
- the right to peaceful assembly, freedom of association, including the right to form and join trade unions
- the right to marry and found a family with free and full consent of the intending spouses
- the right to take part in the conduct of public affairs, directly or through freely chosen representatives
- the right to minorities to enjoy their own culture, profess and practice their own religion, or to use their own language

The Covenant contains specific requirements regarding the conduct of the trial of accused persons. Because of their legal importance, these are set out here:

- the right to a fair and public hearing before a competent, independent and impartial tribunal
- the right of persons charged with offences to be presumed innocent until proven guilty according to law
- the right to be informed promptly and in detail in a suitable language the nature and cause of the charge against him
- adequate time and facilities to prepare a defence and to communicate with counsel of one's own choice
- right to be tried without undue delay and to be present at the trial, as well as being informed of this right
- the right, if necessary, to have free legal assistance assigned to the person accused where the interests of justice so require
- the right to examine witnesses against the person accused and to obtain the attendance of witnesses on his behalf under the same conditions as witnesses against him

- the right to free assistance of an interpreter if needed
- the right not to be compelled to testify against one's self
- the right to have a conviction reviewed by a higher tribunal according to law
- the right to compensation for miscarriages of justice or for new discovered facts in certain circumstances
- the right not be tried for an offence for which the accused has already been either acquitted or convicted

This list, it must be emphasized, is merely a summary. Several of the provisions contain important qualifications and exceptions that are too detailed to set out here, but which might become crucial in a particular case.[8]

Within the context of the United Nations, there are the following treaties or declarations in addition to those discussed above:

- The U.N. Charter
- Universal Declaration on Human Rights, 1948
- Declaration on the Granting of Independence to Colonial Peoples, 1960
- Convention against Torture and other Cruel, Inhuman or Degrading Treatment or Punishment, 1984
- the Convention on the Prevention and Punishment of all Forms of Genocide, 1948
- Convention on Slavery, 1926 as amended
- Convention relating to the Status of Refugees, 1951
- Convention relating to the Status of Stateless Persons, 1954
- Convention on the Political Rights of Women, 1953
- Declaration on the Elimination of All Forms of Intolerance and of Discrimination Based on Religion or Belief, 1981

[8] See further Part III of the International Covenant on Civil and Political Rights. It will also be apparent that this list contains important issues, which will need to be considered when drafting fisheries and related legislation. A basic issue that needs to be addressed is whether these rights are applicable to all trials in a particular country due to the fact that they are covered in some basic law such as the constitution or the criminal code or whether they need to be addressed in each law. While in most countries it is likely that such provisions would be located in such basic laws, the specific answer to this will nonetheless vary considerably from one country to another.

- Convention on the Elimination of all forms of Racial Discrimination, 1979
- Convention on the Elimination of All forms of Discrimination against Women, 1979
- Convention on the Rights of the Child, 1989
- International Convention on the Protection of the Rights of all Migrant Workers and their Families, 1990

In addition to the above instruments, it should be mentioned that there are regional instruments concerning human rights. These are the European Convention on the Protection of Human Rights and Fundamental Freedoms, the Inter-American Convention on Human Rights, and the African Charter on Human Rights adopted by the Organization of African Unity. There is, of course, overlap between all of these instruments, but the list illustrates the breadth of coverage of the term human rights.

The Work of Specialized Agencies

Several of the specialized agencies of the United Nations are also currently involved in work which touches on human rights. It will be immediately apparent from the examples referred to below that the scope here is potentially very wide.

International Labour Organization

The International Labour Organization (ILO) has prepared several basic conventions which have a strong human rights dimension. For example:

- Convention Concerning Forced or Compulsory Labour, 1930
- Convention Concerning the Abolition of Forced Labour, 1937
- Freedom of Association and Protection of the Right to Organize Convention, 1948
- Right to Organize and Collective Bargaining Convention, 1949
- Discrimination (Employment and Occupation) Convention, 1949
- Equality of Treatment (Social security) Convention, 1962
- Social Policy (Basic Aims and Standards) Convention, 1962
- Employment Policy Convention, 1964
- Convention Concerning Indigenous and Tribal Peoples in Independent Countries, 1989

These are just the generally applicable ILO conventions. As well as the above list, there are numerous conventions (over 180) and recommendations (185) that have been prepared by ILO that are capable of applying to employment in the fisheries sector, while some are specifically aimed at the fisheries sector itself.

An important recent ILO convention on employment in the fisheries sector is the Work in Fishing Convention, 2007.[9] The objective of this convention is stated in its preamble to be:

> to ensure that fishers have decent conditions of work on board fishing vessels with regard to minimum requirements for work on board; conditions of service; accommodation and food; occupational safety and health protection; medical care and social security.

The Convention proceeds to elaborate in considerable detail the responsibilities of owners and skippers, and minimum requirements for work on board fishing vessels, including the minimum age, medical examinations, a crew list, a fisher's work agreement, repatriation, recruitment and placement of fishers, the obligations of private employment agencies, payment of fishers, accommodation and food. The Convention also provides for medical care, health protection and social security, occupational health and safety, protection in case of work-related sickness, injury or death, compliance and enforcement. If implemented effectively, the Convention will have a profound impact on the fisheries sector of more direct relevance to the welfare of fishers than the more conventionally described human rights.

International Maritime Organization

The activities of the International Maritime Organization (IMO) also impinge on the rights of seafarers. In recent years this has become a major concern. Once again, while the IMO may not necessarily see the work it is doing as being of a human rights character, nonetheless, the work has that dimension. Some indication of the overlap that exists with the work of the ILO is that a joint working group has been established that led to the joint IMO/ILO Ad Hoc Expert Working Group on Liability

---

[9] *Convention Concerning Work in the Fishing Sector*, 2007 (No.188), adopted 14 June 2007.

and Compensation regarding Claims for Death, Personal Injury and Abandonment of Seafarers.

Following the deliberations of this Joint Expert Working Group in 2001, the IMO Assembly adopted two guidelines which have been summarized as follows on the IMO website:

> A.930 (22) Guidelines on Provision of Financial Security in Cases of Abandonment of Seafarers
> The resolution on provision of financial security in case of abandonment of seafarers states that abandonment of seafarers is a serious problem involving a human and social dimension and recognizes that, given the global nature of the shipping industry, seafarers need special protection. In this context, the resolution notes that the adoption of guidelines is an appropriate interim measure to ensure provision of financial security in case of abandonment of seafarers. The resolution recommends measures to be implemented by shipowners to ensure the provision of an adequate financial security system for seafarers in case of abandonment and includes associated Guidelines which set out the main features and scope of coverage of the financial security system and also contain recommendations for certification of such systems.
>
> A.931 (22) Guidelines on Shipowners' Responsibilities in respect of Contractual Claims for Personal Injury to or Death of Seafarers.
> The resolution on claims for personal injury to or death of seafarers notes a need to recommend minimum international standards for the responsibilities of shipowners in respect of contractual claims in such cases. It expresses the concern that, if shipowners do not have effective insurance cover, or other form of financial security, seafarers are unlikely to obtain full and prompt compensation. It states that putting in place effective arrangements for the payment of compensation is part of the shipowners' responsibilities to provide safe and decent working conditions. The resolution includes associated Guidelines recommending measures to be implemented including certification and a model receipt and release form for claims.

The FAO/IMO/ILO Code of Safety for Fishermen and Fishing Vessels, 2005 has received the concurrent approval of all three agencies involved in it.[10]

Although not put in human rights terms, the above work by the specialized agencies of the United Nations has obvious links with human rights

[10] For a consideration of this Code, see the statement by E. Mitropoulos, Secretary General of IMO, to the FAO Committee on Fisheries, 27th Session, Rome, 5 March 2007, Appendix E, *Report of the twenty-seventh session of the Committee on Fisheries, Rome, 5–9 March 2007*, FAO Fisheries Report No. 830 (Rome: FAO, 2007), pp. 71–74.

issues that underlie the fisheries sector. As is apparent from the above, the range of "human rights" is very extensive. There are also specific rights found in law of the sea instruments which have a human rights character. These are discussed below. Not all of these instruments are directly relevant to the fisheries sector. The main point of listing these instruments is to convey the considerable potential extent of the notion of human rights that could at least indirectly impact on the fisheries sector.

Right to Food

The right to food merits a special subsection because of its link to fisheries. It is not usually addressed in fisheries legislation as such. The right to food is recognized in several important international instruments, the most important being the 1948 Universal Declaration of Human Rights and the 1966 International Covenant on Economic Social and Cultural Rights. The right to food is included in these instruments in the context of the right to an adequate standard of living. This subject was elaborated upon in the 1996 Rome Declaration on World Food Security and the World Food Summit Plan of Action.[11]

According to an FAO study, *Law and Sustainable Development since Rio*,[12] about twenty countries have referred to the right to food in their national constitutions. However, it is difficult to convert the right from an abstract proposition into a workable or functioning right, and not one of the twenty countries which has referred to the right to food has implementing legislation to achieve this goal. Indeed, given the potential breadth of the obligation, for the great majority of countries, the first step would need to be an assessment of their existing legislative powers across a very broad area, leading possibly towards the preparation of framework legislation that sets targets. As the study referred to above points out, the need for such a reference in a constitution is less important in industrialized countries with well developed social security systems than in less developed countries with weak or less developed systems.

---

[11] Adopted in Rome, 13–17 November 1996, <www.fao.org/DOCREP/003/W3613E/W3613E00.HTM>, 28 February 2008.

[12] FAO Legal Office, *Law and Sustainable Development since Rio: Legal Trends in Agriculture and Natural Resource Management*, FAO Legislative Study No. 73 (Rome: FAO, 2002), p. 25.

One important instance of a right similar to the right to food being given judicial enforcement occurred in South Africa with respect to the constitutional right to access to adequate housing. The Constitutional Court, in the *Grootboom Case*,[13] considered the eviction of several hundred squatters from private land. The court evaluated the measures taken by the public authorities in their housing programmes and the steps taken progressively to achieve that right within available resources. The Court issued a declaratory order that the State had the obligation to devise, fund, implement and supervise measures to provide relief to those in desperate need of housing. As the wording in the South African Constitution of the right to adequate housing is not substantially different to that concerning food, it is arguable that the same claim could be raised in that context. However, it would be wrong to leave the impression that the Court would readily make a pronouncement that there had been a failure by the government except where the failure was manifest and was probably beyond the control of the government.

The very nature of a right to food, like the right to development, is perhaps too general and vague to guide any specific decision in the area of technical assistance. However, it could form the basis for the justification of a programme in the area of food security in the fisheries sector.

The 1982 LOS Convention, The U.N. Fish Stocks Agreement, the FAO Compliance Agreement, and General International Law

Against the background of the above catalogue of human rights, it is necessary to consider those provisions of the 1982 LOS Convention, the 1995 U.N. Fish Stocks Agreement, and the FAO Compliance Agreement that have a bearing on the subject of human rights.

The 1982 LOS Convention

Article 73 of the LOS Convention contains provisions that can be equated with human rights in a fisheries context.

---

[13] *Republic of South Africa* v. *Grootboom*. Case No. CCT 11/00. 2000 (11) BCLR 1169.

1. The coastal State may, in the exercise of its sovereign rights to explore, exploit, conserve and manage the living resources in the exclusive economic zone, take such measures, including boarding, inspection, arrest and judicial proceedings, as may be necessary to ensure compliance with the laws and regulations adopted by it in conformity with this Convention.
2. Arrested vessels and their crews shall be promptly released upon the posting of reasonable bond or other security.
3. Coastal State penalties for violations of fisheries laws and regulations in the exclusive economic zone may not include imprisonment, in the absence of agreements to the contrary by the States concerned, or any other form of corporal punishment.
4. In cases of arrest or detention of foreign vessels the coastal State shall promptly notify the flag State, through appropriate channels, of the action taken and of any penalties subsequently imposed.

This article contains in effect two proscriptions of a human rights character: prompt release and non-imposition of imprisonment as a punishment for foreigners. Article 73 does not restrict action taken by the coastal State against its own nationals in the EEZ, thus, it would still be open to a coastal State to imprison its own fishers for fisheries offences committed in that zone. Politically, however, such a situation would obviously be difficult to sustain. It should also be noted that the provision only applies to the EEZ, though not to fishing for sedentary species in the EEZ.[14] In internal waters, archipelagic waters and the territorial sea, the coastal State would be under no such restraints. On the high seas, of course, the situation is different. Except for rights of intervention agreed to by treaty, such as the U.N. Fish Stocks Agreement, and action against pirates, vessels engaged in the slave trade, vessels engaged in unauthorized broadcasting, or vessels without nationality, only the flag State can take action against its own vessels as the freedom of navigation on the high seas applies.

The prompt release upon the posting of a reasonable bond provision of Article 73 has been the subject of judicial decisions by the International Tribunal for the Law of the Sea (ITLOS). There is no need to consider these cases here except to point out that ITLOS has focused primarily on interpreting and applying Article 73. However, a human rights perspective

---

[14] LOS Convention, *supra* note 1, at Art. 68.

underlay certain aspects of the judgments as the following comments in the 2004 *Juno Trader Case* reveal:

> The Tribunal considers that article 73, paragraph 2, must be read in the context of article 73 as a whole. The obligation of prompt release of vessels and crews includes elementary considerations of humanity and due process of law.[15]

This was also expressed very firmly by Judges Mensah and Wolfrum in their separate opinion in the *Juno Trader Case*, where as well as endorsing the above paragraph,[16] they added:

> [T]he tribunal must operate on the basis that the obligation of States... under the convention and general international law, includes the obligation not to deny justice or due process of law, especially in respect of legal and judicial procedures that involve interference with the property rights of aliens.[17]

This position has been reiterated in the latest case with particular reference to confiscation of a foreign fishing vessel.[18] The Tribunal in the *Tomimaru Case* stated:

> A decision to confiscate eliminates the provisional character of the detention of the vessel rendering the procedure for its prompt release without object. Such a decision should not be taken in such a way as to prevent the shipowner from having recourse to available domestic judicial remedies, or as to prevent the flag State from resorting to the prompt release procedure set forth in the Convention; nor should it be taken through proceedings inconsistent with international standards of due process of law. In particular, a confiscation decided in unjustified haste would jeopardize the operation of article 292 of the Convention.[19]

It should be noted that some of the judges in the *Tomimaru Case* appended separate opinions disagreeing with the above proposition.[20]

---

[15] *Juno Trader* (*Saint Vincent and the Grenadines* v. *Guinea Bissau*) Case No. 13, ITLOS, 18 December 2004, para. 77.

[16] *Juno Trader Case*, Joint Separate Opinion of Judges Mensah and Wolfrum, para. 5, <www.itlos.org/case_documents/2004/document_en_253.pdf>, 28 February 2008.

[17] *Ibid.*, para. 6.

[18] *The Tomimaru Case* (*Russian Federation v. Japan*), Case No. 15, ITLOS, 6 August 2007.

[19] *Ibid.*, para. 76.

[20] See Declaration of Judge Nelson, Separate Opinions of Judge Jesus and Judge Lucky, <www.itlos.org/cgi-bin/cases/case_detail.pl?id=15&lang=en>, 28 February 2008. Judge

More problematic is the proscription against imprisonment in Article 73(3). This is not the occasion to discuss the ambiguities underlying this provision, in particular, what is meant by "violations of fisheries laws and regulations," or to address the fact that a number of countries still provide for imprisonment of foreign fishers in their national legislation. Some countries, however, provide for imprisonment but may not actually impose it or they retain the right to imprison foreign fishers, for repeated offences or where there has been an element of violence, for example, in resisting arrest. In the latter instance, the view is taken by some countries that violence is a breach of the criminal law rather than merely a violation of a fisheries law.

It can be noted in passing that Article 230(1) of the LOS Convention also allows only for the imposition of "monetary penalties" in respect of violations of national laws and regulations or applicable rules and standards for the prevention, reduction and control of pollution of the marine environment, committed by vessels beyond the territorial sea. Interestingly, paragraph 3 directs that, in the conduct of proceedings in respect of such violations committed by a foreign vessel which may result in the imposition of penalties, "recognized rights of the accused shall be observed". These recognized rights are not defined but would almost certainly be the more basic rights available to an accused, such as the presumption of innocence, the right to cross examine witnesses, the right to be informed of the charge against her or him, the right to legal representation, the right to challenge the detention before a judge, protection against double jeopardy, and so on.

While there is no equivalent to Article 230 with respect to the fisheries offences in the EEZ under Part V of the LOS Convention, it is very doubtful if the absence of such a clause would mean that such recognized rights could be ignored. The reason for this will become more apparent below.

Yanai noted that, "National prompt release procedure should be based on the principle of due process of law in order to ensure fairness in its implementation" (para. 3, Declaration of Judge Yanai), <www.itlos.org/cgi-bin/cases/case_detail.pl?id=15&lang=en>, 28 February 2008.

The 1995 U.N. Fish Stocks Agreement

The 1995 U.N. Fish Stocks Agreement has only a few clauses that could be said to have a human rights dimension.

Article 21, which deals with subregional and regional cooperation in enforcement, contains a general prohibition against discrimination, though this is directed towards the application of boarding procedures on the high seas and in any event requires procedures to be set up by subregional or regional fisheries management organizations or arrangements. The link here with non-discrimination in the human rights context is very tenuous.

Other provisions of the 1995 Agreement have a more direct human rights link. For example, Article 21(8):

> Where, following boarding and inspection, there are clear grounds for believing that a vessel has committed a serious violation, and the flag State has either failed to respond or failed to take action as required under paragraphs 6 or 7, the inspectors may remain on board and secure evidence and may require the master to assist in further investigation including, where appropriate, by bringing the vessel without delay to the nearest appropriate port, or to such other port as may be specified in procedures established in accordance with paragraph 2.

Article 21(16) also requires that action taken by States, other than the flag State in respect of vessels having engaged in activities contrary to subregional or regional conservation and management measures are to be proportionate to the seriousness of the violation.

Article 23, which deals with port State measures, is an indication of how human rights issues might arise. The article requires that in taking port State measures, there is to be no discrimination in form or in fact against the vessels of any State. This is only in part a human rights issue as it would be a matter for the flag State to take up in the event of discrimination, even though the direct victim is the vessel in question. As with Article 21 referred to above, this is not discrimination in the more traditional human rights sense of the term.

Article 23(3) also refers to the need to for it to be "established" that a catch has been taken in particular manner (i.e., one that undermines the effectiveness of certain high seas conservation measures). In this context, "established" probably indicates the need for at least some minimal level of proof to exist before a prohibition of landing or transshipment is put into effect.

The FAO Compliance Agreement

The FAO Compliance Agreement does not raise any human rights issues directly, though in relation to the collection and exchange of information there could be privacy issues arising under the national laws of individual States. For example, in Article VI(10) states:

> FAO shall circulate promptly the information provided under paragraphs 5, 6, 7, 8 and 9 of this Article to all Parties, and, on request, individually to any Party. FAO shall also, *subject to any restrictions imposed by the Party concerned regarding the distribution of information*, provide such information promptly on request individually to any global, regional or subregional fisheries organization. (emphasis added)

Some indication of how human rights issues could arise indirectly under the FAO Compliance Agreement are indicated in Article III(8):

> Each Party shall take enforcement measures in respect of fishing vessels entitled to fly its flag which act in contravention of the provisions of this Agreement, including, where appropriate, making the contravention of such provisions an offence under national legislation. Sanctions applicable in respect of such contraventions shall be of sufficient gravity as to be effective in securing compliance with the requirements of this Agreement and to deprive offenders of the benefits accruing from their illegal activities. Such sanctions shall, for serious offences, include refusal, suspension or withdrawal of the authorization to fish on the high seas.

It will be apparent that there is at least the potential for conflict between the vigorous application of this provision and respect for certain human rights.

The Application of General International Law to These Treaties

The question arises as to whether the provisions referred to above are the only instances in the fisheries area which have a human rights dimension. In fact, human rights issues are much more pervasive in the three treaties referred to above than the above discussion would suggest. In the case of the LOS Convention and the 1995 U.N. Fish Stocks Agreement, the preamble makes it clear that the treaties are subject to the application of general international law. The 1995 U.N. Fish Stocks Agreement mirrors the preamble of the LOS Convention where it says:

> Affirming that matters not regulated by the Convention or by this Agreement continue to be governed by the rules and principles of general international law.

The FAO Compliance Agreement does not have a clause as explicit as this, but nonetheless makes it clear that the Agreement is intended to be consistent with the provisions of the LOS Convention. Thus, the preamble states:

> Recognizing that all States have the right for their nationals to engage in fishing on the high seas, subject to the relevant rules of international law, as reflected in the United Nations Convention on the Law of the Sea.

As a result of the preamble, it would be reasonable to apply general international law to those aspects of the law of the sea not covered specifically by the three agreements. This would mean that there could be certain minimum international standards of human rights that could be applied in the situations where the treaties in question were silent. This is in effect what was occurring in the *Juno Trader* and the *Tomimaru Cases* discussed above.

It is not the intent here to describe the extent of these minimum international standards except to mention that volumes have been written, and will continue to be written, on this question. Nevertheless, at the risk of oversimplification, there were two views regarding the standard imposed by international law on the treatment of foreigners. The classical, "western," view was that international law imposed some minimum standards below which States could not fall, even if that meant that a foreigner would receive better treatment than a national. The view of the Third World, supported by countries with a Communist or socialist system, was that the foreigner need only be treated like anyone else in the State in question, a national treatment threshold.

This debate, never fully resolved, has been supplanted by the growth of international human rights law. With the exception of the 1948 Universal Declaration on Human Rights, which is not a treaty, the other three instruments, the 1966 Covenant on Civil and Political Rights, the 1996 Covenant on Economic, Social and Cultural Rights, and the 1966 International Convention on the Elimination of all forms of Racial Discrimination, have received high levels of ratifications (approaching 150 States for each).Thus, if a breach of a human right occurred in the context of the law of the sea, and all of the relevant States were parties to the relevant human rights treaty, it would be possible to rely on the breach of the provisions of the human rights treaty, though noting that not all of the agreements provide a mechanism for individuals to bring complaints on their own initiative. Assuming that one of the States to the dispute was not a party to the relevant human rights treaty, it might still be argued that international law now reflects many of the standards contained in these instruments, or, at the very least, those set out in the Universal Declaration. Non-lawyers may well find this conclusion a bit woolly, but that is international law.

Human Rights Aspects Arising Under National Law

Though the situation is far from uniform in some countries, specific fisheries laws are subject to fundamental human rights provisions, and with some exceptions, these fundamental rights are reflected in the law itself for the simple reason that they apply whatever the fisheries law, or any other law, might say.

A question that arises is whether the fundamental rights and freedoms set out in international instruments such as the Covenant on Civil and Political Rights and the Covenant on Economic Social and Cultural Rights automatically apply in national law. This is a very complicated question to answer in general terms as the answer varies according to the particular system of a country. In some countries, for example, those which have the English common law as their basic law, such an instrument has no legal effect until it has been incorporated into the national law by an act of parliament. In this situation, it is the act of the parliament that provides the basis for the operation of the treaty provisions in national law. Other countries have a system whereby the approval of a treaty involves the legislature itself, and once that has occurred, the treaty becomes part of the national law. On the critical question of whether the treaty provisions are unaffected by any subsequent changes in national law, there is no simple universal answer.

Administrative Law Aspects

Depending on the particular legal system under consideration, a person involved in the fisheries sector might be able to obtain redress under administrative law processes. For example, if a decision whether or not to grant a licence was made following criteria not intended to be covered by the fisheries law, it would be possible in most Western countries to bring some type of administrative action. This would probably not be classified as a human rights issue, however, the effect of the action could be the same as if it were.

The link to human rights can be illustrated as follows. If a fisher had their licence removed without a hearing or by an officer who was biased, then the rules of natural justice could be invoked and the decision could be quashed by a court. Likewise, again depending on the particular country, it might be possible for an ombudsman to intervene to provide some protection to the individual fisher against maladministration. In such an instance, we would probably not talk about a violation of human rights, instead, it would be regarded as a straightforward application of administrative law. However, if the country in question had a bill of rights, but lacked an effective system of administrative law, the same result might be achieved, depending on the specific national law, by invoking the provisions of that bill of rights.

In countries that do not having generally applicable human rights provisions, it can be expected that the role of administrative review, including the role of the ombudsman, will play a critical part in protecting human rights. The quality of the administration under consideration and the willingness of courts and tribunals to act independently to protect the rights of individuals are other factors to consider here.

Another aspect to mention briefly is the introduction of a system of civil or administrative penalties for dealing with fisheries offences. The main reason for this approach is that it enables a tribunal to apply a lower standard of proof than is possible in a full criminal trial (proof on the balance of possibilities rather than on the basis of beyond reasonable doubt), it makes possible expedited hearings, and it can also include the possibility of a negotiated settlement of the case.

This method has been adopted in the United States and in several Pacific Island States. Despite the fact that it involves a possible diminution of their legal rights, the system of civil or administrative penalties is often popular with foreign fishers as it enables a speedy resolution of their case and enables them to return quickly to the profitable pursuit of illegal, unregulated or unreported fishing.

The system of civil or administrative penalties is not applicable in all countries as there may be constitutional or legal reasons why such a system cannot be introduced. If, however, it is used and accompanied by adequate safeguards, this approach does not necessarily involve a diminution of human rights of a person accused.

Manuals to Guide Prosecutors, Administrators, and at Sea Enforcement Officials

As noted above, the U.N. Fish Stocks Agreement provides for regionally agreed boarding and inspection measures to be implemented. One aspect of the boarding and inspection provisions that should be noted, and which goes beyond simply giving effect to these provisions, is that considerable work will need to be done by the regional and subregional fisheries bodies for such provisions to be workable. At the national level, there will be a need for detailed instructions to be prepared to assist enforcement officers in how they are to carry out these tasks. These instructions, sometimes described as rules of engagement, give details on when force may be used, provide guidelines on how much force is appropriate, boarding procedures and a host of other issues.

It is often in these documents that there needs to be advice on the human rights implications of certain enforcement actions. This is an area in which technical legal assistance can be very useful, as well as ensuring that human rights issues are promoted by that assistance.

At another level, technical assistance can consist of providing assistance in the drafting of a prosecutions manual in respect of fisheries offences, which is designed to assist prosecutors in preparing and conducting cases against illegal fishing. In such documents there is considerable scope to address human rights aspects of the conduct of prosecutions. Likewise, port State measures will most probably require detailed guidelines for national officials on how to apply these measures. Such guidelines could include aspects concerning the human rights implications of certain action by officials of the port State.

Access to Fisheries Resources

Access to the fisheries resources mainly raises issues of an economic or political nature but it can also have a human rights dimension. In terms of classification, however, problems regarding access to fisheries resources would not normally be considered as a human right, unless there was an argument about equality of treatment before the law, e.g., that one category of persons were receiving discriminatory treatment or preferential treatment over another. Although questions of access to fisheries resources have not traditionally been viewed as a human rights issue, there is no logical reason why it could not become one.

There are, of course, two powerful practical reasons why access is not normally viewed in human right terms. First, the 1982 LOS Convention clearly permits a coastal State to discriminate in favour of its own fishers as against foreigners within the EEZ, territorial sea or internal waters. Second, with fisheries resources under threat in so many parts of the world, the focus is, or should be, instead on the long-term sustainable utilization of the resources rather than creating yet more opportunities for access.

Again, the human rights dimension would not normally be addressed in the fisheries law itself, rather it would be left to the general provisions concerning human rights. Further, it is often the case that the principal fisheries law will not say much, if anything, about the right of access. Leaving aside the question of any right of access which might be granted to foreigners, which international law leaves very much to the discretion of the coastal State; the national law also may say little about access by its own nationals. Typically, national fisheries law allows the minister to allocate licenses, and the policy to be applied to such allocation would be dealt with in policy documents issued by the minister.

There are of course some important exceptions. In the first place, where a system of property rights or individual transferable quota exists, these will often require legislation which sets out in some detail the legal framework for that type of system. In such a system, access may well be seen principally as a political and economic issue, which can be resolved by means of the collection of a resource rent from the quota holders. In other words, the access, or lack of it, to the fisheries resources would not be viewed as a human rights issue. In the second place, if there are special rights accorded to particular groups, this will need to be reflected in legislation. Examples here include the specific regimes concerning the Maori in New Zealand and Canada's First Nations Peoples. In South Africa, legislation provides a basis for redistributing access rights and opportunities in the fisheries sector for previously disadvantaged members of society. Other examples include preferential treatment for certain indigenous groups or community-based fishing rights.

Access under the FAO Code of Conduct[21]

The Code of Conduct for Responsible Fisheries says little about access issues. In the general principles of the Code, there are two provisions that have a bearing on the matter of access. Article 6(13) states:

> States should, to the extent permitted by national laws and regulations, ensure that decision making processes are transparent and achieve timely solutions to urgent matters. States, in accordance with appropriate procedures, should facilitate consultation and the effective participation of industry, fishworkers, environmental and other interested organizations in decision making with respect to the development of laws and policies related to fisheries management, development, international lending and aid.

It is to be noted that this provision does not actually ensure equality of access, rather it advocates that the access processes be transparent and that there should be consultation with the various interest groups. The absence in the Code of a general provision on access is underlined by the second clause in the general principles which focuses solely on artisanal and small-scale fishers.

> Recognizing the important contributions of artisanal and small-scale fisheries to employment, income and food security, States should appropriately protect the rights of fishers and fishworkers, particularly those engaged in subsistence, small-scale and artisanal fisheries, to a secure and just livelihood, as well as preferential access, where appropriate, to traditional fishing grounds and resources in the waters under their national jurisdiction.[22]

The lack of attention to access issues is also brought out in the following passing point in the context of aquaculture development:

> States should ensure that the livelihoods of local communities, and their access to fishing grounds, are not negatively affected by aquaculture developments.[23]

Under integrated coastal area management, there are the following references to access:

> States should develop, as appropriate, institutional and legal frameworks in order to determine the possible uses of coastal resources and to govern

---

21 FAO, *Code of Conduct for Responsible Fisheries* (Rome: FAO, 1995).
22 *Ibid.*, Art. 6(18).
23 *Ibid.*, Art. 9(1)(4).

> access to them taking into account the rights of coastal fishing communities and their customary practices to the extent compatible with sustainable development.[24]

Overall, the Code of Conduct did not adopt a human rights perspective on access questions even though such issues underlie much of what the Code deals with.

The Ethical Dimension

The ethical dimension provides another, indeed novel, way of approaching the subject of fisheries and cuts across traditional approaches to the subject, including the human rights aspect, providing a new perspective on fisheries problems. It is wider in scope than the human rights dimension. The main value of the ethical approach lies in identifying questions or issues which may need to be reconsidered in the fisheries sector. For example, it provides new perspectives on the question of limited access to fisheries, such as the individual transferable quota systems which takes us beyond the predominantly economic explanations often put forward to justify their use. The main parameters of the debate will be outlined.

The institutional framework for an ethical analysis for fisheries is found in instruments such as the Code of Conduct for Responsible Fisheries, the 1982 LOS Convention, the 1993 FAO Compliance Agreement, the 1995 U.N. Fish Stocks Agreement, and the Convention on Biological Diversity.[25] More recently, the 2001 Reykjavik Declaration on Responsible Fisheries in the Marine Ecosystem[26] and the 2002 World Summit on Sustainable Development[27] have provided elements to the ethical dimension of the institutional framework. However, the Code of Conduct is seen as the principal tool in the setting of the institutional framework.

---

[24] *Ibid.*, Art. 10(1)(3).

[25] *Convention on Biological Diversity*, Rio de Janeiro, 22 May 1992, 31 *I.L.M.* 823 (1992).

[26] Adopted by the Reykjavik Conference on Responsible Fisheries in the Marine Ecosystem, 1–4 October 2001, Reykjavik, Iceland.

[27] *Report of the World Summit on Sustainable Development*, Johannesburg, South Africa, 26 August–4 September, 2002, A/CONF.199/20 (New York: United Nations, 2002).

Some indication of the usefulness of the ethical dimension can be gleaned from a consideration of access rights to fisheries. This subject brings out neatly several problems which do not admit of an easy solution. However, the ethical approach pushes into the foreground certain questions which are not answered by a simple economic analysis or a human rights analysis.

Approaching access rights, such as a system of individual transferable quotas, from an ethical point of view raises the question of whether economic efficiency alone is a sufficient basis on which to determine access. The ethical dimension raises the question of balancing out competing interests of who gets access and on what terms. Such an analysis has to take place in a particular context. For example, the right to trade such access rights will rarely exist in an economic vacuum, it will be shaped by the cultural tradition of a particular country or area within a country.

The ethical dimension requires that issues, such as the extent to which such rights confer access to scarce resources and the extent to which the immediate beneficiaries can directly gain the benefits, are addressed. This will bring in a host of issues, such as the rights of disadvantaged groups in a society to gain access to either the resources or the benefits. This in turn will touch upon issues such as whether the rights can be leased or sold, whether they are inheritable, or divisible, etc. and whether there is a place for foreign ownership of such rights. An ethical analysis can also require consideration of issues such as whether a new institutional arrangement will increase inequities or asymmetries in the fisheries sector.

In a wide ranging and important study of an ethical analysis of fisheries, the following conclusion was drawn:

> The moral dimensions of fisheries are manifold, but the main ethical issues concern overfishing, interwoven with those of poverty, food security, food safety and ecosystem degradation. Each of these issues could be broken down into a number of related sub issues, for instance: genetic modification of living organisms, introduction of alien species, protection of endangered or emblematic species, discarding practices, cultural sustainability, knowledge sharing, transboundary impacts, food contamination and safety.[28]

The above discussion is tentative and not intended to address the full range of issues that may arise from an ethical analysis. It is hoped, however, that it demonstrates that the ethical dimension is wider than a human

[28] FAO, *Ethical Issues in Fisheries*, FAO Ethics Series No. 4 (Rome: FAO, 2005), p. 27.

rights based analysis. Just as the ethical dimension has highlighted the limitations of an unequivocal economic rational perspective on questions such as restricted access, so too the ethical dimension can be called upon to supplement the human rights approach. While this is important in itself, the ethical approach is unlikely to provide legal answers to these often difficult issues.

Conclusion

The purpose of this essay has been to explore how human rights issues relate to national legislation in the fisheries sector in the context of technical assistance provided under development assistance. It will be apparent that it is often the case that few human rights aspects are covered in national fisheries legislation. However, human rights considerations apply, or are capable of applying, as a result of the application of treaties or general international law at the international level and at the national level as a result of the application of a bill of rights or other similar instrument.

A major problem in a full analysis is how to define human rights. A simple and safe answer is to define human rights in terms of the principal U.N. instruments, in particular, the Covenant on Civil and Political Rights and the Covenant on Economic Social and Cultural Rights. Such a definition, while safe, is not very helpful, as there are many other instruments that contain human rights provisions. Moreover, in some countries it might be the administrative law which provides the practical remedies. There is also the consideration that a country might have a system of fundamental human rights constitutionally entrenched in its national system, which will largely reflect international human rights standards. It is for this reason that this contribution has avoided drawing a sharp distinction between human rights derived from national law and those derived from international law.

The question of what human rights are of greatest importance to fishers has not been addressed at the international level. However, the work in the IMO and the ILO concerning seafarers, with its obvious implications for fishers suggests that increasing attention is being paid to this subject.

Marine Environmental Protection

Improving the Effectiveness of Environmental Regimes: "Consilience," Science, and Common Sense

Shelley Lexmond*

The last five decades have witnessed the rise of systemic, chronic ecological degradation resulting from a myriad of human activities. Such environmental concerns have forced recognition of the inherent ecological inter-linkages among the Earth's habitats and man's ability to significantly impact the global ecosystem. Largely pursuant to the traditional legal domain of Law of the Sea (LOS) and the newly evolved legal domain of International Environmental Law (IEL), the international community has swiftly responded with a proliferation of global, regional and domestic environmental regimes targeting a myriad of ecological issues.

The international community is now emerging from a phase of environmental legal development that has allowed the proliferation of environmental regimes. However, upon reflection there has been minimal focus on assessing the ecological effectiveness of tools of the regime and their use, such instruments, policy and management tools, decision-making processes and actions related to the above such as implementation. Ecological effectiveness can be measured by the degree of improvement in the health of an ecosystem and the maintenance of its long-term sustainability; the latter being the ultimate goal of environmental regimes. Failure to meet the ultimate objective of improving and sustaining ecological integrity is not only ecologically ineffective, but also economically inefficient as resources are usurped with no ecological gain. To this end, seeking ecological effectiveness of all environmental instruments and related processes and actions is rational.

* Member of the Legal Task Force of the UNEP/GEF Project Reversing Environmental Degradation Trends in the South China Sea and Gulf of Thailand at the time of writing this contribution.

The Laws of Man and the Laws of Nature

The United Nations Convention on the Law of the Sea (LOS Convention),[1] Part XII, is a constitution for environmental protection of not only the seas, but the planet's global ecosystem as, ecologically, the marine environment is inherently interconnected, directly or indirectly, with all terrestrial and freshwater habitats. The ecological interconnections between land and sea have resulted in overlapping legal domains for LOS and IEL as many instruments incorporate principles from each.[2]

As environmental regimes have evolved, LOS and IEL have fundamental, but different purposes in environmental regime-building. LOS, primarily via the LOS Convention, provides States with a legal and jurisdictional framework in which to operate as there is no international and legally-binding equivalent instrument in IEL. Conversely, in implementing the broad principles of the LOS Convention, such as the duties to protect and preserve the marine environment and to cooperate on a global or regional basis,[3] States and other actors draw on the substantive principles and functional actions within IEL framework.[4] Regional seas conventions are prime examples of the melding of IEL and LOS.

IEL has developed relatively swiftly over the previous four decades. Creativity, resourcefulness and diversity mark this journey as IEL has favoured, by necessity, a functional approach to addressing environmental

---

1 *United Nations Convention on the Law of the Sea*, Montego Bay, 10 December 1982, UN Doc. A/Conf.62/122 (1982); 21 *I.L.M.* 1261 (1982) [hereinafter LOS Convention].

2 A few instruments that draw on both LOS and IEL include: *Global Programme of Action for the Protection of the Marine Environment from Land-based Activities,* Washington, 5 December 1995, UNEP (OCA)/LBA/IG.2/7 (1995) [hereinafter GPA]; *Convention on Biological Diversity*, Rio de Janeiro, 5 June 1992, UN Doc. UNEP/Bio. Civ/Conf/L. 2; 31 *I.L.M.* 818 (1992); *Convention on Wetlands of International Importance Especially as Waterfowl Habitat*, Ramsar, 2 February 1971, 996 *U.N.T.S.* 245, 11 *I.L.M.* 963 (1972); and the *Convention for the Protection of the Marine Environment of the North-East Atlantic*, Paris, 22 September 1992, 32 *I.L.M.* 1069 (1993).

3 LOS Convention, *supra* note 1, Articles 192 and 197, respectively.

4 A selection of principles which generally fall within the rubric of IEL includes sustainable development, the concept of precaution, integration (several types and levels), cooperation, intergenerational equity, integrated coastal management, best available technology and best environmental practice.

issues and developed tools accordingly. Examples of the functional nature of IEL include its use of soft-law instruments; regional cooperative initiatives; green technology; education; integrated approaches; and strategies such as including stakeholders, empowering women, addressing poverty, promoting alternative actions and providing economic incentives.

IEL's diverse and functional approach diverged from the formalistic and diplomatic approach of international law generally and LOS specifically. Although many of the early IEL practitioners emanated from the LOS domain in the early decades, the necessity for functional approaches saw the gradual erosion within IEL of the LOS diplomatic mindset of negotiated outcomes and consensual standards. Often reflecting the "lowest common denominator," such a mindset may be acceptable in the international legal arena dictating State (and human) behaviour vis-à-vis each other, but it is inherently unsuitable in the international environmental arena where negotiated outcomes seldom provide the requisite level of environmental protection.

The shift in mindset away from the formalistic LOS mindset is consistent with the fundamental precept governing the effectiveness of IEL and environmental regimes – the laws of nature prevail over the laws of man. There is no negotiating with nature as ecological laws are grounded in immutable scientific principles. Thus, policies and actions must comply with the laws governing natural systems if interventions are to generate ecological improvement, or in other words, be ecologically effective.[5]

Evolution of International Environmental Law

The development of IEL can arguably be divided into three stages. The first stage was influenced by the LOS mindset of politicized outcomes and reflects a lack of critical understanding of environmental issues, particularly

[5] This article uses the word "policy" broadly to include a variety of legal, policy and management initiatives, including global, regional and domestic soft- and hard-law instruments, such as conventions, laws, policies, action plans, management plans, guidelines, declarations, and memoranda of understanding. Likewise, "policy-making" comprises law, policy and decision-making processes and related processes and the term "policy-makers" encompasses the actors involved in the above processes.

in a scientific and ecological context. Accumulating evidence of continuing and chronic degradation and ineffective policies and actions triggered the second stage, a transition phase, in the 1990s. This evolved into the third stage in the early 2000s, which seems to largely focus on effective implementation. This, together with the evolving mindset that recognizes the supremacy of laws of nature, raises questions as to the ecological effectiveness of environmental regimes. Throughout the evolution of IEL, LOS remained relatively stable and provided a solid foundation in which to ground environmental regimes. The three stages are discussed below.

Stage 1 of IEL: Creating an Arsenal of Tools

The first stage in the evolution of IEL commenced with the first macro environmental conference, the 1972 United Nations Conference on the Human Environment (Stockholm),[6] and ended with two macro conferences, 1992 UN Conference on the Environment and Development (UNCED) and the 1995 UN Conference on the Global Programme of Action for the Protection of the Marine Environment from Land-based Activities. This stage focussed largely on developing substantive tools, including legal and management concepts, principles and actions to promote ecological sustainability.[7] These have been entrenched in a prolific array of global, regional and domestic hard- and soft-law instruments.

While this era provided the international community with sufficient and largely philosophically sound tools with which to address environmental issues, weaknesses of stage 1 include the frequently ineffective application and implementation of these tools and instruments and the lack of development and incorporation of mechanisms to assess and review the effectiveness of policies, actions and related processes. Further, there was a general failure to recognize the need for and to infuse sound science into policy-making processes to foster ecological effectiveness in compliance with the laws of nature.

---

[6] United Nations, *Report of the United Nations Conference on the Human Environment,* 5–16 June 1972, A/Conf.48/14/Rev.1 (1973).

[7] Some of the principles are noted above, *supra* note 4.

Stage 2 of IEL: The Transition Stage of Questioning "Effectiveness"

Agenda 21 and the Global Programme of Action for the Protection of the Marine Environment from Land-based Activities (GPA) are products on the cusp of the transition stage.[8] Recognizing the weaknesses of the first era, both of these instruments call for more and better science, encourage improved implementation and highlight the need for "effectiveness."[9] Departing from most legally-binding environmental instruments, both soft-law instruments provide guidance on *how* to improve the overall effectiveness of environmental policies and actions. Although they did not prove to be catalysts for the analyses and revision of policy-making processes, they arguably placed the issue of effectiveness of environmental regimes on the international radar.

Scientists, who have always been vocal,[10] but largely unheard by the international policy community, became more vocal in the 1990s regarding the ecologically unsound application and implementation of largely philosophically sound IEL tools. The scientific voice permeated the international community via reports,[11] articles[12]

---

[8] *Agenda 21*, 13 June 1992, UN Doc. A/Conf.151/4 (1992) and the GPA, *supra* note 2.

[9] See *Agenda 21, ibid.*, Ch. 35, Science for Sustainable Development. The instrument refers to effectiveness over 140 times. The GPA, *ibid.*, refers to the need for science and technical input throughout.

[10] For example: J. Cairns, Jr., "Aquatic Ecosystem Assimilative Capacity," *Fisheries* 2, No. 2 (1977): 5–7, 24; M. Waldichuk, "Control of Marine Pollution: An Essay Review," *Ocean Development and International Law* 4 (1977): 269; R. Carson, *Silent Spring* (New York and Boston: Houghton Mifflin Company, 1962, 1994); Editorial, "Why We Only Accept a Policy if We Know It Will Not Work," *Ecologist Quarterly* 1 (1978): 266; and A. R. D. Stebbing, "Assimilative Capacity," *Marine Pollution Bulletin* 12 (1981): 362.

[11] For example: GESAMP, "A Sea of Troubles," *GESAMP Reports and Studies* No. 70, 2000; GESAMP, "Protecting the Oceans from Land-based Activities: Land-based Sources and Activities Affecting the Quality and Uses of the Marine, Coastal and Associated Freshwater Environment," *GESAMP Reports and Studies* No. 71, 2001; GESAMP, "The State of the Marine Environment," *GESAMP Reports and Studies* No. 39, 1990; GESAMP, "Global Strategies for Marine Environmental Protection," *GESAMP Reports and Studies* No. 45, 1991; and GESAMP, *infra* note 64. For more information, see <www.gesamp.org>, 29 September 2007.

[12] For example: J. Cairns, "Healing the World's Ecological Wounds," *International Journal of Sustainable Development and World Ecology* 8 (2001): 185; J. S. Gray, "Statistics and

and books.[13] A direct approach to the heads of State at UNCED involved the Heidelberg Appeal to the Heads of States and Governments.[14] Other actors involved in environmental regimes, primarily political and social

the Precautionary Principle," *Marine Pollution Bulletin* 21 (1990): 174; L. D. Mee, "Scientific Methods and the Precautionary Principle," in D. Freestone and E. Hey, eds, *The Precautionary Principle in International Law: The Challenge of Implementation* (The Hague, London and Boston: Kluwer Law International, 1996); S. O. Funtowicz and J. R. Ravetz, *infra* note 88; R. G. V. Boelens, "From Policies to Science: Strategies for Marine Environmental Protection," *Marine Pollution Bulletin* 25 (1992): 14; Gray, *infra* note 51; C. Sheppard, "Not Politically Correct," *Marine Pollution Bulletin* 24 (1992): 524; J. C. Hall, W. T. Hall, and C. T. Simmons, "Water Quality Criteria for Copper: A Need for Revisions of the National Standard," *Water Environment and Technology* 9(6) (1997): 45; D. Ludwig, R. Hilborn, and C. Walters, "Uncertainty, Resource Exploitation, and Conservation: Lessons from History," *Science* 260 (1993): 17; T. Wakeford and M. Walters, eds, *Science for the Earth: Can Science Make the World a Better Place?* (Chichester, England: John Wiley & Sons, 1995); M. H. Depledge, "Ecotoxicology: A Science or Management Tool?," *Ambio* 22 (1993): 51; J. A. Hutchings, C. Walters and R. L. Haedrich, "Is Scientific Inquiry Incompatible with Government Information Control?," *Canadian Journal of Fisheries and Aquatic Sciences* 54 (1997): 1198; L. Jeftic, "The Role of Science in Marine Environmental Protection of Regional Seas and Their Coastal Areas: The Experience of the Mediterranean Action Plan," *Marine Pollution Bulletin* 25 (1992): 66; N. Myers, "Environmental Unknowns," *Science* 269 (1995): 358; T. Lovejoy, "Biodiversity: Dismissing Scientific Process," *Scientific American* 286(1) (2002): 67; and J. Rennie, "Misleading Math about the Earth," *Scientific American* 286(1) (2002): 59; Bewers, *infra* note 14; J. S. Gray and J. M. Bewers, *infra* note 58; and Gray, *infra* note 76.

13 C. Sagan, *The Demon-Haunted World: Science as a Candle in the Dark* (New York: Random House, 1995); P. R. Ehrlich and A. H. Ehrlich, *Betrayal of Science and Reason: How Anti-environmental Rhetoric Threatens Our Future* (Washington, D.C.: Covelo, California: Island Press, 1996); Ray and Guzzo, *infra* note 14; Wilson, *infra* note 33; and Capra, *infra* note 54.

14 See J. M. Bewers, "The Declining Influence of Science on Marine Environmental Policy," *Chemistry and Ecology* 10 (1995): 9–23, at p. 21 and Appendix 1, p. 23. Also see D. L. Ray and L. R. Guzzo, *Environmental Overkill: Whatever Happened to Common Sense?* (New York: HarperPerennial, 1993), pp. 6–7. Of the more than 400 signatures, over 250 included the world's leading scientists and 27 Nobel Laureates. Released on 1 June 1992, it noted, *inter alia*, that in two years of UNCED preparation, the involvement of scientific specialists was insignificant. It called for more scientific input and common sense in policy-making. The Heidelberg Appeal was neither acknowledged by the heads of State nor taken into account during the negotiations at UNCED. Ray and Guzzo, p. 7.

scientists, concurred with the scientists.[15] While many discussed the need to foster sound policies and actions for improved ecological effectiveness, only a few acted.

Blatant acknowledgement of the need to improve ecological effectiveness came from the Global Environment Facility (GEF). In the mid-1990s, the GEF required new fund recipients under the International Waters focal area to draft a "Strategic Action Programme" (SAP). The qualifier, "strategic," differentiated the SAP from other action plans as the SAP was required to be grounded in a region-wide Transboundary Diagnostic Analysis (TDA).[16] The TDA, undertaken by scientists, would assess and rank ecological degradation on a regional basis. Subsequently, demonstration sites would be established in accordance with the priority issues identified in the TDA.[17] Ultimately, information collected and lessons learned, primarily from the demonstration sites, would be incorporated into a finalized SAP to be adopted by the relevant parties. Via this process, the SAP would comprise scientifically sound and realistic actions and measures that would result in ecological improvement.

---

15 See, for example, L. K. Caldwell, *Between Two Worlds: Science, the Environmental Movement and Policy Choice* (Cambridge: Cambridge University Press, 1992); Lee, *infra* note 22; Orr, *infra* note 47; O. R. Young, ed., *Global Governance: Drawing Insights from the Environmental Experience* (Cambridge, Massachusetts and London, England: The MIT Press, 1997); R. M. M'Gonigle, T. L. Jamieson, M. K. McAllister, and R. M. Peterman, "Taking Uncertainty Seriously: From Permissive Regulation to Preventative Design in Environmental Decision-making," *Osgoode Hall Law Journal* 32 (1994): 99; N. Robinson, "Legal Systems, Decisionmaking, and the Science of Earth's Systems: Procedural Missing Links," *Ecology Law Quarterly* 27 (2000): 1075; G. E. Brown, "Environmental Science under Siege in the U.S. Congress," *Environment* 39(2) (1997): 12; Hempel, *infra* note 65; Faigman, *infra* note 43; D. Fisk, "Environmental Science and Environmental Law," *Journal of Environmental Law* 10 (1998): 3; P. R. Gross, N. Levitt and M. W. Lewis, eds, *The Flight from Science and Reason* (New York: New York Academy of Sciences, 1996); O. A. Houck, "Are Humans Part of the Ecosystem?," *Environmental Law* 28 (1998): 1; H. Margolis, *Dealing with Risk: Why the Public and the Experts Disagree on Environmental Issues* (Chicago and London: University of Chicago Press, 1996); and S. Rampton and J. Stauber, *Trust Us, We're Experts!: How Industry Manipulates Science and Gambles with Your Future* (New York: Jeremy P. Tarcher/Putnam and the Center for Media and Democracy, 2001).

16 See the GEF website, <www.gefweb.org>, 29 September 2007.

17 *Ibid.*

In a GEF example, the UNEP/GEF South China Sea Project,[18] demonstration sites for each of the identified ecological issues involve scientists working with the policy-makers, law enforcement, politicians, community leaders, industry and other relevant actors to determine effective strategies to reverse trends of degradation. Scientists from each of the seven countries involved meet regularly to determine scientific criteria and other needs to effectively monitor, assess, and address the respective ecological issues. Legal and economic task forces work in conjunction with these groups. Every two years all actors meet to resolve problems and to determine strategies to improve effectiveness further. The draft SAP, which established the sound objectives and detailed relevant criteria and processes, will be finalized towards the end of the project. It will be grounded in sound science, incorporate relevant information from all sectors and disciplines garnered over the life of the project, and establish review mechanisms to ensure the SAP remains current with scientific and other knowledge. Several other regional seas have adopted SAPs, including the Mediterranean, Black, Red and Caspian seas.

Stage 3 of IEL: The Quest for "Effectiveness"

The first macro environmental conference of the new millennium, the 2002 World Summit on Sustainable Development (WSSD), focused global attention on the need for effective implementation.[19] The message of the WSSD departed, quite significantly, from earlier environmental conferences and meetings by highlighting *process* over substantive issues and principles. By doing so, arguably, the WSSD launched IEL into its third evolutionary stage by shifting the focus from "what to do" to "how to do it better."

The focus on improving overall and ecological effectiveness is evident in the frequent use of terms that act as qualifiers, such as "strategic," "effective," "knowledge-based (decision-making),"[20] "science-based (deci-

---

18 UNEP/GEF Project to Reverse Environmental Degradation Trends in the South China Sea and Gulf of Thailand, <www.unepscs.org/>, 29 September 2007.

19 See the *Plan of Implementation of the World Summit on Sustainable Development,* <www.un.org/esa/sustdev/documents/WSSD_POI_PD/English/WSSD_PlanImpl.pdf>, 29 September 2007.

20 UNEP, *A Global Initiative to Strengthen Regional Seas Conventions and Action Plans and Enhance Co-operation: Regional Seas Strategic Directions for 2004–2007*, UNEP(DEC)/RS. 6 (2004), Annex I.

sion-making),"[21] and "adaptive (management)"[22] in instruments and documents. Employing such qualifiers for previously unqualified terms suggests dissatisfaction with previous instruments, actions and outcomes. They offer an avenue to institutionalize pre-meditation, deliberation, calculation, planning, coordination and cooperation to maximize effectiveness. These qualifiers further encourage a shift in mindset to embrace sound policy-making and implementation and, by inference, the need to revise current processes to make them more effective.

"Strategic" and "strategy" gain prominence as the need for greater effectiveness is emphasized. The qualifier "strategic" peppers the 2002 Plan of Implementation of the World Summit on Sustainable Development (Plan of Implementation), appearing in phrases such as "strategic alliances," "strategic approaches," and "strategic plans."[23] Likewise, "strategic partnerships" replace mere partnerships. It has been adopted by both the Strategic Partnership for the Mediterranean Large Marine Ecosystem and the GEF Strategic Partnership on the Black Sea and Danube Basin.[24] The Secretariat for the GPA has formed a "strategic partnership" with the Regional Seas Programmes (RSP) to facilitate coordinated national action within a region in accordance with global directives.[25]

The RSP Members at the 6th Annual Global Meeting in 2004 endorsed a global initiative entitled, in part, Regional Seas Strategic Directions for 2004–2007 (Strategic Directions).[26] In agreeing that a primary role for RSPs should be as facilitators for the domestic and regional implementation of global instruments, Strategic Directions promotes sound processes, including knowledge-based policy-making, education, partnership synergies, coordinated implementation efforts, ecosystem-based integrated management, priority issue focus and innovative tools.[27] In May 2005, the Coordinating Body on the Seas of East Asia (COBSEA) held the

---

[21] Plan of Implementation, *supra* note 19, at para. 109.

[22] See K. N. Lee, *Compass and Gyroscope: Integrating Science and Politics for the Environment* (Washington, D.C. and Covelco, California: Island Press, 1993).

[23] Plan of Implementation, *supra* note 19, at paras. 113, 23(b) and 33(c), respectively.

[24] See respectively, <www.iwlearn.net/iw-projects/Fsp_112799470859> and <webapps01.un.org/dsd/partnerships/public/partnerships/941.html.>, 5 September 2007.

[25] For further information, see the GPA website, <www.gpa.unep.org>, 30 September 2007.

[26] UNEP, *supra* note 20, at Annex I.

[27] *Ibid.*

"Brainstorming Meeting of the National Focal Points on New Strategic Directions of COBSEA."[28]

The terms, "effective," "effectiveness" and "effectively," while found in previous instruments, are used more frequently in the 2002 Plan of Implementation where they appear at least 81 times. References include: "effectively implement," "respond effectively," "deal effectively," "effective participation," "effective governance systems," "operationally effective," "effective measures," "effective science," "effective management frameworks," "effective policy responses," "effective participation," "effective UN system," "promote greater effectiveness of activities," and "enact and enforce clear and effective laws."[29] There is little doubt that the Plan of Implementation is a clarion call to improve effectiveness throughout policy-making processes and environmental regimes.

Additional confirmation of the entrance into stage 3 and the quest to improve ecological effectiveness is the 2002 seminal study entitled *Environmental Regime Effectiveness: Confronting Theory with Evidence.*[30] The authors concluded that although the 14 global and regional regimes studied "by and large make a positive difference, most fail to solve the problems they were designed to solve and often fail by a wide margin."[31] This is an indictment of the failure to focus on ecological effectiveness in the earlier stages of the evolution of IEL.

The third stage in the evolution of IEL marks the philosophical transition of environmental regimes from the era of "command and control" to an era of risk management, as dictated by the amalgam of environmental degradation issues and ecological complexities.[32] Risk management, by definition, is an ongoing, long-term process that requires examining, reviewing and revising to ensure that instruments, actions, measures and

[28] UNEP, *Report of the Brainstorming Meeting of the National Focal Points on New Strategic Directions of COBSEA*, UNEP(DEC)EAS 1.NFP, 25 May 2005.

[29] For the references, search for "effective" in the electronic document, *supra* note 19.

[30] E. D. Miles, A. Underdal, S. Andresen, J. Wettestad, J. B. Skjærseth, and E. M. Carlin, eds, *Environmental Regime Effectiveness: Confronting Theory with Evidence* (Cambridge, Massachusetts and London, England: MIT Press, 2002).

[31] *Ibid.*, at p. 468.

[32] See generally, B. Fischhoff, S. Lichtenstein, P. Slovic, S. L. Derby, and R. L. Keeney, *Acceptable Risk* (Cambridge: University of Cambridge Press, 1981, 1993) and H. Kunreuther and P. Slovic, "Science, Values, and Risk," *The Annals of the American Academy of Political and Social Science* 545 (1996): 116.

initiatives are effective and remain so. The focus is processes and their effectiveness in improving and maintaining ecological integrity.

This evolutionary stage of IEL embodies the quintessential trait of humans – the ability to reason. Common sense is infusing environmental regime development as actors strategise to ensure that efforts produce results, in this case ecological effectiveness.

Two Concepts to Promote Effectiveness

Two concepts, "consilience" and a "grand experiment" mindset, may prove useful for improving the ecological effectiveness of environmental regimes.

"Consilience": A Unifying Concept for Sound Policy-making

"Consilience" is defined as the "'the jumping' together of knowledge by the linking of facts and fact-based theory across disciplines to create a common groundwork of explanation."[33] More simply, consilience fosters an intellectual and comprehensive understanding of environmental events and issues by integrating and unifying all relevant information.[34]

The term, consilience, is not new to the English language, having been introduced by William Whewell in his 1840 publication, *The Philosophy of the Inductive Sciences.*[35] It was recently revived in *Consilience: The Unity of Knowledge*, an illuminating text by an eminent scientist.[36] Consilience has noble roots in the Age of Enlightenment in the 17th and 18th centuries as the three great branches of learning, the arts, humanities, and sciences, were unified as knowledge from one branch was integrated with the other branches.[37] The Enlightenment was dominated by a handful of

---

33 E. O. Wilson, *Consilience: The Unity of Knowledge* (New York: Alfred A. Knopf, Inc., 1998), p. 8.

34 *Ibid.*, pp. 8–13. Also see E. O. Wilson, "Consilience among the Great Branches of Learning," in P. Galison, S. Graubard, and E. Mendelsohn, eds, *Science in Culture* (New Brunswick, N.J.: Transaction Publishers, 2001), pp. 131–150.

35 Wilson, *supra* note 33, at p. 8.

36 *Ibid.*

37 For an encapsulation of "consilience," see E. O. Wilson, "Resuming the Enlightenment Quest," *The Wilson Quarterly* 22 (1998): 16. Also see P. R. Gross, "The Icarian Impulse,"

natural scientists and philosophers, namely Bacon, Descartes, Hobbes, Hume, Locke, Newton, Grotius, Galileo, Kant, and Leibniz.[38] Among the commonalities binding these men, "they believed in the unity of all knowledge and agreed on the power of science to reveal an orderly and understandable universe and thereby lay an enduring base for free and rational discourse."[39]

Consilience is particularly relevant at this juncture in environmental regime-building. Environmental policy-making is increasingly complex, involving systemic, chronic and amorphous issues, such as global climate change, biodiversity loss, habitat degradation, degradation from land-based activities and integrated coastal management (ICM). These issues involve, in addition to scientific complexity and a broad array of socio-economic concerns, fundamental and emotionally charged issues such as sovereignty, values, ethics, cultural and religious beliefs, human rights, and human welfare.

Complex environmental issues require consilience as the latter fosters a *broad knowledge base* for policy-making and a *critical understanding* of the fundamentals of the issue. Currently, policy-makers attempt to create a broad knowledge base by integrating knowledge and information from an array of sources and disciplines. However, it largely commences on an ad hoc, chaotic basis and lacks a master plan or comprehensive strategy for systematically considering and integrating *all* relevant information. Unfortunately, the information gathered is seldom screened to ensure it is sound and reliable and the best information available.

A critical understanding of environmental issues includes not only knowledge relating to the environmental issue, but an understanding of underlying issues, such as ecological principles; scientific challenges (e.g., reservations, uncertainties); hindrances to effectiveness (e.g., societal and cultural biases, media-driven perceptions, viable alternatives); and novel tactical strategies for improving effectiveness (e.g., involving domestic implementing ministries in regional meetings to improve implementation). Possessing a critical understanding does not involve becoming an expert in every area, but involves a sufficient appreciation of the issues to make

The Wilson Quarterly 22 (1998): 39. *Cf.* R. Rorty, "Against Unity," *The Wilson Quarterly* 22 (1998): 28.

[38] Wilson, *supra* note 33, at p. 21.

[39] *Ibid.*

informed, sensible and sound decisions. For example, a large number of actors in environmental regimes suffer a lack of critical understanding of science (which includes ecology) and consequently, they frequently make ecologically unsound decisions.

The premise underlying consilience is that a broad knowledge base and critical understanding empower actors to pose salient questions and reach sound decisions. It embodies systematic, strategic and coherent policy-making processes. To this end, the concept of consilience engenders *effectiveness* in environmental regimes.

The Grand Experiment: A New Mindset

Conceptually viewing environmental regime-building as a grand experiment is an appropriate mindset for fostering effectiveness. Scientific experimentation focuses on sound processes. Experiments are carefully designed, entailing informed thoughts and actions, and assessing and reviewing actions and making necessary adjustments. Adopting an experimental mindset can establish environmental policy-making and regime-building as an on-going process of learning and adaptation. Feedback and review mechanisms, designed to capture successful and unsuccessful strategies and assumptions, the emergence of unexpected phenomena, misstated values and other "missteps," should trigger pre-mediated and rational policy responses to improve effectiveness of policies and actions.[40] Adaptive management, another name for this process, requires that actors be open-minded, flexible, objective, analytical, creative, sceptical (cautious), integrative and persevering (following up inconsistencies) in seeking ecological effectiveness – they must not merely seek expedient or politically popular resolutions.[41]

The mindset of a grand experiment is particularly relevant for complex issues as it fosters a methodological and rational framework to discover and then validate information and to generate sound policies for ecological sustainability. This mindset encourages actors to accept policy-making as a cyclic process of updating policies and actions in accordance with new information. Consistent with the third stage of IEL and consilience,

---

[40] See Lee, *supra* note 22, at pp. 51–86 and GESAMP, *infra* note 64, at pp. 10–12.

[41] Lee, *ibid.*

this mindset focuses attention on processes and improving environmental regime effectiveness.

Science and Consilience

Scientific knowledge is a "universal possession of humanity" and it "comprises what we know of the material world with reasonable certainty."[42] Science is necessary to inform actors of the biological, physical, and chemical implications of human activities; to establish ecological priorities; and to advise on the ecological and economic effectiveness of policy options. To this end, scientific knowledge is at the core of environmental issues as the objective of environmental regimes is ecological sustainability. Without a sound scientific foundation, policies and actions are often ecologically ineffective and economically inefficient.

Science must inform policy, but not dictate, usurp, or hijack the policy-making process. Science can identify sound options and advise whether policy choices are sound, or how they can be adjusted to improve effectiveness. Scientists determine "what *is*" (factual findings) and society, through policy-making processes grounded with sound science, decides "what *ought* to be" given the interests and considerations of the various stakeholders.[43] The key is that options considered in the policy-making process must be scientifically sound. This is consilience in operation.

Declining Influence of Science

Paradoxically, science often fails to influence policy at a time when systemic degradation issues desperately require the input of sound science and at a time when scientists are well placed to provide useful and reliable advice.[44] This is unfortunate, but not surprising as, historically, science

---

[42] Wilson, *supra* note 33, at p. 268.

[43] D. L. Faigman, *Legal Alchemy: The Use and Misuse of Science in Law* (New York: W. H. Freeman and Company, 1999, 2000), p. 33 and P. Castro and M. E. Huber, *Marine Biology*, 2nd edition (Dubuque, Iowa: Wm. C. Brown Publishers, 1997), pp. 14–16.

[44] See Bewers, *supra* note 14. Also see unpublished Ph.D. thesis of S. M. Lexmond, *From Chaos to Consilience: The Need for Science-based Policy to Prevent Marine Degradation from Land-based Activities*, <www.sml.kyl191.net/consilience.pdf>, 30 September 2007, Ch. 5. For examples of improved capacity, see generally P. M. Chapman, "Is Bioaccumulation

has been ignored, neglected and misused in many ways and at every stage of policy-making.[45]

The misuse of science has occurred despite good intentions. Its misuse is also attributable to malfeasance and the intentional manipulation of science for political or other purposes. Misuse includes: failing to consult scientists or to incorporate science; choosing unsound science over sound science; relying on outdated or biased science; relying on inappropriate experts in the targeted issue or those who represent a minority scientific view; ignoring the assimilative capacity of the natural environment; interpreting and applying philosophically sound legal principles, such as the concepts of precaution, best available technology and best environmental practice, in scientifically unsound ways; adopting unsound principles like zero discharge; and inappropriate technology.[46]

Scientific Illiteracy

Scientific illiteracy underlies much of the misuse of science. Policy-makers may be considered "functionally science-illiterate" where their level of critical understanding of science is below the level that is required to make effective decisions about environmental issues and discern sound policy options.[47] Unfortunately, most non-scientific actors have emerged science-illiterate

Useful for Predicting Impacts," *Marine Pollution Bulletin* 34 (1997): 282; L. W. Aarssen, "On the Progress of Ecology," *Oikos* 80 (1997): 177; G. Kullenberg, "Capacity Building in Marine Research and Ocean Observations: A Perspective on Why and How," *Marine Policy* 22 (1998): 185; J. H. Stel, "Editorial – Marine Capacity Building in a Changing Global Setting," *Marine Policy* 22 (1998): 175; E. Okemwa, "The Intergovernmental Oceanographic Commission of UNESCO and Regional Capacity Building," *Marine Policy* 22 (1998): 197; and R. Bailey, ed., *Earth Report 2000: Revisiting the True State of the Planet* (New York: McGraw-Hill, Inc., 2000).

45 Lexmond, *ibid.*, Chs. 2 and 4.

46 See *ibid.*, Ch. 4. Also, many publications noted in footnotes 10–13 and 15, herein, detail misuses.

47 See D. W. Orr, *Ecological Literacy: Education and the Transition to a Postmodern World* (New York: State University of New York Press, 1992); C. P. Snow, *Two Cultures and the Scientific Revolution* (New York: Cambridge University Press, 1959); and G. Hardin, *Filters Against Folly: How to Survive Despite Economists, Ecologists and the Merely Eloquent* (New York: Penguin, 1987). For a more detailed discussion see Lexmond, *ibid.*, Ch. 2.

from schools and tertiary educational institutions as few institutions require that students have even minimal scientific exposure.[48]

The implications of scientific illiteracy can be severe.[49] The most blatant is poor policy choices that result in continuing ecological degradation. Many actors fail to understand fundamentals, such as the extent to which rational policy choices are bound, or should be bound, to science.[50] The science-illiterate make poor choices regarding sound science, suitable experts, policy options and implementation, as noted above. They make fundamental mistakes such as asking the wrong questions or funding inappropriate research.[51] Scientific illiteracy impairs communication, confounds progress and cooperation, generates tension and results in poor choices.[52] Economic inefficiency often ensues as valuable time and resources are used inappropriately, for example, to address low ecological priorities, while high priority issues persist.

Failing to understand the scientific process can result in choosing unqualified or biased experts or unsound science. Misunderstanding scientific terminology, such as the scientific distinction between contamination and pollution and between toxic chemicals and toxic doses, leads to confusion or misunderstandings.[53] Disregarding scientific qualifications can result in false expectations and other outcomes that result in ecological

---

48 Orr, *ibid.*, pp. *x–xii.*

49 These examples in this section are discussed in more detail in Lexmond, *supra* note 44, Ch. 4.

50 See P. E. Allin, "Some Social Aspects of Modern Science: A Point of View," *International Journal of Environment Studies* 3 (1972): 49 at p. 50 and Chapman, *supra* note 44, at p. 227. Also see L. M. Warren, "The Precautionary Principle: Use with Caution," in K. Milton, ed., *Environmentalism: The View from Anthropology* (London and New York: Routledge, 1993), pp. 97–111; Brown, *supra* note 15; Bewers, *supra* note 14; Wilson, *supra* note 33; and F. B. Golley, *A History of the Ecosystem Concept in Ecology: More than the Sum of the Parts* (New Haven and London: Yale University Press, 1993).

51 J. P. Holdren, "Energy: Asking the Wrong Questions," *Scientific American* 286(1) (2002): 63 and J. S. Gray, "Whose Research Is It Anyway?," *New Scientist* 149(2023) (1996): 48, respectively.

52 See generally, Lexmond, *supra* note 44, Chs. 2 and 4.

53 For example, in science, pollution must result in deleterious changes to the environment, whereas contamination is the presence of substances which do not register deleterious changes, as perhaps the concentration is sufficiently low. Thus, contamination can become pollution when deleterious changes are discernible. In law, pollution is most often defined as the introduction of substances that result or may result in deleterious changes. From a science perspective, this comprises both contamination and pollution.

ineffectiveness, unfair loss of scientific credibility and increased tension between the science and policy domains.

Improving Scientific Literacy

Scientific literacy for non-scientists involves a critical understanding of two aspects of science. The first is the processes of science and how scientists operate (scientific method), such as limitations and challenges, scientific reservations, peer review, scientific uncertainties and the incremental process of accumulating scientific knowledge. The second involves ecological literacy, or eco-literacy. In addition to appreciating the complexity of the ecosystem, eco-literacy entails an understanding of the rudiments of ecology, including the five basic principles of ecological organization (inter-dependence; cyclic flow of nutrients, resources, and energy; partnerships; ecosystem dynamism and flexibility; and biological diversity).[54] The aim of scientific literacy is not to create scientific experts, but to generate sophisticated and informed *users* of science who can ask relevant questions and avoid the pitfalls that have led to unsound policy.[55]

The current low levels of scientific literacy can be addressed with scientific education at all levels of society, targeting school children for long-term literacy and offering immediate educational opportunities to key actors, media and the general public.[56] Educating the general public

Consequently, scientists and lawyers discussing "pollution" are talking about two different concepts.

[54] F. Capra, *The Web of Life: A New Scientific Understanding of Living Systems* (New York: Anchor Books, Doubleday, 1996), pp. 298–304. Also see GESAMP, "Protecting the Oceans from Land-based Activities: Land-based Sources and Activities Affecting the Quality and Uses of the Marine, Coastal and Associated Freshwater Environment," *GESAMP Reports and Studies* No. 71, 2001 and E. P. Odum, *Ecology and Our Endangered Life-support System*, 2nd edition (Sunderland, Massachusetts: Sinauer Associates, Inc., 1993).

[55] See generally, Faigman, *supra* note 43. His premise is that actors should have a sufficient understanding of science to be informed users or consumers of science.

[56] See generally, S. Tomkins, "Science for the Earth Starts at School," in Wakeford and Walters, eds, *supra* note 12, pp. 257–276; Orr, *supra* note 47; W. A. Thomas, "A Report from the Workshop on Cross Education of Lawyers and Scientists," *Jurimetrics Journal* 19 (1978): 92; W. A. Thomas, "A Report from the Workshop on Cross Education of Lawyers and Scientists," 19 *Jurimetrics Journal* 92 (1978); H. Schneider, "Concepts and Issues," in OECD, *Environmental Education: An Approach to Sustainable Development*, Prepared under the OECD Development Centre's Research Programme on Coping

helps to change perceptions and behaviour and provides awareness and knowledge to positively influence policy and promote skills for living within the laws of nature.[57]

Actors who are influential in policy-making should become science-literate through dedicated scientific and eco-literacy programmes. Policy-makers, bureaucrats, lawyers, and other government and State actors involved in environmental regimes should be encouraged, if not required, to take a scientific literacy exam, or enrol in a government-sponsored scientific literacy course with accreditation or incentives given to those who upgrade their skills. It is essential that the actors in decision-making positions be able to cut through emotionally charged issues and be empowered to make wise choices utilizing the best available science.[58]

A motivational tool is a twelve-step self-help programme entitled, "Becoming a Sophisticated Consumer of Science."[59] Programme participants commence with an admission of scientific illiteracy, acknowledge the need for scientific literacy to fulfil professional obligations, commit to seeking sound scientific advice, and agree that, by following the steps, they will become sophisticated consumers of science.[60] The programme may sound somewhat pedantic, but it holds promise as a tool to encourage decision-makers to improve their levels of scientific understanding as part of the process of improving regime effectiveness. While other factors, such as socio-economic issues, must guide environmental decisions, the premise is that the options must be grounded in sound science to ensure wise decisions and that the actions chosen will positively impact the ecosystem in accordance with the objectives and sound use of financial and other resources.

---

with Environmental Threats in Developing Countries (Paris: OECD, 1992), pp. 25–33; and O. S. Dijksterhuis, "Environmental Education: A Tool for Coastal Management? A Study of the Caribbean Region," *Coastal Management* 24 (1996): 339.

[57] E. W. Weidner, "Educational Aspects of Environmental Issues," *International Journal of Environmental Studies* 2(1) (1972): 301–308, at p. 308. For a discussion on the need for behavioural fixes in addition to technological fixes, see R. W. Kates, "Success, Strain, and Surprise," *Issues in Science and Technology* 2 (1985–1986): 46.

[58] See J. S. Gray and J. M. Bewers, "Towards a Scientific Definition of the Precautionary Principle," *Marine Pollution Bulletin* 32 (1996): 768, at p. 771.

[59] Faigman, *supra* note 43, at p. 198.

[60] *Ibid.*

A key group of actors deserving special and immediate scientific education is the media, who, in turn, can become educators of the general public.[61] There is a move towards responsible and balanced reporting, particularly as bodies such as the Union of Concerned Scientists, based in the United States, speak out against inaccurate or misleading science reporting on environmental issues.[62] The contrarian reports and open debates in the media must continue, and they should be encouraged. However, clarity and honesty on behalf of reporters is necessary, for example, to distinguish minority scientific opinions from those of the mainstream scientific body or to confirm whether a cited expert has expertise in the relevant area. The media needs to fully inform the public to assist the latter in making rational choices.

Scientific literacy can be addressed on a global (perhaps within the United Nations) or regional level via a multi-disciplinary task force mandated to conceptualize, design, and implement novel ways of making science less daunting and more interesting for non-scientific actors while creating materials to support scientific literacy. Such a task force requires the participation of numerous disciplines, including social scientists and psychologists who understand factors influencing mindsets and perceptions and consequently policy choices, as well as cultural and religious biases, information processing styles and fear and risk perception.[63]

Institutionalizing a Role for Science

Scientists have long recognized that science requires a defined role in environmental regime development. The Group of Experts on Scientific

---

61 See L. S. Susskind, *Environmental Diplomacy: Negotiating More Effective Global Agreements* (New York: Oxford: Oxford University Press), 1994, p. 136 and Ray and Guzzo, *supra* note 14, at pp. 171–187. See H. A. Cohl, *Are We Scaring Ourselves to Death?: How Pessimism, Paranoia, and a Misguided Media are Leading Us Toward Disaster* (New York: St. Martin's Griffin Press, 1997).

62 The Union of Concerned Scientists started in the 1960s in response to concerns about the social uses of scientific discovery. Its members encourage the sound use of sound science, particularly in publicly debated environmental concerns such as global climate change. See the Union of Concerned Scientists, <www.ucsusa.org>, 29 September 2007.

63 For example, see D. Goleman, *Emotional Intelligence* (New York: Bantam Books, 1995); S. Pinker, *How the Mind Works* (New York: W. W. Norton, 1997); and Wilson, *supra* note 33.

Aspects of Marine Environmental Protection (GESAMP), a body established to advise specific United Nations-related bodies, published in 1996 "The Contributions of Science to Integrated Coastal Management."[64] They foresaw a two-pronged role for science. The first role entails scientists offering scientific advice. They can offer the best available scientific knowledge and advice to tailor policies and actions and provide a sound scientific framework in which to discuss socio-economic and political concerns. This will involve sound scientific knowledge about the natural world, assessment of scientific impacts of human activities and assessment of various options for ecological effectiveness (to ensure resources are not wasted pursuing actions that are ecologically ineffective). Where scientific uncertainty is high, scientists can advise on feasible options and suggest feedback mechanisms to regularly assess the effectiveness of the policies and actions. In the second role, scientists are educators, improving scientific literacy and communication and building confidence in science. Scientists can educate, direct, guide, advise, challenge, and support actors in an effort to foster sound policy and to discourage or eliminate misperceptions, myths, prejudices, abuse, neglect, misunderstanding, and confusion.

GESAMP, in their publication, has detailed a role for science through five consecutive stages of an iterative, feedback-enhanced policy-making process for integrated coastal management.[65] GESAMP's five stages[66] are equally relevant to all environmental policy-making processes and they detail a two-fold role for science at each stage as follows:

Stage 1: Issue Identification and Assessment

In this initial stage, scientists should collect sufficient information to guide a multi-disciplinary team in subsequent stages of policy-making. Social and natural scientists and other relevant actors take the first step towards

---

[64] GESAMP, "The Contributions of Science to Integrated Coastal Management," *GESAMP Reports and Studies* No. 61, 1996.

[65] *Ibid.*, pp. 5–12 and 20–21. Also see L. C. Hempel, *Environmental Governance: The Global Challenge* (Washington D.C. and Covelo, California: Island Press, 1996), pp. 134–147. Further see Susskind, *supra* note 61, at pp. 76–78, which identifies five roles for science, namely trend spotters, theory-builders, theory-testers, communicators, and policy analysts.

[66] See GESAMP, *ibid.*, pp. 5–12. Further, see Hempel, *ibid.*, for a very insightful discussion of each stage of the policy-making process, complete with pitfalls, constraints, and needs.

consilience, as economic, social (e.g., values and interests), and ecological information establishes a basis for goal- and priority-setting. This entails compiling, integrating, and prioritizing information to identify and assess ecological and social issues, and developmental, institutional, cultural, and other relevant contexts in which a policy must operate.[67]

Natural scientists must assume scientific and educational roles in roughly equal measure in this stage. They must identify and communicate the scientific priorities (e.g., ecological and research), trends, linkages, gaps in scientific knowledge and scientific uncertainty and the implications of their findings.[68] Their role as educators is fundamental to the initial stage, as it is here that misperceptions, prejudices, biases, and other types of misunderstanding foment.[69]

Stage 2: Planning and Programme Preparation

This organizational stage utilizes information from the assessment stage as a basis for consultation and planning a detailed operational strategy. Programme objectives, environmental quality objectives, and competing values must be concisely articulated.[70] The economic, ecological, and social costs of various options must be analyzed and debated. The participation of diverse disciplines and the acknowledgment of the validity of diverse interests are fundamental at this stage.

The scientific role in this stage includes research, monitoring, identifying characteristics or conditions of concern, elaborating on natural and anthropocentric influences, and collecting and providing to other actors the necessary scientific information.[71] Cooperation among scientists and other actors to formulate specific questions for scientific investigation greatly enhances the probability of success.[72] Scientists can highlight alternative strategies and the ecological effectiveness of various options. By educating actors about the ecological priorities, scientists can work with other actors to tailor policy options that will provide sufficient environmental protection without undue economic or social costs.

---

[67] GESAMP, *ibid.*, p. 5.

[68] *Ibid.*, p. 7.

[69] *Ibid.*, p. 21.

[70] *Ibid.*, p. 7.

[71] *Ibid.*, pp. 7–9.

[72] *Ibid.*, p. 9.

Scientists should emphasize the need for scientific feedback mechanisms to indicate whether the stated environmental quality objectives are being met and to assess the level of environmental protection secured via the chosen options. Adaptive management strategies to revise policies accordingly should be determined by the relevant actors.

Stage 3: Adoption and Funding

This is a crucial stage as policy choices must be marketed to politicians (most often ministers, heads of agencies, or even heads of government), funding agencies, and the public.[73] A well-organized, explicit, scientifically sound, accurate proposal with a detailed and substantiated budget for each stage will attract interest and encourage constructive dialogue and debate. Regardless of how soundly science was applied at the initial stages, this stage presents the greatest risk of neglect or failure to incorporate sound science as this stage, more than any other, is one of negotiation, bargaining, and accommodation of interests across a broad array of actors.[74]

The primary role for scientists in this stage is to educate, identify, explain, and if necessary, defend the policies as scientifically sound and economically efficient.[75] As issues are debated, scientists need to clarify and advise on ecological repercussions of a range of actions, or inaction, ensuring that the scientific/ecological issues are understood and not overlooked in negotiations and that the ecological pros and cons of various policy options are understood. Scientists must ensure that the best available scientific expertise is utilized and identify unsound proposals.[76] Negotiation and accommodation are legitimate and necessary, as society must set priorities, but the options must be grounded in sound science to maximize ecological effectiveness and wise use of scarce financial and other resources.

---

[73] *Ibid.*

[74] *Ibid.*

[75] *Ibid.*, p. 10. As nature's interpreters, scientists are the only actors competent to explain and clarify scientific aspects and ecological implications of management options.

[76] See J. S. Gray, "Integrating Precautionary Scientific Methods into Decision-making," in D. Freestone and E. Hey, eds, *The Precautionary Principle in International Law: The Challenge of Implementation* (The Hague, London and Boston: Kluwer Law International, 1996), p. 146.

Stage 4: Implementation

Effective implementation is best attained through a pre-meditated implementation strategy. Facets of such a strategy may include controls, regulations, incentives, monitoring programmes, feedback mechanisms, enforcement mechanisms and institutional arrangements.[77] Issues to be addressed both before and during implementation include conflict resolution, inter-agency coordination, infrastructure development, public education, personnel training, and contemplation of new problems.[78]

The scientific role entails continuous assessment of policy performance, revision and adaptation of techniques, experimental designs, monitoring programmes, and related scientific endeavours.[79] Mechanisms are required to enhance communication between scientists and other actors and to incorporate scientific feedback into policy-making and implementation efforts.

The scientist's role as educator and advisor is essential in this stage. As policies become operational, the involvement of additional actors increases the risk of poor utilization of scientific information, perpetuation of misperceptions and poor use of scarce resources.

Stage 5: Review/Evaluation and Adjustment/Adaptation

The evaluation of policy performance is the only way to measure the effectiveness of policies and actions. Paradoxically, this is the most frequently omitted or poorly addressed stage in environmental policy-making.[80] Pre-determined feedback mechanisms, assessments and periodic reviews are crucial to identifying new information, inefficiencies, ineffectiveness, over-effectiveness (which wastes scarce resources) and successes.

The review process should trigger a pre-determined revision process for adapting policies and actions to reflect the findings of the assessments and reviews. Embracing the cyclical nature of policy-making processes, the revision process should trigger a return to stage 1 (assessment), re-initiating

[77] GESAMP, *supra* note 64, at p. 10.

[78] *Ibid.*, p. 10.

[79] *Ibid.*

[80] *Ibid.* This stage is either omitted or superficially addressed in ICM, *ibid.* Likewise, multilateral environmental agreements and domestic policy often lack effective review mechanisms.

the five-stage process to ensure policy revisions are regular, rational and sound.

These five stages proposed by GESAMP are consistent with consilience of science with other information and the "grand experiment" mindset embracing adaptive management.[81] Adopting the five-stage policy-making process will foster strategic and effective regime development. This is consistent with the current stage in the evolution of IEL.

Encouraging Consilience

Encouraging consilience, by definition, should improve the ecological effectiveness of environmental regimes. Consilience infers strategic planning utilizing all relevant information and, in particular, science. Existing, but under-utilized options for encouraging consilience, include epistemic communities, post-normal science and the use of facilitators. Creative and lateral thinking can produce novel options, for example, in the form of instruments, tools, principles and actions. Various options are discussed below.

An environmental epistemic community is a network of individuals with expertise or competence in various domains or sectors, who, sharing a common outlook (e.g., ecological improvement), come together to influence policy-making processes.[82] Epistemic communities are most successful where they offer science-based, consensual knowledge on priority issues and sound measures.[83] Epistemic communities have successfully influenced policy related to the Mediterranean RSP,[84] the 1987 Montreal

[81] *Ibid.* and Lee, *supra* note 22, at pp. 51–86. Lee discusses in detail the strategy of "adaptive management."

[82] P. M. Haas, "Do Regimes Matter?: Epistemic Communities and Mediterranean Pollution Control," *International Organizations* 43 (1989): 377; P. M. Haas, ed., *Knowledge, Power, and International Policy Coordination, International Organizations* (Special Issue) 46 (1992); P. M. Haas, "Introduction: Epistemic Communities and International Policy Coordination," *International Organizations* (Special Issue) 46 (1992): 1; H. Breitmeier, "International Organizations and the Creation of Environmental Regimes," in Young, *supra* note 15, at pp. 90–92; O. S. Stokke, "Regimes as Governance Systems," in Young, *supra* note 15, at pp. 57–58; M. Valiante, P. Muldoon, and L. Botts, "Ecosystem Governance: Lessons from the Great Lakes," in Young, *supra* note 15, at pp. 197–225; and Hempel, *supra* note 65, at pp. 124–128.

[83] See generally, Haas, 1989, *ibid.*

[84] Haas, 1989 and 1992, *supra* note 82.

Ozone Protocol,[85] the management regime for the Great Lakes boundary waters between Canada and the United States,[86] and the Parliamentarians of the Arctic Region.[87]

"Post-normal science" is a rational methodology that draws on an extended peer community to provide "extended facts" in situations where scientific uncertainty is high, the issues are complex, the ecological or health stakes are high, values are in dispute and policy decisions are urgent.[88] The extended facts may include observations and experiences of relevant actors (e.g., birders, fishers, scuba divers, indigenous community, cottage industries, etc.).[89] Utilizing extended facts gathered from outside the traditional scientific community to confirm the accuracy and validity of scientific predictions, assumptions, and models and drawing on actors that have relevant information provides a rational, iterative and dynamic process to manage the various types of scientific uncertainty inherent in contemporary environmental issues.[90]

The use of facilitators at negotiations and meetings encourages consilience and overall effectiveness.[91] Facilitators are "specialized generalists" who have a critical understanding of the most relevant disciplines and cross-disciplinary appreciation of the types of issues that derail cooperation and agreement. Their role in policy-making processes and regime-building are to facilitate communication, cooperation and problem-solving by rephrasing and explaining positions in terms understood by the actors. Facilitators act as both interpreters and psychologists to fathom misunderstandings

---

[85] Haas, 1989, *ibid.*, p. 402.

[86] Valiante, et al., *supra* note 82, at pp. 218 and 224–225.

[87] See Minister of Public Works and Government Services Canada, *Report of the Second Conference of Parliamentarians of the Arctic Region, Yellowknife, Northwest Territories, Canada, 13–14 March 1996* (Ottawa: Minister of Public Works and Government Services Canada, 1996).

[88] S. O. Funtowicz and J. R. Ravetz, "A New Scientific Methodology for Global Environmental Issues," in R. Costanza, ed., *Ecological Economics: The Science and Management of Sustainability* (New York: Columbia University Press, 1991), p. 139 and S. O. Funtowicz and J. R. Ravetz, "Uncertainty, Complexity and Post-normal Science," *Environmental Toxicology and Chemistry* 13 (1994): 1881.

[89] *Ibid.*

[90] Funtowitz and Ravetz, 1994, *ibid.*, pp. 1883–1885 and Funtowitz and Ravetz, 1991, *ibid.*, pp. 148–150.

[91] See T. Justice and D. W. Jamieson, *The Facilitator's Handbook* (New York: American Management Association, 1999).

and misperceptions, foster understanding and acceptance among actors and ensure meaningful inclusion of all relevant actors.

Creative and lateral thinking can produce unlimited novel options for encouraging consilience and improving ecological effectiveness. To this end, an international think tank mandated with rethinking current approaches and devising novel options and strategies would be welcome. An example of a novel diplomatic strategy could involve an international (or regional) declaration that shifts the focuses from *what to do* to *how to do it*.[92] Its primary theme should be strategies to improve regime effectiveness. Provisions can include improving science-literacy, infusing sound science into policy-making processes, promoting consilience, enumerating the lessons learned from the Regime Effectiveness Study,[93] and strategies such as adaptive management, epistemic communities, post-normal science and procedural improvements (such as including domestic policy implementers in regional policy-setting meetings and endorsement of policies by heads of State to increase credibility). This instrument could synthesize efforts in the current ad hoc quest for greater regime effectiveness and place environmental regime development on a strategic path forward.

Conclusion

The international environmental arena has evolved into a sophisticated array of environmental regimes comprising principles of IEL and LOS, instruments and related entities and actions. Likewise, the perception of environmental issues and the related mindset are evolving as actions ineffective at resolving environmental issues highlight the need for the laws of man to heed the laws of nature. The era of command and control is past and resolving environmental issues is increasingly accepted as an exercise in risk management – a grand experiment in policy-making.

This current stage of IEL is increasingly recognizing that environmental regimes must be effective and dynamic – cycles of incremental knowledge-building and policy adaptation consistent with the long-term objective of ecological sustainability. There is growing emphasis on shifting to strategic means to improve ecological effectiveness and ensure that

[92] See Lexmond, *supra* note 44, at Ch. 5.

[93] Miles, et al., *supra* note 30.

environmental actions become learning opportunities via feedback and review mechanisms.

The international environmental community is shifting in positive directions. It is hoped that the next step will be to acknowledge that the only avenue to ecological effectiveness is to embrace sound science and infuse it into policy-making processes to encourage the systematic collection and rational use of reliable information and highlight the most ecologically effective and economically efficient options available. Commensurate with this is the need to empower environmental actors to make, via improved levels of scientific literacy, effective choices. While other aspects of environmental regimes require attention, science and scientific literacy for scientific non-actors are critical components as nature is governed by the immutable laws of science. The growing emphasis on strategy and effectiveness is leading the actors toward science, holding promise for ecologically effective environmental regimes embodying consilience and common sense.

Actors positioned to advance effective regime development should possess attributes that mirror those of the late Douglas M. Johnston, an individual who was the embodiment of consilience. These attributes include a diverse knowledge base, multi-disciplinary perspective, science-literacy, creativity, and open-mindedness. In addition, he or she should be a good communicator (in particular, a good listener), problem solver, lateral thinker, visionary, and grounded in common sense.

International Law and Ocean Dumping: Steering a Precautionary Course Aboard the 1996 London Protocol, but Still an Unfinished Voyage*

David L. VanderZwaag** and Anne Daniel***

Introduction

The oceans cover some 70 per cent of this planet. We are increasingly coming to understand their value and their vulnerability – and *our* vulnerability if we fail to maintain them in a reasonable condition.[1] This contribution examines a global treaty designed to protect the oceans from one source of marine pollution and assesses its contribution to protect our oceans' health.

---

* This contribution attempts to be accurate regarding law and policy developments up to 1 December 2007.

** Director, Marine & Environmental Law Institute, Canada Research Chair in Ocean Law & Governance, and Professor of Law, Dalhousie University, Halifax, Canada. The authors gratefully acknowledge the research assistance of Rebecca Moore, LL.B. 2008, Dalhousie Law School. Professor VanderZwaag also acknowledges the research support of the Social Sciences and Humanities Research Council of Canada.

*** General Counsel, Department of Justice, Ottawa, Canada. The views expressed are those of the author alone and do not reflect the views of the Government of Canada.

1 For a recent description of the challenges facing the oceans, see the Report of the Secretary General, *Oceans and the Law of the Sea*, A/62/66, 12 March 2007. At paragraph 361, he states: "Also as noted in the present report, despite heightened efforts at the global level, coastal and marine ecosystems continue to deteriorate as a result of pressures from human development. Consequently, priority needs to be given to the management of human activities that adversely impact marine ecosystems in order to ensure conservation, sustainable use and development of ocean resources in the interests of present and future generations."

Industrialization in the 1800s and 1900s contributed greatly to the pollution of oceans, and as one author has put it: "The perception of the sea as a convenient and limitless waste dump persists even today amongst many sectors of the global community."[2] Dredging in the early 20th century, increasing use mid-century of the oceans to dump waste, especially military waste after World War II, and rapid industrialization through to the 1970s led to substantial increases in ocean dumping. This involved increased sewage sludge from rapidly growing cities, increasing amounts of dredged material as ports were expanded, toxic industrial wastes, and radioactive wastes from an expanding nuclear industry. During the 1960s a specialized industry grew up focused on the loading of wastes on ships for purposes of dumping at sea.[3] Dredges were eventually designed so that they could both dredge and directly dump the dredged material at sea. Again in the 1960s specialized ships "were designed and built for the specific purpose of incinerating... these highly toxic materials at sea."[4]

Although only 10 per cent of ocean pollution is estimated to come from ocean dumping,[5] the deliberate disposal of wastes into the oceans still remains an ongoing management challenge. Large amounts of waste continue to be deposited into marine waters with about 80 to 90 per cent of the material resulting from dredging.[6] Some 100 to 150 million tonnes of dredged materials are dumped annually into internal waters of States, and between 150 and 400 million tonnes are dumped in marine areas beyond internal waters.[7] In 2004, China reported issuing permits authorizing disposal of 190,593,000 tonnes of dredged material in East Asian Seas.[8]

---

[2] *"Barriers to Compliance" Project Final Report*, LC 29/INF.2, p. 1.

[3] *Ibid.*, pp. 1–2.

[4] *Ibid.*, p. 2.

[5] *UN Atlas of the Oceans*, "Dumping at Sea," <www.oceansatlas.com/unatlas/oceandumping-wastes/dumping/dumping_at_sea.htm>. The International Maritime Organization, *The London Dumping Convention: The First Decade and Beyond* (London: IMO, 1991), p. 44.

[6] *UN Atlas of the Oceans*, "Ocean Dumping and Ship Wastes," <www.oceansatlas.org/html/workuses.jsp>.

[7] "Dumping at Sea," *supra* note 5.

[8] IMO, Scientific Group of the London Convention – 30th Meeting and Scientific Group of the London Protocol – 1st Meeting, *Review of Reports on Dumping Permits Issued and Efforts to Improve Reporting: Review and Improvement of Reporting*, LC/SG 30/5 (18 May 2007), Annex, Table 2.2.

The number of ocean dumping permits issued around the globe each year, while difficult to estimate in light of only partial reporting by countries to the International Maritime Organization (IMO),[9] is substantial. For example, IMO's 2004 summary report on issuance of ocean dumping permits documented 1,369 total permits with countries responsible for most permits including China (587), Hong Kong China (127), United Kingdom (125), Canada (85), United States (78), Norway (55), and France (51).[10]

International law and policy responses to ocean dumping might be summarized as progressing through five main steps.[11] First was a narrow and general international focus on the dumping of radioactive wastes. The 1958 Convention on the High Seas[12] through Article 25(1), simply called upon States to "take measures to prevent pollution of the seas from the dumping of radio-active wastes taking into account any standards and regulations which may be formulated by the competent international organizations."

A second step was international recognition on the need for a broader approach to ocean dumping and marine environmental protection. The 1972 Stockholm Declaration[13] in Principle 7 urged States to "take all possible steps to prevent pollution of the seas by substances that are liable to create hazards to human health, to harm living resources and marine life, to damage amenities or to interfere with other legitimate uses of the sea." The Stockholm Action Plan for the Human Environment[14] called upon governments to control ocean dumping by their nationals anywhere or by any person in areas under their jurisdiction. The Action Plan also urged

---

[9] For example, as of 18 May 2007, 45 parties to the global instruments governing ocean dumping had not yet reported on their dumping activities for the year 2004. *Ibid.*, p. 2.

[10] *Ibid.*, Annex, Table 1.

[11] A review of regional ocean dumping agreements is beyond the scope of this chapter. For a partial review, see S. Marr, *The Precautionary Principle in the Law of the Sea: Modern Decision Making in International Law* (The Hague: Martinus Nijhoff, 2003), pp. 124–125.

[12] *Convention on the High Seas*, Geneva, 29 April 1958, 450 *U.N.T.S.* 82.

[13] H. Hohmann, ed., *Basic Documents of International Environmental Law*, Vol. 1 (London: Graham & Trotman, 1992), p. 21.

[14] *Ibid.*, p. 26.

governments to complete and bring into force an over-all instrument for the control of ocean dumping.[15]

The third progression was actual adoption of the London Dumping Convention in 1972[16] and its subsequent implementation where the international approach might be described as following a permissive and assimilative philosophy.[17] Although the Convention prohibits dumping of wastes listed on Annex I, often referred to as the "black list" which includes particularly threatening wastes,[18] the Convention allows various exceptions. Contracting parties can still permit dumping of some Annex I wastes if the substances "are rapidly rendered harmless by physical, chemical or biological processes in the sea provided they do not... make edible marine organisms unpalatable, or... endanger human health or that or domestic animals."[19] Parties can also allow dumping of many black listed wastes, if they are contained as trace contaminants in other materials such as sewage or dredged materials.[20] Even dumping of radioactive wastes may be permitted if they contain *de minimis* levels of radioactivity.[21] A "grey list" set out in Annex II of the Convention includes wastes such as lead, nickel, zinc, cyanides and fluorides, where only special permits should be issued requiring particular care regarding locations and control

---

15 *Ibid.*, p. 44, Recommendation 86.

16 *Convention on the Prevention of Marine Pollution by Dumping of Wastes and Other Matter*, London, 29 December 1972, 11 *I.L.M.* 1291 (1972) [hereinafter London Convention].

17 For further reviews see J. W. Kindt, "Ocean Dumping," *Denver Journal of International Law & Policy* 13 (1983–1984): 335; E. L. Hughes, "Ocean Dumping and Its Regulation in Canada," *Canadian Yearbook of International Law* 26 (1988): 155; and F. F. Richards III, "Ocean Dumping: An International and Domestic Perspective," *Journal of Legislation* 17 (1990–1991): 287.

18 Listed wastes include: organohalogen compounds, mercury and mercury compounds, cadimum and cadimum compounds, persistent plastics and other persistent synthetic materials, crude oil and its wastes, radioactive wastes, biological and chemical warfare materials, incineration at sea of industrial waste and sewage, sludge and industrial waste (as from 1 January 1996).

19 London Convention, *supra* note 16, Annex I, para. 8. The exception may be applied to six categories of waste listed, namely, organohalogen compounds, mercury and mercury compounds, cadimum and cadimum compounds, persistent plastics, crude oil wastes and biological/chemical warfare materials.

20 *Ibid.*, Annex I, para. 9, allows the trace contaminant exception to be applied to the first five listed categories of wastes.

21 *Ibid.*, Annex I, para. 9.

measures.[22] All other wastes may be dumped subject to general permits which shall only be issued after considering all the factors listed in Annex III such as toxicity, persistence, tainting potential and possible effects on other uses of the sea. An underlying premise is that almost any waste may be dumped in the ocean, except where the waste because of its toxicity and concentration would exceed the assimilative capacity of the marine environment.[23]

A fourth step forward was inclusion of Article 210 in the 1982 United Nations Convention on the Law of the Sea[24] (LOS Convention) which cast a potentially wide legal net. All parties to the Convention must adopt laws and regulations to control ocean dumping through a permitting process. National laws and measures are to be no less effective than global rules and standards for ocean dumping. Thus, parties to LOS Convention would be implicitly bound by London Convention standards for ocean dumping even if not a contracting party to that specific Convention.[25] The same may hold true for the Protocol, something that might be of significance until the Protocol attracts a more universal membership.

The fifth transition is the embracement of precautionary and pollution prevention approaches to ocean dumping through the 1996 Protocol to the London Convention[26] which is the focal point of this essay. The next section provides an overview of the precautionary thrusts of the Protocol supported through a "reverse listing approach" whereby only those wastes included on a global "safe list" may be considered for ocean dumping.

The following section of this contribution highlights the "sea of issues" that still must be navigated in the Protocol's implementation. Those issues include: exclusions and interpretation of the Protocol; clarification through

---

22 *Ibid.*, Art. IV(1)(b).

23 That said, the ethos of the Convention gradually changed over time and the dumping of radioactive waste and incineration at sea were eventually prohibited. See L. de La Fayette, "The London Convention 1972: Preparing for the Future," *International Journal of Marine and Coastal Law* 13 (1998): 515 at 517.

24 *United Nations Convention on the Law of the Sea*, Montego Bay, 10 December 1982, UN Doc.A/Conf. 62/122; 21 *I.L.M.* 1261 (1982).

25 O. Schrane Stokke, "Beyond Dumping? The Effectiveness of the London Convention," *Yearbook of International Co-operation on Environment and Development* (1998–1999): 39 at 40.

26 *Protocol to the Convention on the Prevention of Marine Pollution by Dumping of Wastes and Other Matter*, London, 7 November 1996, 36 *I.L.M.* 1 (1996) [hereinafter London Protocol].

additional guidance documents; regulation of "new uses" such as carbon sequestration in the seabed and ocean fertilization; provision for liability and compensation for harms resulting from ocean dumping; ensuring compliance; promotion of technical and financial assistance to developing countries; and achievement of universal acceptance of the Protocol. The Conclusion summarizes the strengths and challenges of the 1996 Protocol and suggests how the Protocol could be a model for addressing other international environmental challenges such as the further regulation of toxic chemicals.

The 1996 Protocol to the London Convention: Steering a Precautionary Course

Adopted by the Special Meeting of Contracting Parties to the London Convention 1972 on 7 November 1996, the Protocol promotes a precautionary approach to ocean dumping in three key ways. First, the Protocol imposes a general obligation on all parties to follow a precautionary approach in the control of ocean dumping. Article 3(1) provides:

> In implementing this Protocol, Contracting Parties shall apply a precautionary approach to environmental protection from dumping of wastes or other matter whereby appropriate preventative measures are taken when there is reason to believe that wastes or other matter introduced into the marine environment are likely to cause harm even when there is no conclusive evidence to prove a causal relation between inputs and their effects.

A second precautionary thrust is adoption of a reverse listing approach to ocean dumping. Article 4(1) requires parties to prohibit dumping of any wastes or other matter except for those wastes/materials listed under Annex I. The original Annex I list of what may be considered as acceptable for dumping included seven categories:

- dredged material;
- sewage sludge;
- fish waste, or material resulting form industrial fish processing operations;
- vessels and platforms or other man-made structures at sea;
- inert, inorganic geological material;
- organic material of natural origin; and

- bulky items primarily comprising iron, steel, concrete and similarly non-harmful materials for which the concern is physical impact, and limited to those circumstances where such wastes are generated at locations, such as small islands with isolated communities, having no practicable access to disposal options other than dumping.

At the first Meeting of Contracting Parties to the London Protocol in November 2006, Annex I was amended to also allow carbon dioxide sequestration in geological formations under the seabed.[27] The amendment entered into force on 10 February 2007.[28]

A third avenue towards precaution is imposition of precautionary waste assessment responsibilities even for those wastes on the Annex I "safe list." Annex II, entitled "Assessment of Wastes or Other Matter That May Be Considered for Dumping," urges parties to require ocean dumping applicants to undertake waste prevention audits to identify opportunities for waste prevention at source and to formulate waste prevention strategies.[29] Permitting authorities are required to refuse issuing waste dumping permits if appropriate opportunities exist to re-use, recycle or treat the waste without undue risks to human health or the environment or disproportionate costs.[30] Permits are also to be refused where a waste is so poorly characterized that proper assessment cannot be made of its potential impacts on human health and the environment.[31]

The Protocol also contains a new requirement for the parties to establish procedures and mechanisms to assess and promote compliance with the Protocol and to establish a subsidiary body to evaluate new reports that are required on Protocol implementing measures and their effectiveness.[32] The Protocol also enables developing countries that want to ratify the

---

[27] IMO, *Report of the Twenty-eighth Consultative Meeting and the First Meeting of Contracting Parties*, LC 28/15 (6 December 2006), Annex 6 [hereinafter Report of the 28th Consultative Meeting].

[28] Except 29 January 2007 for Canada. IMO, *Status of the London Convention and Protocol*, LC 29/2/1 (10 September 2007).

[29] London Convention, *supra* note 16, Annex II, ss. 2–3.

[30] *Ibid.*, Annex II, s. 6.

[31] *Ibid.*, Annex II, s. 7.

[32] London Protocol, *supra* note 26, Arts. 11 and 9.5 respectively.

Protocol and cannot immediately comply with its provisions to ratify without having to immediately meet all of their obligations.[33]

Still an Unfinished Voyage: A Sea of Issues

Exclusions and Interpretation of the Protocol

Interpretation of three categories of ocean disposal excluded from Protocol coverage has been problematic. Those exclusions include: placement for a purpose other than mere disposal; wastes incidental to the normal operations of a ship; and industrial waste and inert, inorganic geological material.

Placement for a Purpose Other Than Disposal

The 24th through 27th Consultative Meetings discussed the matter of the exclusion from dumping, found in both the Convention and Protocol, of "placement of matter for a purpose other than the mere disposal thereof, provided that such placement is not contrary to the aims of the" Convention or Protocol, as the case may be. No consensus could be achieved, and the parties had to settle for agreement on a number of elements of policy guidance: (1) placement should not be used as an excuse for disposal at sea of waste materials; (2) placement should not be contrary to the aims of the Convention; (3) information on placement activities by parties should be provided to the Secretariat as available; and (4) materials used for placement activities should be assessed in accordance with the relevant Specific Guidelines.[34]

At the 27th Consultative Meeting, it was agreed that the Scientific Group of the London Convention would continue to keep placement on the agenda, with a focus on specific guidance for placement of artificial reefs on the basis of information-gathering by the United Nations Environment Programme (UNEP) on current national and regional artificial reef

[33] *Ibid.*, Art. 26. See de La Fayette, *supra* note 23, for a description of the key features of the 1996 Protocol.

[34] The discussion was contained in para. 155 of the Report of the 28th Consultative Meeting, *supra* note 27.

guidelines, practices and cases studies. Spain is now leading this work, and the guidelines are scheduled to be proposed for adoption in 2008.[35]

As an illustration of the kind of issues involved, at the 28th Consultative Meeting Greenpeace International raised the question of the "placement" of a former U.S. Navy ship 24 miles off the coast of Florida, which it alleged had not been properly cleaned of PCBs. Greenpeace was of the view that such reefing was contrary to the aims of the Convention. The United States made a detailed reply, indicating that extensive clean-up of the ship had taken place and there was ongoing monitoring in place. The PCBs left in place in solid materials, such as electrical cable and gaskets, would not pose an unreasonable risk of injury to human health or the environment. However, the United States also reminded the meeting of the 13th Consultative Meeting where the general view was that it is up to each party to determine, on a case-by-case basis, whether an activity would be contrary to the aims of the Convention. Further, the United States disagreed with Greenpeace International's view that the phrase "contrary to the aims of the Convention" imposed on parties undertaking placement activities the equivalent of the substantive obligations placed on ocean dumping activities.[36]

The primary difficulty with the placement exemption – apart from some of the confused discussion around its interpretation – is that it has the potential to be a loophole that swallows both the Convention and the Protocol. At the national level, it is the permitting authority that has to be consulted by regulatees to ensure they obtain permits for dumping when required. While regulatees can argue that certain activities are placement, it will of course be up to the national permitting authority not to allow placement that is not in accordance with the aims of the Convention or Protocol. However, there may be some activities about which the governing bodies can generate consensus on what is or is not contrary to the aims of the Protocol, in order to establish consistent collective approaches to placement activities, given their exclusionary nature. This has already generated discussion in the context of iron fertilization of the oceans, discussed below.[37]

---

[35] The draft guidelines are contained in, *Development of Global Guidance for the Placement of Artificial Reefs*, LC/SG/30/4.

[36] See paras. 161–170, Report of the 28th Consultative Meeting, *supra* note 27.

[37] See "Ocean Fertilization," *infra*.

Normal Operations of a Ship

The exclusion from the definition of "dumping" commonly referred to as the "normal operations of a ship,"[38] was inserted to help draft an intelligible line between the scopes of the Convention/Protocol and the International Convention for the Prevention of Pollution from Ships (MARPOL),[39] assuming proper regulation by each instrument in its sphere. Through its annexes, MARPOL controls oil, noxious substances, sewage and, in Annex V, garbage.[40] However, over the years, there have been examples of situations where ships have claimed as "normal operations" increasingly egregious examples, such as the throwing overboard of old furniture as a ship heads into port to be refurbished.[41]

The 26th Consultative Meeting considered two issues related to the "boundary" between the London Convention/Protocol and MARPOL and agreed to draw these to the attention of the Marine Environment Protection Committee (MEPC) of the IMO. As a result of a positive response from MEPC 53 a joint correspondence group was established under the lead of Canada. As a result of this work, the group concluded that the approach to manage spoilt cargo would in most cases fall under the London Convention and London Protocol, and the group is proceeding to develop practical guidance for mariners to manage spoilt cargoes. This working group is liaising closely with the MEPC review of MARPOL Annex V.

Among the concerns with spoilt cargo that the Working Group has considered are the differences in requirements under the London

---

38 Under the Convention and Protocol, dumping does not include "the disposal into the sea of wastes or other matter incidental to, or derived from the normal operations of vessels, aircraft, platforms or other man-made structures at sea and their equipment, other than wastes or other matter transported by or to vessels, aircraft, platforms or other man-made structures at sea, operating for the purpose of disposal of such matter or derived from the treatment of such wastes or other matter on such vessels, aircraft, platforms or other man-made structures" (Art. 1.4.2.1 of the Protocol, *supra* note 26, and Art. III(1)(b)(i) of the Convention, *supra* note 16).

39 *International Convention for the Prevention of Pollution by Ships as modified by the Protocol of 1978*, London, 2 November 1973 <//sedac.ciesin.columbia.edu/entri/texts/pollution.from.ships.1973.htm> [hereinafter MARPOL].

40 *Ibid., Annex V: Regulations for the Prevention of Pollution by Garbage from Ships.*

41 See: Evaluation of the exemption for discharges occurring as part of the normal operations of vessels, platforms, etc. under Art. III(1)(b)(i) of the London Convention 1972 and Art. 1.4.2.1 of the 1996 Protocol, submitted by Canada, LC26/6/2, at para. 2.

Convention/Protocol and MARPOL for managing spoilt cargo and cargo residues; specific materials that could be considered for disposal as spoilt cargo and measures for managing small quantities; measures for managing temporary storage, should the ship be unsuitable for carrying out dumping of the spoilt cargo; mitigating measures to address invasive species and pathogenic organisms; options available should a permit not be allowed; who should apply for and receive a permit, as well as considerations for pre-arranging a permit; and reviewing a permit and setting its conditions, taking into account Annex II of the Protocol and the Specific Guidelines for Assessment of Organic Material of Natural Origin.[42]

The question of how to treat wastes resulting from hull scrapings, which can yield substantial amounts of toxic substances, has become of serious concern due to the entry into force (17 September 2008) of the International Convention on the Control of Harmful Anti-fouling Systems on Ships,[43] which requires the replacement or overcoating of existing organotin-based anti-fouling systems.[44] An intersessional correspondence group has been preparing advice for consideration of the Scientific Groups of the Convention and Protocol on the management of waste streams resulting from the removal of such anti-fouling systems from ships.[45]

The importance of ongoing monitoring of the use of this exception cannot be understated. If practices are argued to fall under MARPOL, but are not regulated by MARPOL, this can create a very large gap between it and the two ocean disposal treaties. The implications are even further reaching when one considers that the Basel Convention on the Transboundary Movement of Hazardous Waste and their Disposal (Basel Convention) also excludes from its application wastes derived from the normal operations of a ship.[46]

---

[42] Discussed below under "Specific Guidelines."

[43] *International Convention on the Control of Harmful Anti-fouling Systems on Ships*, London, 5 October 2001 <//www.environment.gov.au/coasts/pollution/antifouling/pubs/convention.pdf>.

[44] *Report of the Twenty-ninth Consultative Meeting and the Second Meeting of Contracting Parties*, LC 29/17 (14 December 2007), para. 7.6 [hereinafter Report of the 29th Consultative Meeting].

[45] *Ibid.*, paras. 7.8–7.10.

[46] Article 1.4 of the Basel Convention provides: "Wastes which derive from the normal operations of a ship, the discharge of which is covered by another international instrument, are excluded from the scope of this Convention." *Basel Convention on the*

Industrial Waste

The London Convention has prohibited the dumping of industrial waste since January 1, 1996. It provides in Annex I that the term "industrial waste means waste materials generated by manufacturing or processing operations and does not apply to," among other things, "uncontaminated inert geological materials the chemical constituents of which are unlikely to be released into the marine environment." The Protocol prohibits all dumping of wastes or other matter except for those items listed on the reverse list. The reverse list includes "inert, inorganic geological material."

From 1999 to 2005, the Consultative Meeting discussed interpretations of "industrial waste," and conditions under which materials exempted from the definition were eligible for dumping under the Convention. Japan in particular had been continuously questioned about its dumping of bauxite residues or "red mud," as most other parties considered it, as industrial waste, to be ineligible for dumping. However, as complete consensus was not reached on the interpretation of industrial waste in this context, in order to try and move forward on the issue it was agreed that the Scientific Group would develop pre-screening criteria to assess inert materials as part of the Specific Guidelines for Assessment of Inert, Inorganic, Geological Materials.[47] In the meantime, Japan had committed to the cessation of dumping of bauxite residues by 2015. However, they reported to the 2007 Scientific Group on progress and indicated that they are examining the possibility of exporting some of the red mud to ensure it meets this deadline, as it cannot do so solely by the development of alternative uses of the bauxite residues.[48]

Controversy has also arisen over the decision of the Cook Islands government to permit the dumping of asbestos-laden building materials into the ocean. The government allowed the scuttling at sea of a ship reportedly filled with 300 tonnes of cement sheeting containing asbestos in the belief that the disposal would meet the uncontaminated, inert geological material exception.[49]

Transboundary Movement of Hazardous Wastes and Their Disposal, Basel, 22 March 1989 <www.basel.int/text/documents.html>.

47 For a further discussion, see "Specific Guidelines," *infra*.

48 Report of the 30th Meeting of the Scientific Group, LC/SG/30/14, paras. 3.73–3.77.

49 "Cook Islands Defends Asbestos Dumping Decision," *Pacific Magazine*, 7 November 2007, <www.pacificmagazine.net/news/2007/11/07/Cook-Islands-defends-asbestos-dumping-decision>.

Clarification through Guidance Documents

Besides a generic waste assessment guidance document adopted in 1997,[50] a set of eight specific guidelines were developed by the Scientific Group during the period 1997–2001. They include: Specific Guidelines for Assessment of Dredged Material; Specific Guidelines for Assessment of Sewage Sludge; Specific Guidelines for Assessment of Fish Waste, or Material Resulting from Fish Processing Operations; Specific Guidelines for Assessment of Vessels; Specific Guidelines for Assessment of Platforms or Other Man-made Structures at Sea; Specific Guidelines for Assessment of Inert, Inorganic Geological Material; Specific Guidelines for Assessment of Organic Material of Natural Origin; and Specific Guidelines for Assessment of Bulky Items Primarily Comprising Iron, Steel, Concrete, etc.[51]

Given that many materials contain natural radionuclides and are often contaminated with artificial radionuclides such as from nuclear testing fallout, the need arose to define *de minimis* levels of radionuclides that could be disposed of in the oceans. Guidelines for the Application of the *De Minimis* Concept under the London Convention 1972 were adopted in 1999 and amended in 2003.[52]

The various guidelines are meant to be "living documents" subject to review and updating and the door is open for development of further guidelines. At the 29th Consultative Meeting of Parties to the London Convention and the Second Meeting of Parties to the 1996 Protocol in November 2007, parties adopted Specific Guidelines for Assessment of Carbon Dioxide Streams for Disposal into Sub-seabed Geological

---

[50] "Guidelines for the Assessment of Wastes or Other Matter That May Be Considered for Dumping," <www.imo.org/dynamic/mainframe.esp?topic_id=1503&doc_id=7550> [hereinafter Generic Guidelines].

[51] The Guidelines may be found in Annexes 3–10 respectively in the Report of the Twenty-Fourth Meeting of the Scientific Group, LC/SG 24/11 (24 July 2001) [hereinafter Report of the 24th SG Meeting]. The status of all the Specific Guidelines was confirmed by the First Meeting of Contracting Parties to the 1996 Protocol and the guidelines were formally adopted under the London Protocol. Report of the 28th Consultative Meeting, *supra* note 27, at para. 238.

[52] "Guidelines for the Application of the *De Minimis* Concept under the London Convention 1972," <www.imo.org/home.asp?topic_id=1488>.

Formations.[53] Guidelines for the Placement of Artificial Reefs are in the process of development.[54]

Generic Guidelines

The Guidelines for the Assessment of Wastes or Other Matter That May Be Considered for Dumping are meant to guide national authorities in evaluating applications for dumping wastes not prohibited under the London Convention or the 1996 Protocol. The Guidelines repeat much of the text from Annex II to the 1996 Protocol with some further details on such issues as dump-site selection, assessment of potential effects and permitting. A checklist of factors which should influence dump-site selection is set out, such as proximity to shoreline and bathing beaches, fishing areas, migration routes, seasonal and critical habitats, shipping lanes and military exclusion zones.[55]

Additional guidance on assessment of potential effects reemphasizes the need for precautionary and preventive approaches and encourages assessment of cumulative effects. Permitting authorities are urged to give preference to waste management techniques preventing the input of contaminants into the environment.[56] Authorities are also urged to consider cumulative effects of repeated dumping at a site and possible interactions with other waste dumping practices in the area.[57]

Guidance on permitting leaves considerable discretion to national authorities. Public review and participation in the permitting process is merely recommended.[58] Regulators are urged to keep environmental changes "as far below the limits of allowable environmental change as practicable, taking into account technological capabilities as well as economic, social and political concerns."[59] Reviewing permits "at regular intervals" is also called for.[60]

---

[53] Report of the 29th Consultative Meeting, *supra* note 44, at para. 4.13, Annex 4.
[54] *Ibid.*, paras. 8.1–8.10.
[55] Generic Guidelines, *supra* note 50, at para. 19.
[56] *Ibid.*, para. 29.
[57] *Ibid.*, para. 36.
[58] *Ibid.*, para. 47.
[59] *Ibid.*, para. 48.
[60] *Ibid.*, para. 49.

Specific Guidelines

While the nine specific guidance documents adopted to date have many common elements including substantial text drawn from the generic guidelines, each guidance document contains specific recommendations and directions for the particular waste being addressed. Specific Guidelines for Assessment of Dredged Material suggest various beneficial uses for dredged material which should be considered as alternatives to ocean dumping, such as land creation, beach nourishment and construction material.[61] Capping contaminated dredged material with clean sediment is suggested as one of many management techniques.[62]

Specific Guidelines for Assessment of Sewage Sludge clarify that the intent is to cover dumping of residues remaining from the treatment of municipal sewage.[63] Permitting authorities are encouraged to consider alternatives to dumping, such as use in agriculture or silviculture, production of energy and disposal on land.[64]

Guidelines for Assessment of Fish Waste clarify the intent to address only primary and secondary fish processing wastes and not discharges from ships actively fishing.[65] Applicants for ocean dumping permits are required to consider other options for waste management such as reprocessing into fishmeal, use as food for domestic animals, transformation in biochemical industry products and use as fertilizer in land farming.[66]

Guidelines for the Assessment of Vessels are meant to cover proposed dumping of "any waterborne or airborne craft of any type whatsoever"[67] but permitting authorities have discretion to determine the minimum vessel size to which the Guidelines will apply.[68] Consideration of alternative waste management options to dumping is encouraged, for example, re-use of a vessel, recycling and disposal on land.[69] Development of a pollution prevention plan prior to dumping is also called for to assure that wastes,

---

61 Report of the 24th SG Meeting, *supra* note 51, at Annex 3, para. 3.2.
62 *Ibid.*, para. 3.4.
63 Report of the 24th SG Meeting, *supra* note 51, at Annex 4, para. 15.
64 *Ibid.*, para. 3.2.
65 Report of the 24th SG Meeting, *supra* note 51, at Annex 5, para. 1.4.
66 *Ibid.*, para. 3.1.
67 Report of the 24th SG Meeting, *supra* note 51, at Annex 6, para. 1.4.
68 *Ibid.*, para. 1.6.
69 *Ibid.*, para. 3.1.

such as oils, chemicals, sludges and materials capable of creating floating debris, have been removed to the "maximum extent" in order to avoid pollution of the marine environment.[70]

Guidelines for Assessment of Platforms or other Structures at Sea are primarily aimed at disposal of oil and gas platforms[71] and once again highlight the importance of considering waste management options other than ocean dumping. Those options include re-use at sea or on shore, recycling and final disposal on land.[72] Regulators are to ensure platforms/structures to be disposed at sea have been cleaned of petroleum hydrocarbons or other substances likely to cause harm to the marine environment to the "maximum extent" (within technical and economic feasibility and taking into consideration the safety of workers).[73] Materials capable of creating floating debris are also to be removed.[74]

Guidelines for Assessment of Inert, Inorganic Geological Material leave considerable leeway for what materials might fall into the category. National authorities are urged to specify the basis on which a material is characterized as geological and inert in the marine environment. The category is meant to cover those geological materials where the only effects will be due to physical properties rather than chemical in nature.[75] Alternatives to ocean dumping are urged for consideration including re-use, recycling and disposal on land.[76]

Specific Guidelines for Assessment of Organic Material of Natural Origin are intended to apply to animal and vegetable matter predominantly of agricultural origin but also to the dumping of spoilt organic cargoes at sea such as beef, potatoes, rice, corn and bananas.[77] Applicants for ocean dumping permits are required to consider waste management alternatives such as composting, use as animal food and disposal on land.[78]

Guidelines for Assessment of Bulky Items are meant to cover items primarily comprising iron, steel, concrete and similarly non-harmful materials

[70] *Ibid.*, para. 4.1.
[71] Report of the 24th SG Meeting, *supra* note 51, at Annex 7, para. 1.4.
[72] *Ibid.*, para. 7.2.
[73] *Ibid.*, para. 5.2.
[74] *Ibid.*
[75] Report of the 24th SG Meeting, *supra* note 51, at Annex 8, para. 4.1.
[76] *Ibid.*, para. 3.1.
[77] Report of the 24th SG Meeting, *supra* note 51, at Annex 9, para. 1.4.
[78] *Ibid.*, para. 3.1.

for which concern is physical impact.[79] The Guidelines are limited to where such wastes are generated at locations, such as small islands with isolated communities, having no practicable access disposal options other than dumping.[80] Permitting authorities are urged to consider ways to reduce bulk and the possible need for treatment to reduce or remove hazardous constituents.[81] To ensure bulky items reach the sea floor relatively rapidly, the specific gravity of such materials should exceed 1.2 when allowance of the ingress of water into internal and void spaces has been made.[82]

Guidelines for the Assessment of Carbon Dioxide Streams for Sub-seabed Disposal, which were approved at the Second Meeting of Contracting Parties, highlight the risks associated with carbon dioxide sequestration in geological formations including leakage of carbon dioxide into the marine environment and possibly other substances, such as heavy metals, in or mobilized by the carbon dioxide stream.[83] Consideration of waste management alternatives is urged, such as greenhouse gas emission reductions and land-based underground storage.[84] Carbon dioxide streams must consist "overwhelmingly" of carbon dioxide consistent with the purpose of reducing greenhouse emissions.[85] A geological risk assessment is to be required in order to establish the feasibility and integrity of the injection site.[86]

Other Guidance

Guidelines for the Application of the *De Minimis* Concept under the London Convention 1972 provide advice on what categories of materials containing radionuclides might be eligible for ocean dumping. Some materials, such as natural radionuclides in raw materials and widely-distributed radionuclides resulting from global fallout and accidents, may be automatically defined as *de minimis* while other materials may require a specific assessment to determine if radiological criteria to protect human

---

79 Report of the 24th SG Meeting, *supra* note 51, at Annex 10, para. 1.4.
80 *Ibid.*
81 *Ibid.*, para. 3.1.
82 *Ibid.*, para. 4.1.
83 Report of the 29th Meeting, *supra* note 44, at Annex 4.
84 *Ibid.*, para. 3.2.
85 *Ibid.*, para. 5.4.
86 *Ibid.*, paras. 7.1–7.14.

health are met.[87] Since no international radiological criteria have been developed for the protection of flora and fauna, permitting authorities are urged to use a precautionary approach in conducting assessments of the potential impacts on the marine environment.[88]

Guidance on the National Implementation of the 1996 Protocol[89] has also been prepared. The guidance document provides a checklist by key elements that need to be considered for implementation in national legislation and/or regulations, such as establishing a permit-system instituting procedures for emergency dumping, setting out offences and penalties, facilitating scientific and technical research and establishing monitoring requirements.

Regulation of New Uses

Carbon Sequestration

Through the sustained efforts of the United Kingdom, Norway and Australia in particular, an amendment to the Protocol was adopted after a vote at the First Meeting of Contracting Parties. This amendment adds to the reverse list "Carbon dioxide streams from carbon dioxide capture processes for sequestration." Additional text provides that carbon dioxide streams may only be considered for dumping if (1) disposal is into a sub-seabed geological formation and (2) they consist "overwhelmingly" of carbon dioxide. They may contain "incidental" associated substances derived from the source material and the capture and sequestration processes used, and (3) no wastes or other matter are added for the purpose of disposing of those wastes or other matter.[90]

There was intense, lengthy debate at the First Meeting of Contracting Parties to the Protocol, both in plenary and later in a working group. The three main proponents pushed hard to have their proposal adopted, while other delegations argued that it was premature and more work was needed before any such adoption.[91]

---

87 De Minimis Guidelines, *supra* note 52, at para. 2.3.

88 *Ibid.*, para. 3.3.

89 Report of the 28th Meeting, *supra* note 27, at Annex 12.

90 Report of the 28th Consultative Meeting, *supra* note 27, at Annex 6. This added a new paragraph 1.8 to paragraph 1 of Annex 1 and a new paragraph 4 with the additional conditions.

91 *Ibid.*, para. 81.

The main arguments by the proponents were that this was just one option in a portfolio of options for combating climate change, that many projects were under consideration around the world, the technology was promising, and any further development of it would not occur without the amendment – and that it was consistent with the precautionary approach of the Protocol.[92]

Those against expressed the view that there were still too many scientific uncertainties regarding site selection, acceptable leakage rate, long-term monitoring and issues regarding the purity of the captured carbon dioxide. It was also argued that the specific guidelines for assessment of permits for such sequestration should be developed before any decision could be made on the proposed amendment – and waiting for this would be in line with the precautionary approach.[93]

There was considerable debate on whether and how to define the word "overwhelmingly," with Greenpeace arguing that 99.9 per cent should be utilized.[94] This led to disagreement about the scientific basis for selecting any particular percentage,[95] and ultimately no percentage was adopted in the amendment.

Resolution LP.1(1) provided not only for the amendment of the Protocol, but requested the Scientific Group of the Protocol to develop specific guidance on the application of Annex 2 to the Protocol to carbon dioxide sequestration in sub-seabed geological formations for consideration at the Second Meeting of Protocol Parties. A number of delegations were concerned that adopting the amendment before the related guidance was finalized would make it more difficult to resolve questions around "overwhelmingly" in a manner that was sufficiently rigorous.[96]

Similarly, paragraph 3 of the resolution invited parties that issued permits before the specific guidance was completed, to use the best available guidance relevant to carbon dioxide sequestration in sub-seabed geological formations to ensure the protection of the marine environment, while also recognizing the requirements of Annex 2 to the London Protocol. Earlier

92 *Ibid.*, para. 82.
93 *Ibid.*, para. 83.
94 *Ibid.*, para. 80; the paper was LP1/6/2.
95 *Ibid.*, para. 89.3.
96 See comments by Denmark, *ibid.*, para. 95.

language had been stronger, but did not survive. Nevertheless, any parties issuing permits were put on notice.

The intense debate in the Working Group carried into the plenary, where it became clear that there would be no consensus on the matter. Australia called for a vote, and then Denmark insisted on a roll call. Because there were only 29 parties to the Protocol at that time, a quorum consisted of 17. When the vote was called, 12 countries supported the resolution, and 5 abstained. Given the voting rule requiring a two-thirds majority, the proponents were indeed fortunate that other parties abstained instead of voting against the proposal.[97] The amendment automatically entered into force for each party to the Protocol on 10 February 2007, except for Canada, which notified its acceptance earlier, to which the amendment applied as of 29 January 2007.

Ocean Fertilization

Originally raised at the 2007 Scientific Group as an issue of iron fertilization, the issue had broadened at the Consultative Meeting/Meeting of Contracting Parties to be expressed as a concern about ocean fertilization in general.

Iron fertilization of the oceans to stimulate phytoplankton growth in order to sequester carbon dioxide has been of recent commercial interest due to carbon markets. The concern of the Scientific Group was stimulated by the existence of a proposal that came to light through the United States about a for-profit company planning to dissolve iron dust in a 100 by 100 kilometre area of ocean off the Galapagos Islands. The Scientific Groups of the Convention and Protocol advised "caution" in relation to large-scale iron fertilization and issued a Statement of Concern. They declared that knowledge about the effectiveness and potential environmental impacts of ocean iron fertilization are currently insufficient to justify large-scale operations. Further, they noted that the Intergovernmental Panel on Climate Change (IPCC) has indicated that fertilization of the oceans may offer a potential strategy for removing carbon dioxide from the atmosphere by stimulating the growth of phytoplankton and thereby sequestering the carbon dioxide in the form of particulate organic

[97] *Ibid.*, para. 101, in support: Australia, Canada, France, Germany, Mexico, New Zealand, Norway, Saudi Arabia, Spain, Sweden, United Kingdom, and Vanuatu. Abstaining: Belgium, China, Denmark, Egypt, and South Africa.

carbon. However, the Scientific Groups also noted the IPCC's statement that ocean iron fertilization remains largely speculative, with many of the environmental impacts yet to be assessed.[98]

The Scientific Groups noted the potential for large-scale ocean iron fertilization to have negative impacts on the marine environment and human health. They therefore recommended that any such operations be evaluated carefully to ensure, among other things, that such operations were not contrary to the aims of the London Convention and Protocol.[99]

The Scientific Groups also suggested that the following need to be taken into account in evaluating such operations: the estimated amounts and potential impacts of iron and other materials that may be released with the iron; the potential impacts of gases that may be produced by the expected phytoplankton blooms or by bacteria decomposing the dead phytoplankton; the estimated extent and potential impacts of bacterial decay of the expected phytoplankton blooms, including reduced oxygen concentrations; the types of phytoplankton that are expected to bloom and the potential impacts of any harmful algal blooms that may develop; the nature and extent of potential impacts on the marine ecosystem including naturally occurring marine species and communities; the estimated amounts and timescales of carbon sequestration, taking account of partitioning between sediments and the water; and the estimated carbon mass balance for the operation.[100]

The two Scientific Groups requested the 2007 meetings of parties to the Convention and Protocol to consider the issue of large-scale ocean iron fertilization operations with a view to ensuring adequate regulation of such operations. In particular, they asked parties to focus on: (1) the purposes and circumstances of proposed large-scale ocean iron fertilization operations and whether these are compatible with the aims of the Convention and the Protocol; (2) the need, and potential mechanisms, for regulation of such operations; and (3) the desirability of bringing to the attention of other international instruments and institutions proposals for such operations. In the meantime, they invited parties to the Convention and

---

98 IMO Circular 14, 13 July 2007, "Convention on the Prevention of Marine Pollution by Dumping of Wastes and other Matter, 1972 and Its 1996 Protocol, Statement of Concern regarding iron fertilization of the oceans to sequester carbon dioxide," para. 1.

99 *Ibid.*

100 *Ibid.*, para. 2.

Protocol to take this statement of concern into account when considering experimental or large-scale ocean iron fertilization to sequester carbon dioxide, and to provide further information on such operations to the secretariat and to the Scientific Groups when it becomes available.[101]

The parties to both instruments responded equally cautiously. Since the issuance of the June Statement of Concern, a scientific symposium had been held at the Woods Hole Oceanographic Institution in the United States, and a meeting held by the Member States of the Permanent Commission for the South Pacific (La Comisión Permanente del Pacífico Sur – CPPS) (Chile, Colombia, Ecuador, Panama and Peru).[102] At the meetings of the governing bodies, this information was made available. Vanuatu also made a strong statement that the focus should be on reducing carbon dioxide emissions at source instead of looking for additional and potentially dangerous interim carbon dioxide sequestration solutions.

The governing bodies: (1) endorsed the Statement of Concern of the Scientific Groups; (2) agreed that the scope of work of the London Convention/Protocol include ocean fertilization, as well as iron fertilization; (3) agreed that the London Convention/Protocol are competent to address this issue due to their general objective to protect and preserve the marine environment from all sources (Article 1 of the Convention, Article 2 of the Protocol); (4) agreed that they would further study the issue from a scientific and legal perspective with a view to its regulation; and (5) recognized that it was within the purview of each State to consider proposals on a case-by-case basis in accordance with the London Convention and Protocol and urged States to use the utmost caution when considering proposals for large-scale ocean fertilization operations. The governing bodies also took the view that, given the present state of knowledge regarding ocean fertilization, such large-scale operations were currently not justified.

It is important to note that after discussion, and other proposals being discussed, the statements being made were broadened to ocean fertilization

101 *Ibid.*, para. 4. For a review of the various iron fertilization experiments proposed around the globe, see E. Kintisch, "Carbon Sequestration: Should Oceanographers Pump Iron?" *Science* 318: 5855 (30 November 2007): 1368.

102 Report of the 29th Consultative Meeting, *supra* note 44, at paras. 4.17 and 4.18. CPPS implements the "Action Plan for the Protection of the Marine Environment and Coastal Area of the South-east Pacific."

more generally.[103] The reference to "large-scale" was intended to reflect the fact that appropriate research could still go forward. The statement in (5) was crafted with a "case-by-case" emphasis as a result of concern on the part of some regulators that they have to evaluate each proposal for a dumping permit on the basis of domestic law and could not agree based on the current level of information that no permits should be allowed while the issue is being studied further.

The intersessional study by the Scientific Groups and an intersessional legal correspondence group will be based on several fact scenarios, resulting in reports to be considered by the governing bodies in 2008. Legal issues that need to be canvassed, based on specific fact situations are as follows: (1) whether the application of the Convention and Protocol will be different; (2) whether this is dumping at the first instance; (3) if so, whether the exclusion of "placement" applies; (4) in discussing that exclusion, one question is whether seeding the oceans in this manner is even an act of "placement"; (5) if it is placement, whether it is contrary to the aims of either the Convention or Protocol, given Articles 1 and 2 respectively, and given Article 3 in particular of the Protocol; and (6) the relevance of the LOS Convention, particularly for parties thereto who are not parties to the Protocol.

It will be interesting to see whether, at the next meetings of the governing bodies, parties will be prepared to reach a collective decision on whether the placement exclusion should be taken off the table. The overarching tension regarding placement, as noted elsewhere in this contribution, is that some parties believe that placement is a decision that should be left to national governments to decide on a case-by-case basis, while others are of the view that there are circumstances in which all parties should agree that certain activities are contrary to the aims of the instruments.[104]

---

[103] An example of a broader ocean fertilization issue is the proposed dumping of nitrogen-rich urea by an Australian company in the Sulu Sea off the Philippines. See J. Aning, "WWF opposes dumping of urea in Sulu Sea," <newsinfo.inquirer.net/breakingnews/nation/view_article.php?article_id=106151>.

[104] For further review of the potential role of the London Convention/Protocol in controlling ocean fertilization and the related uncertainties, see R. Rayfuse, M. G. Lawrence and K. M. Gjerde, "Ocean Fertilization and Climate Change: The Need to Regulate Emerging High Seas Uses," *International Journal of Marine and Coastal Law* 23 (2008): 297–326.

Provision for Liability and Compensation

Both the London Convention and the 1996 Protocol contain a provision on liability, but little has been done under either instrument. Under the Convention, interest initially arose in this subject around the time when the radioactive wastes issue became of collective concern. Resolution LDC.21(9) adopted at the 9th Consultative Meeting in 1985, suspended the dumping of low-level radioactive wastes and other radioactive matter at sea. The resolution also "called upon contracting Parties to develop, as envisaged in Article X, procedures for the assessment of liability. . . ." Article 15 of the Protocol provides that, in accordance with the principles of international law regarding State responsibility for damage to the environment of other States or to any other area of the environment, the contracting parties undertake to develop procedures regarding liability arising from the dumping or incineration at sea of wastes or other matter.

At the Eleventh Consultative Meeting, a small legal task group was asked to examine the liability question, and the team's report was adopted by the 12th Consultative Meeting. The report questioned whether this was a matter of priority in the London Convention and further questioned whether any regime that would be elaborated would be likely to achieve wide acceptance by contracting parties. More legal discussion was held thereafter, with the 14th Consultative Meeting ultimately concluding that the issue was only a moderate priority for the Convention and that it would not embark upon the development of a liability regime.[105] Subsequently, the attention of the parties became focused on amending the Convention, and no further work was done on liability under the Convention.

The liability issue arose again under the Protocol in 2006 at the first Meeting of Contracting Parties to the London Protocol in the context of the debate on whether to amend the reverse list to add carbon dioxide in order to allow permits to be given for sequestration of carbon dioxide in the deep seabed. With a view to exploring liability questions under the Protocol, including in the context of carbon dioxide sequestration, Germany proposed that the secretariat prepare an overview of developments in recent years regarding liability issues under multilateral environmental agreements relevant for the purpose of the London Protocol.[106] The Meet-

[105] LDC 14/16, para. 5.10.

[106] Report of the 28th Consultative Meeting, *supra* note 26, at para. 104.

ing of Contracting Parties agreed, but couched the charge as "with the aim to explore liability questions relating to carbon dioxide sequestration in the context of Article 15 of the Protocol."[107]

When the issue came for consideration before the 2nd Meeting of Contracting Parties in 2007, little had changed about the general liability context since the 14th Consultative Meeting: liability treaties are still being negotiated, but are still not generating sufficient support to enter into force.[108] One of the considerations in the Convention legal discussions was that most dumping occurs within the territorial or internal waters of States, and that such dumping would therefore be covered by national liability rules. The Protocol has, in Article 6, the additional prohibition of exports for disposal to another country, thus reducing even further possible transboundary situations of concern.

It should be noted that no action has been taken under the LOS Convention to develop liability rules relating to Article 235(3).[109] Guidelines on liability and compensation have recently been developed under the Barcelona Regional Seas Convention by legal experts for subsequent adoption by the parties. Currently, due to the sensitive nature of the topic, the

---

[107] *Ibid.*, para. 105.

[108] In its papers for the November 2007 discussions at the Meeting of Contracting Parties, the secretariat utilized information prepared by the secretariat of the Convention on Biological Diversity, prepared for the Open-ended Ad Hoc Working Group of Legal and Technical Experts on Liability and Redress in the Context of the Cartagena Protocol on Biosafety at its third meeting: LC29/INF.5 and LC29/9. These papers also illustrate the long list of treaties still not in force. Some authors have probed the circumstances in which the development of strict liability treaty regimes represents good environmental policy. See, R. R. Churchill, "Facilitating (Transnational) Civil Liability Litigation for Environmental Damage by Means of Treaties: Progress, Problems, and Prospects," *Yearbook of International Environmental Law* 12 (2001): 3; A. Daniel, "Liability Regimes as a Complement to Multilateral Environmental Agreements: Sound International Policy or False Comfort?," *Review of European Community and International Environmental Law* 12(3) (2003): 225. More generally see, R. Wolfrum, C. Langenfeld and P. Minnerop, *Environmental Protection by Means of International Liability Law* (Berlin: Erich Schmidt Verlag, 1999).

[109] This provides in part that "States shall co-operate in the implementation of existing international law and the further development of international law relating to responsibility and liability for the assessment of and compensation for damage and the settlement of related disputes, as well as, where appropriate, development of criteria and procedures for payment of adequate compensation, such as compulsory insurance or compensation funds."

guidelines focus on providing guidance for the development of national liability rules.

Further, the adoption of compliance procedures under the Protocol at the 2nd Meeting of Contracting Parties in November 2007, although intended to be facilitative in nature, may also help reduce dumping not sanctioned by the Protocol. Interestingly, few delegations spoke on this issue at the meeting. There was no interest expressed in developing general liability rules for ocean disposal, at least one delegation indicating that the nature of most disposal being within parties' territories effectively taking care of the issue. However, notwithstanding concerns expressed at the previous meeting by some parties, there was also no support for developing specialized liability rules for seabed sequestration. Nevertheless, parties agreed to provide information to the secretariat on existing domestic liability rules for ocean disposal, and the secretariat was requested to continue to update the Meeting of Contracting Parties on developments in international environmental liability rules. Other than that, the parties took no further action. This may have been affected by the fact that earlier in the meeting the guidance for carbon dioxide sequestration was agreed, and concerns about transboundary issues will be addressed through a Working Group. Although liability issues were not specifically listed within the mandate of the Working Group, they will no doubt arise.

Given the nature of most ocean disposal, the decision by the parties appears to be sensible. However, notification, consent and accountability issues related to transboundary sequestration will need to be adequately addressed in the Working Group, or parties who are not proponents of such sequestration may develop more interest in specialized liability rules. Further, an area that has been generally ignored is the question of the global commons, although at the moment, it may not be feasible for sequestration to occur so far out at sea.[110]

110 State responsibility for past dumping of large amounts of radioactive wastes at sea is also a global commons issue, and although there was little interest expressed at the 2nd Meeting of Contracting Parties in discussing issues of State responsibility and liability, a failure to address accountability issues in the carbon dioxide context might generate more interest in all issues of liability. For a review of past dumping activities, see: *The London Dumping Convention: The First Decade and Beyond, supra* note 5, at pp. 56–64.

Ensuring Compliance

Since 1998,[111] Canada has led discussions on compliance in the London Convention Consultative Meetings, based on the concerns of the then chair of the Scientific Group, Canadian John Karau, about the levels of non-compliance that had become evident in Scientific Group technical discussions.[112] At the time, Canada noted that the parties to an agreement cannot evaluate its effectiveness in meeting its goals unless there is a good sense of compliance levels with the instrument. There was also the outstanding issue of meeting the requirements of Article 11 of the 1996 Protocol, which required the establishment of "procedures and mechanisms necessary to assess and promote compliance" with the Protocol. Canada proposed that Article 11 procedures could be developed in anticipation of the entry into force of the Protocol, and the same procedures be adopted under the Convention.[113] From that time, an Ad Hoc Working Group on reporting and compliance has been struck at each Consultative Meeting to address a now standing agenda item on compliance.

Over the years, the Working Group has made recommendations to the Consultative Meeting on a range of compliance matters, including on compliance procedures for the Convention,[114] the development of national guidance on implementing the Protocol, and most recently in the development of compliance procedures under Article 11. The development of the Article 11 procedures was kicked off by parties after another nudge

---

[111] Canada: Compliance under the London Convention 1972 and the 1996 Protocol, document LC 20/5, Twentieth Consultative Meeting of Contracting Parties to the Convention on the Prevention of Marine Pollution by Dumping of Wastes and other Matter, December 14–18, 1998.

[112] Mr. Karau later became the chair of the Consultative Meeting, a post he held for four years. Co-author Anne Daniel has led the compliance work under the Protocol since 1998, including chairing the negotiations of the compliance procedures until their adoption.

[113] See LC 20/5, which also pointed out that the Protocol required the parties to establish a body under Article 9.5 and this could be done under the same initiative.

[114] In 1999 the 21st Consultative Meeting, based on another Canadian paper outlining six common non-compliance scenarios, confirmed how compliance issues should be resolved under the Convention, basically confirming the role of existing institutions such as the Scientific Group, the secretariat and the Consultative Meeting, with the role assigned to the Ad Hoc Working Group on Reporting and Compliance of advice to the Consultative Meeting on overcoming barriers to compliance: Report of the Twentieth Consultative Meeting, LC 21/13, para. 4.12.

by Canada in 2002, but discussions progressed in a fairly leisurely manner, given the low number of ratifications of the Protocol, with the pace heating up after entry into force in 2006. One of the challenges faced in the negotiations was the fact that many delegations were represented by transport departments, or their country's permanent representative to the IMO, and were not familiar with the approach being taken over the last two decades to compliance procedures in multilateral environmental agreements (MEAs), while other delegations were represented by environment departments which wanted to see the MEA approach replicated.

The United States wanted the procedures to be limited in that the committee being proposed would be *ad hoc* only and could not act unless it had a matter referred to it by the Meeting of Contracting Parties. Few other delegations supported this perspective, but it made the early negotiations complex and resulted in additional text in the procedures not found in other MEAs. However, at the 1st Meeting of Contracting Parties, the parties decided that the Compliance Group would be a limited, standing committee of 15 members, which could have a number of issues directly referred to it.

The 1st Meeting of Contracting Parties also decided, partly in reaction to the lack of time devoted to compliance at that meeting due to time-consuming discussions on the proposed amendment for carbon dioxide sequestration, and the impending deadline in Article 11, that a two-day meeting of the Working Group would be held in November 2007 to make further progress in advance of the 2nd Meeting of Contracting Parties. Interestingly, a number of the most active players in earlier Convention discussions to prepare Protocol compliance procedures are not parties to the Protocol (e.g., the United States and the Netherlands). Among the more controversial issues were who could trigger the procedures, how the committee would be constituted, and what measures they could recommend to the Meeting of Contracting Parties. Not controversial, but nevertheless difficult to straighten out, was the interlinkage between the committee and the Scientific Group.

Negotiators considered which among proposed trigger options should be retained: the Meeting of Contracting Parties, subsidiary bodies, a party about itself, one party regarding another party, and organizations with observer status. The self-trigger option was never an issue, but without substantial technical and financial assistance available, it is not likely that this will be used as frequently as other mechanisms. There was little support for subsidiary bodies to trigger the mechanism – the only one in existence being the Scientific Group – as this could politicize their functions. The Meeting of Contracting Parties, although not controversial as such, is not typically listed in other compliance mechanisms as a trigger and originally appeared in the text due to the U.S. desire that it be the only trigger.

The party-to-party trigger, which has been a difficult one to negotiate in several MEAs, is qualified so as to limit one party's complaint about another to situations where the complaining party is affected or likely to be affected by the possible non-compliance. This qualification was the price of getting agreement on this trigger, which most delegations consider to be fundamental to such procedures. In addition, it is noted that a party intending to make a submission of this nature should first undertake meaningful consultations with the party whose compliance is in question with the aim of resolving the matter.

A trigger that had been proposed for observer organizations did not receive sufficient support to be retained, and is not standard in other compliance procedures. However, it was understood that the Meeting of Contracting Parties could trigger the procedures as a result of matters being drawn to its attention in accordance with the rules of procedure. Over the years, Greenpeace International has raised several issues directly in the plenary.

The Compliance Group would also be given the power to examine issues of systemic non-compliance. Given that parties are typically reluctant to invoke traditional MEA compliance procedures against each other, this work at the outset can often prove to be the most helpful to the instrument. For example, under the Basel Convention, the Compliance Committee initially examined the systemic issues of failure to appoint national focal points and competent authorities, failure to report, and failure to adopt implementing laws.[115] The procedures provide that the Compliance Group can initiate discussions of systemic issues in order to make recommendations to the Meeting of Contracting Parties of systemic issues to pursue.[116]

---

[115] See the *Report of the Third Session of the Basel Convention Implementation and Compliance Committee*, UNEP/CHW/CC/3/8, pp. 2–4.

[116] Report of the 28th Consultative Meeting, *supra* note 27, at para. 136.

The Scientific Group is specifically acknowledged in the procedures to have a role in reviewing dumping reports and that will continue, and it will advise the Compliance Group as appropriate. However, it is also acknowledged that the Compliance Group should also have a role in reviewing those reports from a different perspective. In addition, the Meeting of Contracting Parties agreed that the Compliance Group would implement the requirement under Article 9.5 of the Protocol to establish a subsidiary body to review reports on administrative and legislative measures taken to implement the Protocol, as well as the effectiveness of such measures and any problems encountered in their application.[117]

The innovative Article 26 of the Protocol, which is intended to encourage ratification by developing countries by allowing for a transition period after ratification in which not all obligations have to be met, can be tracked through the Compliance Group.[118] The process involves party progress reports to be submitted to the Meeting of Contracting Parties through the secretariat for each meeting held during the transitional period. The Meeting of Contracting Parties can then decide if it wishes to refer the reports to the Compliance Group or the Scientific Group.

The Compliance Group can recommend to the Meeting of Contracting Parties one or more of the following: (1) the provision of advice and recommendations; (2) the facilitation of cooperation and assistance; (3) the elaboration, with the cooperation of the party or parties concerned, of compliance action plans, including targets and timelines; (4) and the issuing of a formal statement of concern regarding a party's compliance situation.[119] A measure involving the invitation of the non-compliant party to appear before the Meeting of Contracting Parties and outline the action it would take to bring itself into compliance and report back to the next Meeting of Contracting Parties was considered too "punitive" by some parties. The measure was removed and transformed into a procedural step which provides that where the Meeting of Contracting Parties has decided on any of the measures outlined, a party may make a statement to the Meeting of Contracting Parties on its situation.

The Meeting of Contracting Parties is given the clear authority to make the final decision regarding any measures to be taken in response to a

[117] *Ibid.*, para. 6.3.
[118] *Ibid.*, para. 6.5.
[119] *Ibid.*, para. 5.1.

party's possible non-compliance. It may also consider additional measures within its mandate, as appropriate, to facilitate compliance by the party concerned.

The qualities and establishment of the Compliance Group itself was typically difficult. In MEA compliance procedure negotiations, there are usually a group of States which argue that the members represent parties, while others argue that to remain independent they should serve in their personal capacity. This has been resolved in other recent mechanisms – as it was here – by honing in on the quality of the participation through a simple statement that "Members shall serve objectively and in the interest of promoting compliance with the Protocol," coupled with their nomination by a party and their election by the Meeting of Parties.

It has been agreed that parties not member of the Compliance Group and non-party observers may attend meetings of the Compliance Group, except when individual situations of compliance are under consideration. Such meetings shall be closed upon the request of the party whose compliance is in question.

With the adoption of these procedures, the Protocol will enter into a new era[120] in which compliance with Protocol obligations will have a standing forum for discussion, and which the various provisions of the Protocol, such as Articles 9.5, 11 and 26, as well as the role of the Scientific Group, are unified into procedures and mechanisms that should provide a sound basis for monitoring parties' compliance in the years to come. However, to complete the picture on compliance promotion it is necessary to have technical and financial assistance available to the parties, a matter addressed below.

Technical and Financial Assistance

One of the keys to successful compliance promotion is the availability of technical assistance to developing countries. The London Convention and Protocol have traditionally had little financial assistance available to offer to those Parties who needed assistance in meeting their obligations. Although Article 26 can provide some incentive to developing countries

[120] "In effect, the London Convention has become a non-dumping convention whose only real challenge is to ensure that compliance is more fully monitored and effectively controlled." P. Birnie and A. Boyle, *International Law and the Environment*, 2nd ed. (Oxford: Oxford University Press, 2002), p. 428.

to ratify the Protocol, it has not been invoked to date. Concerns about the inability of the Convention and Protocol to improve implementation of the two instruments and to attract new parties resulted in the adoption of a Long-Term Strategy for Technical Co-operation and Assistance under the London Convention 1972 in 2001 and its subsequent updating in 2006.[121]

The priorities for the Strategy are to promote compliance and implementation of the two treaties, to promote membership in the London Protocol, the development of capacity for the prevention and management of marine pollution and protection of marine resources generally, and to facilitate cooperation with other organizations.[122] The most recent Strategy particularly stressed the need to promote compliance with the parties' obligations. A project focused on this, through joint IMO/UNEP funding, is the "Barriers to Compliance" project. Through a survey of parties and non-parties alike, this project has identified several barriers to compliance and implementation, resulting in a report which the parties considered at the November 2007 meetings.[123] The report identifies legislative and policy, institutional, technical and socio-economic barriers.[124] It proposes a US$1.7 million action plan over three years to remove barriers to compliance.

Among the recommendations of the project are for the IMO to further develop and promulgate the Guidance on National Implementation of the 1996 Protocol to the London Convention 1972, to develop model legislation in all U.N. languages, and the use of national consultancies to undertake national legislative reviews and to develop national legal "road maps" to London Protocol accession/implementation.[125] It also proposes to develop a Global State of Sea Dumping report in order to better understand where dumping occurs and whether it is outside the jurisdiction of parties to either the Convention or Protocol.

In 2007, the governing bodies approved a Strategic Approach to Implementation of the Barriers project that included creation of a Work-

[121] The original strategy can be found at LC 23/16, annex 7. The 2006 update can be found at Report of the 28th Meeting of Consultative Parties, *supra* note 27.

[122] *Ibid.*, Annex 10, section 6.

[123] *Supra* note 2.

[124] *Ibid.*, pp. 15–22.

[125] *Ibid.*, see chart at p. viii.

ing Group to lead the planning and implementation of the project. The Working Group is to initiate work for which funding has been secured, organize work that is of a general nature, identify existing tools and support mechanisms available to advance this work, and identify regional leads to advance work to identify and engage new members.

Another key contribution to improving parties' capabilities in complying with ocean disposal rules includes the Waste Assessment Guidance Tutorial (WAG), several years in development, which includes both instructor and participant manuals in hard copy and on CD, and is available on the members part of the website.[126] Regional workshops, such as that held in China in 2006, in coordination with the Regional Organization for the Protection of the Marine Environment (ROPME) in the Middle East, engender further interest in the Protocol and requests for assistance from developing countries.

Much has been done over the years with minimal resources. There have been substantial efforts to provide one-on-one advice to developing countries, or lead countries volunteering to develop the WAG Tutorial, the Guidance on National Implementation of the 1996 Protocol, and individual specific waste assessment guidelines. A lot of progress has relied on the volunteer efforts of dedicated individuals from numerous parties. The secretariat has been equally efficient in its wise use of limited funds.

However, until there is a fairly substantial and sustained level of funding made available, it will be difficult to increase membership of developing countries in the Protocol and to enhance compliance with the Protocol and Convention. Furthermore, the Compliance Group will be frustrated in its work, as much non-compliance is a result of lack of capacity. The establishment of that Group needs to be coupled with resources that can be used to address non-compliance that is identified.

At the 2007 meetings, several financial offers totalling approximately US$400,000 were made to begin funding the Barriers project. While this was nowhere near what the project authors proposed, it is a substantial sum. There is recognition that member countries and other organizations may be asked to provide in-kind support, and Greenpeace International volunteered to assist in this regard. In addition, the Compliance Group and the Scientific Groups may also be able to assist in implementing the

[126] Report of the 29th Consultative Meeting, *supra* note 44, at paras. 6.13–6.17.

project.[127] Parties need to maintain this momentum as technical assistance is going to be one of the key items that will dictate the future success of the 1996 Protocol.

Achievement of Universal Acceptance of the Protocol

Global acceptance and implementation of the 1996 Protocol is still a distant goal. As of 9 November 2007, the Protocol only had 32 Contracting States while the "mothership agreement," the London Convention 1972, had 82 contracting parties.[128] Various constraints stand in the way of expedited and widespread adoption of the 1996 Protocol's precautionary approach to ocean dumping. Although Article 210 of the LOS Convention requires national laws to be no less effective than global rules in controlling ocean dumping pollution, whether the 1996 Protocol's reverse listing approach or the London Convention's more permissive measures represent global standards may be subject to debate. Article 26 of the 1996 Protocol allows any State not a contracting party to the London Convention before 31 December 1996 to notify the Secretary-General of its inability to meet Protocol obligations, other than prohibiting incineration at sea or dumping of radioactive wastes, for a period up to five years. Political will to adopt the Protocol has been especially lacking in Africa, Asia and Latin/South America. Egypt and South Africa are the only African States that have accepted the Protocol. Only China and Japan have ratified the Protocol among East Asian States. Mexico is the sole Latin American country to formally embrace the Protocol, and Suriname is the only South American contracting State.[129]

Conclusion

The 1996 Protocol to the London Convention 1972 has fundamentally altered course in the control of ocean dumping. While the London Convention permitted a permissible approach to deliberate ocean disposals that was slowly tightened over time, the Protocol embraces a precautionary

---

[127] *Ibid.*, paras. 6.28 et seq.

[128] IMO, "Status of Conventions Summary," <www.imo.org/Conventions/Mainframe.asp?topic_id=247>.

[129] Status of the London Convention and Protocol, *supra* note 28, at Annex I.

stance primarily through a "reverse listing" approach whereby only waste materials on a very short "safe list" may be considered for dumping at sea.

The Protocol's adoption and subsequent implementation efforts and debates demonstrate some of the major international environmental law challenges facing the global community as highlighted in a co-authored article by Douglas M. Johnston in the year 2000.[130] Those challenges include limited political will, particularly in developing States, to "buy into" a strong environmental agenda, divergent national capacities to implement international environmental agreements and ethical debates over whether activities threatening the marine environmental should be prohibited or regulated.

The Protocol also exemplifies the "living nature" of multilateral environmental agreements. Over ten guidance documents have been developed to fill-in textual generalities and the guidelines will almost certainly be subject to periodic reviews. Annual Meetings of the Parties provide a venue for addressing new ocean disposal/placement threats and for discussing implementation challenges such as technical and financial assistance to developing countries, and broader ratification. The recently agreed to compliance mechanism promises to place a spotlight on at least some of the major causes of national implementation lethargy.

The 1996 Protocol may be viewed as a "beacon of hope" for guiding future law and policy developments in other sectors of international environmental law. A "reverse listing" approach and development of a global consensus on what wastes/materials should be allowed for ocean dumping might be a model extended to other global challenges, such as the control of toxic chemicals and substances. While two global chemical agreements have thus far been forged, the 2001 Stockholm Convention on Persistent Organic Pollutants[131] and the 1998 Rotterdam Convention,[132] these

---

130 D. M. Johnston and D. L. VanderZwaag, "The Ocean and International Environmental Law: Swimming, Sinking, and Treading Water at the Millennium," *Ocean & Coastal Management* 43 (2000): 141.

131 *Stockholm Convention on Persistent Organic Pollutants*, Stockholm, 22 May 2001, <chm.pops.int/Convention/tabiol/54/language/en-US/Default.aspx#convtext>.

132 *Convention on the Prior Informed Consent Procedure for Certain Hazardous Chemicals and Pesticides in International Trade*, Rotterdam, 11 September 1998, <www.pic.int/en/ConventionText/ONU-GB.pdf>.

agreements are extremely limited in the number of chemicals addressed[133] and both suffer from the rather tedious and slow mechanisms for adding chemicals for regulatory controls.[134] With some 100,000 chemicals available for commercial use and over 1,000 new chemicals reportedly produced per year,[135] a "losing regulatory battle" seems obvious. A global "reverse listing" approach modelled on the 1996 Protocol stands as a political and ethical guide that the world might eventually follow.[136]

The 1996 Protocol is certainly steering a precautionary course in the global approach to ocean dumping but its implementation voyage has not yet reached final shores. It also remains to be seen whether political will and sufficient ethical, social and academic forces can be mustered to transfer the strong marine environmental protection "reverse listing" approach to other realms.[137]

---

133 The Stockholm Convention initially addressed only 12 chemicals referred to as the "dirty dozen" while the Rotterdam Convention has subjected just 39 chemicals listed in Annex III to the prior informed consent procedure. See "Rotterdam Convention Overview," <www.pic.int/home/php?type=t&id=5&sid=16>.

134 For example, under the Stockholm Convention a party must first propose listing a chemical and provide detailed information as set out in Annex D and the Persistent Organic Pollutants Review Committee must then develop a risk profile based on Annex E followed by a risk management evaluation based on socio-economic information specified in Annex F before a chemical may be recommended to the Conference of Parties for listing. See "Chemicals in review process," <www.pops.int/documents/meetings/poprc/chem_review.htm>.

135 *Projet de Sociéte: Canada and Agenda 21* (Winnipeg: International Institute for Sustainable Development, 1995), <iisd.ca/worldsd/Canada/project/c19.htm>.

136 See D. VanderZwaag, R. Huebert and S. Ferrara, "The Arctic Environmental Protection Strategy, Arctic Council and Multilateral Initiatives: Tinkering While the Arctic Marine Environment Totters," *Denver Journal of International Law and Policy* 30(2) (2002): 131 at 168. At present the world, of course, seems content to follow a further "soft path" through the voluntary Strategic Approach to International Chemicals Management (SAICM) comprising a declaration, an overarching policy strategy and a global plan of action. For a recent critique, see A. Logomasini, "Strategic Approach to International Chemicals Management: Lack of Interest Belies Importance," *Sustainable Development Law & Policy* 6(3) (Spring 2006): 33.

137 For example, a reverse listing approach might eventually be applied to the growing problem of land-based marine pollution through a new global instrument or through regional embraces. For a discussion of existing global and regional approaches, see D. Hassan, *Protecting the Marine Environment from Land-based Sources of Pollution: Towards Effective International Cooperation* (Burlington, VT: Ashgate Publishing Co., 2006).

Regulation of the Dumping of Wastes at Sea: The Chinese Practice

Zou Keyuan*

Introduction

Marine environmental protection remains an important priority for human beings in the 21st century despite the existing extensive efforts that have been undertaken at the global, regional and national levels. According to a recent report published in *Science*, many fish species will be extinct by 2050,[1] and the global ocean remains threatened by natural and human environmental impacts. Dumping of wastes, although not a primary threat to ocean ecological systems, is a source of pollution that needs to be tackled to better protect the marine environment.

According to statistics, there are three main sources of marine pollution in China: (1) land-based, the largest source of pollution, accounting for almost half of the total waste water discharged into the sea and comprising about 80 per cent of marine pollutants; (2) discharge of waste waters and dumping of wastes into the sea by marine activities; and (3) atmospheric deposition of pollutants.[2] It was common along China's coastal region to deposit waste, both domestic and industrial, into the sea at random and without care. This has caused severe environmental problems, not only

---

* Harris Professor of International Law, University of Central Lancashire, Preston, United Kingdom.

1 Reported in "Salt-Water Fish Extinction Seen By 2048: Study by Ecologists, Economists Predicts Collapse of World Ocean Ecology," <www.cbsnews.com/stories/2006/11/02/health/webmd/main2147223.shtml>, 23 November 2006. See B. Worm, et al., "Impacts of Biodiversity Loss on Ocean Ecosystem Services," *Science* 314 (3 November 2006): 787–790.

2 See Compiling Committee of the Series of the Chinese Natural Resources (CCSCNR), *The Series of the Chinese Natural Resources: Ocean* (Beijing: China Environmental Science Press, 1995), pp. 59–60 (in Chinese).

damaging the marine environment, but also harming coastal residents. Unfortunately, before 1982, dumping of wastes offshore was not regulated in China. Only dredge materials from ports were managed by the harbour superintendent administration.[3] Regulations concerning dumping at sea in China were formally introduced in 1982 when the basic law on marine environmental protection, i.e., the Marine Environmental Protection Law (MEPL), came into being. MEPL entered into force on 1 March 1983 and was substantially amended in December 1999.[4] In addition to this basic law, relevant regulations include the 1985 Regulations Concerning Dumping of Wastes at Sea and the 1990 Measures for the Implementation of the Regulations Concerning the Dumping of Wastes at Sea. These regulations are the principal national legal instruments governing ocean dumping activities in China.

The regulation of dumping of wastes at sea is an integral part of the comprehensive marine environmental protection system in China. Currently, China has been attempting to establish a harmonious society for its people, within which harmony between nature and human beings is critical. Thus, effective control of dumping of wastes at sea is a necessary component in efforts to achieve that goal. Since China is the largest country and has the longest mainland coastline in East Asia, its marine environmental protection laws and practices affect the entire East Asian seas area.

This essay will address the ocean dumping issue from a legal perspective, analyzing and assessing the relevant laws and regulations and their implementation and enforcement in China. In addition, it will compare Chinese regulations and their implementation with the relevant international law to see whether or not they are compatible. To better explain China's law and practice, relevant case law regarding dumping of waste will be reviewed as well.

---

[3] Fang Shaoming, "The 1972 London Convention and China's Marine Waste Dumping Management," *Ocean Development and Management* 2 (1995), p. 72 (in Chinese).

[4] English text of the 1999 amended Law is available in Office of Policy, Law and Regulations, State Oceanic Administration, ed., *Collection of the Sea Laws and Regulations of the People's Republic of China*, 3rd edition (Beijing: Ocean Press, 2001), 216–250 [hereinafter MEPL].

General International Law

Dumping of wastes at sea has been governed by international law since the 1970s. The first international legal instrument in this respect is the 1972 Convention on the Prevention of Marine Pollution by Dumping of Wastes at Sea and Other Matter (the London Dumping Convention), which entered into force on 30 August 1975. It governs the dumping at sea of the wastes generated from land and the prevention of such activities from polluting the marine environment. The Convention takes a so-called "black- and grey-list" approach. The black list contains substances, the dumping of which is prohibited. The grey list contains substances the dumping of which is only permitted under strict control and provided certain conditions are met.[5]

The 1996 Protocol to the London Dumping Convention, which came into force on 24 March 2006, has substantially altered the 1972 Convention. The 1996 Protocol involves

> a major change of approach to the question of how to regulate the use of the sea as a depository for waste materials in that, in essence, dumping is prohibited, except for materials on an approved list. This contrasts with the 1972 Convention which permitted dumping of wastes at sea, except for those materials on a banned list.[6]

The wastes on the approved list include: dredged material; sewage sludge; fish wastes; vessels and platforms; inert, inorganic geological material (e.g., mining wastes); organic material of natural origin; and bulky items primarily comprising iron, steel and concrete. Based on the new Protocol, guidelines for the assessment of wastes or other matter that may be considered for dumping have been prepared.

The 1996 Protocol redefines the term dumping from how it was defined in the 1972 Convention to mean

1. any deliberate disposal into the sea of wastes or other matter from vessels, aircraft, platforms or other man-made structures at sea;

---

[5] "Background," <www.londonconvention.org/London_Convention.htm>, 1 December 2006.

[6] "International Rules on Dumping of Wastes at Sea to be Strengthened with Entry into Force of 1996 Protocol," *IMO Briefing*, 22 February 2006, <www.londonconvention.org/documents/lc72/LP press release feb 06.doc>, 13 November 2006.

2. any deliberate disposal into the sea of vessels, aircraft, platforms or other man-made structures at sea;
3. any storage of wastes or other matter in the seabed and the subsoil thereof from vessels, aircraft, platforms or other man-made structures at sea; and
4. any abandonment or toppling at site of platforms or other man-made structures at sea, for the sole purpose of deliberate disposal.[7]

However, "dumping" does not include:

1. the disposal into the sea of wastes or other matter incidental to, or derived from the normal operations of vessels, aircraft, platforms or other man-made structures at sea and their equipment, other than wastes or other matter transported by or to vessels, aircraft, platforms or other man-made structures at sea, operating for the purpose of disposal of such matter or derived from the treatment of such wastes or other matter on such vessels, aircraft, platforms or other man-made structures;
2. placement of matter for a purpose other than the mere disposal thereof, provided that such placement is not contrary to the aims of this Protocol; and
3. notwithstanding paragraph 4.1.4, abandonment in the sea of matter (e.g., cables, pipelines and marine research devices) placed for a purpose other than the mere disposal thereof.[8]

The new definition is more detailed than the definition of dumping provided in the London Dumping Convention.

The third important international instrument concerning ocean dumping is the 1982 United Nations Convention on the Law of the Sea (LOS Convention), which came into force on 16 November 1994. Part XII, Protection and Preservation of the Marine Environment, lists six sources of marine pollution, including ocean dumping of waste. The LOS Convention calls on States to adopt laws and regulations to prevent, reduce and control pollution by dumping and to ensure that no dumping is carried

[7] *Protocol to the Convention on the Prevention of Marine Pollution by Dumping of Wastes and Other Matter*, London, 7 November 1996, 36 *I.L.M.* 1 (1996), Art. 1(4.1).
[8] *Ibid.*, Art. 1(4.2).

out without the permission of the competent authorities of States.[9] The Convention encourages State parties to adopt national laws and regulations to support global rules and standards. It should be noted that the LOS Convention uses the same definition of dumping as the 1972 London Dumping Convention.

China acceded to the 1972 London Dumping Convention in November 1985. On 29 June 2006 China ratified the 1996 Protocol to the London Dumping Convention.[10] Since 1984, China has attended the annual Consultative Meeting of the Treaty Parties to the London Dumping Convention.[11] China has called for international, regional and bilateral cooperation to coordinate the prevention and control of marine pollution and disposal and transport of wastes at the global level.[12] The competent authority in charge of dumping at sea in China is the State Oceanic Administration (SOA), which reports annually to the Secretariat of the London Dumping Convention about the categories and quantity of dumped materials.

At the 16th Consultative Meeting of the Parties to the London Dumping Convention in 1993, three resolutions concerning disposal at sea of radioactive wastes, incineration at sea of industrial wastes and sewage sludge and sea disposal of industrial wastes were adopted as amendments to the annexes to the London Dumping Convention.[13] The Chinese delegation has stated that the dumping of radioactive material and other hazardous substances at sea should be banned, and that China has never dumped such materials at sea nor does it have such a dumping plan. Nevertheless, China had to modify its domestic regulations to bring them in line with these amendments. For example, industrial wastes without hazards or low hazards are permitted to be dumped at sea pursuant to the Chinese

---

[9] *United Nations Convention on the Law of the Sea*, Montego Bay, 10 December 1982, UN Doc.A/Conf. 62/122; 21 *I.L.M.* 1261 (1982), Art. 210.

[10] Text of the Ratification Decision by the Standing Committee of the National People's Congress is available at *Gazette of the Standing Committee of the National People's Congress*, No. 6, 2006, p. 458.

[11] During the period of 1984–1985, China participated as an observer. Since China acceded to the Convention, it has participated as a State party.

[12] Yang Wenhe, ed., *London Convention: 25 Years* (Beijing: Ocean Press, 1999), 235–236 (in Chinese).

[13] Doc LC.51 (16) (1993); Doc LC.50 (16) (1993); and Doc LC.49 (16) (1993). These amendments came into force on 20 February 1994.

Regulations Concerning the Dumping of Wastes at Sea (1985). The acceptance of these amendments by the Chinese government means that it will be necessary to revise the relevant domestic laws and regulations.[14]

Following the United Nations Conference on Environment and Development (UNCED), held in Rio de Janeiro in 1992, environmental protection at the global level has been strengthened, as has the protection of the marine environment. UNCED triggered the revision of the London Dumping Convention through the negotiation of the 1996 Protocol. Thus, dumping of wastes at sea has been further limited at the international level. China was an active participant in these negotiations, and its suggestions were reflected in the final text of the Protocol. For example, China preferred the use of the term "precautionary approach" instead of "precautionary measures."[15] As a result of compromise, the 1996 Protocol requires State parties to apply

> a precautionary approach to environmental protection from dumping of wastes or other matter whereby appropriate preventive measures are taken when there is reason to believe that wastes or other matter introduced into the marine environment are likely to cause harm even when there is no conclusive evidence to prove a causal relation between inputs and their effects.[16]

The second area where China exerted its influence was with regard to whether internal waters should be included into the definition of "sea" in the Protocol. China opined that dumping within the internal waters should be subject solely to domestic laws of the coastal State. The final definition of "sea" does not include internal waters, but the Protocol contains an optional clause stipulating that States parties may at their discretion either apply the provisions of the Protocol or adopt other effective measures to control the deliberate disposal of wastes in "marine internal waters."[17] Finally, China opposed the idea of the "reverse list," preferring to maintain the existing lists provided for in the London Dumping Convention. The Chinese delegation stated that it was not necessary to change the list mechanism that had proven to be practical and effective for the last two decades.[18] However, when the majority of the delegations

---

[14] Yang, *supra* note 12, at p. 239.

[15] *Ibid.*, pp. 246–250.

[16] *Supra* note 7, at Art. 3(1).

[17] *Ibid.*, Art. 7.

[18] Yang, *supra* note 12, p. 254.

favoured the reverse list, China retreated from its original position and made concessions after determining that the materials contained in the reverse list would not be contrary to the interests of China and that the reverse list would not be a barrier to China's economic development.[19] China's change of position paved the way to the adoption of the reverse list during the final minutes of the negotiations.

In addition to its involvement in international efforts to protect the marine environment, China attended UNCED, which produced Agenda 21, and prepared its national Agenda 21 in 1994[20] and Ocean Agenda 21 in 1996.[21] China is also a party to the 1989 Basel Convention on the Control of Transboundary Movements of Hazardous Wastes and Their Disposal[22] and MARPOL73/78.[23] All of these undertakings indicate that China is determined to strengthen the international regime to protect the marine environment and preserve marine resources so as to realize the ultimate goal of sustainable development in the use of oceans.

Chinese Legislation

To implement the London Dumping Convention, China was obliged to adopt appropriate domestic laws and regulations and establish and operate enforcement authorities. Since 1982, China has adopted a series of laws and regulations governing and managing the protection of the marine environment (see Table 1). While it is not possible to review details of all such laws and regulations, the basic marine environment law, i.e., MEPL, is the most relevant to ocean dumping.

---

19 *Ibid.*, p. 256.

20 China's national Agenda 21 was adopted on 25 March 1994 by the State Council. The whole White Paper was published by the China Environmental Science Press in 1994 and entitled *China Agenda 21.*

21 The *China Ocean Agenda 21* was adopted by the State Oceanic Administration and later published by the Ocean Press in 1996 in both Chinese and English.

22 *Basel Convention on the Control of Transboundary Movements of Hazardous Wastes and their Disposal*, Basel, 22 March 1989, 28 *I.L.M.* 657 (1989).

23 *International Convention for the Prevention of Pollution from Ships*, London, 2 November 1973, 1340 *U.N.T.S.* 184, as amended by *Protocol of 1978 relating to the International Convention for the Prevention of Pollution from Ships of 1973*, 17 February 1978, 1340 *U.N.T.S.* 61 [hereinafter MARPOL 73/78].

Table 1. China's laws and regulations relating to the protection of the marine environment

Laws/Regulations	Date of Promulgation
Law on Marine Environmental Protection	23 August 1982; amended 25 December 1999
Regulations Concerning Environmental Protection in Offshore Oil Exploration and Exploitation	29 December 1983
Regulations Concerning the Prevention of Pollution of Sea Areas by Vessels	29 December 1983
Regulations Concerning the Dumping of Wastes at Sea	6 March 1985
Regulations Concerning the Prevention of Pollution Damage to the Marine Environment by Coastal Construction Projects	25 May 1990
Regulations Concerning the Prevention of Pollution Damage to the Marine Environment by Land-based Pollutants	25 May 1990
Regulations Concerning the Prevention of Environmental Pollution by Ship-breaking	18 May 1988
Implementing Regulations on the Protection of Aquatic Wild Animals	5 October 1993
Regulations of Natural Protected Reserves	9 October 1994
Regulations for the Protection of Wild Plants	30 September 1996
Measures for the Implementation of the Regulations Concerning Environmental Protection in Offshore Oil Exploration and Exploitation	20 September 1992
Measures of the Management of Marine Nature Reserves	29 May 1995
Measures on the Protection of Nature Reserves of Aquatic Fauna and Flora	17 October 1997
Provisions on the Procedure for Investigation and Handling of Accidents of Pollution in Fishing Areas	26 March 1997
Measures for the Implementation of the Regulations Concerning the Dumping of Wastes at Sea	25 September 1990
Provisions Governing the Management of Coastal Forest Belts under Special State Protection	9 December 1996

Table 1. (*cont.*)

Laws/Regulations	**Date of Promulgation**
Law on the Management of Sea Area Use	27 October 2001
Provisional Provisions on the Management of Abandoned Offshore Oil Platforms	24 June 2002
Provisional Measures on the Management of Dumping Sites	14 November 2003
Provisions on the Management of the Protection and Utilization of Uninhabited Islands	23 June 2003
Regulations on the Management of the Prevention of Pollution Damage to the Marine Environment by Marine Construction Projects	30 August 2006
Provisions on the Management of Sea Area Use Rights	13 October 2006
Provisions on the Management of Sea Functional Zoning	30 July 2007

Source: Compiled by the author.

Marine Environmental Protection Law

The purposes of MEPL are as follows: to protect the marine environment and resources, prevent pollution damage, maintain ecological balance, protect people's health, and promote marine understanding. The Law covers five sources of pollution: coastal construction projects, offshore oil exploration and exploitation, land-based pollutants, vessel source pollution, and dumping of wastes. In order to implement this law, a series of regulations were put in place to govern the various sources of marine pollution.

MEPL applies to China's internal waters and territorial sea, exclusive economic zone (EEZ) and continental shelf, and all other sea areas under China's jurisdiction. While the wording in the Law concerning "internal waters and the territorial sea," the EEZ and the continental shelf is clear, the term "other sea areas under China's jurisdiction" is not clear since it is not specified as to what particular sea areas are included besides the above mentioned maritime zones.

MEPL grants implementing power to the relevant governmental agencies, including the State Environmental Protection Administration (SEPA), the State Oceanic Administration (SOA), the Harbour Superintendent

Administration, Fishery Administrative and Fishing Port Superintendent Agency, the environmental protection department of the armed forces, and environmental protection departments of the coastal provinces.[24] The power to regulate dumping of wastes at sea is delegated to the SOA.

MEPL provides for three kinds of liability – administrative, civil and criminal. In the first case, the relevant supervising departments have the right to order a law-breaker to remedy the damages caused and to impose fines. In comparison with the schedule of fines set down by other Chinese environmental laws, the maximum fine for damage to the marine environment is much higher.[25] The reasons given for this are as follows: (1) marine pollution may cause much more severe consequence than any other pollution; and (2) foreign vessels or enterprises are usually involved in marine pollution.[26] Regarding civil liability, all violators will be held strictly liable for any damage to the Chinese marine environment, except those caused by "acts of war," "irresistible natural calamities" or "negligence or other wrongful acts on the part of the departments responsible for the maintenance of beacons or other navigational aids in exercising their functions."[27] Strict liability, or non-fault liability, has become an accepted principle in China's environmental legal system.[28] Instituting criminal liability was controversial because there was no provision in China's criminal law that dealt with environmental damage at the time MEPL came into being. However, this has been remedied in the revised Criminal Law where a new section containing nine provisions deals with environmental crimes, ranging from causing severe pollution accidents, to illegal import of hazardous solid wastes, illegal catching of aquatic products, illegal catching and selling of endangered wildlife, illegal mining,

---

[24] MEPL, *supra* note 4, at pp. 217–218.

[25] In general the fines are 1,000 RMB. However, in the marine environmental protection field, the maximum fine is 100,000 RMB.

[26] Zhang Kunmin and Jin Ruilin, *Talks on Environmental Protection Law* (Beijing: Qinghua University Press, 1990), p. 160 (in Chinese).

[27] The 1982 MEPL is reprinted in Office of Policy, Law and Regulations, State Oceanic Administration, ed., *Collection of the Sea Laws and Regulations of the People's Republic of China*, revised edition (Beijing: Ocean Press, 1998), pp. 195–208. Art. 43 of the 1982 MEPL.

[28] For details, see Han Depei, ed., *Textbook on Environmental Protection Law*, 2nd edition (Beijing: Law Press, 1991), pp. 224–226 (in Chinese).

and illegal logging.[29] Thus, the criminal liability provisions in MEPL can be imposed upon violators in accordance with the revised Criminal Law. As a consequence, environmental law enforcement has been improved considerably.

In 1999, MEPL was substantially amended and expanded from the original eight chapters to ten and from 48 clauses to 98. Compared with the old Law, the new Law contains a number of principles enshrined in general international environmental law. First, the 1999 Law incorporates the principle of sustainable development in the purpose of the Law (Article 1). Based on this principle, the 1999 Law establishes the mechanisms for controlling the total quantity of pollution discharged into important sea areas. The Law prescribes the maximum amounts of the main pollutants that may be discharged into the sea, and allocates the discharge quantity among the main pollution sources (Article 3). It should be noted that the old Law controlled discharges by density, which left a loophole whereby some polluters diluted the pollutants before they discharged. The concept of sustainability is also reflected in the environmental impact assessment requirements for coastal and marine construction projects. Further, the 1999 Law calls for a mechanism for the preparation of contingency plans for potential grave pollution accidents (Article 18). The second principle enshrined in the 1999 Law is that of polluter pays. Anyone who discharges listed pollutants into the sea must pay discharge fees (Article 11). Fines will be imposed for discharges of other substances that are not allowed under the new MEPL. This has changed the former legal arrangement under which fees could only be collected for discharges beyond environmental standards.

Dumping Regulations

In 1985, when China acceded to the London Dumping Convention, it adopted the Regulations Concerning Dumping of Wastes at Sea. On 25 November 1990, the Measures for the Implementation of the Regulations Concerning the Dumping of Wastes at Sea were adopted. These regulations

[29] *Criminal Law of the People's Republic of China*, adopted on 1 July 1979, amended on 14 March 1997, in *Gazette of the Standing Committee of the National People's Congress of the People's Republic of China*, No. 2 (1997), pp. 198–200, Section 6, Arts. 338–346 (in Chinese).

were followed by other new regulations, such as, the 2003 Provisional Measures on the Management of Dumping Sites.

Definition of Dumping

The regulations define dumping as follows: the disposal of wastes or other matter from vessels, aircraft, platforms or other vehicles at sea; the disposal of vessels, aircraft, platforms or other man-made structures at sea; and the disposal at sea of wastes or other matter arising from, or related to the exploration and exploitation of seabed mineral resources and offshore processing related thereto. Dumping does not include the discharge of wastes arising from normal operation of vessels, aircraft, or other vehicles and facilities. This definition includes waste disposal "arising from or related to the exploration and exploitation of seabed mineral resources and offshore processing related thereto," which is not included in the definition in the London Dumping Convention.

In the 1990 Implementing Measures, wastes are divided into three categories in accordance with factors such as toxicity, harmful substance content, impact upon the marine environment, etc. Accordingly, dumping of Category A substances is generally prohibited unless their disposal on land will severely harm human health and dumping at sea is the only way to prevent such health threat. Category B refers to substances that can be rapidly rendered harmless. Their dumping requires a prior special permit. Category C includes low toxic and harmless materials, dumping of which needs a prior general permit.

China prohibits incineration of wastes and disposal of radioactive wastes at sea. The State Council has the authority to determine the exemption density of radioactivity.[30]

Dumping Permits and Fees

According to the relevant law and regulations, any person or entity which intends to dump wastes at sea must obtain a permit from the competent authority. The application for a dumping permit should include a test paper on the characteristics and composition of the wastes.[31] The SOA is

[30] MEPL, *supra* note 4, at Art. 61.

[31] 1985 Dumping Regulations, Art. 5. The English text is reprinted in Office of Policy, Law and Regulations, State Oceanic Administration, ed., *Collection of the Sea Laws and Regulations of the People's Republic of China*, revised edition (Beijing: Ocean Press, 1998), pp. 362–370.

to examine the application within two months from the date of receiving the application and decide on whether the application is approved. The permit for dumping should clearly indicate the waste dumping entity, terms of validity, quantity and category of the wastes, and the method of dumping. The issuing authority "may change or revoke a permit in view of the variation of the marine ecological environment and the development of science and technology."[32] Since wastes are divided into three categories in accordance with factors such as toxicity, harmful substance content, and impact on the marine environment, the permits are accordingly divided into three types: (a) an emergency permit, which is issued for the dumping of prohibited wastes in emergency situations; (b) a prior special permit, which is issued for dumping of wastes list in Annex II to the 1985 Dumping Regulations, including wastes containing arsenic, lead, copper, zinc, cyanides, fluorides, beryllium, chromium, nickel, vanadium, organosilicon, and their compounds; and (c) a prior general permit, which is used for the dumping of low-toxic and non-toxic wastes not included in Annexes I and II to the regulations.[33] The first type of permit is to be approved by the SOA and can only be used once, while the other permits can be issued by various branches of the SOA and are valid for six months or one year.[34]

Those who have obtained permits are to notify the SOA for verification at the time of loading of the wastes to be dumped. If the SOA finds the wastes loaded are not the ones specified in the permit, it has the right to stop the shipment. In serious cases of breach, the permit can be suspended or revoked. Where the dumping of wastes is to be conducted from vessels, the Harbour Superintendent Administration of the port of departure is to be notified for verification as well.[35] Any entity which has obtained a permit is to dump its wastes in the designated site within the time limit and according to the requirements as specified in the permit. The entity is to complete the form of detailed records of dumping and submit it to the competent authority and the details of dumping are also to be recorded in the log book.[36]

---

32 *Ibid.*, Art. 10.

33 *Ibid.*, Art. 11.

34 1990 Implementing Measures, Arts. 13–14, <www.mem.gov.cn/docs/statute/statute-Content3.jsp>, 3 March 2008.

35 1985 Dumping Regulations, *supra* note 31, at Art. 12.

36 *Ibid.*, Art. 14.

Article 11 of MEPL provides that those who conduct ocean dumping are to pay dumping fees. The fees levied should be used for the prevention and control of marine environmental pollution and not be appropriated for any other purpose. Charges are also levied on applying for dumping permits. In 1992, the SOA prepared the criteria of charges for wastes dumping. These criteria only apply to dredged materials and industrial wastes. Because of these two factors, the criteria are being reviewed and revised. As others have observed, if the fees are too low, polluters would rather pollute the marine environment than reduce or eliminate pollutants at the source.[37]

Selection and Designation of Dumping Sites

According to MEPL, dumping sites should be selected and designated in accordance with scientific, economic and safety principles. Environmental principles and standards were later added to the considerations through the implementing regulations for ocean dumping site selection. Once the SOA has selected and mapped out the dumping sites, they are submitted to the State Environmental Protection Administration (SEPA) for review and to the State Council for final approval.[38] The SOA is to supervise and manage the use of dumping sites and carry out environmental monitoring. Once a dumping site is no longer suitable for use, the SOA is to close it down and report this to the State Council.

Ocean dumping sites are divided into five categories: separate sites for designated Categories A, B and C wastes; an experimental dumping site; and a temporary dumping site. The designated waste sites correspond to the wastes listed in the categories contained in the Annexes to the 1985 Dumping Regulations. Experimental dumping sites are designated for experimental use only and are not to be used for more than two years. Temporary dumping sites are designated for engineering projects' special needs and are used only once.

The 2003 Provisional Measures on the Management of Dumping Sites also provides that dumping sites include marine dumping sites and temporary marine dumping sites. The former are designated for long-term use

[37] Xu Lina, "Certain Issues Relating to the Criteria of Levying Fees on Ocean Wastes Dumping," *Ocean Development and Management* 3 (2006), p. 17 (in Chinese).

[38] MEPL, *supra* note 4, at Art. 57.

while the latter are for temporary purposes (limited to three years with a possible extension of one year).

Dumping operators should expect surveillance and inspection by the China Marine Surveillance. The State Council has approved 38 dumping sites for dredged materials. As of October 2006, 98 ocean dumping sites had been designated.[39]

Regulations Related to Dumping

There are several other regulations concerning marine environmental protection that have a connection to the special regulations on dumping at sea. These include, the 2006 Regulations on the Management of the Prevention of Pollution Damage to the Marine Environment by Marine Construction Projects,[40] which requires that wastes including oily waste water, plastic products, and toxic dredges should not be discharged directly into the sea, and the 2002 Provisional Provisions on the Management of Abandoned Offshore Oil Platforms,[41] which apply to abandoning activities of all platforms including fixed and mobile platforms, their appended appliances and other floating instruments within China's jurisdictional waters. The 2002 Provisions provide that any abandoning activity must comply with the environmental requirements so as to prevent the marine environment from being harmed, and that the owner of the platform is to demolish all devices and equipment which could pollute the marine environment during the course of abandoning activities (Article 4).

Management and Law Enforcement

The SOA is the competent authority in China for managing and law enforcement concerning dumping of wastes at sea. The SOA has three branch bureaus for North China, the East China Sea and the South China Sea. Through these branches, the SOA undertakes management and inspection. Since 1990, coastal provinces and cities that have established

---

39 See "Ocean Dumping Increases Year by Year in Our Country," *People's Daily*, 12 October 2006, p. 2 (in Chinese).

40 Text is available at <www.soa.gov.cn/law/061008law.htm>, 15 November 2006.

41 Text is available in *China Ocean News*, 23 July 2002, p. 3 (in Chinese).

Offices of Marine Affairs also have management responsibilities for dumping of wastes at sea.

A new system was introduced on 1 January 1998, in accordance with a circular issued by the SOA in August 1996, for the installation of dumping recorders on self-navigating vessels used for dumping of wastes. In January 2002, SOA issued another circular requesting non-self-navigating vessels to be equipped with such devices. Vessels without such equipment are not eligible to apply for dumping permits. In order to improve this new scheme, an inspection is necessary to determine whether the vessels have been equipped with the devices.[42]

According to the May 2006 Monthly Bulletin, China used 37 dumping sites including 29 temporary ones. Approximately 15 million cubic meters of dredged materials were dumped. Forty-one dumping permits were issued and about 2.6 million RMB of dumping fees were collected. At the local level, 27 permits were issued and 239,282 RMB of dumping fees were collected.

Under the SOA, the permanent maritime law enforcement agency, China Marine Surveillance, carries out inspection functions in China's jurisdictional waters to find and deal with violations of Chinese laws in respect of marine rights and interests, use of sea areas, marine environment and resources, damage of installations at sea and the disturbance of order at sea.[43] Monitoring dumping of wastes at sea falls within the jurisdiction of the agency. Inspection and monitoring is to be conducted to determine: (a) whether the dumping entity holds a valid permit; (b) whether the dumping vessel has a verified pass from the Harbour Superintendent Department; (c) whether the loaded wastes conform with the contents indicated in the permit; (d) whether the dumping record form is appropriate; (e) whether the dumping is conducted in the place and under the conditions prescribed in the permit; and (f) other related matters.[44]

From 1985 to 1998, the China Marine Surveillance undertook air patrols covering one million kilometres and sea patrols of 800,000 kilometres. More than 1,800 inspections on board vessels dumping wastes were conducted and illegal dumping activities were punished. Meanwhile,

---

[42] See Ding Jingzhao, "Further Strengthening the Management of the Use of Ocean Dumping Devices," *China Ocean News*, 21 November 2006 (in Chinese).

[43] *China Ocean News*, 20 August 1999, p. 1 (in Chinese).

[44] Yang, *supra* note 12, at pp. 278–279.

SOA branches also carried out more than 4,000 inspections of vessels engaged in dumping activities and imposed administrative punishment in about 300 cases.[45] In 2004, the China Marine Surveillance made 600 inspections and found two cases of illegal dumping.[46]

According to MEPL, dumping wastes without a permit is to result in a fine of not less than 30,000 RMB, but no more than 200,000 RMB.[47] This monetary punishment gives some degree of discretion to law enforcers to decide on the exact amount for comparable illegal dumping incidents. Illegal dumping can take several forms: (a) dumping in a dumping or non-dumping site without a permit; (b) dumping without the permit having been reviewed or with an expired permit; (c) dumping before obtaining a permit or while the application is being processed; and (d) dumping for more than two engineering operations but with only one permit.[48] The China Marine Surveillance and SOA local branches consider the following factors when they impose fines for illegal dumping: (a) the quantity of dumped wastes and the duration of dumping; (b) the behaviour of law-breakers, i.e., whether they cooperated with law enforcers during the investigation, made timely corrections, or remediated harmful consequences; and (c) if the dumping is conducted deliberately or repeatedly.[49] In addition, environmental considerations are also important. In 2003, the SOA prepared the Procedure on the Classification and Assessment of Dredged Materials for Ocean Dumping, which divides dredges into three categories: (a) clean; (b) dusted; and (c) polluted. Different dredges will have different marine environmental consequences, thus the corresponding administrative punishment should also be differentiated.[50]

There are recent reports of illegal dumping and punishment being imposed by the China Marine Surveillance. On 27 March 2005, the 5th Brigade, during a law enforcement patrol, found a ship (*Hu Huan Ni* 506) belonging to the Shanghai Waste Disposal Company dumping

---

[45] *Ibid.*, p. 308.

[46] "Marine Law Enforcement and Monitoring," *China Ocean Yearbook 2005* (Beijing: Ocean Press, 2006), p. 192 (in Chinese).

[47] MEPL, *supra* note 4, at Art. 73.

[48] Shi Xingping, You Zhiyong, et al., "Study on the Application of Rational Discretion in the Administrative Punishment for Ocean Dumping without a Permit," *Ocean Development and Management* (in Chinese), No. 4, 2005, p. 50 (in Chinese).

[49] *Ibid.*, p. 50.

[50] *Ibid.*, p. 51.

dredged materials in the dumping site north of the mouth of the Wusong River. Law enforcement officers visited the ship. During the investigation, it was determined that the ship had dumped dredged materials six times between 24 to 27 March, without filling in the form required by MEPL. The company admitted that its ship broke the law, and an administrative punishment was imposed.[51] In September 2006, the Lianyungang Brigade found a ship that was rented by an engineering company for dredging operations dumping sludge that was a Category C waste. The ship did not have a permit or a dumping record. The Lianyungang Bureau of Marine and Fishery Affairs ordered the shipowner to stop dumping and imposed a fine of 30,000 RMB.[52]

Between 21–27 November 2006 the China Marine Surveillance North Sea General Brigade and the General Brigade of Liaoning Province carried out their first joint law enforcement inspection for waste dumping activities in the sea areas adjacent to Liaoning. The targets included dumping operations approved by the North Sea Branch of the SOA and the Office of Marine and Fishery Affairs of Liaoning Province. More than 50 dumping vessels were examined.[53]

In April 2004, the Implementing Program of Comprehensively Pushing forward Administration in Accordance with Law was issued,[54] which provided a guidance document for law-based administration that includes enhancement of transparency and accountability. Governments at all levels are required to act in accordance with Law. The goal of law-based administration is also reflected in the government's management of dumping activities. For example, in May 2004, the East China Sea Branch Bureau of the SOA implemented a system of notification and committed itself to simplifying the application procedure.[55] Although MEPL prohibited wastes from outside China from being dumped into the sea area under

---

[51] "Shanghai Waste Disposal Company was Investigated for its Breach to Dump Wastes into Sea," *China Ocean News*, 28 June 2005 (in Chinese).

[52] "Ship Owner in Lianyungang Fined 30,000 *Yuan* for his Illegal Dumping at Sea," *China Ocean News*, 12 September 2006 (in Chinese).

[53] See "China Marine Surveillance North Sea General Brigade and the General Brigade of Liaoning Province Jointly Launched Special Law Enforcement Inspection for Wastes Dumping," *China Ocean News*, 1 December 2006, p. 1 (in Chinese).

[54] Text is available in *People's Daily*, 21 April 2004, p. 6 (in Chinese).

[55] "Marine Environmental Protection," *China Ocean Yearbook 2005* (Beijing: Ocean Press, 2006), p. 182 (in Chinese).

China's jurisdiction, China does permit dumping of wastes from Hong Kong and Macao by special agreement. For example, in March 2004, the SOA signed cooperative agreements with its counterparts in Hong Kong and Macao regarding the dumping of wastes in offshore areas of mainland China.[56]

Conclusion

The regulation and management of dumping of wastes at sea in China has improved in recent years. However, there is much room for improvement and a need to adopt more effective measures.

It is clear that the environmental situation in China is reaching a critical point as an increasing number of environmental problems are being identified. China will face a serious environmental challenge by 2020.[57] The same is true with the marine environment. Recent reports reveal that marine environmental quality continues to decline despite the efforts of the Chinese government. A recent survey revealed that about 80 per cent of the sea water in the Yangtze River estuary and surrounding sea areas (38,000 km^2) is of very poor quality and that there were no sedentary species living on one-third of the seabed.[58] With rapid economic growth and the increased number of offshore projects, it is expected that waste-dumping activities and the quantity of dumped wastes will increase.[59] For example, the quantity of Category C wastes dumped in the East China Sea increased from 42,666,000 cubic metres in 2003 to 89,800,000 cubic metres in 2006.[60] Strengthening the management of the dumping of ocean wastes is an important task in the overall efforts to protect China's marine environment.

---

[56] "SOA and Hong Kong and Macao Signed Agreements on Ocean Wastes Dumping," *China Ocean News*, 2 April 2004 (in Chinese).

[57] "Environmental Situation at 'Critical Point'," *China Daily*, 13 November 2006, p. 1.

[58] "Pollution in the Yangtze River Estuary and Surrounding Areas is Shocking," *People's Daily*, 5 December 2006, p. 6 (in Chinese).

[59] "Ocean Dumping Increases Year by Year in Our Country," *People's Daily*, 12 October 2006, p. 2 (in Chinese).

[60] "SOA East China Sea Bureau Tightens its Supervision and Management of Wastes Dumping at Sea," *China Ocean News*, 13 July 2007, p. 1 (in Chinese).

Regulations governing the dumping of wastes at sea were adopted as early as 1985. More than two decades have passed, and there have been significant changes at both the international and domestic levels. At the international level, numerous regulations and measures concerning the ocean dumping of wastes have been adopted. The most significant is the 1996 Protocol to the 1972 London Dumping Convention. China, as a party to both the London Dumping Convention and the 1996 Protocol, must accommodate the new legal norms and rules into its domestic legislation. At the domestic level, the 1982 MEPL has been amended substantially, and new laws and regulations governing other areas of the marine environmental protection have been adopted. However, the 1985 Dumping Regulations have remained substantially unchanged. Thus, doubts exist as to whether these regulations can effectively govern the management of dumping of wastes at sea. It is expected that the regulations will be revised in the near future, and there are suggestions within China as to how to amend and improve the regulations.[61]

China has adopted straight baselines to measure its territorial sea. Although only part of its baselines and basepoints were publicized in 1996, the size of China's internal waters is immense. The existing dumping sites are predominantly located in these internal waters rather than in the territorial sea or EEZ. These sites are not subject to regulation under the 1996 Protocol, although China can adopt domestic laws to reflect the spirit and letter of the Protocol and has the obligation to provide the competent international organization with "information on legislation and institutional mechanisms regarding implementation, compliance and enforcement in marine internal waters."[62] There are doubts as to whether China's regulations are sufficient and effective in governing the waste-dumping activities in its internal waters.

Overall, there is a general trend towards more stringent rules and standards concerning the dumping of wastes at sea. This was reflected in the shift of the definition of dumping from the 1972 Convention to the 1996 Protocol with the new definition encompassing more activities. The creation of the reverse list is also significant because it only allows a few categories of wastes to be dumped at sea. The tightening process continues,

[61] Ding Jizhao, "Issues and Suggestions on the Revising of the Dumping Regulations," *China Ocean News*, 6 March 2007, p. 3 (in Chinese).

[62] 1996 Protocol, *supra* note 7, at Art. 7(3).

and some waste substances currently included in the reverse list may be removed in the future. The International Maritime Organization recently issued a news release stating that storage of carbon dioxide (CO_2) under the seabed was to be permitted from 10 February 2007 under amendments to the London Dumping Convention. It is expected that the competent international organization, i.e., IMO, will take further measures to develop and improve the existing legal regime on the dumping of wastes at sea. China must be actively involved in these negotiations and, at the same time, be prepared at the domestic level to respond timely and harmoniously to the new legal developments.

The Designation of Particularly Sensitive Sea Areas: A New Layer in the Regime for Marine Environmental Protection from International Shipping*

Aldo Chircop**

Introduction

Like other contributors to this book, this author had the privilege of working closely with Douglas M. Johnston over a long period, and benefited from and is indebted to his functionalist approach to the analysis of regime-building in the international law of the sea and international environmental law. Professor Johnston's numerous scholarly contributions provide insights on the close relationship between legal and political processes, the relationship between unilateral and cooperative action in international arenas, and the role of functionalist values of efficiency, effectiveness and equity in regime-building. Like other policy analysts of the Yale school, Professor Johnston saw the international legal process as a process of political claim and counter-claim in pursuit of foreign policy values and interests. The current international discourse on particularly sensitive sea areas (PSSAs) is similarly reflective of claims and counter-claims in the international arena, frequently fuelled by strong advocacy of new tools to promote marine environmentalism on the one hand, and equally strong support of protection of the sacrosanct right of international navigation.

---

* Parts of this contribution were published prior to the amendment of the IMO PSSA Guidelines in December 2005 in "Particularly Sensitive Sea Areas and International Navigation Rights: Trends, Controversies and Emerging Issues," in Iwan Davies, ed., *Issues in International Commercial Law* (Aldershot: Ashgate Publishing, 2005), pp. 217–243. The author is grateful to Iwan Davies for permission to use the earlier work.

** Professor of Law and Associate, Marine & Environmental Law Institute, Dalhousie University, Halifax, Canada.

Since the adoption of the International Convention for the Prevention of Pollution from Ships, 1973 (as amended in 1978; MARPOL 73/78)[1] and the United Nations Convention on the Law of the Sea, 1982 (LOS Convention),[2] there have emerged regulatory ocean zoning practices aimed at the protection of the marine environment from the impact of vessel operations. These have developed under the auspices of the International Maritime Organization (IMO), which is the principal international arena for the regulation of safety at sea and marine environmental protection from shipping and navigation. A first type of zoning commenced with MARPOL 73/78 and concerns the designation of marine spaces known as "special areas," within which there is a higher level of protection from specific types of vessel-source pollution than the norm. A second type is the PSSA, the subject of this essay. Emerging in 1991, the PSSA is a relatively new type of ocean zoning applying to marine areas of various sizes and within which a range of navigational safety and environmental protection tools may be applied.

Between 1991 and 2000 there were only two PSSAs designated, at the behest of Australia and Cuba, and following technical consideration by the IMO.[3] Since then many more PSSAs have been designated. PSSAs have been triggered in part by widely-publicized oil pollution casualties and underlying concerns over the safety, environmental, social, and eco-

[1] *International Convention for the Prevention of Pollution from Ships, 1973, as modified by the Protocol of 1978 relating thereto*, London, 17 February 1978, *U.N.T.S.* 1340 (1983): 62 [hereinafter MARPOL 73/78]. The convention was further amended by the *Protocol of 1997 to amend the International Convention for the Prevention of Pollution from Ships, 1973, as modified by the Protocol of 1978 relating thereto*, London, 26 September 1997, which added Annex VI. There have been numerous amendments to the annexes. For information on the latest amendments to the annexes, see IMO, online: <www.imo .org/Conventions/contents.asp?doc_id=678&topic_id=258#2004sewage>, accessed January 2008.

[2] *United Nations Convention on the Law of the Sea*, Montego Bay, 10 December 1982, in *The Law of the Sea: Official Text of the United Nations Convention on the Law of the Sea* (New York: United Nations, 1983) [hereinafter the LOS Convention].

[3] "Identification of the Great Barrier Reef Region as a Particularly Sensitive Area," IMO/ MEPC.44(30), 16 November 1990, which was extended to include the Torres Strait; "Report of the Marine Environment Protection Committee on its Forty-Ninth Session," IMO/MEPC 49/22, 8 August 2003; "Identification of the Archipelago of Sabana-Camaguey as a Particularly Sensitive Sea Area," IMO/MEPC.74(40), 25 September 1997.

nomic threats posed by ageing or substandard shipping in coastal waters and exclusive economic zones (EEZs) and the vulnerable ecosystems within them.

The designation of PSSAs has not always been a smooth technical process. Some PSSAs have been proposed in the context of contentious political and legal debate, within and outside of the IMO, suggesting to some that re-naming this zone as a "politically sensitive sea area" might not be altogether inappropriate. The concerns expressed are more than mere technical concerns over combating vessel-source pollution. At the heart of the discourse is genuine regard for the need to protect sensitive coastal and marine ecosystems at a time when there is heightened concern over the growing loss of marine life and habitats and traditional concerns over coastal State encroachment and potential impediments to international navigation rights. The IMO has been the central arena for PSSA claims and counter-claims, because the decisions of this body provide legitimacy for coastal State measures in relation to international shipping, which would not otherwise be the case if those measures were undertaken unilaterally by States.

Regime Context

To understand the potential of PSSAs as a modern ocean zoning practice, it is useful to conceptualize their role in the scheme of international shipping regulation. Over the last three decades, the global protection of the marine environment from vessel-source pollution has evolved into a complex multi-level protective regime within the LOS Convention framework and several specialized international legal instruments. The regime can be envisaged as consisting of four major levels of protection. The first three are discussed in depth below, while the fourth level is not considered in detail because it is peripheral to the PSSA discussions. The four levels are as follows:

- Level 1 consists of a general global obligation for States to protect the marine environment from vessel-source pollution, set out principally in the LOS Convention and the general standards set out in international maritime law conventions, notably MARPOL 73/78.
- Level 2 provides a higher protection level for certain marine regions in the form of more stringent vessel discharge standards in special areas

designated by the IMO within the MARPOL 73/78 framework and the guidelines adopted by the IMO for the designation of special areas.
- Level 3 protection derives from Article 211(6) of the LOS Convention and the guidelines adopted by the IMO to facilitate the designation of PSSAs where the existing legal and management regime is deemed insufficient to meet their protection from international shipping.
- Level 4 concerns exceptional and ad hoc situations where coastal States, motivated by serious concerns for self-protection, are empowered to intervene in relation to shipping casualties "to prevent, mitigate or eliminate grave and imminent danger to their coastline or related interests from pollution" or the threat thereof under the International Convention relating to Intervention on the High Seas in Cases of Oil Pollution Casualties, 1969[4] and the LOS Convention.[5]

Although not considered here, it should be pointed out that in several marine regions there are additional protections initiated in regional, rather than global fora, and within the global framework set out above. Regional initiatives cut across the first three protection levels, serving to regionalize global obligations and thereby reinforce special regional seas protection. There are also marine protected areas (MPAs) that are national or transboundary which are of concern to international shipping. These are not addressed by this contribution, except to a limited reference where a particular PSSA is adopted to provide further protection for an existing MPA.

Level 1: Duty to Protect the Marine Environment from Vessel-Source Pollution

Parties to the LOS Convention have a fundamental legal obligation to protect and preserve the marine environment.[6] This includes a duty to harmonize their policies and to take individual or collective measures

---

[4] *International Convention relating to Intervention on the High Seas in Cases of Oil Pollution Casualties*, Brussels, 29 November 1969, 970 *U.N.T.S.* 211 (1975), Art. 1.

[5] LOS Convention, *supra* note 2, at Art. 221.

[6] *Ibid.*, Art. 192. The performance of this obligation is further developed through a multilateral obligation to "co-operate on a global basis and, as appropriate, on a regional basis, directly or through competent international organizations, in formulating and elaborating international rules, standards and recommended practices and procedures." *Ibid.*, Art. 197.

to prevent, reduce and control pollution.[7] Specifically in relation to vessel-source pollution, this general duty translates into a specific duty to take measures to minimize both operational and accidental pollution.[8] The measures taken by a State must not unjustifiably interfere with the exercise of legitimate rights by other State parties to the LOS Convention.[9] Such measures "include those necessary to protect and preserve rare or fragile ecosystems as well as the habitat of depleted, threatened or endangered species and other forms of marine life."[10] Thus, the basic principle is that while coastal States are empowered, and indeed have the duty to take protective measures against vessel-source pollution, they must do so in a manner that does not interfere unjustifiably with international navigation rights. In general, if there is to be interference, it must be in accordance with international rules and standards. This applies to the territorial sea,[11] international straits,[12] and the EEZ.[13] In fact, the protective action to be taken must be in accordance with international rules and standards adopted through the IMO and may include routeing schemes to minimize threats to local interests.[14] Moreover, the process of adoption

---

[7] *Ibid.*, Art. 194(1).

[8] Specifically, "pollution from vessels, in particular measures for preventing accidents and dealing with emergencies, ensuring the safety of operations at sea, preventing intentional and unintentional discharges, and regulating the design, construction, equipment, operation and manning of vessels." *Ibid.*, Art. 194(3)(b).

[9] *Ibid.*, Art. 194(4).

[10] *Ibid.*, Art. 194(5).

[11] "Coastal States may, in the exercise of their sovereignty within their territorial sea, adopt laws and regulations for the prevention, reduction and control of marine pollution from foreign vessels, including vessels exercising the right of innocent passage. Such laws and regulations shall, in accordance with Part. 11, section 3, not hamper innocent passage of foreign vessels." *Ibid.*, Art. 211(4); in relation to this provision see also Art. 24. Art. 21 concerns, *inter alia*, the coastal State's legislative jurisdiction for safety of navigation and marine environment protection in the territorial sea. Art. 221(2) provides that "[Such] laws and regulations shall not apply to the design, construction, manning or equipment of foreign ships unless they are giving effect to generally accepted international rules and standards." See also Art. 22 in which the coastal State may designate or prescribe sea lanes and traffic separation schemes for ships exercising innocent passage in the territorial sea. In doing so, it is required to take into account IMO recommendations.

[12] *Ibid.*, Arts. 41–42.

[13] *Ibid.*, Art. 211(5). See also Art. 56(1)(b)(iii) concerning the allocation of marine environmental jurisdiction to the coastal State and Art. 58 on the rights and duties of other States in the EEZ.

[14] *Ibid.*, Art. 211(1).

Table 1. Status of MARPOL 73/78

Convention & Annexes	Entry into force	Parties	Percentage of world tonnage
Convention	2 October 1983	146	99
Annex I/II	2 October 1983	146	99
Annex III	1 July 1992	128	95.07
Annex IV	27 September 2003	119	80.74
Annex V	31 December 1988	134	96.85
Annex VI	19 May 2005	51	80.36

Source: IMO, Summary of Status of Conventions (31 May 2008), online: http://www.imo.org/conventions/mainframe.asp?topic_id=247 (accessed 30 June 2008)

of a measure that interferes with international navigation requires not only justification of the measure, but also approval through the IMO. In providing the framework for protective action, the LOS Convention crafts a delicate balance between coastal State interests and international navigation rights. The LOS Convention framework and general principles set out above are then nourished by several specialized instruments addressing aspects of vessel-source impact on the marine environment, chief of which is MARPOL 73/78. Boasting a wide membership (Table 1), MARPOL 73/78 provides in-depth treatment of Level 1 responsibilities for the prevention of operational vessel-source pollution.

Thus at Level 1, while the LOS Convention provides the general framework, MARPOL 73/78 provides the substance and regulatory teeth for the prevention of pollution from ships through a standard-setting approach and a compliance system.

Level 2: Protection through MARPOL 73/78 Special Areas

In addition to the general standards for vessel-source pollution at Level 1, MARPOL 73/78 also provides for a higher level of protection for certain areas, technically known as "special areas."[15] A special area is defined as "...a sea area where for recognized technical reasons in relation to its oceanographical and ecological conditions and to the particular character

[15] Even earlier, OILPOL, the precursor of MARPOL 73/78, introduced no-discharge zones for oily wastes. *International Convention for the Prevention of Pollution of the Sea by Oil*, London, 12 May 1954, 327 *U.N.T.S.* 3(1959). This instrument was amended in 1962, 1969 and 1971, and is largely superseded by MARPOL 73/78.

Table 2. MARPOL 73/78 special areas and atmospheric emission control areas

Annex I: Oil pollution	**Annex II: Noxious liquid substances**	**Annex V: Garbage**	**Annex VI: Atmospheric emission control area (SOx)**
– Mediterranean Sea – Baltic Sea – Black Sea – Red Sea – "Gulfs" area* – Gulf of Aden – Antarctic area (south of 60° latitude) (Antarctic Treaty Area) – Northwest European waters (comprising the North Sea, Irish Sea, Celtic Sea, English Channel and part of the Northeast Atlantic west of Ireland) – Oman area of Arabian Sea – Southern South African Waters	– Baltic Sea – Black Sea – Antarctic area (Antarctic Treaty Area)	– Mediterranean Sea – Baltic Sea – Black Sea – Red Sea – "Gulfs" area – Antarctic area (Antarctic Treaty Area) – North Sea (including the Skagerrak and English Channel) – Wider Caribbean	– Baltic Sea – North Sea

* "Gulfs" area is how the Persian/Arabian Gulf are referred to by the IMO for special area purposes.

Source: Adapted from IMO, Special Areas under MARPOL, online: http:/www.imo.org/Environment/mainframe.asp?topic_id=760 (accessed 8 January 2008).

of its traffic, the adoption of special mandatory methods for the prevention of sea pollution by oil, noxious liquid substances, or garbage, as applicable, is required."[16] Special areas deserve a higher level of protection than the general norm. The seas or marine areas that have been so designated in the relevant MARPOL 73/78 annexes are listed in Table 2.

[16] "Guidelines for the Designation of Special Areas under MARPOL 73/78," IMO Doc A 22/Res 927, 15 January 2002 [hereafter Special Area Guidelines]. The Special Area Guidelines were formulated for Annexes I, II, and V, and do not concern emission control areas in Annex VI.

Special areas can be established under MARPOL 73/78 Annexes I (oily wastes), II (noxious liquid substances), and V (garbage), and they subject qualifying ships in those areas to more stringent restrictions than the general rules on vessel discharges. Annex VI further provides for the designation of emission control areas insofar as atmospheric pollution from ships is concerned, namely sulphur and nitrogen oxides (respectively SOx and NOx). As the various annexes suggest, special areas are designated with reference to a particular type of waste generated by ships. The IMO adopts special areas on the basis of amendments to the pertinent annexes of MARPOL 73/78. To facilitate designation, the IMO has developed Special Area Guidelines for submissions of special area proposals.[17] While designed to assist, the intention within the IMO is to avoid a proliferation of such areas.[18] By the time the Guidelines were adopted, most of the special areas were already designated. The Guidelines have thus been applied only to the Antarctic (Annexes I, II and V), northwest European waters (Annex I), Wider Caribbean (Annex V), Gulf of Oman area of the Arabian Sea (Annex I), and southern South African waters (Annex I).

MARPOL 73/78 does not impose *a priori* limits on the size of a special area. The size of each special area depends on the geographical space that needs protection. The designations to date relate to large marine areas, that is, entire seas, EEZs, or parts of EEZs that satisfy three key criteria, namely oceanographic, ecological and vessel traffic conditions. Concerning oceanographic conditions, proponent States have to show that the physical conditions are such that they "may cause the concentration of harmful substances in the waters or sediments of the area."[19] These could include, for example, restricted hydrology or extreme ice conditions. In the case of the Baltic, for example, it takes 30 years to recycle the waters. Ecological conditions include threats to marine species; protection of high biological productivity areas, nurseries and habitats; protection of migratory routes for marine avifauna and mammals; and protection of rare or fragile coastal and marine ecosystems.[20] As for vessel traffic conditions, "[T]he sea area is used by ships to an extent that the discharge of harmful substances by

[17] *Ibid.*

[18] IMO/MEPC 30/19/1, 17 August 1990, p. 24.

[19] Special Area Guidelines, *supra* note 16, at para. 2.4.

[20] *Ibid.*, para. 2.5.

ships when operating in accordance with the requirements of MARPOL 73/78 for areas other than special areas would be unacceptable in the light of the existing oceanographic and ecological conditions in the area."[21]

In addition to these three criteria, arguments that may strengthen a special area proposal include threats to amenities (e.g., coastal values), threats by other sources of marine pollution, and the existing management regime.[22] These arguments are important because consideration is given to cumulative impact; at the same time, other measures planned or taken to combat other pollution sources strengthen a proposal. Further, a special area proposal may need to be coordinated with other regional management initiatives, such as the United Nations Environment Programme (UNEP) Regional Seas Programme applicable in the area, and national integrated coastal and ocean management initiatives. The proposal requirements are detailed and rigorous, and in addition to placing the burden of proof as to why a marine area should be designated a special area on the proponent(s), there is the further onus to provide information on the availability of reception facilities.[23] As will be seen below, this is a critical requirement.

As can be expected, while operating at sea ships generate various types of waste, discharge of which is mostly banned or controlled. When and where a controlled discharge is permitted, the discharge tends to be at a regulated low level. With very few exceptions,[24] in special areas the discharge level is at or close to zero release. An important element in relation to wastes that can be discharged at sea is the distance from the shore, i.e., "the nearest land." Although no special areas have been designated under MARPOL 73/78 Annex IV on sewage, the term "from the nearest land" for discharge purposes was re-defined in the case of the coast of northeastern Australia, in relation to the Great Barrier Reef, to enable

---

[21] *Ibid.*, para. 2.6.

[22] *Ibid.*, paras. 2.7, 2.8 and 2.10.

[23] *Ibid.*, Part 3.

[24] MARPOL 73/78, *supra* note 1, at Annex 1, Reg. 11, concerning discharges in cases of *force majeure*, as a result of damage to ship or equipment and where there is approval of a national maritime administration to combat a pollution incident.

authorized releases further out at sea than would otherwise be the case.[25] Identical terms are also found in Annexes I, II, and V.[26]

Level 2 protection has soft spots. When a State becomes a party to MARPOL 73/78, only Annexes I and II are compulsory. Annexes III–VI are optional, so that parties to the main Convention may choose not to be bound by these annexes. Clearly, this is also a weakness for Level I insofar as the regulation of pollution from liquid noxious substances, sewage, and garbage from ships are concerned. For Level II protection, clearly special areas adopted in relation to liquid noxious substances and garbage would not bind non-parties to Annexes III and V. This concern could and has been addressed through port State control inspection and enforcement under Port State Control Memoranda of Understanding (MOU), which enable member maritime administrations to apply international safety and environmental standards also in relation to ships flying the flags of non-party States.[27] Another weakness in the past was that the annexes did not enter into force until a long period after adoption. Annexes I–V were adopted in 1973 (and re-incorporated into the 1978 Protocol); but Annex I only came into force in 1983, Annex II in 1987, Annex III in 1992, Annex IV in 2003 and Annex V in 1988. Annex VI, which was introduced into MARPOL 73/78 as a result of an amendment in 1997, only entered into force in 2005.[28]

A more significant and enduring concern is the extent to which the conditions for a special area are being complied with by affected coastal States. Coastal States geographically located adjacent to special areas have an obligation to provide adequate reception facilities for oily residues and

---

25 MARPOL 73/78, *ibid.*, Annex IV: Regulations for the Prevention of Pollution by Sewage, Reg. 1(5).

26 The adjusted definition of "nearest land" in the case of Australia had been introduced earlier in an amendment to OILPOL, *supra* note 15, at Art. 1, by IMCO (now IMO) Assembly Resolution A 232 (VII), 12 October 1971.

27 In inspecting ships visiting their ports, port States apply conventions designated in the MOU(s) to which the flags of visiting ships may not be parties. On port State control see G. Kasoulides, *Port State Control and Jurisdiction: Evolution of the Port State Regime* (Dordrecht: Martinus Nijhoff, 1993).

28 Annex VI came into force when the 1997 Protocol entered into force on 19 May 2005 after 15 State parties to the Convention representing 50 per cent of the world gross tonnage became parties to the Protocol. 1997 Protocol, *supra* note 1, at Art. 6.

mixtures,[29] noxious liquid substances,[30] sewage,[31] and garbage[32] and to inform IMO accordingly. Although by 2003 there were 128 State parties to MARPOL 73/78,[33] at that time only 57 of those States actually communicated information on reception facilities for oily residues and mixtures to IMO and many of the submitted reports did not demonstrate a capability to accept all the expected oil wastes.[34] This is important because special areas under the annexes become effective after the necessary reception facilities are established and this information is communicated to IMO. As a result, it appears that not all the designated special areas are actually in effect.[35] The special area practice to date appears to be mixed and its effectiveness in some regions remains to be assessed.

---

29 MARPOL 73/78, Annex I, *supra* note 1, at Regs. 10(7)(8) and 12.

30 MARPOL 73/78, *ibid.*, Annex II, Reg. 7.

31 MARPOL 73/78, *ibid.*, Annex IV, Reg. 10. It is appropriate to note that despite the late entry into force of this annex, many State parties had already implemented its provisions into national law (for example, Denmark and Canada).

32 MARPOL 73/78, *ibid.*, Annex V, Reg. 5(4), (5).

33 IMO, "Summary of Status of Conventions" (31 July 2004), online: <www.imo.org/Conventions/mainframe.asp?topic_id=247>, 3 September 2004.

34 The wastes at issue are dirty ballast water, tank-washing slops, oily mixtures containing chemicals, scale and sludge from tank cleaning, oily bilge water, and sludge from fuel oil purifier. See "Facilities in Ports for the Reception of Oily Wastes from Ships," IMO/MEPC 3/Circ 4, 18 November 2003. At its 12th Session in March 2004, the IMO Sub-Committee on Flag State Implementation embarked on an initiative to address the problem of low level reporting by disseminating a questionnaire. See IMO/MEPC/Circ 417, 7 July 2004. See also, IMO, *Comprehensive Manual on Port Reception Facilities*, 2nd ed. (London: IMO, 1999).

35 For example in relation to Annex I, see Reg. 10(7)(a)(i) and (iii). Also, "[t]he requirements of a special area designation can only become effective when adequate reception facilities are provided for ships in accordance with the provisions of MARPOL 73/78." Special Area Guidelines, *supra* note 16, at para. 2.7. As of 2007, the following special areas became effective: *Annex I* – Mediterranean, Baltic and Black Sea areas on 2 October 1983; Antarctic on 17 March 1992; Northwest European waters on 1 August 1999; Southern South Africa waters is expected to become effective on 1 August 2008; the Gulfs Area is expected to become effective on 1 August 2008; *Annex II* – Baltic on 6 April 1987 and Antarctic on 1 July 1994; *Annex V* – Baltic on 1 October 1989, North Sea on 18 February 1991, Antarctic on 17 March 1992 and Wider Caribbean on 4 April 1993; "Gulfs Area" is expected to become effective on 1 August 2008 and Mediterranean on 1 May 2009. Annex VI emission control areas now in effect are the Baltic Sea (as of 19 May 2006) and North Sea (on 22 November 2007). Thus, the following special areas are not in effect or have no date of projected effectivity: Annex I – Red

Level 3: LOS Convention Article 211(6) and PSSAs

Level 3 protection is of most direct concern for the establishment of PSSAs and the level that has raised some of the most contentious claims and counter-claims. The starting point is Article 211(6) of the LOS Convention because there are different views in the international maritime community on the meaning of this provision and also because there appears to be a relationship between this provision, PSSAs, and MARPOL 73/78 special areas which calls for clarification.

LOS Convention Article 211(6): Special Mandatory Measures

Article 211(6) of the LOS Convention provides the following:

a) Where the international rules and standards referred to in paragraph 1 are inadequate to meet special circumstances and coastal States have reasonable grounds for believing that a particular, clearly defined area of their respective exclusive economic zones is an area where the adoption of special mandatory measures for the prevention of pollution from vessels is required for recognized technical reasons in relation to its oceanographical and ecological conditions, as well as its utilization or the protection of its resources and the particular character of its traffic, the coastal States, after appropriate consultations through the competent international organization with any other States concerned, may, for that area, direct a communication to that organization, submitting scientific and technical evidence in support and information on necessary reception facilities. Within 12 months after receiving such a communication, the organization shall determine whether the conditions in that area correspond to the requirements set out above. If the organization so determines, the coastal States may, for that area, adopt laws and regulations for the prevention, reduction and control of pollution from vessels implementing such international rules and

---

Sea, Gulf of Aden and the recent Oman Sea; Annex II – Black Sea; Annex V – Black Sea, Red Sea and Wider Caribbean. See IMO, "Special Areas under MARPOL," online: <www.imo.org/Environment/mainframe.asp?topic_id=760>, 8 January 2008. See also, "Status of Multilateral Conventions and Instruments in Respect of Which the IMO or its Secretary-General Performs Depositary or other Functions," as at 31 December 2005, online: <www.imo.org/includes/blastDataOnly.asp/data_id%3D14744/9193.pdf>, 8 January 2008.

standards or navigational practices as are made applicable, through the organization, for special areas. These laws and regulations shall not become applicable to foreign vessels until 15 months after the submission of the communication to the organization.

b) The coastal States shall publish the limits of any such particular, clearly defined area.

c) If the coastal States intend to adopt additional laws and regulations for the same area for the prevention, reduction and control of pollution from vessels, they shall, when submitting the aforesaid communication, at the same time notify the organization thereof. Such additional laws and regulations may relate to discharges or navigational practices but shall not require foreign vessels to observe design, construction, manning or equipment standards other than generally accepted international rules and standards; they shall become applicable to foreign vessels 15 months after the submission of the communication to the organization, provided that the organization agrees within 12 months after the submission of the communication.

The first observation is that this provision does not provide for the designation of special areas or PSSAs per se, but rather it provides for the adoption of special mandatory measures in special circumstances. Second, the reference to "special area" in this provision is distinct from MARPOL 73/78 special areas.[36] It has been suggested that Article 211(6) was modelled on the MARPOL 73/78 special area provisions, although this view also holds that the LOS Convention provision provides for a much wider application than the MARPOL 73/78 equivalents.[37]

The idea of special protection, that is, referring to particular marine areas that deserve a higher level of protection from shipping activities, predates both the LOS Convention and MARPOL 73/78. The 1954 International Convention for the Prevention of Pollution of the Sea by Oil (OILPOL),[38] the precursor to MARPOL 73/78, provided for no-discharge

36 M. Nordquist, ed., *United Nations Convention on the Law of the Sea 1982: A Commentary*, Vol. 4 (S. Rosenne and A. Yankov, vol. eds.) (Dordrecht: Martinus Nijhoff, 1991), p. 181 [Rosenne and Yankov]; E. Molenaar, *Coastal State Jurisdiction over Vessel-Source Pollution* (The Hague: Kluwer, 1998), p. 403.

37 Molenaar, *ibid.*, at p. 403.

38 OILPOL, *supra* note 15.

zones. Canada's submission in 1972 to the Seabed Committee preparing the agenda for the Third United Nations Conference on the Law of the Sea (UNCLOS III) referred to the principle of "special authority of the coastal state in areas adjacent to its territorial waters" and, significantly, "a specific and limited exercise of residual authority by the coastal state to ensure compliance with internationally agreed standards or with special local standards."[39] This contribution was probably influenced by the 1970 unilateral initiative taken by Canada to establish special protection in the Arctic through the *Arctic Waters Pollution Prevention Act*.[40] Some of these ideas made it into the International Conference on Marine Pollution that produced the MARPOL Convention in 1973.

The diplomatic history of UNCLOS III sheds useful light on the origins and intentions of Article 211(6), for which many proposals were submitted over several conference sessions. In 1974, Kenya advanced a proposal for "appropriate standards for special situations."[41] Later that year, a proposal by ten States referred to special circumstances for which coastal States could adopt laws and regulations in addition to or more stringent than international rules and standards, including stricter standards justified by exceptional hazards and special environmental vulnerability.[42] A "softer" text and one that encapsulated some of the key concepts to eventually appear in the provision (that is, special area, recognized technical justification, special mandatory measures, application to the IMO) was advanced in 1975 by nine States.[43] In 1975, the Informal Single Negotiating Text captured most of these ideas, but created as a precondition the inexistence or inadequacy of international rules and standards to deal with the special circumstances, as well as the coastal State's reasonable belief concerning the

---

39 Rosenne and Yankov, *supra* note 36, at pp. 183–184.

40 R.S. 1985, c. A-12. This initiative eventually influenced the adoption of Article 234 of the LOS Convention on the protection of ice-covered areas. On the Canadian view of the relationship of this provision to MARPOL 73/78, and reactions from the United States and several European States, see Status of Multilateral Conventions, *supra* note 35, at p. 96.

41 Rosenne and Yankov, *supra* note 36, at p. 186.

42 Canada, Fiji, Ghana, Guyana, Iceland, India, Iran, New Zealand, the Philippines, and Spain. *Ibid.*, at pp. 186–187.

43 Belgium, Bulgaria, Denmark, German Democratic Republic, Federal Republic of Germany, Greece, Netherlands, Poland, and the United Kingdom. *Ibid.*, at p. 187.

needed measures. The zone would be "a particular area" of the EEZ.[44] By and large, and with adjusted text, the same ideas were reproduced in the Revised Single Negotiating Text in 1976. In this text, the terms "special" and "particular" appear more than once. This would also carry through in the Informal Composite Negotiating Text (ICNT) in 1977, which further developed and refined the text and clearly set out that the determination of special areas belonged to the IMO, not to the proponent State, and that only after such a determination could the coastal State proceed with domestic action.[45] This was necessary if coastal State measures were to be legislated and enforced against international navigation. In effect, the coastal State would be enforcing international rules and standards. As a result of the *Amoco Cadiz* disaster in 1978, France introduced new proposals to emphasize preventive measures, such as routeing systems and notification of coastal States.[46] Following much debate, a compromise proposal was submitted by Tanzania, included in the second revision of the ICNT, and eventually in the LOS Convention.[47]

It was clear from the negotiations of Article 211(6) that what was being sought was the possibility to impose stricter measures than the norm with reference to "a particular, clearly defined area" of the EEZ. Special mandatory measures meant that international standards for special areas would be raised or made more stringent. There is not anything that suggests that an entire EEZ could be captured under Article 211(6) as requiring special measures, hence the distinction that was maintained with special areas in MARPOL 73/78, which evidently could include entire EEZs in semi-enclosed seas. It must be recalled that the intention of the LOS Convention negotiators was to "replace a plethora of conflicting claims by coastal states with agreed limits" for the maritime zones, and that the freedom of navigation was facilitated by, *inter alia*, the compromise reached on the status of the EEZ.[48] Hence, to justify that a PSSA (including the connotation of further restrictions on international navigation) could encompass an entire EEZ (as will be seen

[44] *Ibid.*, at p. 189.

[45] *Ibid.*, at pp. 192–193.

[46] *Ibid.*, at pp. 193–194.

[47] *Ibid.*, at pp. 196–199.

[48] "A Constitution for the Oceans," Remarks by Tommy T. B. Koh, Singapore, President of the Third United Nations Conference on the Law of the Sea, in *Law of the Sea, supra* note 2, at p. xxxiii.

below), the PSSA concept has to be considered separate from the special area contemplated in Article 211(6), because it is not derived from this provision. Otherwise the PSSA concept would appear to be inconsistent with the apparent spatial constraint for special areas under Article 211(6) which reflects the delicate balance that the LOS Convention negotiators attempted to reach and produced.

It is against this backdrop that the conditions in Article 211(6) should be interpreted. The conditions justifying special mandatory measures under this Article are

(1) The normally applicable international rules and standards are inadequate to meet special circumstances (at a minimum, this means that Level 1 protection is not sufficient, but arguably applicable may also refer to Level 2 protection); and
(2) The coastal State has reasonable grounds for believing that a particular, clearly defined area of its EEZ is an area where the adoption of special mandatory measures for the prevention of pollution from vessels is required for:
    (a) recognized technical reasons in relation to the area's oceanographical and ecological conditions;
    (b) the utilization or the protection of the area's resources; and
    (c) the particular character of the area's traffic.

The burden of proof rests with the coastal State requesting a special designation and the procedure applicable is as follows:

(1) The coastal State is required to conduct appropriate consultations with other concerned States through the IMO. "Concerned States" may well be more than States that actually express a concern and can include any IMO member State that expresses an interest. Basically, this is a duty to consult.
(2) After such consultations, the coastal State "may," suggesting discretion, submit supporting scientific and technical evidence and information on reception facilities for the area concerned. The coastal State has to formally submit a well-founded proposal.

If a coastal State makes such a submission, the IMO has 12 months within which to deliberate on whether the area concerned does indeed satisfy the above conditions. The coastal State has to persuade other IMO member

States of its proposal. Should the IMO, in effect the IMO member States, presumably participating in the deliberations of the Marine Environment Protection Committee (MEPC), decide that this is the case, the coastal State may proceed to legislate international rules and standards that the IMO has adopted as being applicable for such areas. Note the restrictions on the content of that legislation in Article 211(6)(c) – the coastal State's legislation becomes applicable to international navigation, at the earliest, 15 months after the formal submission to the IMO.

The IMO has not developed Article 211(6)-specific guidelines, although it has been suggested that IMO should do so.[49] There is a practical relationship between, on the one hand, the provisions of this article and, on the other, PSSA designation and adoption of associated protective measures (APMs) in accordance with the PSSA Guidelines that will be explored below. APMs are restrictions on international shipping authorized by the IMO for the proponent State to implement in a PSSA. A measure pursuant to Article 211(6) could be advanced as a legal basis for a proposed associated protective measure. Article 211(6) could provide an even higher level of protection in PSSAs (Level 3b-type) because it authorizes *special mandatory measures*, whereas associated protective measures adopted under the current PSSA Guidelines may include *non-mandatory* measures.

Particularly Sensitive Sea Areas

Background and Rationale

In 1978 the IMO was called upon to commence studies to produce an inventory of sea areas in special need of protection from vessel-source pollution and dumping.[50] The call was for an assessment of the extent of the need for such areas and what action would be appropriate to deal

---

[49] A proposal of the United States; see "Proposed Amendments to Assembly Resolution A 927(22) to Strengthen and Clarify the Guidelines for the Identification and Designation of Particularly Sensitive Sea Areas," Submitted by the United States, IMO/MEPC 52/8, 9 July 2004 [hereinafter U.S. Amendments], 3 and 11, note 12.

[50] In the International Conference on Tanker Safety and Pollution Prevention, Resolution 9, February 1978. For an overview of the early history, see K. Gjerde and D. Freestone, "Particularly Sensitive Sea Areas – An Important Environmental Concept at a Turning Point," *International Journal of Maritime & Coastal Law* 9 (1994): 431–468; G. Peet, "Particularly Sensitive Sea Areas – A Documentary History," *International Journal of Maritime & Coastal Law* 9 (1994): 469–506.

with the issue. Action in this regard commenced at the 23rd session of the MEPC in 1986. The early discussions did not result in the acceptance of any one term or concept to capture the nature of the proposed protective zone, but did raise questions on the relationship between MARPOL 73/78 special areas, ice-covered areas (Article 234 of the LOS Convention), and generally the special area reference in Article 211(6) of the LOS Convention. In addition to PSSA as a term of art, other terms were employed to capture the same idea, including "special area," "particularly sensitive area" (which is the Great Barrier Reef's official designation), and "ecologically sensitive area." At the 23rd session it was recognized that the focus should be on a concept for the protection of sea areas, distinct from the MARPOL 73/78 special area concept, that were considered to be particularly sensitive to vessel-source pollution. The Great Barrier Reef was often used as an example. It was recognized that there were a number of international conventions that also called for special protection for specific types of areas.[51]

There was concern over the possibility of national standard-setting for vessel-source pollution occurring outside the global framework provided by the IMO. If stricter vessel standards were to be adopted, it was thought that there should first be consideration on whether such higher standards should be adopted on a global basis. If the need was for something less than a global approach, then, it was thought that MARPOL 73/78 provided an opportunity for the designation of stricter standards for large sea areas.[52] If additional protection was needed for even smaller areas, it was conceivable to simply limit navigation in such areas and to use buffer zones. The International Convention for the Safety of Life at Sea, 1974 (SOLAS) already provided for routeing and the designation of areas to be

[51] For example, *Convention on Wetlands of International Importance Especially as Waterfowl Habitat*, Ramsar, 2 February 1971, online: <www.ramsar.org/key_conv_e.htm>; *Convention for the Protection of the World Cultural and Natural Heritage*, Paris, 23 November 1972, online: <whc.unesco.org/world_he.htm>; *Convention on the Conservation of Migratory Species of Wild Animals*, Bonn, 23 June 1979, online: <fletcher.tufts.edu/multi/texts/BH752.txt>; and several regional conventions. Also, at the time, UNESCO's Man and Biosphere Programme called for the designation of biosphere reserves. IMO/MEPC 30/19/1, 17 August 1990, pp. 14–18.

[52] IMO/MEPC 30/19/1, *ibid.*, at p. 20.

avoided and other measures for safe navigation.[53] Thus, the IMO could entertain requests for additional protection requested by member States, on a case-by-case basis, and prescribe measures from within its arsenal as may be proposed by those States.

These ideas were further developed between 1986 and 1990 and led to the adoption of the PSSA Guidelines in 1991. The Guidelines were amended in 1999, 2001 and more recently by IMO Assembly Resolution A.982(24) on 1 December 2005.[54] The discourse leading to the 2005 amendments has been contentious mostly in relation to the Baltic and Western European PSSAs, which had provided the impetus for the latest revision of the Guidelines. Major flag States, such as Liberia, Panama and the Russian Federation, supported by industry organizations, suggested that a moratorium on new PSSA designations should be adopted until the Guidelines had been reviewed and amended. There was division between those States that thought revisions were needed and those that thought otherwise.[55] The moratorium idea was not accepted with the result that consideration of PSSAs-in-principle (which included the Baltic Sea PSSA) continued. There was agreement on the need for review of the Guidelines,[56] and several proposals were submitted for consideration at the

[53] *International Convention for the Safety of Life at Sea, 1974*, London, 1 November 1974, 1184 *U.N.T.S.* 278 (1980), in particular Chapter V (Navigation). As an example, see "General Provisions on Ships' Routeing," IMO Assembly Resolution A 572(14), 20 November 1985.

[54] IMO Assembly Resolution A 720(17), 6 November 1991, as amended by A 885(21), 25 November 1999. In addition to the 2001 amendments, the MEPC also approved a "Guidance Document for Submission of PSSA Proposals to IMO," IMO/MEPC 48/7/1, 30 April 2002 [hereafter PSSA Guidance Document]. The current version is "Revised Guidelines for the Identification and Designation of Particularly Sensitive Sea Areas," adopted by IMO Assembly Resolution A.982(24), 1 December 2005 [hereafter PSSA Guidelines].

[55] "Report of the Marine Environment Protection Committee on its Fifty-First Session," IMO/MEPC 51/22, 22 April 2004 [hereafter MEPC 51 Report], pp. 29–32. PSSA proliferation was a concern expressed by the Russian Federation, Panama, Liberia, and several industry bodies. See "'Politically Sensitive Sea Areas' Agreed in Principle," *Fairplay*, 8 April 2004, 6.

[56] It was felt by Norway, and the MEPC agreed, that the institution of a moratorium was outside the MEPC mandate and would have to be proposed to the Assembly. IMO/MEPC 51/22, *ibid.*, p. 32.

MEPC's 52nd session in October 2004.[57] The review was undertaken by an inter-sessional group led by the United States and the MEPC's 53rd session adopted the group's recommendations.[58]

All along the distinction between special areas and PSSAs has been maintained. Whereas special areas are direct creatures of a multilateral convention, PSSAs are not, but their designation must be consistent with the provisions of the LOS Convention, and in particular Article 211(6). It has been stated that a "PSSA has no legal significance in and of itself," it is the associated measures that carry legal significance.[59]

According to the U.N. Department of Ocean Affairs and the Law of the Sea, the PSSA Guidelines "derive from the provisions" of Article 211(6) of the LOS Convention, although they are "far more detailed and 'liberal' in their approach" than the LOS Convention provision.[60] In proceeding with the development and eventual amendment of the PSSA Guidelines, the IMO member States do not appear to have derived the idea or content

[57] In particular, U.S. Amendments, *supra* note 49; "Proposed Amendments to Guidelines for the Identification and Designation of Particularly Sensitive Sea Areas (Annex 2 of IMO Assembly Resolution A 927(22))," Submitted by the Russian Federation, IMO/MEPC 52/8/1, 6 August 2004 [hereafter Russian Federation Amendments]; "Proposed Amendments to Resolution A 927(22) on the Identification and Designation of Particularly Sensitive Sea Areas (PSSA)," Submitted by the International Chamber of Shipping (ICS) and INTERTANKO, IMO/MEPC 52/8/2, 6 August 2004; "Proposed Amendments to Resolution A 927(22) on the Identification and Designation of Particularly Sensitive Sea Areas (PSSA) – PSSA Proposal Review Form," Submitted by ICS and INTERTANKO, IMO/MEPC 52/8/3, 6 August 2004; and "Proposed Amendments to Assembly resolution A 927(22) to Strengthen and Clarify the Guidelines for the Identification and Designation of Particularly Sensitive Sea Areas (PSSAs) – Comments on MEPC 52/8," Submitted by WWF, IMO/MEPC 52/8/4, 18 August 2004 [hereafter WWF Comments].

[58] A major player, the United States proposed review of the criteria for PSSA designation, establishment of a link between vulnerability and a proposed measure, legal basis for protective measures and procedural matters. U.S. Amendments, *supra* note 49. The Russian Federation suggested amendments that emphasized the exceptional nature of a PSSA as Level 3 protection, in comparison to Level 2 protection under MARPOL 73/78 special areas or some other regional regime. Russian Federation Amendments, *supra* note 56, at p. 2.

[59] U.S. Amendments, *supra* note 49, at p. 3.

[60] "Comments made by the Division for Ocean Affairs and the Law of the Sea of the United Nations (DOALOS) in connection with Issues Raised in Document LEG 87/16/1," IMO LEG 87/WP 3, 13 October 2003 [hereafter DOALOS Comments].

of the Guidelines from Article 211(6). Rather, the IMO member States invoked mandates in IMO conventions to proceed with the development and adoption of the PSSA Guidelines.[61] As noted earlier, Article 211(6) of the LOS Convention plays a limited role in the PSSA Guidelines, namely in providing the legal basis for special mandatory measures as associated protective measures within a given PSSA.

PSSAs Designated

A PSSA is defined as a marine area

> that needs special protection through action by IMO because of its significance for recognized ecological, socio-economic, or scientific attributes where such attributes may be vulnerable to damage by international shipping activities.[62]

It has been described as a comprehensive management tool, but in reality the management measures to be adopted are selective tools and are specifically in relation to international shipping activities.[63] In the first decade after the adoption of the PSSA Guidelines in 1991, there were only two proposals that led to designations. Since 2002, however, there have been several proposals submitted to the IMO that have led to PSSA designations, together with the adopted associated measures (see Table 3).

Conditions and Sensitivity Criteria

The key conditions to be satisfied both for PSSA designation and the adoption of related measures are in three steps: (1) the area's environmental conditions, (2) its vulnerability to international shipping and (3) availability of measures within the IMO's competence.[64] It should be emphasized that the PSSA designation is a problem-oriented mechanism to enable the adoption of measures to protect against damage to such areas from

---

[61] The IMO's general authority in this regard stems from its own constitutive instrument. *Convention on the International Maritime Organization*, 3 September 1976, 1143 *U.N.T.S. 105* (1982), Arts. 1(a) and 15(j). According to DOALOS, the PSSA "Guidelines conform to the requirements of article 237 [of the LOS Convention], as a subsequent agreement adopted by IMO Assembly in furtherance of the general principles set forth in the Convention, in particular, those in Part XII." DOALOS Comments, *ibid.*, note 59.

[62] PSSA Guidelines, *supra* note 54, at para. 1.2.

[63] PSSA Guidance Document, *supra* note 54, at para. 1.2.

[64] PSSA Guidelines, *supra* note 54, at para. 1.5.

Table 3. PSSAs designated

PSSA	Year adopted	Proponent State(s)	Appropriate protective measures adopted
Great Barrier Reef	1990	Australia	Compulsory pilotage IMO-recommended pilotage Mandatory reporting
Sabana-Camaguey Archipelago	1997	Cuba	Traffic separation schemes Area to be avoided Prohibited discharges
Wadden Sea	2002	Denmark, Germany, Netherlands	No new additional measures; only existing measures invoked (compulsory reporting and traffic surveillance, routeing systems)
Florida Keys sea area	2002	United States	Areas to be avoided (new and amendment of existing) No anchoring area
Malpelo Island	2002	Colombia	Area to be avoided
Paracas National Reserve	2003	Peru	Tanker traffic restrictions Prohibited discharges Traffic separation schemes
Western European area	2004	Belgium, France, Ireland, Portugal, Spain and United Kingdom	Mandatory ship-reporting system
Torres Strait	2005	Australia and Papua New Guinea	Two-way shipping route Extension of Great Barrier Reef pilotage
Canary Islands	2005	Spain	Traffic separation schemes (including inshore traffic zones and precautionary areas) Areas to be avoided Mandatory ship reporting
Galapagos Archipelago	2005	Ecuador	Area to be avoided (as a buffer zone around the PSSA)

Table 3. (*cont.*)

PSSA	**Year adopted**	**Proponent State(s)**	**Appropriate protective measures adopted**
Baltic Sea (less Russian waters)	2005	Denmark, Estonia, Finland, Germany, Latvia, Lithuania, Poland and Sweden	New and amended traffic separation schemes and associated routing measures (including inshore traffic zones and precautionary areas) in three areas
Papahānaumokuākea Marine National Monument (Hawaii)	2008	United States	Areas to be avoided Ship reporting system (mandatory for ships passing through PSSA to or from a U.S. port; voluntary for all others)

Sources: Identification and Protection of Special Areas and Particularly Sensitive Sea Areas: Report of the Technical Group on Particularly Sensitive Sea Areas (PSSAs), IMO MEPC MEPC 53/WP.15, 20 July 2005; IMO Briefing 13/2008, 9 April 2008; NOAA Press Release, Papahānaumokuākea Marine National Monument Designated a Particularly Sensitive Sea Area, <hawaiireef.noaa.gov/news/releases/pdfs/FINAL_PSSA040708.pdf>.

international navigation and not from other ocean uses. Other threats to the same areas, for example, dumping, are covered by other regimes.

The step 1 conditions share some substantive similarities with those for MARPOL 73/78 and Article 211(6) special areas and include clusters of sensitivity criteria in the form of ecological, social, cultural and economic, and scientific and educational criteria (see Table 4).

Only one of the sensitivity criteria within any one of the three clusters needs to be satisfied before proceeding with consideration of the Step 2 condition.[65] In the past this minimalist requirement was the cause of significant concern because virtually any internationally navigable area of the oceans could qualify and, as a result, the effect of providing special protection for very special places, that is, particularly sensitive, could be undermined. Several States advocated change in the PSSA Guidelines on this point in anticipation of the 2005 amendments. The Russian Federation

[65] *Ibid.*, para. 4.4.

Table 4. PSSA Sensitivity Criteria

Ecological cluster	**Social, cultural and economic cluster**	**Scientific and educational cluster**
(1) Uniqueness or rarity (2) Critical habitat (3) Dependency (4) Representativeness (5) Diversity (6) Productivity (7) Spawning or breeding grounds (8) Naturalness (9) Integrity (10) Vulnerability (11) Bio-geographical importance	(12) Economic benefit (13) Recreation (14) Human dependency	(15) Research (16) Baseline and monitoring studies (17) Education

wanted to see a prioritization of criteria, with the highest emphasis to be given to ecological criteria, but this was not adopted.[66] The United States was of the view that "the criteria are not as clear and meaningful as they should or could be, thus making it difficult to conduct a meaningful review of proposed PSSAs."[67] However, a more flexible view advanced by European Union member States held sway, perhaps best articulated by the World Wildlife Fund (WWF) when it recommended that the precautionary character of the PSSA criteria should be maintained.[68] It was felt that a tightening of the criteria in the U.S. proposal to emphasize scientifically-recognized as an element in the criteria could have conflicted with the needs of precaution.[69]

Motivated by its experience with the Baltic PSSA, the Russian Federation unsuccessfully proposed an additional criterion which would have required PSSA proponents to take into account the interests of other States in the proposed area. The implication was that in the case of enclosed or semi-enclosed seas, the PSSA designation should be made on the basis

[66] *Ibid.*

[67] U.S. Amendments, *supra* note 49, at p. 2.

[68] WWF Comments, *supra* note 57, at p. 2.

[69] WWF Comments, *ibid.*, pp. 3–4.

of consensus of all the regional States.[70] It was difficult to find a legal, rather than a political basis for this proposal. While Article 123 and Part XII of the LOS Convention do encourage regional cooperation, there is no requirement for a consensus-based approach to regional cooperation, such that one objecting State could impede an otherwise broadly supported regional initiative. A related Russian view that emphasis should be on local ecosystems, perhaps to justify national MPA protection in the first place, also was not adopted.[71] The feeling here was that "local" may be somewhat constraining when the marine ecosystem to be protected was spatially large.

Nature of Areas to be Protected

Since the designation of the Great Barrier Reef and the Sabana-Camaguey Archipelago as PSSAs, much of the practice has been related to demonstrably sensitive coastal and marine ecosystems that probably satisfied more than one criterion. They have tended to be very special places. Malpelo Island's waters are biologically very productive, biogeographically significant and are home to a unique Pacific coral reef community. The waters had already been designated as a national MPA.[72] Even so, the MEPC required re-submission of Colombia's first proposal to designate Malpelo Island as a PSSA, and it was not so designated before additional information was provided.[73] Also sensitive to commercial and recreational shipping activities, the Florida Keys have a different but equally fragile system, this time a barrier coral reef (the third largest in the world), which is also a national marine sanctuary.[74] The Paracas National Reserve was designated a Ramsar Convention site because of its great biodiversity and

[70] Russian Federation Amendments, *supra* note 57, at p. 2; also suggested by ICS and INTERTANKO, *supra* note 57.

[71] Russian Federation Amendments, *ibid.*, p. 2.

[72] "Additional Information for the Designation of Malpelo Island as a 'Particularly Sensitive Sea Area'," IMO/MEPC 46/6/3, 16 February 2001 [hereafter Malpelo Island Submission].

[73] *Ibid.* This occurred during consideration at the MEPC's 44th session, 6–13 March 2000. The Colombian proposal, although based on the 1991 PSSA Guidelines, lacked the information required by the 1999 amendments.

[74] "Designation of the Marine Area around the Florida Keys as a Particularly Sensitive Sea Area," Submitted by the United States, IMO/MEPC 46/6/2, 19 January 2001 [hereafter Florida Keys Submission].

a regional reserve for migratory birds, thus making it unique, exceptional and highly sensitive.[75] Peru was able to argue that there was a trend in increased shipping through or in the vicinity of the reserve, as well as accidents and operational pollution in the area. The Wadden Sea, on the other hand, has a very different sensitive coastal ecosystem in the form of a highly dynamic tidal ecosystem shared by three countries and deserving of special protection.[76] Similar situations and considerations of sensitivity were weighed in the case of the Canary Islands, and Galapagos PSSA proposals, and more recently, for the protection of the fragile coral reef system in the Papahānaumokuākea Marine National Monument (Hawaii) PSSA proposal.[77] Generally, the natural value of the sites concerned is such that PSSAs tend to be considered as areas of global and/or regional importance. The proposed areas are also actively under national protection, and therefore provide a persuasive argument for further protection from international navigation activities that they could not otherwise receive through domestic regulation alone.

If any of the 17 criteria in Table 4 is satisfied, a Step 2 assessment requires that "the area should be at risk from international shipping activities."[78] This entails consideration of vessel traffic characteristics (operation factors, vessel types, traffic characteristics, harmful substances) and natural factors (hydrographical, meteorological, oceanographical).[79] Additional information such as impact of shipping activities, past casualties, foreseeable incidents, other environmental stresses and existing measures (and their impact) facilitates risk assessment and the determination of appropriate protective measures.[80] For example, in the case of Malpelo Island, the threats posed included illegal fishing, diving activities from

[75] "Designation of the Marine Area of the Paracas National Reserve as a 'Particularly Sensitive Sea Area'," Submitted by Peru, IMO/MEPC 48/7, 18 April 2002.

[76] "Designation of the Wadden Sea as a Particularly Sensitive Sea Area," Submitted by Denmark, Germany and The Netherlands, IMO/MEPC 48/7/2, 28 June 2002.

[77] "Designation of the Waters of the Canary Islands as a Particularly Sensitive Sea Area," Proposal Submitted by Spain, IMO/MEPC 51/8, 24 October 2003; "Designation of the Galapagos Archipelago as a Particularly Sensitive Sea Area," Submitted by Ecuador, IMO/MEPC 51/8/2, 24 December 2003 [hereafter Galapagos Submission]; Designation of the Papahānaumokuākea Marine National Monument as a Particularly Sensitive Sea Area, IMO Doc. MEPC 56/8, 5 April 2007.

[78] PSSA Guidelines, *supra* note 54, at para. 5.1.

[79] *Ibid.*, Part 5 – Vulnerability to Impacts from International Shipping.

[80] *Ibid.*

cruise ships, and major international commercial traffic in the vicinity of the island.[81] The threats to the Florida Keys included, among others, the high risk posed by one of the most heavily utilized sea lanes in the world and physical damage of the coral reef area from ship groundings and anchoring.[82]

Size of Area

There does not appear to be a limit on the size of a PSSA, and an area may include a buffer zone.[83] On size, a number of States have taken the view that a PSSA should be a local area within the territorial sea and/or an EEZ, and therefore should not encompass an entire EEZ, or EEZs, or whole seas. It is argued that size should be related to an ecosystem(s) that need additional protection. It is the MEPC that determines whether the size of a proposed area is commensurate with that necessary to address the identified need,[84] and that the need is necessarily the sensitivity of the area with reference to the threat posed by international navigation. Reactions to size of PSSAs have been variable. The very large area covered by the Great Barrier Reef was never in dispute in that the whole marine ecosystem in question required Level 3 protection. However, size was an issue with the Baltic[85] and Western European[86] PSSA proposals. To

[81] Malpelo Island Submission, *supra* note 72.

[82] Florida Keys Submission, *supra* note 74, at pp. 9–10.

[83] For example, the Galapagos submission proposed an area to be avoided that would function as "a separation zone or contiguous zone for the proposed PSSA," justified on the basis of currents, logistical concerns in tackling spills, and correspondence of buffer area to the existing national marine protected area. Galapagos Submission, *supra* note 77, at pp. 20–21.

[84] PSSA Guidelines, *supra* note 54, at para. 8.2.3.

[85] "Designation of the Baltic Sea Area as a Particularly Sensitive Sea Area," Submitted by Denmark, Estonia, Finland, Germany, Latvia, Lithuania, Poland and Sweden, IMO/MEPC 51/8/1, 19 December 2003. See Ylva Uggla, "Environmental Protection and the Freedom of the High Seas: The Baltic Sea as a PSSA from a Swedish Perspective," *Marine Policy* 31 (2007): 251–257.

[86] "Designation of a Western European Particularly Sensitive Sea Area," Submitted by Belgium, France, Ireland, Portugal, Spain and the United Kingdom, IMO/MEPC 49/8/1, 11 April 2003 [hereafter Western European Submission]. See also, J. Roberts, M. Tsamenyi, T. Workman and L. Johnson, "The Western European PSSA Proposal: A 'Politically Sensitive Sea Area'," *Marine Policy* 29 (2005): 431–440; Markus Detjen, "The Western European PSSA – Testing a Unique International Concept to Protect Imperilled Marine Ecosystems," *Marine Policy* 30 (2006): 442–453.

their critics, both of these proposals appeared different in geographical and ecosystemic scope from earlier PSSA designations, which tended to focus on local sensitivities, perhaps mostly domestic (with the exception of the Wadden Sea).

A strong collective argument was made by the Baltic States (Denmark, Estonia, Finland, Germany, Latvia, Lithuania, Poland and Sweden) to justify Level 3 protection for the Baltic Sea as a unique ecosystem. A weakness in their submission was that it was not immediately apparent that the various MARPOL 73/78 special area designations, i.e., Level 2 protection, were insufficient to protect the Baltic; and if they were insufficient, whether strengthened implementation of MARPOL 73/78 annexes would have addressed the concerns behind the submission. Basically, States opposing the PSSA proposal for the Baltic did not see that the Level 2 protection possibilities had been exhausted without satisfactory results so as to then justify more restrictive Level 3 protection. In particular, the Russian Federation, an important maritime and Baltic coastal State, argued that the proposal did not provide added value to Level 2 MARPOL 73/78 special and emission control areas protection and the Helsinki Commission efforts in the region.[87] The Russian Federation is a major oil exporter increasingly using the Baltic as a principal export route for its Primorsk production. At the time the IMO deliberated on the Baltic PSSA proposal, originally a designation-in-principle, the majority of Baltic States were members of the European Union and several boasted strong environmental records. There was a political willingness to proceed without the Russian Federation. The consequence was that Russian waters were excluded from the PSSA, even though the purpose of the PSSA was to protect the entire sea. There were also other major flag States, in particular Liberia and Panama, that sympathized with the Russian concerns. It remains to be seen whether opposing States remain outside the PSSA since, if they do, the purpose and effectiveness of the Baltic PSSA may be greatly reduced or remain restricted to relatively uncontroversial albeit beneficial protective measures.

The Western European PSSA was similarly a collective initiative of six European Union member States, namely Belgium, France, Ireland, Portugal, Spain, and the United Kingdom. Consisting to a great extent

87 "Report of the Marine Environment Protection Committee on its Fifty-First Session," IMO MEPC 51/22, 22 April 2004 [hereafter MEPC 51 Report], pp. 34–35 and Annex 8.

of the EEZs of the proponent States, the area covered is approximately of a comparable size to the Great Barrier Reef and is one of the most densely navigated areas in the world. The proponents argued that this area probably experienced more maritime casualties than any other area in the world. The massive casualties of the *Torrey Canyon, Amoco Cadiz, Erika*, and *Prestige* are evidence of this. The Western European PSSA is home not to one, but rather to several marine ecosystems captured within the EEZs of the proponent States. Perhaps deeper concerns than in the case of the Baltic PSSA were expressed in relation to the consistency of a PSSA of this size with Level 3 protection, as well as with the LOS Convention.[88] It was counter-claimed that a more defensible approach would have been to designate a system of PSSAs, including buffer zones, that were more local ecosystem-specific. The critics argued that the large size was likely to push more shipping into the open ocean and towards neighbouring States, for example towards Norway. In effect, this was considered tantamount to exporting the risk to other maritime zones. The northern limit of the proposed PSSA had to be adjusted to respond to this particular concern. Further, the concern was raised that if more shipping is pushed to the high seas, the feasibility of search and rescue and provision of places of refuge could be put in question.[89]

Designations in Principle

It is possible for a PSSA to be designated in principle pending final adoption by the IMO. Under the 2001 Guidelines there was significant flexibility for the designation of PSSAs in principle. PSSAs could be submitted even though no firm protective measures were proposed. Protective measures could be proposed subsequently. At the most, there was an expectation that the types of measures contemplated be indicated at the time of submission.[90] The MEPC could designate a PSSA in principle

---

[88] "Designation of a Western European Particularly Sensitive Sea Area," Submitted by Liberia, Panama, the Russian Federation, BIMCO, ICS, INTERCARGO, INTERTANKO and IPTA, IMO/LEG 87/16/1, 12 September 2003.

[89] "Report of the Marine Environment Protection Committee on its Forty-Ninth Session," IMO/MEPC 49/22, 8 August 2003, 47. A similar concern was raised by Brazil in relation to the Baltic and Canary Islands proposals. MEPC 51 Report, *supra* note 87.

[90] PSSA Guidance Document, *supra* note 54, at Part. 3.5.

with a two-year period within which at least one appropriate measure had to be proposed and adopted through the IMO and, failing such submission, the designation lapsed. In the run-up to the latest review of the PSSA Guidelines, the United States proposed that the designation or approval in principle pending submission of appropriate measures be discontinued.[91] It was thought that consideration of the proposed PSSA needed to have the full picture of demonstrated vulnerabilities and precise measures, including the legality of such measures.[92] However, others thought that a designation in principle was consistent with the precautionary approach in responding in the interim to a recognized risk or threat and until more substantive action could be taken.[93] The 2005 Guidelines tightened designations in principle so that the MEPC can still designate a PSSA in principle "if appropriate following its assessment."[94] At the time of submission of a PSSA proposal, the proponent also has to propose measures and indicate how "they provide the needed protection from the threats of damage posed by international maritime activities occurring in and around the area."[95]

Protective Measures

The teeth of Level 3 protection consist of the protective measures that may be adopted by the IMO for the area. Protective measures consist of restrictions on international navigation and/or related activities adopted through the IMO.[96] There are various navigation measures that the IMO

---

91 U.S. Amendments, *supra* note 49, at p. 3.

92 *Ibid.*

93 WWF Comments, *supra* note 57, at p. 4.

94 PSSA Guidelines, *supra* note 54, at para. 8.3.2.

95 *Ibid.*, para. 7.5.2. Despite its concern with designations-in-principle, the United States sought and achieved such a designation of the Papahānaumokuākea Marine National Monument (Hawaii) PSSA at the 56th MEPC Session, before its final PSSA designation at the 57th MEPC Session in 2008. *Supra*, note 77.

96 Potential measures could include "designated anchorage areas and methods; no anchorage zones; closure of routes to certain types of vessels or cargoes; speed restrictions; compulsory pilotage or tug escort to ensure safe navigation in or near PSSAs; prohibitions/restrictions on cargo transfer; required submission of pre-filed passage plans and adherence to time schedules; special under-keel clearance restrictions; regulation of offshore bunkering; prohibition of intentional discharges, including ballast waters; seasonal closures to protect migrating marine mammals." "Additional Protection for Particularly Sensitive Sea Areas (PSSAs)," IMO/MEPC 46/6/1, 19 January 2001 [hereafter

member States are competent to endorse, including routeing measures,[97] reporting systems,[98] and other forms of navigation assistance.[99] In implementing these measures, the coastal State must not require new design, construction, manning or equipment standards for vessels, unless they are IMO-adopted standards and are endorsed as appropriate measures for the area concerned. Otherwise there would be a risk of violating international navigation rights. Also, as indicated earlier, such measures may be adopted by virtue of Article 211(6) of the LOS Convention.[100] By referring to the latter, the PSSA Guidelines in effect maintain a distinction with the Article 211(6) special mandatory measures, while at the same time establishing a link to this provision. In general, the IMO favours the least restrictive options in the interests of commerce.[101] For instance, a measure creating a no-anchoring area may be more appropriate than a measure indicating that an area is to be avoided, if the principal threat is from anchoring and not from the navigation activity.[102] The measures may apply to particular categories of ships. The assessment of a measure will include examination of the potential impact of the measure on navigation safety and efficiency.[103] Should there be non-compliance with IMO-adopted measures, the coastal State concerned would be in a position to undertake actions to ensure compliance, but in doing so must act consistently with the LOS Convention. Proposed national compliance and control action needs to be communicated to the MEPC along with the proposal for associated protective measures.[104]

PSSA Additional Protection], para. 2.4.10. See also, "Draft Guidance Document on Associated Protective Measures for Particularly Sensitive Sea Areas," Submitted by WWF, IMO/MEPC 49/8/2, 8 May 2003. One interesting idea advanced in the latter document is air pollution emission limitations which could require exclusive utilization of high quality low sulphur fuels in or near PSSAs.

[97] For example, areas to be avoided, traffic separation schemes, inshore traffic zones, deep water routes, precautionary areas, recommended routes and vessel traffic service systems.

[98] For example, radar surveillance and interactive ship/shore reporting.

[99] For example, pilotage and towage.

[100] PSSA Guidelines, *supra* note 54, at para. 7.5.2.3(iii).

[101] PSSA Additional Protection, *supra* note 96, at para. 2.1.3.

[102] This was the case in the Florida Keys. See Florida Keys Submission, *supra* note 74, and U.S. Amendments, *supra* note 49, at p. 2.

[103] PSSA Guidelines, *supra* note 54, at para. 7.6.

[104] *Ibid.*, para. 7.9. See also, PSSA Guidance Document, *supra* note 54, at para. 3.5.3.

When the Baltic PSSA was adopted in principle, no measures had been proposed.[105] Some Baltic States had hoped that compulsory pilotage, among other measures, would be considered, but major maritime powers objected to such a measure. The measures that were eventually adopted were less controversial and consisted of a deep water route for the oil trade. The original Western European PSSA submission proposed an associated protective measure that would have banned single-hull traffic in the area and a mandatory 48-hour notice requirement for all laden double-hulls entering or exiting the reporting area.[106] The single-hull ban appeared to be inconsistent with Article 211(6)(c) because it addressed design and construction standards. A contentious debate in the MEPC forced the withdrawal of the single-hull ban, but the 48-hour notice area in accordance with SOLAS (Chapter V/11) was adopted as an approved protective measure.[107] Outside the PSSA discourse at the time, there was already agreement between the European Union and IMO to accelerate the phase-out of single hulls by bringing forward the phase-out date from 2015 to 2010.[108]

Where IMO-approved measures already exist in an area, a PSSA proposal does not need to renew these measures. It may be questioned as to what is the purpose of a PSSA proposal for an area that is already protected by existing measures. One answer is that PSSAs and their associated protective measures are designated on navigation charts in accordance with

[105] "Designation of the Baltic Sea Area as a Particularly Sensitive Sea Area," Submitted by Denmark, Estonia, Finland, Germany, Latvia, Lithuania, Poland and Sweden, IMO/MEPC 51/8/1, 19 December 2003. However, the proposal indicated that there a number of protective measures already in place, namely compulsory reporting in some areas and traffic surveillance, routeing systems in certain areas for big draught ships, local pilotage and the MARPOL 73/78 special area and emission control area. Also, the proposal indicates that new measures that will be considered including common monitoring system based on compulsory reporting and traffic surveillance, routeing systems, escort and escorting tugs, pilotage and areas to be avoided. *Ibid.* See also, "HELCOM Expert Working Groups," Submitted by Denmark, Germany and Sweden, IMO/NAV 50/INF.5, 29 April 2004.

[106] Western European Submission, *supra* note 86.

[107] "Mandatory Ship Reporting System for the Western European Particularly Sensitive Sea Area," Submitted by Belgium, France, Ireland, Portugal, Spain and the United Kingdom, IMO/NAV 50/3/6, 2 April 2004.

[108] See "IMO Saves Face and EU Claims Victory," *Fairplay*, 11 December 2003, 4.

International Hydrographic Organization (IHO) symbols,[109] thus ensuring that they are brought to the attention of seafarers navigating in or in the vicinity of a PSSA. Also, flag States have a responsibility to ensure compliance by their ships with the protective measures.[110]

In the discussion leading to the 2005 Guidelines, it was felt that PSSA proponent States did not always sufficiently demonstrate the nexus between the identified vulnerabilities of an area and the associated protective measures proposed, such that area size (spatial dimension), propriety, and sufficiency of the proposed measure could be ascertained and the least restrictive measure approved. The United States proposed that there should be a greater emphasis on the purpose of the proposed measure in addressing identified vulnerabilities, in particular in relating the international shipping threat more directly to the vulnerabilities of the area, the degree of actual and/or potential harm, and reasonable forseeability.[111]

In some PSSA proposals, the legal basis of a proposed measure has not always been apparent. A case in point was the original Western European PSSA proposal restriction on single-hull tankers. This proposed restriction would, in effect, lead to restrictions on international navigation rights. For such restrictions to be legitimate, and not tantamount to encroachments on the right of international navigation, it is necessary to have appropriate legal justification. Thus, the U.S. position during the review of the Guidelines emphasized the legality test for special mandatory measures with reference to the "particular, localized circumstances of the proposed area."[112] Under the 2005 Guidelines, the proposal for a measure has to be consistent with existing legal instruments approved or adopted by IMO to prevent, reduce or eliminate the vulnerability addressed by the PSSA.[113] One measure that raised legality concerns was compulsory pilotage. Both the United States and Russian Federation successfully proposed that this possible measure be excluded from the 2005 Guidelines.[114] However controversial compulsory pilotage as protective measure might be, safety and common sense might still dictate its usefulness in certain areas where

---

109 PSSA Guidelines, *supra* note 54, at para. 9.1.
110 *Ibid.*, para. 9.3.
111 U.S. Amendments, *supra* note 49, at Annex, pp. 7–8.
112 *Ibid.*, p. 11.
113 *Ibid.*, Annex 2.
114 Russian Federation Amendments, *supra* note 57. WWF argued otherwise. See WWF Comments, *supra* note 57, at p. 3.

the navigation conditions are such that its application can considerably reduce the risks sensitive ecosystems might otherwise be exposed to. A clear and convincing case that leaves no better alternatives would have to be made.

Conclusion

PSSAs constitute a significant Level 3 global marine environmental protection layer that legitimizes regional and national efforts for higher standards of protection from shipping for eligible marine ecosystems. There is a presumption that proponent States have the political and management will to act. At the same time, however, the argument that proponent States should endeavour to better implement existing Level 1 and 2 responsibilities to justify additional encroachments on international navigation rights through Level 3 action still holds. This is especially true for those States that have sought and succeeded in achieving MARPOL 73/78 special area status for their marine regions, but have not followed up on their commitment to provide reception facilities to enable those areas to take effect.

There is relatively little practice to indicate how effective PSSAs really are, partly because the majority of PSSAs have been designated only since 2002. Effectiveness will depend on the nature of associated protective measures and how they are implemented by the coastal State(s) concerned. Recently, Australia reported that since pilotage was introduced in 1991, the rate of accidents within the Great Barrier Reef was reduced by more than 50 per cent. This is a significant achievement, even though accidents continue to happen.[115] The value of such a measure is even more evident when it is considered that despite the recommended pilotage for the Torres Strait, prior to the PSSA-extension approval in principle and in accordance with an earlier IMO resolution, Australia reported declining compliance.[116] There appears to be a clear cause-effect relationship to justify a PSSA and

115 "Torres Strait PSSA Associated Protective Measure – Compulsory Pilotage," Submitted by Australia and Papua New Guinea, IMO/MEPC/NAV 50/3, 22 March 2004. See also, "Results of a Safety of Navigation Assessment Conducted for the Torres Strait," Submitted by Australia, IMO/NAV 50/INF 2, 2 April 2004.

116 NAV 50/3, *ibid.*

associated protective measures for the area concerned, which may need to be updated to respond to evolving problems.

The need to avoid a proliferation of MARPOL 73/78 special areas was addressed successfully, at least judging by the fewer number of special area proposals submitted after the adoption of the Special Area Guidelines. Is there a continued risk of PSSA proliferation, as some suggest? Despite their recent revision, the PSSA Guidelines place a relatively light onus on proponent States to justify Level 3 protection, rather than Level 1 or 2, and also why the utilization of specific SOLAS (Chapter V) measures, rather than PSSA designation, is not sufficient. The international maritime and environmental communities cannot possibly be well served by the designation of areas that are stronger on optics than on effective management responses to genuine particularly sensitive areas. As industry has cautioned, there is a danger of watering-down the special character and value of PSSAs unless there is effective implementation of existing protective measures to which a PSSA adds further value.

The PSSA discourse in the IMO is likely to remain contentious and, given the perceived stakes by those States that represent minority but significant views, it may well be that majority decision-making could have unhappy consequences. Unfortunately, majority decision-making substituted consensus, despite the sharp differences over the Western European and Baltic PSSA applications. The concern here is particularly accentuated when differences are not necessarily purely political or economic, but may also involve important issues of principle, such as whether the international law of the sea permits the restriction of international navigation rights in an entire EEZ (or series of connected EEZs) or a semi-enclosed sea. A consequent danger is that the PSSA designation and associated protective measures may not be recognized by objecting States, and in turn, this could raise difficulties with compliance.[117]

On a concluding note, the PSSA discourse with its animated arguments by all sides on how best to protect the marine environment within national

[117] The Russian Federation has already provided such notice: "The Russian Federation does not agree with this decision to grant PSSA status to the whole Baltic Sea, does not associate itself with this decision and, should the Organization approve any other recommendations, regulations or action for adoption on the basis of the Baltic Sea's designation as a PSSA, reserves its right not to give effect to any such decision." MEPC 51 Report, *supra* note 77, at Annex 8.

jurisdiction without imposing unnecessary encroachments on international navigation rights may well serve to enhance procedures for coastal State transparency in this regard. Those States that wish to exercise greater marine environmental authority have to demonstrate that they are fulfilling their basic international legal responsibilities in the first place.

Dispute Settlement and Regional Cooperation

Third States in Maritime Delimitation Cases: Too Big a Role, Too Small a Role, or Both?

Alex G. Oude Elferink*

Introduction

The delimitation of maritime boundaries between neighbouring States is primarily a bilateral process. At the same time, in many instances more than two States may have an interest in the delimitation of a specific area. The existence of a claim of a third State invariable will be a factor in bilateral negotiations or the adjudication of a bilateral delimitation dispute. Douglas Johnston amply considered this issue in his work on maritime boundary delimitation, including in one of his most significant contributions to the law of the sea, *The Theory and History of Ocean Boundary-Making* published in 1988. This monograph presents a functionalist approach to ocean boundary-making[1] and this extends to his views on the position of third States in maritime boundary delimitation. These views are succinctly summed-up in a comment on the *Libya/Malta Continental Shelf Case*.[2] In this Case, the impact of the position of a third State on the judgment on the merits was criticized in the opinions of a number of judges.[3] After reviewing those criticisms, Johnston observed:

> On the other hand, functionalists are likely to welcome the Court's treatment of the third-party-interest problem as an ingenious effort to provide a truly effective facilitative service to the parties. A facilitative judgment, it may be argued, is wholly prospective and problem-oriented in character, and should not be constrained unduly by the bilateral, or bifurcatory, frame

* Senior Research Associate, Netherlands Institute for the Law of the Sea, Utrecht University, The Netherlands.

[1] For a detailed elaboration of that approach see especially, D. M. Johnston, *The Theory and History of Ocean Boundary-Making* (Kingston and Montreal: McGill-Queen's University Press, 1988), Part III, pp. 223–284.

[2] *Continental Shelf (Libyan Arab Jamahiriya/Malta)*, Judgment, [1985] *I.C.J. Reports*, 13.

[3] *Ibid.* For a discussion of this aspect of the judgment see also *infra*.

> of reference traditionally associated with adversarial proceedings. [...] For functionalists, it seems a step forward, not backward, for a tribunal to give weight to the regional (or subregional) characteristics of the physical situation with which it must deal and to acknowledge the relevance of third-party interests in that region without prejudgment of other disputes that should be settled, preferably through negotiation.[4]

The functionalist approach to the establishment of maritime boundaries between neighbouring States accords particular weight to regional circumstances. Starting from the premise that the overriding obligation under international law is the requirement to produce an equitable solution,[5] the appropriate principles, criteria and methods should be selected by a tribunal in the light of the following elements: (1) the general (macrogeographical) characteristics within the appropriate relevant area; (2) the special (microgeographical) characteristics of the area in dispute; (3) the opportunities and problems of ocean development and management in the disputed area and adjacent waters; and (4) the recent history of ocean uses and relevant relationships in the region.[6] Johnston welcomed the fact that tribunals in the period preceding the publication of *The Theory and History of Ocean Boundary-Making* had consistently emphasized the particularistic orientation of their decisions and had warned not to try to extract general principles from them.[7] It was further observed that certain counsel and commentators did not heed those warnings, assuming that the consolidation of legal norms is the primary function of any tribunal.[8]

The present essay revisits the views expounded in *The Theory and History of Ocean Boundary-Making* on the regional context of adjudicated bilateral delimitations in the light of the developments that have taken place in the 20 years since its publication. The recent case law displays a tendency to downplay the importance of the wider regional context

[4] Johnston, *supra* note 1, at pp. 209–210 (footnote omitted).

[5] *Ibid.*, pp. 282–283. As is observed the term "solution" should be interpreted as acknowledging the existence of a "problem" that is likely to be larger that the associated legal or technical issues (*ibid.*, p. 283). Secondary-level legal norms should not be given undue saliency in the problem-solving process in respect of maritime boundary delimitation (*ibid.*).

[6] *Ibid.*

[7] *Ibid.*, p. 284.

[8] *Ibid.*

of bilateral delimitation disputes and has been considered to have led to a less particularistic, more structured approach, giving more weight to general rules and less to the individualization of the specific case.[9] In order to properly appreciate these matters the essay will not only discuss recent case law, but will also look at the case law antedating *The Theory and History of Ocean Boundary-Making.*

Third States play a role in the decisions of courts and tribunals in a number of ways. There may be a claim of a third State to the maritime areas that the parties to a case have asked a court or tribunal to delimit. Such a claim involves questions of legal principle – a third State has not consented to its claims being considered – which will have to be taken into account in establishing the endpoint(s) of the boundary between the parties to the proceedings. Then, there are instances in which the presence of a third State is one of the relevant circumstances for the delimitation between the parties to the case. Finally, the reasoning of a court or tribunal may have implications for the future maritime delimitations of the parties with third States or future maritime delimitations between third States in the same region. The case law illustrates that the development of maritime delimitation law has grappled with these issues and has not always come up with consistent answers. Whether this reflects a functionalist approach leading to facilitative judgments is another matter.

The following sections of this essay look at the way in which courts and tribunals have dealt with claims of third States at the level of legal principle and the establishment of the endpoint(s) of the maritime boundary between the parties to the proceedings, the presence of third States as a relevant circumstance and the possible implications of the reasoning of decisions for future delimitations in the same region. This discussion will be followed by some concluding observations.

---

[9] See, e.g., M. D. Evans, "Maritime Boundary Delimitation: Where Do We Go From Here?" in D. Freestone, R. Barnes and D. Ong, eds, *The Law of the Sea; Progress and Prospects* (Oxford: Oxford University Press, 2006), pp. 137–160, at p. 160; Y. Tanaka, "Reflections on Maritime Delimitation in the *Cameroon/Nigeria* Case," *International and Comparative Law Quarterly* 53 (2004): 369–406, at p. 405.

Claims of Third States – Legal Issues

The delimitation of maritime boundaries is primarily a bilateral process. This is the case for negotiations as well as the settlement of maritime boundaries through third party adjudication. As far as third party adjudication is concerned, this accords with the bilateral cast in which that mode of dispute settlement is generally framed.[10] At the same time, public international law is not only concerned with regulating the bilateral legal relationships between States. The decision of a court or tribunal carries the risk of affecting the legal interests of third States. That fact is recognized by Articles 59, 62 and 63 of the Statute of the International Court of Justice.[11] Article 59 provides that a decision of the Court has no binding force except between the parties and in respect of the particular case. Articles 62 and 63 are respectively concerned with the intervention of third States where they consider that they have an interest of a legal nature which may be affected by the decision in the case and the intervention of third States whenever the construction of a convention to which States other than those concerned in the case are parties is in question.

These provisions of the Statute of the International Court reflect the consensual nature of third party dispute settlement. That consensual nature leads to an inherent tension in the adjudicatory process in public international law. On the one hand, a court or tribunal will have to respect the autonomy of the parties before it to define the issues they want to submit. On the other hand, the absence of consent of a third party requires that a court or tribunal not pass judgment on its claims.[12] This tension is particularly visible in maritime delimitation cases.[13] States formulate their

---

10 For a further discussion see, e.g., C. Chinkin, *Third Parties in International Law* (Oxford: Clarendon Press, 1993), pp. 147–149.

11 Similar provisions are contained in Arts. 31, 32 and 33(2) of the Statute of the International Tribunal for the Law of the Sea.

12 For a further discussion of this matter see, e.g., Chinkin, *supra* note 10, at pp. 147–149.

13 See also, E. Jouannet "L'impossible protection des droits du tiers par la Cour Internationale de Justice dans les affaires de délimitation maritime," in V. Coussirat-Coustère, Y. Daudet, P.-M. Dupuy, P. M. Eisemann, and M. Voelckel, eds, *La mer et son droit: mélanges offerts à Laurent Lucchini et Jean-Pierre Quéneudec* (Paris: Éditions A. Pedone, 2003), pp. 315–341, at pp. 315 and 322; R. Kolb, *Case Law on Equitable Maritime Delimitation: Digest and Commentaries* (The Hague: Martinus Nijhoff Publishers, 2003),

claims by reference to specific areas or lines, which may lead to a spatial overlap between the claims of the parties and a third State.

The International Court first had to consider the significance of the presence of a third State as a relevant circumstance for a bilateral delimitation in the *North Sea Continental Shelf Cases*. These Cases are distinct from all later maritime delimitation decisions in one important respect. The third State in both bilateral delimitations involving Germany was a party before the Court. The question has been raised if the Court could have reached the conclusions it did if it would only have had two of the parties concerned before it. At the oral hearings, Judge Petrén specifically addressed that issue in a question to Denmark and the Netherlands.[14] The Court's judgment deals with this issue in a single rather intricately worded paragraph.[15] The Court observed that Denmark and the Netherlands had acted in close cooperation, such that "to this extent" the two cases could be treated as one and, even though either case might have been brought separately, that did not alter the problem with which the Court was actually faced having regard to the manner in which the parties had brought the matter before the Court.[16] At the same time, the Court observed that the cases themselves remained separate and that two separate delimitations were involved, which, however, gave rise to a single situation.[17] Rather than indicating that it was the joining of the cases which allowed the Court to reach its conclusions, paragraph 11 of the Judgment indicates that the two cases remained separate, but that their joining had consequences for the structure of the Judgment.[18] The applicable law led to the single

p. 101; L. Lucchini and M. Voelckel, *Droit de la Mer; Tome 2 Volume 1 Délimitation* (Paris: Éditions A. Pedone, 1996), p. 278.

14 International Court of Justice, *North Sea Continental Shelf Cases (Federal Republic of Germany/Denmark; Federal Republic of Germany/The Netherlands)*, Pleadings, Oral Arguments, Documents, Vol. II, p. 266. The Agent for the Netherlands, Riphagen categorically rejected that the principles and rules of international law permitted that the location of either bilateral boundary of Germany be determined or influenced in law by the other bilateral boundary (*ibid.*, pp. 283–284; see also the argument of Sir Humphrey Waldock at *ibid.*, pp. 269–270).

15 *North Sea Continental Shelf Cases (Federal Republic of Germany/Denmark; Federal Republic of Germany/The Netherlands)*, Judgment [1969] *I.C.J. Reports*, p. 19, para. 11.

16 *Ibid.*

17 *Ibid.*

18 The issue addressed in paragraph 11 was considered in the separate opinion of Judge Padilla Nervo ([1969] *I.C.J. Reports*, pp. 89–90) and the dissenting opinions of Judges

situation involving the three States, not the joining of the cases.[19] The Court could have taken two approaches had only two of the parties been before it. The Judgment could have dealt with the merits or the Court could have indicated the absence of a necessary third party.[20] The latter outcome would also have revealed that equidistance was not applicable in the instant case because of the single situation resulting from the two separate delimitations.[21]

In the *Anglo-French Arbitration*, the Court of Arbitration had to consider the position of Ireland as a third State. The Court observed that it would not pronounce itself in any way on the delimitation between Ireland and the United Kingdom as that matter was outside its competence.[22] At the same time, the Court made it clear that the legal interest of Ireland did not imply a limitation on the competence of the Court to decide the dispute between the parties, even though it was acknowledged that a potential boundary between Ireland and the United Kingdom might meet a boundary between France and the United Kingdom to the east of the 1,000 metre isobath on which the parties had asked the Court to

---

Tanaka (*ibid.*, pp. 191–192) and Morelli (*ibid.*, pp. 209–212). Judges Padilla Nervo and Morelli observed that the combined effect of the two bilateral delimitations applying equidistance led to an inequitable result. Judge Padilla Nervo expressed the belief that the parties by submitting the matter to the Court in the way they did, recognized that the matter constituted an integral whole (*ibid.*, p. 90). On the other hand, Judge Morelli observed that it would have been quite possible for the Court to reach the conclusion it did if it would only have had two of the three States concerned before it (*ibid.*, p. 211). Judge Tanaka reached exactly the opposite conclusion (*ibid.*, p. 192). This difference is accompanied and explained by a different view on the procedural and substantive rules of maritime delimitation law. Judge Tanaka's opinion stresses the bilateral nature of that process and the centrality of equidistance. Judge Padilla Nervo and Morelli indicate – as does the Judgment of the Court – that the equitable nature of a bilateral delimitation has to be assessed in a broader context (on the relation between the substantive rules of delimitation law and the role of third States see also *infra*, especially the discussion of the *Cameroon v. Nigeria Case*).

19 *Contra* Evans, *supra* note 9, at p. 157.

20 The former approach accords with that of courts and tribunals in subsequent cases in which the coast of third States was taken into account as a relevant circumstance (see also *infra*).

21 See further *infra*.

22 *Arbitration between the United Kingdom and France on the Delimitation of the Continental Shelf*, Decision of 30 June 1977 (ILR Vol. 54, p. 11; UNRIAA Vol. XVIII, p. 3), paras. 27–28.

locate the terminus of their bilateral boundary.[23] The Court's decision would only be binding on the parties before it and would be *res inter alios acta* for Ireland.[24]

The approach of the Court of Arbitration has not been followed by the International Court and tribunals in subsequent cases. In the International Court's view, Article 59 of the Statute of the Court did not offer sufficient protection to the rights of third States.[25] To protect the legal interest of third States, the case law subsequent to the *Anglo-French Arbitration* generally has taken the position that a decision should leave this legal interest unaffected. That position has in practice implied that courts and tribunals have refrained from establishing a boundary in areas that are claimed by a third State.[26] In the *Eritrea/Yemen Arbitration*, the Tribunal justified that choice as follows:

---

23 *Ibid.*, paras. 28 and 25.

24 *Ibid.*, para. 28.

25 *Continental Shelf (Libyan Arab Jamahiriya/Malta)*, Application to Intervene, Judgment [1984] *I.C.J. Reports*, pp. 26–27, para. 43; *Land and Maritime Boundary between Cameroon and Nigeria (Cameroon v. Nigeria: Equatorial Guinea intervening)*, Judgment [2002] *I.C.J. Reports*, p. 421, para. 238. The former judgment, however, also indicates that Article 59 of the Statute offers a large measure of protection to a third State in a maritime delimitation case (see [1984] *I.C.J. Reports*, p. 26, para. 42).

26 On the specific approaches that have been followed in this respect see further *infra*. The award of the Tribunal in the *Nova Scotia and Labrador/Newfoundland Arbitration* (*Arbitration between Newfoundland and Labrador and Nova Scotia concerning Portions of the Limits of their Offshore Areas as defined in the Canada-Nova Scotia Offshore Petroleum Resources Accord Implementation Act and the Canada-Newfoundland Offshore Petroleum Resources Accord Implementation Act*, Award of the Tribunal in the Second Phase, 26 March 2002) forms an exception as far as the deference to third party claims is concerned. The continental shelf boundary between Nova Scotia and Labrador/Newfoundland established by the Tribunal passes close to the maritime boundary between Canada and the French islands of Saint-Pierre and Miquelon, which ends at 200 nautical miles from the baseline of the islands, and cuts across the area in which the continental shelf beyond 200 nautical miles of Saint-Pierre and Miquelon is located. The Tribunal's approach in the *Nova Scotia and Labrador/Newfoundland Arbitration* was premised on the assumption that the maritime claims of France had been definitively settled (*ibid.*, para. 4.26). However, the basis for that assumption is debatable, to say the least. The Tribunal recalled that the Court of Arbitration, which established a single maritime boundary between Saint-Pierre and Miquelon and Canada in 1992, declined to rule on whether France had any continental shelf areas beyond 200 nautical miles (*ibid.*, para. 4.27; information concerning France's project on the preparation of its submission to the Commission on the Limits of Continental Shelf under the

> The Tribunal has the competence and the authority according to the Arbitration Agreement to decide the maritime boundary between the two Parties. But it has neither competence nor authority to decide on any of the boundaries between either of the two Parties and neighbouring States. It will therefore be necessary to terminate either end of the boundary line in such a way as to avoid trespassing upon an area where other claims might fall to be considered.[27]

The International Court in the *Libya/Malta Continental Shelf Case* made a similar link between the protection of the legal interests of third States and the area they claim. The Court observed that limiting its decision to the area not claimed by Italy

> signifies simply that the Court has not been endowed with jurisdiction to determine what principles and rules govern delimitations with third States, or whether the claims of the Parties outside that area prevail over the claims of those third States in the region.[28]

United Nations Convention on the Law of the Sea (hereinafter LOS Convention) (10 December 1982, in force 16 November 1994, 1833 *U.N.T.S.* 396) indicates that France is carrying out investigations in respect of the continental shelf beyond 200 nautical miles of Saint-Pierre and Miquelon (<www.extraplac.fr/FR/extensions/StPierre.php>, 23 August 2007). Nonetheless, the Tribunal decided to proceed on the basis that the maritime areas of France were those contained within the limits defined in 1992 (*Nova Scotia and Labrador/Newfoundland Arbitration*; Award of the Tribunal in the Second Phase, 26 March 2002, para 4.27). To support this approach the Tribunal referred to paragraph 34 of the judgment on the merits of the *Libya/Malta Continental Shelf Case*. However, paragraph 34 of that judgment addresses the relation between the criteria of distance and natural prolongation in Article 76 of the LOS Convention in areas within 200 nautical miles. It does not consider the delimitation of the continental shelf in areas located within 200 nautical miles of one State but beyond 200 nautical miles of another State. The Tribunal's cavalier approach of a third party claim contrasts with the handling of this matter in other recent decisions. The *Nova Scotia and Labrador/Newfoundland Arbitration* was, as the Tribunal itself pointed out in discussing the delimitation of the continental shelf beyond 200 nautical miles, in quite a different position as the Court of Arbitration deciding the maritime boundaries between Canada and France. The reasons set out in that connection (*ibid.*, para. 2.31) may have also contributed to the Tribunal's approach in relation to France.

[27] *In the Matter of an Arbitration Pursuant to an Agreement to Arbitrate Dated 3 October 1996 between the Government of the State of Eritrea and the Government of the Republic of Yemen*; Award of the Arbitral Tribunal in the Second Stage of the Proceedings (Maritime Delimitation), para. 136.

[28] [1985] *I.C.J. Reports*, p. 26, para. 21; see also, [2002] *I.C.J. Reports*, p. 421, para. 238.

This focus on the area claimed by a third State to achieve that the legal interests of that State are left unaffected raises two questions. First, this approach is based on the assumption that a ruling on the boundary between the parties implies a decision on a) the boundary with a neighbouring State, or b) the claim of neighbouring State. Is that assumption justified? Second, is it defendable to presume that the delimitation of a boundary between two States outside an area claimed by a third State in principle will not affect the legal interests of that third State?

Answering these questions requires a closer look at the role of the coasts of third States in maritime delimitation. The delimitation of a boundary between neighbouring States requires an assessment of all the relevant circumstances in order to arrive at an equitable result. As the case law indicates, the regional context, i.e., the coasts of third States, may be relevant to the delimitation of a bilateral boundary.[29] Courts and tribunals have never considered themselves to be debarred from ruling on a bilateral delimitation because its regional context was one of the relevant circumstances. This matter was explicitly addressed in connection with Nicaragua's intervention in the 1985 *Land, Island and Frontier Dispute Case* between Honduras and El Salvador. The Chamber found that Nicaragua had no legal interest in the maritime delimitation between Honduras and El Salvador:

> the question is whether a legal interest of Nicaragua would be "affected" by such maritime delimitation. It occurs frequently in practice that a delimitation between two States involves taking account of the coast of a third State; but the taking into account of all the coasts and coastal relationships within the Gulf as a geographical fact for the purpose of effecting an eventual delimitation as between two riparian States – El Salvador and Honduras in the instant case – in no way signifies that by such an operation itself the legal interest of a third riparian State of the Gulf, Nicaragua, may be affected.[30]

If this finding is accepted it would seem that the legal interest of a third State will never be affected by the establishment of a bilateral boundary

[29] See further *infra*.

[30] *Land, Island and Maritime Frontier Dispute (El Salvador/Honduras)*, Application to Intervene, Judgment, [1990] *I.C.J. Reports*, p. 124, para. 77. See also, Jouannet, *supra* note 13, at p. 318.

outside an area which is claimed by it.[31] However, this finding also suggests that serious doubts may exist as to the possibility that a third State's legal interests may be affected by a delimitation that extends into an area that is claimed by it. Are there, apart from the taking into account of the coast of a third State, other issues in a bilateral delimitation that might affect the legal interests of a third State? Clearly, there are. The establishment of the terminus of the boundary between the two States and a third State or the indication of the approximate location of boundaries with a third State to assess the maritime areas attributed to the parties would prejudice the delimitation of the maritime boundaries of the third State. However, the establishment of a maritime boundary between the parties does not require that a finding is made on such issues. The fact that no finding need be made on the claims of a third State to assess the claims of the parties also indicates that the latter assessment does not indicate or create a precedence of the latter claims over the former.[32]

Apart from the question of whether a decision that establishes a bilateral boundary between two States affects the legal interests of a third State in the same area, a second issue is whether this "affect test" itself is an appropriate standard. In dealing with Nigeria's eighth preliminary objection in the *Cameroon v. Nigeria Case* the International Court recalled, referring to the 1995 *East Timor (Portugal v. Australia) Case*,[33] that "one of the fundamental principles of its Statute is that it cannot decide a dispute between States without the consent of those States to its jurisdiction," but that the Court was not necessarily prevented from adjudicating when the judgment it is was asked to give might affect the legal interests of a State which is not a party to the case.[34] The Court also recalled that in the *Certain Phosphate Lands in Nauru (Nauru v. Australia) Case*,[35] it had taken the approach that

---

[31] For a further discussion of this matter see also *infra*.

[32] See also, *Continental Shelf (Tunisia/ Libyan Arab Jamahiriya)*, Application to Intervene, Judgment [1981] *I.C.J. Reports*, p. 19, para. 33.

[33] *East Timor (Portugal v. Australia)*, Judgment, [1995] *I.C.J. Reports*, p. 101, para. 26.

[34] *Land and Maritime Boundary between Cameroon and Nigeria*, Preliminary Objections, Judgment, [1998] *I.C.J. Reports*, p. 324, para. 116.

[35] *Certain Phosphate Lands in Nauru (Nauru v. Australia)*, Preliminary Objections, [1992] *I.C.J. Reports*, p. 240.

> a finding by the Court regarding the existence or the content of the responsibility attributed to Australia by Nauru might well have implications for the legal situation of the two other States concerned, but no finding in respect of that legal situation will be needed as a basis for the Court's decision on Nauru's claims against Australia [...] (I.C.J. Reports 1992, pp. 261–262, para. 55.).[36]

These observations fit the situation of maritime boundary delimitation involving third State claims. No finding on third State claims is in principle required to delimit a bilateral boundary between the parties to a case.[37] In respect of the maritime boundary issue submitted by Cameroon, the Court, in ruling on Nigeria's preliminary objections, left open the possibility that it could decide on Cameroon's claim even though the rights and interests of third States might be involved.[38] In the Judgment on the merits, the Court considered this matter further. The Court did not seek to define the circumstances in which "the impact of the judgment required by Cameroon on the rights and interests of the third States could be such that the Court would be prevented from rendering it in the absence of these States,"[39] but instead retreated to the position that it could not rule on Cameroon's claims in so far as they might affect the rights of third States.[40]

The impact of a decision of a court or tribunal on the legal interest of a third State would potentially seem to be most significant if the presence of a third State is considered to be a relevant circumstance, even if the boundary is not established in an area that is also claimed by that third State. For instance, the conclusion that the equidistance method does not lead to an equitable result for a State lying between two other States on a concave coasts has implications for both neighbouring States. In that circumstance, the assessment that taking into account the coast of a third State is only a finding of fact, as was indicated by the Court in the *Land, Island and Maritime Frontier Dispute Case* between Honduras and El Salvador,[41] seems to be untenable. The delimitation between the

---

36 *Land and Maritime Boundary between Cameroon and Nigeria, supra* note 34.

37 See also, [1981] *I.C.J. Reports*, pp. 8–9, paras. 12–13; pp. 12–14, paras. 19–23; and pp. 19–20, paras. 33–35.

38 [1998] *I.C.J. Reports*, p. 324, para. 116.

39 See *ibid.*

40 [2002] *I.C.J. Reports*, p. 421, para. 238.

41 See *supra* the quotation at note 30.

parties also implies an assessment of the position of the third State and as such affects its legal interests.[42]

Should a court or tribunal, which finds that the delimitation between the parties affects the legal interests of a third State refrain from exercising its jurisdiction over the dispute submitted by the parties? Two options exist for a court or tribunal being faced with a delimitation in which the coast of a third State is found to be part of the relevant geography for the delimitation between the parties. The adjudicative body could refrain from establishing a boundary between the parties in the area concerned or it could establish a boundary while including appropriate statements safeguarding the legal interests of the third State. Both approaches affect the legal interests of the third State. At first glance it might seem that the latter approach has more profound implications. However, a number of arguments indicate that the difference between the two approaches may not be particularly significant. The limited determinacy of the law on maritime delimitation and the significance of the factual context of each individual case imply that the choice for the location of one specific boundary in an area does not determine the location of other boundaries

[42] The interest involved in this instance would seem to fall squarely under the test formulated by the International Court in the *South West Africa* cases:

> it may be said that a legal right or interest need not necessarily relate to anything material or "tangible," and can be infringed even though no prejudice of a material kind has been suffered. [...] [S]uch rights or interest, in order to exist must be clearly vested in those who claim them, by some text or instrument or rule of law (*South West Africa cases (Ethiopia v. South Africa; Liberia v. South Africa) second phase*, Judgment of 18 July 1966; [1966] *I.C.J. Reports* 3, at para. 44).

In maritime delimitation cases two approaches can be discerned in assessing whether the legal interests of a third State may be affected. The Court has taken a restrictive approach in assessing whether a legal interest was involved in considering the application for intervention of Malta in the *Tunisia/Libya Continental Shelf Case* and of Italy in the *Libya/Malta Continental Shelf Case* (for a discussion see Chinkin, *supra* note 10, at pp. 160–163). That approach is also apparent in the way in which the case law has dealt with the presence of third States as a relevant circumstance (see further *infra*). On other hand, in its judgments on the merits in *Tunisia/Libya Continental Shelf Case*, the *Libya/Malta Continental Shelf Case*, and the *Cameroon v. Nigeria Case*, the Court adopted a broader view, limiting the area in which it established a boundary by reference to the claims and interests of a third State.

in the same area.[43] A court or tribunal would not have to decide on the actual delimitation involving a third State to assess the impact of the presence of that States as a relevant circumstance. This suggests that the rights of a third State would be similarly affected if an adjudicative body would either deal or refrain from dealing with the request of the parties before it because of the finding that the regional context of a bilateral delimitation is relevant to its determination.[44] In these circumstances, the balance should arguably be tipped towards dealing with the request of the parties before the court or tribunal to the fullest extent possible, while including the necessary safeguards for third party rights.

Claims of Third States and the Establishment of the Endpoint(s) of the Maritime Boundary Between the Parties to Proceedings

Courts and tribunals have taken a number of approaches to define the extent of a maritime boundary between the parties in the face of third party claims. The parties to a case may themselves indicate the area to which the decision is to be limited by reference to third party claims or existing maritime boundary agreements. In such cases, a court or tribunal may not have to address the spatial impact of third party claims directly. However, in most cases involving claims of third parties, this easy way out is not offered by the litigants or the factual circumstances. Several solutions have been adopted by adjudicative bodies to deal with the claims of third States. One option is to define the boundary as a line running in a specified direction without the indication of a final point. This solution guarantees that the decision addresses the request of the parties in full. Whether a court or tribunal will find it possible to adopt this solution will depend on the extent of the area in which claims of third States exist

---

43 See also, N. Marques Antunes, *Towards the Conceptualisation of Maritime Delimitation: Legal and Technical Aspects of a Political Process* (Leiden: Martinus Nijhoff Publishers, 2003), p. 371.

44 In this specific case, a court or tribunal would probably not be in a position to limit the scope of its decision by reference to the area claimed by a third State, as that claim is unlikely to be affected by the delimitation between the parties. The *North Sea Continental Shelf Cases* are a case in point. The position of Germany implied that its bilateral boundary with the Netherlands would be located in an area that was not claimed by Denmark. The same applies to Germany's bilateral delimitation with Denmark.

and how an adjudicative body deals with the existence of such claims, and the method of delimitation adopted to delimit the boundary between the parties.[45] This approach does point to the inherent difficulties involved in dealing in full with the request of the parties, while at the same time not affecting the claims and legal interests of third States. The fact that the boundary extends in a direction after a specified point implies that the same circumstances remain relevant for the determination of the boundary up to the point where it will eventually link up with other maritime boundaries in the same area. In other words, this approach identifies the circumstances that are relevant to the establishment of that point, which involves a third State, to the exclusion of other circumstances.

A second option in the face of a third party claim employed by courts and tribunals is to leave all or part of the boundary affected by the claim of a third State undefined. In the *Libya/Malta Continental Shelf Case* the parties had requested the Court to consider the delimitation of the areas of continental shelf appertaining to each party.[46] However, the Court decided to limit its decision to an area that lay outside the claims that Italy had presented during the hearings on its application to intervene, leaving a large part of the potential boundary between Libya and Malta undefined. The Judgment observes that the decision did not imply a finding that the principles and rules it had formulated were not applicable to the area outside the area it would delimit. The Court took the view that it had not been endowed with jurisdiction to determine the principles and rules governing delimitations with third States.[47] The principles and rules applicable to the delimitation between Libya and Malta to the east of the boundary identified by the Court may indeed be different. For one thing, it would seem difficult to maintain that all of the relevant coast of Libya, as identified by the International Court, is also relevant to the determination of that boundary.[48] The *Libya/Malta Continental Shelf*

[45] See further *infra* for the approach of the International Court in the *Libya/Malta Continental Shelf Case* and of the Tribunal in the *Eritrea/Yemen Arbitration*.

[46] The question submitted by the parties to the Court is contained in Article I of the Special Agreement reproduced in [1985] *I.C.J. Reports*, p. 16, para. 2; see also, *ibid.*, p. 25, para. 21.

[47] *Ibid.*, p. 25, para. 21.

[48] Most of the relevant coast of Libya identified by the Court is beyond 200 nautical miles from the eastern part of an equidistance line between Malta and Libya, which is controlled by Libyan basepoints to the east of the Gulf of Sirte.

Case illustrates that limiting a decision to the area that is not claimed by a third State may leave no other option than to refrain from addressing a significant part of the request of the parties. The International Court only determined the continental shelf boundary between Libya and Malta between the meridians of 13° 50' E and 15° 10' E. The provisional equidistance line between Libya and Malta bounded by these two meridians is approximately 125 kilometres. The provisional equidistance line between the parties to the east of that area measures over 280 kilometres.

In the *Eritrea/Yemen Arbitration*, the Tribunal was faced with the third State issue at the northern and southern end of the maritime boundary between the parties. In view of the Tribunal's conclusion that it had to avoid trespassing upon an area where other claims might fall to be considered,[49] the Tribunal adopted a northern and southern terminal point which it believed to be "well short of where the boundary line might be disputed by any third State."[50] The approach of the Tribunal can be explained in terms of the geography of the area to be delimited, the method of delimitation the Tribunal adopted in view of that geography and its views concerning the rights of third States. The maritime boundary established by the Tribunal takes the median line as its starting point. That line is controlled by numerous basepoints on the coasts of the parties and has many turning points. In those circumstances, the definition of an open-ended boundary is fraught with difficulty. If the last turning point of the boundary is too far from a potential trijunction point the boundary can be expected to veer off course as it does not take into account further controlling basepoints of the median line. Clearly, there is no solution to this problem if it is considered that determining a boundary in an area that is also claimed by a third State affects the legal interests of that State or implies a decision on its claims.

The Tribunal's actual choice of the northern end point of the boundary between Eritrea and Yemen suggests that it sought to balance the interest of the parties in the determination of their bilateral boundary with those of the third State concerned, instead of adopting the view of the third State. Saudi Arabia had written to the Registrar of the Tribunal to point out that its boundaries with Yemen were disputed. It had suggested that

---

49 Award of the Arbitral Tribunal in the Second Stage of the Proceedings (Maritime Delimitation), *supra* note 27, at para. 136.

50 *Ibid.*, para. 164.

the Tribunal should restrict its decision to the area south of the line of latitude through the most northern point of the island of Jabal al-Tayr. Yemen wished the delimitation to extend to the latitude of 16° north. Eritrea did not object to the Saudi Arabian proposal.[51] The northern terminus of the boundary between Eritrea and Yemen established by the Tribunal is to the north of the line of latitude indicated by Saudi Arabia, but to the south of the latitude of 16° north and even further to the south of the maritime boundary that was agreed between Yemen and Saudi Arabia in 2000.[52]

In the *Anglo-French Arbitration*, the Court of Arbitration did not limit its decision by reference to the presence of Ireland as a third State. That approach is explained – as is the case for the examples discussed above – by the way the Court perceived the impact of its decision on that third State.[53] The Tribunal in the *Nova Scotia and Labrador/Newfoundland Arbitration* also did not take into account the maritime zones of a third State in establishing the boundary between the parties. The Tribunal, after a cursory review, concluded that no claim of a third State existed.[54]

The Presence of Third States as a Relevant Circumstance

The *North Sea Continental Shelf Cases*, the first case concerned with maritime delimitation beyond the territorial sea, signified a highpoint for the presence of third States as a relevant circumstance. The presence of a third State can be considered the main consideration – arguably even the only consideration – underlying the conclusion that the equidistance method did not result in an equitable result. The Court's decision indicates that the consensual nature of its jurisdiction did not debar it from taking into account the presence of a third State to the extent this was required to decide a bilateral delimitation.[55]

---

[51] *Ibid.*, para. 44.

[52] *Ibid.*

[53] See *supra* the discussion at note 22.

[54] See further *supra* the discussion in note 26.

[55] See further *supra* the discussion in respect of paragraph 11 of the Court's judgment at note 15.

The Court of Arbitration in the *Anglo-French Arbitration* had to consider the argument of the parties concerning the presence of Ireland for a part of the delimitation it had been requested to effect by the parties. The Court rejected that the fact that the United Kingdom was situated between France and Ireland was a relevant circumstance for the delimitation between France and the United Kingdom in Atlantic region. This approach has been considered as a welcome movement away from that taken in the *North Sea Continental Shelf Cases*.[56] The Court of Arbitration observed that the delimitation of a boundary between France and the United Kingdom did not depend on "any nice calculations of proportionality based on conjectures as to the course of the prospective boundary between the United Kingdom and Ireland,"[57] suggesting that it had a different view on proportionality and third States as a relevant circumstance than that expressed by the International Court in the *North Sea Continental Shelf Cases*. The delimitation the Court of Arbitration effected depended solely on the assessment that the Scilly Isles, viewed in their relation to the mainland coast of the United Kingdom and the relation of the United Kingdom coast to the French coast, constituted a special circumstance warranting an adjustment of the median line.[58]

However, the distinction between the *North Sea Continental Shelf Cases* and the *Anglo-French Arbitration* may in the end be more apparent than real. The Court of Arbitration observed that two successive delimitations in the Atlantic region "*may* result in *some* overlapping" of the continental shelf.[59] This observation suggests that the Arbitral Court was aware of the relationship of the coasts of the three States abutting on the Atlantic region, but found that there was only a limited amount of overlap of their natural prolongations.[60] The Court's reference to "any nice

---

[56] See E. D. Brown, *Sea-bed Energy and Minerals: The International Legal Regime* (The Hague: Martinus Nijhoff Publishers, 1992), pp. 124–126; P. Weil, *The Law of Maritime Delimitation: Reflections* (Cambridge: Grotius Publications, 1989), pp. 253–254.

[57] *Anglo-French Arbitration*, *supra* note 22, at para. 27.

[58] *Ibid.*, paras. 248–253. That assessment of the Scilly Isles does constitute recognition of the fact that the weight of any geographical feature in a maritime delimitation has to be assessed in its wider context. The delimitation the Court effected in the area involving the Channel Islands (see *ibid.*, paras. 180–202) is a further case in point.

[59] *Ibid.*, para. 28 (emphasis provided).

[60] At the end of paragraph 28 of the Award, the Court indicates that the overlap was most likely to occur beyond the terminus of the continental shelf boundary the parties

calculations of proportionality" also suggests this point.[61] Moreover, to reach that conclusion the Court of Arbitration distinguished the geography of the three States involved in the *North Sea Continental Shelf Cases* from the Atlantic region with which it was concerned.[62] That geography is indeed completely different. The relevant coasts of Denmark, Germany and the Netherlands, although one may have different views as to their exact definition, are all of considerable length. In the Atlantic region under consideration in the *Anglo-French Arbitration* the approximate length of the relevant coastlines would be more difficult to assess. What seems beyond doubt is that all of the southern coast of Ireland faces that area. As far as the coasts of the parties are concerned, the Court of Arbitration limited itself to observing that the coasts abutting on the area to be delimited were relatively short.[63] It further noted that those coasts were formed by Finistère and Ushant on the French side and the Scilly Island and Cornwall on the U.K. side, without providing any further specification.[64] Any definition of these coasts would result in coastal lengths that are less than half of the Irish coast facing the same area. That might suggest the existence of a disproportion that can be easily ascertained without any nice calculations. However, the British and Irish coasts relevant for their delimitation in the Celtic Sea and Atlantic Ocean are comparable in length and the only special or relevant circumstance in this case are the Scilly Isles, suggesting that an equidistance line giving limited effect to the Scilly Isles results in an equitable solution.[65] This assessment illustrates that an approach similar to that in the *North Sea Continental Shelf Cases* would have required

had imposed on the Court through the Special Agreement (see also the answer of the parties to the first question of the Court as reported in para. 25 of the Award).

[61] As the Court of Arbitration also indicated (*ibid.*, para. 99) the International Court in the *North Sea Continental Shelf Cases* also did not suggest "any nice calculations of proportionality" but rather considered that there should be a "reasonable degree of proportionality."

[62] *Ibid.*, paras. 250 and 99–101.

[63] *Ibid.*, para. 238.

[64] *Ibid.*, para. 248.

[65] The continental shelf boundary that has been agreed between the United Kingdom and Ireland is a stepped line, which is more advantageous to Ireland than an equidistance line giving full effect to the Scilly Isles (see the map in D. H. Anderson, "Ireland-United Kingdom," in J. I. Charney and L. M. Alexander, *International Maritime Boundaries*, Vol. 1 (Dordrecht: Martinus Nijhoff Publishers, 1993), pp. 1767–1779, at p. 1772. It has been reported that this boundary follows the precedent of the *Anglo-French*

addressing the bilateral delimitation between the United Kingdom and Ireland, a matter outside the jurisdiction of the Court. The fact that the relevant coast of the United Kingdom abutting the Atlantic region under any definition is shorter than the Irish and French coasts indicates that a delimitation that gives a more limited area of continental shelf to the United Kingdom than the other two States is equitable.

The *Libya/Malta Continental Shelf Case* and the *Guinea/Guinea-Bissau Arbitration* both identified the presence of third States as a relevant circumstance. The treatment of third States in these cases also raises questions concerning the possible impact of decisions on future delimitations in the same region. These cases will therefore be further discussed in the next section.

In the *Cameroon v. Nigeria Case*, the International Court did not consider that the presence of a third State constituted a relevant circumstance for the delimitation it had been requested to effect. The Gulf of Guinea is characterized by a highly complex political geography. Cameroon is situated between Nigeria and the mainland coast of Equatorial Guinea. In front of the coast of Cameroon lies the island of Bioko, which is part of Equatorial Guinea. To the southwest of Bioko lies the island State of São Tomé and Príncipe. In the *Cameroon v. Nigeria Case*, the Court had to consider the consequences of the mainland geography and the presence of Bioko for the maritime boundary between Cameroon and Nigeria. Cameroon had argued that an equidistance line was inequitable for a number of reasons, including the concavity of the relevant mainland coasts and the location of Bioko in front of Cameroon's coast.[66] The Court found that the arguments of Cameroon did not justify any adjustment of the provisional equidistance line.[67]

The delimitation effected by the Court in the *Cameroon v. Nigeria Case* indicates a clear break with the earlier case law assessing similar situations, in which it was found that equidistance did not lead to an equitable result. A comparison of the geography in the North Sea and the Gulf of Guinea and its impact on maritime delimitation makes it difficult to view the outcome of the *North Sea Continental Shelf Cases* as

Arbitration and other cases in giving less than full effect to islands lying near the coast (*ibid.*, pp. 1769–1770).

66 *Cameroon v. Nigeria Case*, *supra* note 25, at pp. 433–434, paras. 272–275.

67 *Ibid.*, pp. 445–447, paras. 296–301.

anything but completely opposite to the *Cameroon v. Nigeria Case*.[68] In the *North Sea Continental Shelf Cases*, the Court found that the coastal geography of Denmark, Germany and the Netherlands had to be viewed in conjunction to assess the effects of the equidistance method for the bilateral delimitations of the three States. The same assessment of the three coasts involved can be made in the case of the mainland coasts of Cameroon and Nigeria and the coast of Bioko. The concavity at the back of these three coasts is much more pronounced than the concavity of the coasts of Denmark, Germany and the Netherlands. Moreover, there can be little doubt that the Court's finding in the *North Sea Continental Shelf Cases* concerning the broad equality of the coastlines of the three States[69] is also applicable to the coasts Cameroon and Bioko.[70]

The Court in the *Cameroon v. Nigeria Case* did not consider the geographical relationship between the three coasts, but instead looked separately at the relationship between the coasts of Cameroon and Nigeria and the coasts of Cameroon and Bioko.[71] The Court concluded that the effect of Bioko on the seaward projection of the coastal front of Cameroon was an issue between Cameroon and Equatorial Guinea and was not relevant to the delimitation before the Court.[72] The Court's bilateralized approach, explained by its reliance on equidistance to define the relevant area,[73] effectively excluded an assessment of the impact of the regional geography outside of that area.[74]

---

68 See also, Evans, *supra* note 9, at p. 146.

69 *North Sea Continental Shelf Cases, supra* note 15, at pp. 49–50, para. 91.

70 The Court considered that the relevant coast of Cameroon ran from East Point on the Bakassi Peninsula to Debundsha (*Cameroon v. Nigeria Case, supra* note 25, at pp. 442–443, para. 291). That coast is comparable in length to the west facing coast of Bioko. Especially in a broader regional context, it could be argued that a larger part of Nigeria's coast is relevant. However, that longer coast of Nigeria would get a much larger share of maritime zones than the shorter coast of Cameroon if the equidistance line is adjusted to some extent.

71 *Cameroon v. Nigeria Case, supra* note 25, at pp. 445–446, paras. 296–299. It should be noted that Cameroon had placed the issue before the Court mostly in terms of two bilateral relationships (see, e.g., *Cameroon v. Nigeria* case CR2002/6 paras. 40 *et seq.*).

72 *Ibid.*, p. 446, para. 299.

73 See further *infra*.

74 As is observed by Evans "[h]aving limited itself in this fashion it would have been peculiar if the Court *had* decided to adjust the equidistance line that it went on to

It has been questioned whether the presence of third States should be considered to be a relevant circumstance.[75] One argument is that there is a fundamental objection to the inclusion of the interests of third States in the category of relevant circumstances. Only the projections of coasts abutting on an area meet and overlap and it would thus be inconceivable for a court to widen its field of consideration to coasts that have nothing to do with the delimitation in question.[76] That argument is not convincing. There are instances in which the coasts of three (or more) States abut on the same area and the projections of all those States overlap.

Evans's rejection of the presence of third States as a relevant circumstance is in line with the reasoning of the International Court in the *Cameroon v. Nigeria Case*. Evans views the bilateral delimitation between Nigeria and Cameroon and the mainland geography as one matter and the bilateral delimitation between Cameroon and Equatorial Guinea (Bioko) as another ("the superaddition of an additional factor").[77] If that view is adopted "[t]he area to be delimited *becomes* the 'region' in which the delimitation it to take effect. [...] The point is simple: the general geographical situation in the *region* in which the delimitation takes place has no real bearing on the outcome at all."[78] Is it really that simple? The answer hinges on the role of equidistance in the delimitation process. In the *Cameroon v. Nigeria Case*, the Court defined the maritime area and coastlines it took into account by reference to the equidistance lines between the parties and between the parties and the third State concerned.[79] In the limited area thus defined, equidistance did not lead to an inequitable result. The problem involved in this approach is that equidistance was effectively used to asses the equitableness of equidistance.

A further justification for a rejection of the presence of third States as a relevant circumstance is the subjectivity involved in assessing this

identify between Cameroon and Nigeria by reference to [Bioko]" (Evans, *supra* note 9, at p. 149; emphasis in the original).

75 Evans, *supra* note 9, at pp. 149–160; Lucchini and Voelckel, *supra* note 13, at pp. 280–281; Weil, *supra* note 56, at p. 253.

76 Weil, *supra* note 56, at p. 253; see also, Lucchini and Voelckel, *supra* note 13, at pp. 280–281.

77 Evans, *supra* note 9, at p. 150.

78 *Ibid.*, p. 150 (emphasis in the original).

79 *Cameroon v. Nigeria Case*, *supra* note 25, at p. 443, para. 292.

circumstance.[80] It is problematical to rely on the perceived subjectivity of a circumstance as a reason to discard it if it forms part of the applicable law. Secondly, whether excluding this particular circumstance from the category of special/relevant circumstances actually leads to greater objectivity may be doubted. Small islands have consistently been mentioned as a prime example of a circumstance justifying an adjustment of the provisional equidistance line. An assessment of a specific island will always have to be made in the light of the other circumstances of the specific case, as is illustrated by the reasoning in respect of the Scilly Isles and the Channel Islands in the *Anglo-French Arbitration*. A hypothetical example may serve to further illustrate the difficulties that may be involved. In the North Sea, the equidistance method gives Germany a smaller area of continental shelf than its neighbours, Denmark and the Netherlands. Would the presence of a small island in front of the coast of Germany that would deflect the equidistance line significantly to the advantage of Germany have to be considered as a special circumstance requiring the adjustment of the equidistance line? Viewed in isolation from other circumstances, the answer would likely be affirmative. If the regional context were to be taken into account, which would indicate that discounting the island would give Germany a considerably smaller part of the North Sea than its two neighbours, that likelihood diminishes. Giving full effect to the small island in this hypothetical case would contribute to one of the perceived merits of the equidistance method, namely that it in principle leads to an equal division of the area of overlapping continental shelves.[81]

In the end, since equidistance is not a mandatory method, subjectivity is inevitable. This should, in itself, be a caution against stressing the virtues of a "more structured approach" to maritime delimitation with a central role for the equidistance method.[82] In view of the large number of pending maritime boundary delimitations, in widely diverging geographical contexts, there is much to be said for maintaining the flexibility that has characterized the case law.[83] This would seem to be the understanding that

[80] Evans, *supra* note 9, at pp. 159–160; Lucchini and Voelckel, *supra* note 13, at pp. 280–281; Weil, *supra* note 56, at p. 253.

[81] See, e.g., Weil, *supra* note 56, at p. 257.

[82] As Evans observes, a more robust approach to the scope of relevant/special circumstances would probably only serve to shift the debate to new questions in what appear to be more relevant areas (Evans, *supra* note 9, at p. 159).

[83] See also, Evans, *supra* note 9, at p. 160.

inspired the compromise concerning Articles 74 and 83 of the United Nations Convention on the Law of the Sea.[84] That is not to say that equidistance should not continue to play a significant role in maritime delimitation law. However, equidistance should not become the driving force for identifying the relevant geography and the area to be delimited, and specific special/relevant circumstances should not be assessed solely in the light of their effect on the provisional equidistance line, but should be assessed in the overall context of the situation under consideration.

The Possible Impact of Decisions on Future Delimitations in the Same Region

A decision of a court or tribunal only has binding force between the parties in respect of the particular case concerned. Notwithstanding this fundamental principle, the decisions of courts or tribunals may have an impact on future delimitations. As was noted above, the potential impact of a decision on the legal interest of a third State or for the construction of a treaty is recognized in the Statutes of the International Court and the International Tribunal for the Law of the Sea. This impact is particularly noticeable in the case of maritime delimitation, which generally is considered to have been elaborated to a large extent through the case law. This is illustrated by the numerous references by courts and tribunals to earlier decisions. Such reliance is in line with the requirement to decide the case before an adjudicative body on the basis of the applicable law. The content of a rule of law contained in a convention or customary international law does not depend on the parties before the court or tribunal.

The application of the rules of delimitation law in a specific case may have implications for delimitations involving States that are not a party to the proceedings. At times it may be inevitable to rule on matters that also have implications for third States. As was argued earlier in respect of the *North Sea Continental Shelf Cases*, the fact that the International Court considered that the configuration of the German coast implied that equidistance did not lead to an equitable result because of the combined effect of the delimitations with Denmark and the Netherlands inevitably

---

84 See L. D. M. Nelson, "The Roles of Equity in Maritime Delimitation," *American Journal of International Law* 84 (1990): 837–859, at p. 858.

meant that the findings on one of the bilateral delimitations had consequences for the other delimitation. In these circumstances, as was argued above, an adjudicative body should not refrain from pronouncing itself on the delimitation of the bilateral boundary of the parties before it. It is difficult to envisage that the Court could have identified (other) relevant circumstances of a purely bilateral nature that would have justified the conclusion that equidistance did not lead to an equitable result.

What guidelines can be used in assessing to what extent an adjudicative body should address issues that may have implications for future delimitations involving third States? A couple of examples from the case law may be helpful to illustrate some aspects of this matter. In the *Cameroon v. Nigeria Case*, the International Court considered the argument of Cameroon that the concavity of its coastline constituted a relevant circumstance. The Court observed that the coastline of Cameroon it had identified as relevant "exhibit[ed] no particular concavity."[85] The Court went on to note that "the concavity of Cameroon's coastline is apparent primarily in the sector where it faces Bioko."[86] The Judgment obviously does not pronounce itself on the implications of this concave nature of the coast of Cameroon facing Bioko. As a matter of fact, there may be no implications for the delimitation between Cameroon and Equatorial Guinea at all. Nonetheless, it is to be expected that such an observation would be picked up by a State involved in bilateral negotiations. The finding that the relevant coast of Cameroon for the area to be delimited with Nigeria did not display a marked concavity would have sufficed to decide the issue before the Court.

The *Guinea/Guinea-Bissau Arbitration* and the *Libya/Malta Continental Shelf Case* employed a reasoning that has much more significant implications for future delimitations than the Court's pronouncement on the concavity of Cameroon's coastline where it faces Bioko. In the *Guinea/Guinea-Bissau Arbitration*, the Tribunal concluded that it had to assess the boundary to be established in a regional context. The Tribunal observed that if Sierra Leone were taken into consideration there are three adjacent States along an adjacent coastline and that equidistance has the drawback

[85] *Cameroon v. Nigeria Case, supra* note 25, at p. 445, para. 297.

[86] *Ibid.*

of the middle country being enclaved.[87] In order to establish the method of delimitation to be employed, the Tribunal found it appropriate to look at the whole coastline of West Africa, which would lead to a delimitation integrated into the present and future delimitations as a whole.[88] To achieve this result the Tribunal drew a line representing the general configuration of the coastline between Almadies Point in Senegal and Cape Shilling in Sierra Leone.[89] Most of the maritime boundary between Guinea and Guinea-Bissau established by the Tribunal is perpendicular to that line. The use of the Tribunal's general direction line to draw perpendicular maritime boundaries of the other States in the region has remarkable results. Such perpendicular lines would deviate considerably from the two parallel maritime boundaries agreed between The Gambia and Senegal.[90] In the case of Guinea and Sierra Leone, a perpendicular line is advantageous to Guinea in comparison to an equidistance line throughout the length of their maritime boundary, including the nearshore area, in which case the presence of Guinea-Bissau more than 270 kilometres to the northwest by no stretch of the imagination can be considered to be a relevant circumstance. Of course, the Tribunal did not assert that the perpendicular formed an equitable maritime boundary between Guinea and Sierra Leone. However, its emphasis on the regional context of the delimitation and the specific line selected gives to Guinea a detailed argument backed up by the authority of an international tribunal. It would no doubt be possible to explain away the relevance of the Tribunal's decision for at least part of the boundary between Guinea and Sierra Leone.[91] Still, the Award implies a burdening of the negotiations between Guinea and Sierra Leone, and it cannot be excluded altogether that it would have

87 *Guinea/Guinea-Bissau Maritime Delimitation Maritime Delimitation Arbitration*, Decision of 14 February 1985 (International Boundary Cases, Vol. II, p. 1301), para. 104.

88 *Ibid.*, para. 108.

89 *Ibid.*, para. 110. For a depiction of this line see Brown, *supra* note 56, at p. 296, figure 18.

90 *Agreement between The Gambia and the Republic of Senegal*, 4 June 1975; in force 27 August 1976; Charney and Alexander, *supra* note 65, at pp. 854–855.

91 In the *Guinea/Guinea-Bissau* arbitration the Tribunal also did not apply the perpendicular line for the first segment for the boundary because of the presence of other circumstances that were found to be relevant for that part of the maritime boundary.

implications for a settlement by a third party.[92] Unnecessarily at that, as the relevant geography characterized by the presence of numerous islands along the coast, especially of Guinea-Bissau, offered the Tribunal ample opportunity to arrive at a similar boundary without explicitly addressing the regional context. The Award includes all the necessary elements for that approach. The Tribunal observed that the coastline of Guinea-Bissau is convex if the Bijagos Islands, lying in front of Guinea-Bissau's coast, are taken into account and the coast of Guinea is concave.[93] The Tribunal distinguished the Bijagos Islands from coastal islands. According to the Tribunal, the latter "must be considered as forming an integral part of the continent."[94] By implication, the Bijagos were not considered to be an integral part of the African continent.[95]

In the *Libya/Malta Continental Shelf Case*, the International Court considered that the difference in the length of the relevant coasts justified an adjustment of the median line so as to attribute a larger area of continental shelf to Libya.[96] To establish the extent of this shift, the Court looked at the broader geographical context in which the delimitation had to be effected and concluded that the maximum shift of a median line was indicated by the median line between the coasts of Libya and the Italian island of Sicily located to the north of Malta.[97] An implication of this finding is that, according to the Court, there was no difference in coastal length between the relevant coasts of Libya and Sicily which could justify the adjustment of a median line between those two coasts. That implication is open to serious questioning, as is argued in the joint separate opinion of Judges Ruda, Bedjaoui and Jiménez de Aréchaga.[98] Depending on the definition of the relevant coast of Sicily, there is a considerable difference in the lengths of the relevant coast of Sicily and

92 See also, M. D. Evans, *Relevant Circumstances and Maritime Delimitation* (Oxford: Clarendon Press, 1989), pp. 237–238; B. H. Oxman, "Political, strategic, and historical considerations," in Charney and Alexander, *supra* note 65, pp. 3–40 at p. 18.

93 *Guinea/Guinea-Bissau Maritime Delimitation Maritime Delimitation Arbitration*, Decision of 14 February 1985, para. 103.

94 *Ibid.*, para. 95.

95 See also *ibid.*

96 *Libya/Malta Continental Shelf Case, supra* note 2, at p. 50, para. 68.

97 *Ibid.*, pp. 51–52, para. 72.

98 *Separate Opinion of Judges Ruda, Bedjaoui and Jiménez de Aréchaga*, [1985] *I.C.J. Reports*, p. 76.

the relevant coast of Libya as defined in the judgment.[99] Judges Ruda, Bedjaoui and Jiménez de Aréchaga provide two definitions of the relevant coast of Sicily, one between Cape Passero and Gela and the other between Cape Passero and Marina di Ragusa, just to the east of Cape Scaramia. The ratio of those coasts and the relevant coast of Libya is given as respectively 1 to 1.55 and 1 to 3.5.[100] Although the conclusion that the former ratio would make a median line inequitable[101] may be open to question, it is clear that the conclusion that in the hypothetical case of a delimitation between Italy and Libya the length of the relevant coasts of Sicily and Libya does not require any adjustment of the median line is a shaky foundation on which to build the reasoning for shifting the provisional median line between Libya and Malta.

The implications of the International Court's approach of taking into account Sicily are obvious for any delimitation between Italy and Malta. If the Court's reasoning is followed, the coast of Sicily that was taken into account by the Court also is the relevant coast for the delimitation between Italy and Malta in at least the area directly between the two States and arguably also in the area to the west of it.[102]

The approach of the Court is all the more debatable because there were no apparent grounds to evaluate the delimitation between Libya and Malta in a regional context.[103] There is no compelling logic to the reasoning supporting the shift of the median line by the Court.[104] In view of the amount of discretion that courts and tribunals possess in applying maritime delimitation law, a similar shift of the median line could have

[99] See *Libya/Malta Continental Shelf Case*, *supra* note 2, at p. 50, para. 68.

[100] *Ibid.* Joint sep. op., pp. 81–82, para. 18. The method for measuring the length of the coast used for the second ratio is not provided. If these coasts are measured as straight line segments, the ratio seems to be from around 1:5.6 to 1:5.8.

[101] *Ibid.*, Joint sep. op., p. 82, para. 18.

[102] However, if the area to the north of Malta is considered, the coast of Sicily between Marina di Ragusa and Cape Passero identified in the joint separate opinion of Judges Ruda, Bedjaoui and Jiménez de Aréchaga could well be argued to be the relevant coast. Whether there is any disproportion justifying a shift of a median line between that coast and the relevant coast of Malta is questionable.

[103] See also, B. Conforti, "L'arrêt de la cour internationale de justice dans l'affaire de la délimitation du plateau continental entre la Libye et Malte" *Revue générale de droit international public* 90 (1986): 313–343, at pp. 331–332.

[104] See also, *Libya/Malta Continental Shelf Case*, Joint sep. op., *supra* note 2, at p. 82, para. 19; *ibid.*, Schwebel, diss. op., pp. 180–182, esp. at p. 181.

been justified on the basis of a broad reference to the disparity of the relevant coasts of the parties.[105]

States will obviously be attracted by arguments that can be backed up by references to the case law. Despite the political context in which negotiations on maritime delimitation take place, legal arguments invariably are an important factor. The stature of decisions of the International Court and tribunals makes their arguments on the law compelling. Even though maritime delimitation law is open-textured and it can always be argued that the factual situation of the case at hand does not justify the application of the reasoning contained in a decision made in a different factual context, a State that has to adopt this approach will be at a tactical disadvantage. This will especially be the case if a previous decision has addressed a factual scenario that is part of the later negotiating context and specific legal consequences have been drawn from it. As was argued above, in some cases it may be inevitable to evaluate the consequences of the presence of the third State on the maritime delimitation before a court or tribunal. However, as the discussion of the case law indicates, there are also instances in which a decision can be based on legal arguments that only need to take into account the bilateral relationship between the States concerned to achieve a similar result. It is suggested that adjudicative bodies should frame their decisions to the largest extent possible on the latter type of argument.

Concluding Observations

The case law undoubtedly has moved in the opposite direction from the trend that Douglas Johnston welcomed in *The Theory and History of Ocean Boundary-Making*. Instead of an emphasis on the particularistic of any delimitation, the recent case law has contributed to the consolidation of legal norms, emphasizing the centrality of equidistance in the delimitation process. In the *Cameroon v. Nigeria Case* this development was coupled with a rejection of the relevance of the regional context for considering the bilateral delimitation between Cameroon and Nigeria. It is difficult to see that approach as anything other than a complete break with the

[105] See also, *ibid.*, Sette Camara, sep. op., pp. 74–75; *ibid.*, Joint sep. op., pp. 91–92, paras. 39–40.

previous case law.[106] If the *Cameroon v. Nigeria Case* is a portent of the future of maritime delimitation law, any further role for third States as a relevant circumstance seems to be excluded.[107] Whether this is a desirable development, in view of the complexity and variety of issues a court or tribunal may be faced with, is questionable. Using the location of equidistance lines as the primary consideration for identifying the area to be delimited has the attractiveness of simplicity and apparent objectivity, but also means that equidistance is used to set up the parameters for its own evaluation in terms of the equitableness of the delimitation that is to be achieved. To avoid this result a broader area should be taken into consideration to make that evaluation. Depending on the circumstances of the specific case that approach may lead to also considering the coast of a third State as part of the relevant circumstances of that case.

Following the *Anglo-French Arbitration*, courts and tribunals have shown a particular concern that the delimitation they were requested to effect might affect the legal interest of third States. Whether this approach is justified is questionable, in view of the options that are open to a court or tribunal. In the face of a third party claim an adjudicative body could refrain from establishing a boundary between the parties in the area concerned or it could establish a boundary and at the same time include appropriate pronouncements concerning the legal interest of the third State. As was argued, both approaches affect the legal interests of the third State in not dissimilar ways. In these circumstances, the balance should be tipped towards dealing with the request of the parties before the court or tribunal to the fullest extent possible, while including the necessary safeguards for third party rights. This approach would also seem to be preferable from a public order perspective.

---

[106] It seems slightly ironic that the *Cameroon v. Nigeria Case*, which downplayed the regional context of the delimitation it had to effect, was the first maritime delimitation case in which the intervention of a third State was allowed, whereas the Court rejected intervention in two cases (*Tunisia/Libya Continental Shelf Case* and the *Libya/Malta Continental Shelf Case*) at a time when the particularism of the case law and the evaluation of bilateral delimitations in a regional context were at a high point.

[107] The fact that the maritime delimitation effected by the Court took place in a very complex factual matrix involving among others the territorial dispute over the Bakassi peninsula should possibly caution against drawing too much on that case for establishing the future direction of maritime delimitation law.

The taking into account of the claims of third States by courts and tribunals is translated to a specific approach to the delimitation of the boundary between the parties in or near areas that are also claimed by third States. Depending on the circumstances, it may still be possible to address the delimitation of the entire boundary between the parties. However, as was observed, this may imply an implicit evaluation of circumstances relevant to the delimitation with the third State. At other times, the extent of the claims of a third party that is taken into account, the geography concerned or the method of delimitation that is chosen prevent this approach.

The reticence to deal with areas involving the claims of third States, even in circumstances in which it would not lead to any qualification of the legal rights of third States, has not prevented courts and tribunals from making pronouncements that arguably have affected the legal rights of third States. This point illustrates that one should be careful to equate a larger role to the regional context of a delimitation with a facilitative approach to ocean boundary-making. Depending of the way in which an argument is structured, it may complicate the future delimitation of other boundaries in the area.

At times it may be necessary to deal with arguments that have implications for the legal interests of third States, either in establishing a bilateral boundary or in refraining from deciding that issue. However, in cases in which the law and the circumstances of the case leave various options to achieve an equitable result, the option that is built on a reasoning that has the least consequences for third States should be preferred. It is suggested that a facilitative approach to maritime boundary delimitation at times may be best realized by taking cognizance of the relevant circumstances of a regional nature without explicitly stating those circumstances in the decision.

The magnitude of the impact of the presence of a third State on a decision on a maritime boundary and the consequent impact of that decision on the legal rights of a third State in the first place depends on the content of maritime delimitation law. These impacts can be said to be the inverse of the place that is accorded to equidistance in maritime delimitation law. As illustrated by the *Cameroon v. Nigeria Case*, defining the relevant area by reference to equidistance lines indicating the extent of the claims of a third State signified that the presence of a third State in practice was excluded from the category of relevant circumstances.

Whether this approach will always do justice to the expectations of States seeking an equitable solution to their delimitation disputes is a different question.[108] The great variety of factual situations that may confront the judge or arbitrator may be the best guarantee against too rigid a reliance on a "more structured approach" to maritime delimitation. If that expectation is confirmed, the regional context will continue to play a role in maritime delimitation. In that case, it will remain necessary for courts and tribunals to find a balance between the legal rights of third States and deciding the issues that the parties have placed before them. The flexibility that has characterized most of the development of maritime delimitation law should make it possible to deal with both issues simultaneously in a satisfactory manner.

[108] See also, Nelson, *supra* note 84, at p. 858.

Brokering Cooperation Amidst Competing Maritime Claims: Preventative Diplomacy in the Gulf of Thailand and South China Sea

Clive Schofield* and Ian Townsend-Gault**

Introduction

One of the hallmarks of Douglas Johnston's approach to international law was his insistence that norms should be viewed in their context: a law regarded as autonomous from its sphere of application was likely to be of little or no utility. Furthermore, it could actually constitute a hazard to the achievement of the goal it was seeking to advance. This thinking shaped much of his attitude to the law of the sea, and he communicated it vividly to his younger colleagues as they started to forge their careers. It is for reasons like this that the present writers, from their different disciplinary viewpoints, have always tried to situate the law of the sea and the policies that lie behind it in the specific context of modern marine management.

Also characteristic of Johnston's approach to the study of ocean affairs was his willingness to, indeed, great enthusiasm for, tackling especially sensitive and challenging issues of great significance to good oceans governance and, ultimately, to international peace and security. The region under examination here, the South China Sea/Gulf of Thailand, clearly accords with these distinguishing criteria as these maritime spaces and the claims made to them are undoubtedly geographically and legally complex and politically contentious and problematic.[1]

---

* QEII Research Fellow, Australian National Centre for Ocean Resources and Security, University of Wollongong, Australia. Dr. Schofield is the recipient of an Australian Research Council QEII Fellowship (DP0666273).

** Associate Professor of Law and Director, Southeast Asian Legal Studies, University of British Columbia, Vancouver, Canada.

[1] Although the distinct terms South China Sea and Gulf of Thailand are employed in this essay, these maritime spaces should be considered, particularly in geographical and environmental terms, as one region.

It is also notable that Johnston was deeply involved in efforts to foster enhanced cooperation and oceans management practice in these regions and this has informed our choice of topic. In keeping with the spirit and substance of the approach outlined above, this has meant, *inter alia*, trying to tackle the root causes of outmoded attitudes to ocean law and policy that promised little and often delivered nothing of value. This was especially necessary where arid debates on sovereignty were allowed to dominate at the expense of, for example, a marine ecosystem facing irreversible deterioration because the countries concerned could not agree who was to manage it, and accordingly refused to work together. Even when this hurdle was overcome, Johnston always insisted that the main task was not simply the contextualized application of rules, but rather a consideration of norms from the functional point of view, with the purposive approach very much in mind.

In this contribution, we examine some of the issues and challenges that arise in the context of maritime cooperation in the South China Sea and the Gulf of Thailand. An overview of the geographical and geopolitical context is offered, together with a brief assessment of the key maritime interests at stake. There then follows an assessment of the baselines, claims to maritime jurisdiction, and related agreements, including those involving maritime joint development, existing in the South China Sea and Gulf of Thailand. Claims inconsistent with the relevant provisions of the United Nations Convention on the Law of the Sea (LOS Convention)[2] are highlighted, in particular claims based on alleged historic rights. The problematic issue of dealing with islands in the South China Sea context is also explored. Efforts towards enhancing maritime cooperation, management and dispute resolution are then considered. In particular initiatives in preventative diplomacy designed to forestall conflict and foster regional cooperation, specifically the South China Sea workshops process and Southeast Asian Programme in Ocean Law (SEAPOL) Gulf of Thailand project, are addressed and some thoughts as to the way forward offered.

---

[2] *United Nations Convention on the Law of the Sea*, Montego Bay, 10 December 1982 (entered into force on 16 November 1994), UN/Doc. A/CONF.62/122, 7 October 1982, 33 *I.L.M.* 1309 [hereinafter LOS Convention].

Geographical and Geopolitical Context

The limits of the South China Sea have been defined as extending southwards from the Strait of Taiwan to the 3° South parallel of latitude.[3] Djalal has suggested that the 1° North parallel of latitude may be a more appropriate southern limit. If the latter definition is taken, the total surface area of the South China Sea (including the Gulf of Thailand) has been calculated at approximately three million square kilometres.[4] Accordingly, eight States, Brunei, Cambodia, China, Indonesia, Malaysia, the Philippines, Singapore and Viet Nam, together with Taiwan, have coastal fronts on the South China Sea. The Gulf of Thailand represents an arm of the South China Sea encompassing a total surface area of approximately 82,715 square nautical miles (nm^2) (283,700 km^2), bounded by Malaysia, Thailand, Cambodia and Viet Nam.[5]

The coastal geography of both the South China Sea and Gulf of Thailand is characterized and complicated by the presence of numerous islands, islets, rocks and reefs. Of particular note in the South China Sea context are two disputed archipelagos of islands and reefs generally known respectively as the Spratly and Paracel islands (see below).[6] The semi-enclosed, tropical environment of these maritime spaces has been termed "an area of globally significant biological diversity."[7] Similarly, the Gulf of Thailand has been identified as one of the richest biological resources in the world.[8] This exceptionally productive marine environment

---

[3] According to the International Hydrographic Organization. Referred to in H. Djalal, "South China Sea Island Disputes," in M. H. Nordquist and J. N. Moore, eds, *Security Flashpoints: Oil, Islands, Sea Access and Military Confrontation* (The Hague: Martinus Nijhoff Publishers, 1998), pp. 109–133, at p. 109.

[4] *Ibid.*

[5] Assuming a closing line for the Gulf of a straight line joining Mui Ca Mau at the southern tip of the Vietnamese mainland to a point on the Malaysian coast in the vicinity of Tumpat. See J. R. V. Prescott, *The Gulf of Thailand* (Kuala Lumpur: Maritime Institute of Malaysia, 1998), p. 10.

[6] The littoral States give one or both of these island groups names in their respective vernaculars: for the sake of clarity, we will adhere to the (no doubt lamentable but arguably neutral) English usage.

[7] See, UNEP/GEF, *Reversing Environmental Degradation Trends in the South China Sea and Gulf of Thailand* project website at <www.unepscs.org>.

[8] H. M. Ibrahim, "Overfishing in the Gulf of Thailand: Issues and Resolution," in D. M. Johnston, ed., *SEAPOL Integrated Studies of the Gulf of Thailand*, Vol. 2 (Bangkok: SEAPOL, 1998), pp. 55–93, at p. 62.

supports a fishery that accounts for 10 per cent of the world's landed catch and plays a critical role in regional food security.[9] Access to the waters of the South China Sea and Gulf of Thailand in order to exploit these abundant living resources represents a traditional, and enduring, maritime concern of the littoral States.

The environment, biological diversity and living resources in question are, however, under serious threat. It has been suggested that over 80 per cent of reefs are at great risk and will collapse within 20 years unless unsustainable practices are abated; 70 per cent of mangrove cover has been lost in the last 70 years and at current rates of habitat loss, the remainder will be lost by 2030; 20–60 per cent of seagrass beds have similarly disappeared over the last 50 years and those still in existence are also threatened with destruction.[10] With regard to the Gulf of Thailand, although in the late 1990s it was concluded that its marine environment was "not seriously polluted,"[11] the Gulf's role as a "concentration area" for marine pollution represents a major threat.[12]

The other key maritime resource interest of the coastal States is access to potential seabed hydrocarbon reserves. Commercially viable oil and, particularly, gas reserves have been discovered in the margins of the South China Sea. Indonesia's Natuna gas fields, Brunei and Malaysia's developments offshore Borneo, and the Philippines Malampaya project off the coast of Palawan all represent good examples in this regard. For its part, the Gulf of Thailand has proved to be an established and relatively abundant source of seabed hydrocarbons. However, for all the talk of the oil-related riches of the South China Sea, and particularly references to the (in)famously "oil rich" Spratly Islands, there is considerable uncertainty

[9] M. D. Fortes, "Reversing Environmental Degradation of the South China Sea," unpublished paper given at conference on The South China Sea: Towards a Cooperative Management Regime (Singapore: S. Rajaratnam School of International Studies (RSIS), Nanyang Technological University, May 2007).

[10] *Ibid.*

[11] D. M. Johnston and D. VanderZwaag, "Towards the Management of the Gulf of Thailand: Charting the Course of Cooperation," in D. M. Johnston, ed., *SEAPOL Integrated Studies of the Gulf of Thailand*, Vol. 1 (Bangkok: SEAPOL, 1998), pp. 69–135 at p. 74.

[12] C. Ake-uru, "Thailand and the Law of the Sea," in C. Park and J. K. Park, eds, *The Law of the Sea: Problems from the East Asian Perspective* (Honolulu: Law of the Sea Institute, University of Hawaii, 1987), pp. 414–425, at p. 418.

over the presence of seabed resources in these areas, such that estimates as to the potential oil wealth to be derived from the South China Sea vary greatly.[13] While seabed resources may indeed underlie the South China Sea, expectations of a second Persian Gulf appear far-fetched. Nonetheless, the abiding perception among the competing littoral States that such resources do indeed exist remains an important underlying issue in the South China Sea disputes. Indeed, such unproven assertions as to the hydrocarbon potential of the South China Sea have served to fuel an already difficult jurisdictional dispute.

The hydrocarbon factor needs to be set in the context of escalating regional energy security concerns. Countries across the region are facing serious supply challenges in the face of spiralling domestic energy demands. This scenario has served to sharpen interest in the potential seabed resources of the South China Sea and Gulf of Thailand and added urgency to efforts to facilitate exploration and exploitation of these suspected reserves. While this provides a potent rationale for cooperation in areas of overlapping maritime claims, concerns have simultaneously been raised that competition for energy resources could become the catalyst for dispute and conflict.

Heightened unease over access to energy resources also serves to reinforce longstanding coastal State interests in ensuring freedom of navigation through the South China Sea and Gulf of Thailand. As over 90 per cent of world trade is transported by sea, freedom of navigation and seaborne trade is crucial to the economies of all the littoral States of the region as well as to an increasingly globalized and interdependent world economy as a whole.[14] With regard to energy supplies, the South China Sea hosts key sea lanes of communication (SLOCs) which constitute the primary

---

[13] For example, while the Energy Information Administration (EIA) of the United States estimates proven oil reserves for the South China Sea region to be of the order of seven billion barrels, it is also noted that Chinese estimates as to the area's seabed resource potential fall in the range of 105–213 billion barrels potential oil resources See EIA, "South China Sea," *Country Analysis Briefs*, <www.eia.doe.gov/cabs/South_China_Sea/Background.html>. See also, C. H. Schofield, "Dangerous Ground: A Geopolitical Overview of the South China Sea," in S. Bateman, J. Chen and R. Emmers, eds, *Security and International Politics in the South China Sea: Towards a Co-operative Management Regime* (London: Routledge, 2008), pp. 7–25, at pp. 14–17.

[14] International Chamber of Shipping and International Shipping Federation, "Overview of the International Shipping Industry," <www.marisec.org>.

maritime link between the Indian Ocean and East Asia. More than a quarter of the world's trade passes along these SLOCs, including 70 per cent of Japan's energy needs and 65 per cent of China's.[15] As regional energy imports appear set to rise, the importance of these SLOCs will also grow.

Competing Claims to Maritime Jurisdiction

It is hardly surprising that there are overlapping claims to maritime jurisdiction in the South China Sea and Gulf of Thailand. The former appears to be almost zone-locked,[16] the latter entirely so. Prompted by the maritime interests outlined above, the littoral States have proven to be enthusiastic in claiming maritime jurisdictional zones.[17] The potential extent of their maritime entitlements is, however, uneven. Some States, such as Indonesia and the Philippines, benefit from extensive maritime areas by any standard. Others, such as Brunei, Cambodia, and Singapore, are not so fortunate.

The distinguishing features of the patchwork of claims in the region are their great complexity and diversity, as the coastal States have shown remarkable powers of creativity to maximize their claimed maritime entitlements. Claims to baselines have been advanced on the basis of remarkably liberal interpretations of established rules of international law. Some maritime claims are based on history, or modern interpretations thereof. Others are the result of the application of the basic rules of maritime jurisdiction set forth in the LOS Convention. In some instances, there are distinct traces of irredentism or the desire to right perceived past wrongs

15 See C. H. Schofield and I. Storey, "Energy Security and Southeast Asia: The Impact of Maritime Boundary and Territorial Disputes," *Harvard Asia Quarterly* 9(4) (Fall 2005): 36–46, at p. 42.

16 Final determination of this point will depend on a detailed (and objective) application of Art. 121(3) of the LOS Convention (see below). On the other hand, there is a broad assumption on the part of some experts in the region and beyond that the whole of the South China Sea will be apportioned between the littoral States.

17 For a global summary of claims to maritime jurisdiction See Division for Ocean Affairs and the Law of the Sea, Office of Legal Affairs, United Nations, *Table of Claims to Maritime Jurisdiction* (2007), <www.un.org/Depts/los/LEGISLATIONANDTREATIES/PDFFILES/table_summary_of_claims.pdf>.

inflicted on the modern States by the actions or careless inactions of former colonial powers. History and irredentism, and the righting of wrongs, are powerful forces that complicate an already tense jurisdictional picture. National pride and notions of lingering injustice are hard to dispel, not least by arguments based on the cold logic of international law.

The South China Sea and Gulf of Thailand States and the Law of the Sea

Brunei, China, Indonesia, Malaysia, the Philippines, Singapore and Viet Nam have all signed and ratified the LOS Convention. Cambodia and Thailand have signed the Convention but have yet to ratify it.[18] Taiwan, as a non-UN member, is not eligible to become a party, though there is little doubt that it would sign and ratify the Convention if it could. In light of the strong support for the LOS Convention by the South China Sea and Gulf of Thailand coastal States, it seems appropriate to use the Convention as a framework for analysis of the littoral States competing maritime claims. Furthermore, substantial portions of LOS Convention, including its jurisdictional-related provisions, are generally considered to have, through acquiescence to their terms over time, become part of customary international law. Nonetheless, it is also worth recalling that the LOS Convention contains often deliberate ambiguities, leading to significant differences of view between States as to the proper interpretation of certain provisions.

Baselines

Baselines are fundamental to claims to maritime jurisdiction, as it is essential to determine the points from which the breadths of maritime zones are measured. A coastal State's "normal" baselines, in accordance with the LOS Convention, Article 5, are "the low-water line along the coast as marked on large-scale charts officially recognized by the coastal

---

[18] It is understood that Thailand is preparing to ratify the Convention and that, in any case, the fact that both Cambodia and Thailand have signed the LOS Convention indicates acceptance in principle.

State." The choice of low-water line is left to individual coastal States and is dependent on choice of vertical datum.[19]

With the exceptions of Brunei and Singapore, all the South China Sea and Gulf of Thailand littoral States have, however, departed from the use of these normal baselines in favour of straight or archipelagic baselines for all or part of their coastal fronts. The majority of these claims are problematic. Indeed, the straight baseline claims of Cambodia, China, and Viet Nam,[20] have all been subject to international protests.[21] Article 7 of the LOS Convention provides for the application of straight baselines in localities where "the coastline is deeply indented and cut into, or if there is a fringe of islands along the coast in its immediate vicinity." Most of these States' straight baselines have been subject to protests because they appear to depart from this provision. They frequently define systems of straight baselines along essentially smooth coastlines, devoid of major embayments, or utilize basepoints on a few small islands located both distant from one

---

[19] As no particular vertical datum or "zero line" is specified under the LOS Convention, jurisdictional disputes can arise as a result of States opting for different vertical datums. Choice of vertical datum will also impact on the status of certain insular features, for example, whether a particular feature is an island or a low-tide elevation and this, in turn, can have significant implications in terms of the capacity of a particular feature to generate extensive maritime claims to jurisdiction. See C. Carleton and C. H. Schofield, *Developments in the Technical Determination of Maritime Space: Charts, Datums, Baselines, Maritime Zones and Limits*, Maritime Briefing 3(3) (Durham: International Boundaries Research Unit, University of Durham, 2001), pp. 21–25. See also, International Hydrographic Organization (with the International Oceanographic Commission and the International Association of Geodesy), *A Manual on Technical Aspects of the United Nations Convention on the Law of the Sea, 1982*, Special Publication No. 51, 4th ed. (Monaco: International Hydrographic Bureau, 2006), chapter 2, pp. 16–20. Available at <www.iho.shom.fr/>.

[20] Malaysia is not mentioned in this list because, although it appears to have defined straight baselines around its coast, these have not been claimed or publicized formally, and consequently they have not been subject to diplomatic protest notes (at least in the public domain). The location of Malaysia's straight baselines can, however, be deduced, and commentators have been critical of this inferred baseline system. Similarly, Taiwan has advanced what appear to be excessive straight baselines that have also provoked adverse comment from analysts.

[21] Notably from the United States which, as the world's pre-eminent maritime power, routinely protests any practice that it deems to be excessive or contrary to the provisions of the LOS Convention, or, more specifically, the United States' interpretation of those provisions, through its Freedom of Navigation Program. See J. A. Roach and R. W. Smith, *United States Responses to Excessive Maritime Claims* (The Hague: Martinus Nijhoff Publishers, 1996), pp. 74–145.

another and substantially offshore. The latter practice makes it difficult to characterize such islands as either constituting a fringe of islands or being located in the "immediate vicinity" of the coast. Consequently, the sea areas enclosed are frequently too expansive and distant from the mainland to be genuinely considered as capable of qualifying as internal waters. It is important to note, however, that such seemingly excessive baseline claims need not prove an insurmountable obstacle to reaching bilateral agreements on overlapping claims where sufficient political will is present.[22] Additionally, two of the States under consideration, Indonesia and the Philippines, claim archipelagic status. While the fundamentally archipelagic nature of these States is not in doubt, their baselines present some problems.[23]

The South China Sea Disputes

Six coastal States (or in Taiwan's case, a State-like entity) are generally considered to be parties to the South China Sea jurisdictional disputes: Brunei, China/Taiwan, Malaysia, the Philippines, and Viet Nam. A distinction can be made between the dispute between China/Taiwan and Viet Nam over the Paracel Islands and the dispute over the Spratly archipelago involving all the above littoral parties. With regard to the disputes over the Spratly Islands, it is also worth noting that only China/Taiwan and

[22] For example, despite the fact that both Thailand and Viet Nam had advanced arguably highly problematic straight baseline claims in the Gulf of Thailand, the two States were nevertheless able to reach agreement on the delimitation of the maritime boundary between them in 1997 (see below). This agreement was apparently achieved through the simple expedient of both sides agreeing to ignore their respective straight baseline claims for delimitation purposes.

[23] Indonesia made its first archipelagic baselines claim in 1960. These baselines, which pre-dated the LOS Convention but were generally not viewed as excessive, were revised in 1996 (with coordinates issued through supplementary regulations of 2002 and 2008). However, the new Indonesian baseline system appears to enclose East Timor's Oecussi region within Indonesia's archipelagic waters. With regard to the Philippines, its claimed baselines, which also pre-date the LOS Convention, include a 140 nautical mile straight line across the mouth of the Gulf of Moro which, at 140 nautical miles, exceeds the LOS Convention Art. 47 restriction that no archipelagic baseline segment exceeds 125 nautical miles. The claims of the Philippines are also problematic because straight baselines enclosing internal waters rather than archipelagic baselines enclosing archipelagic waters are claimed under its legislation dating from 1961 and 1968.

Viet Nam claim all of the islands in question on the grounds of discovery, historical usage and occupation. Both Brunei and Malaysia appear to base their claims on the basis that the insular features concerned lie on their claimed continental shelves,[24] whilst the Philippines claim is founded on proximity allied to relatively recent historical usage and claims. Also of note is the fact that all of the claimants to the Spratly Islands, with the exception of Brunei, have occupied at least one insular feature in the Spratly group.

An exhaustive examination of the claims of each of the parties is unnecessary; suffice it to say that all of the sovereignty claims concerned have been appraised as weak by international legal authorities.[25] However, two key issues of particular note in the South China Sea context will be explored in further detail. First, historic claims will be considered in the context of the provisions of the LOS Convention. Second, the contentious issue of the classification of islands and their capacity to generate claims to maritime jurisdiction will be assessed.

Historic Claims

The notorious "dotted line" claim formally made by the Government of China in 1947 and subsequently adopted by that of the People's Republic (PRC) must be one of the most extraordinary assertions of jurisdiction anywhere. Comprised of a series of discontinuous dashes,[26] starting (for these purposes) in the northeast, the first of nine segments is drawn roughly midway between Taiwan and the Philippine Island of Luzon in the Luzon Strait. The second to fourth are close to the coast of the archipelago itself, the fifth and sixth just off the Malaysian provinces of Sabah and Sarawak, the seventh midway between Indonesia's Natuna Besar and the Mekong Delta, and the final two just off the coast of Viet Nam.

---

[24] It has, however, been observed that while Brunei appears to claim Louisa Reef by virtue of the fact that it occurs on its continental shelf, the Bruneian authorities do not appear to view Louisa Reef as an island capable of being subjected to a sovereignty claim. See D. J. Dzurek, *The Spratly Islands: Who's On First?* (Durham: International Boundaries Research Unit, University of Durham, 1996), p. 22.

[25] *Ibid.*, 47–55; and M. J. Valencia, J. M. Van Dyke, and N. A. Ludwig, *Sharing the Resources of the South China Sea* (The Hague: Martinus Nijhoff, 1997), pp. 17–40.

[26] Some maps have joined the nine segments. Both Beijing and Taipei maintain that this has no official status.

The rationale for this claim is that Chinese navigators and fishers have used these waters and the associated islands since time immemorial, thus giving China a claim to both. There does not appear to have been an authoritative statement as to what exactly is being claimed, although the second author has heard senior Chinese experts suggest that ownership over all islands "within" the segments is being asserted. But could a claim to almost the whole of the South China Sea based on historic waters have ever been viable? This point is addressed below.

Viet Nam for its part has issued a map of the South China Sea showing a line running due east from the median point of the Gulf of Tonkin and terminating close to the island of Luzon. It proceeds south, roughly paralleling the course of the Chinese line. So far as Hanoi is concerned, the historical record of Vietnamese activity in the South China Sea is no less significant than that of China, allowing it to assert jurisdiction accordingly. The map was issued in connection with potential oil and gas exploration permits, so at the least this is a claim to resource jurisdiction. Viet Nam's claim to the Spratly and Paracel island groups also has an historical basis derived from continuous use by sailors and fishers since time immemorial. Viet Nam furthermore asserts that France asserted a claim to the islands during the colonial period.

The roots of the Philippine claim lie in history, but very much more specifically and recently than that of China or Viet Nam. In this connection it should not be forgotten, assertions to the contrary notwithstanding, most of the features making up the Spratly Islands group are closer to the Philippines than any other country. The argument has been advanced that the colonial powers, Spain and subsequently the United States, appeared to regard the archipelago as an integral part of their possessions, though no convincing evidence for this has been adduced. In May 1956, the Philippine fishing magnate Tomas Cloma claimed most of the Spratly Islands (but not Spratly Island itself); the Government of the Philippines asserted a claim by Presidential Decree in 1978.[27]

What can be said of the three tenaciously asserted historical claims more than a decade after the entry into force of the LOS Convention? It has

---

27 Presidential Decree No. 1596, Declaring Certain Area Part of the Philippine Territory and Providing for their Government and Administration, 11 June 1978. Curiously enough, the map of the country available on the website of the Government of the Philippines does not show the Kalayaan, see <www.gov.ph/aboutphil/philmap.asp>.

long been accepted that there is a hierarchy between the two sources of international law whereby treaties have primacy over rules derived from custom. The rationale for this is that treaties are formal, in written form, and consensual. It is possible this last element, the fact that the obligations accepted voluntarily, is what justifies their status. In any case, logic dictates that there must be a hierarchy: if States were able to elide treaty obligations on the grounds that "general customary international law" had developed in a different substantive direction, then any semblance of legal certainty would be removed, and the basic foundations of the international legal order would be undermined. Treaty rules are to be applied in preference to rules of international customary law. States are not free to effect *de facto* amendments of their treaty rights by State practice.

The question is this: is it possible to interpret the LOS Convention so as to allow countries to maintain positions regarding jurisdiction which are essentially derived from history and ignore the rules on the generation of the zones of maritime jurisdiction set forth in the 1982 Convention, now accepted as general international law? This is possible only if one adopts a concept of interpretation that is not merely liberal, but one that goes beyond the recognition of virtually any constraining factors.

Such an approach presupposes that international law presents States with a range of choices from which they are free to choose to construct their claims to maritime jurisdiction. But no such variety of choice is available under the LOS Convention. The language of the Convention is totally prescriptive with respect to the nature and extent of the rules of maritime jurisdiction. International lawyers have long agreed that rights such as those flowing from the regime of the territorial sea and the continental shelf are inherent in that they do not depend on proclamation, occupation, or indeed any act of the coastal State.[28] The question of the regime to be deployed within these zones governing activities there is another matter, but the absence of, say, a petroleum law does not signal open season for other States or their nationals.

Accordingly, in our view, the historical claims to maritime jurisdiction in the South China Sea have been superseded by concepts of jurisdiction enunciated in the LOS Convention because the States concerned have, of their own volition, ratified the agreement. This is despite the fact

[28] For example, Art. 77(3) of the LOS Convention states, "the rights of the coastal State over the continental shelf do not depend on occupation, effective or notional, or on any express proclamation."

that the LOS Convention itself is silent on the issue of historic waters[29] and notwithstanding declarations made on signature/ratification of the LOS Convention, which cannot change the substance of the obligations assumed. It is very difficult for States to retreat from jurisdictional positions they have taken, but if the above argument is correct, then China/Taiwan, the Philippines, and Viet Nam have, by implication at least, abandoned their historical claims, at least in the sense of historic-based claims to maritime spaces in the South China Sea. Having made this observation, it is important to note that many States appear to be quite capable of maintaining apparently illogical or contradictory positions and that, in this context, there often appears to be a profound disconnect between the implications of international law and the requirements of *real politik*. Furthermore, the implicit demise of these historical claims has no implications whatsoever for claims to sovereignty over islands in the South China Sea. The point is that the historical lines have no significance in terms of maritime jurisdiction per se: territorial sovereignty is another matter.

It is arguable that the studied ambiguity of the exact nature of the China/Taiwan claim would provoke a contumelious response from other littoral States, and so it has proved: although the exact course of the boundary agreed by Indonesia and Viet Nam has yet to be revealed,[30] it is clear that neither country paid the slighted heed to the "dotted line" in the course of negotiations. Indeed, Dr. Ali Alatas, Foreign Minister of Indonesia for much of the 1990s (and one of the foremost supporters of the Managing Potential Conflicts initiative), told the second author that China had offered to commence maritime boundary discussions with his country, only to be informed that the two States did not exercise jurisdiction over adjacent ocean spaces, and in consequence, there was nothing to discuss.[31] It is in ways such as this that Southeast Asian countries have resisted what some have seen as the "inevitability" of Chinese hegemony over the South China Sea, at least insofar as that phrase means that China sets the agenda and the other parties tamely concur with it. The same approach can be applied to the other historic claims made at the margins

---

29 The concept of "so-called 'historic' bays" is, however, mentioned at Art. 10(6) of the LOS Convention.

30 While it is understood that the treaty had now been ratified by both States, at the time of writing its terms had not entered the public domain.

31 This was immediately after the ASEAN joint representation on Mischief Reef.

of the South China Sea. While Viet Nam's longstanding claim to a maritime boundary based on the Sino-French Frontier Convention of 1887 has now been abandoned,[32] Manila's claim to the so-called "Philippines Box" remains another highly problematic historically based maritime claim.[33]

Issues with Islands

Since the jurisdiction over ocean space flows from sovereignty over land, ownership of the various islands must first be decided. Much has been said about the island sovereignty question. It comes as no surprise that the features have been used by fishers and sailors of the littoral States. One would not expect otherwise. Accordingly, it is difficult to see that there would be much utility in assembling evidence of occasional occupation, although in the nature of things, it would be logical to expect a higher degree of use and occasional occupation on the part of the inhabitants of what is now the Philippines or the Malaysian state of Sarawak, being the most proximate major centres of human habitation to the Spratly Islands, than with fishers from further away. As previously noted, however, all the sovereignty claims to the Spratly Islands are equally weak (or, alternatively, equally strong) in international law terms.

Looming beyond the sovereignty question though are the intertwined issues of the classification of islands under the LOS Convention and their treatment in the context of maritime boundary delimitation. These are clearly crucial issues throughout the South China Sea/Gulf of Thailand. While the Gulf of Thailand is mercifully free of sovereignty disputes over islands,[34] much of the overlap between the maritime claims of the coastal

---

32 Viet Nam claimed that this Convention, which was largely devoted to defining the parties' land boundary, also defined the maritime boundary in the Gulf of Tonkin (known in China as the Beibu Gulf and in Viet Nam as the Bac Bo Gulf) as coincident with longitude 108° 3' 13" E. China rejected this position resulting in dispute which was finally resolved in December 2001, when China and Viet Nam signed an equidistance-line based maritime boundary delimitation agreement. See D. A. Colson and R. W. Smith, *International Maritime Boundaries*, Vol. V (Leiden: Martinus Nijhoff, 2005), pp. 3755–3758.

33 The Philippines claims territorial seas from its baselines to the limits established by the Spanish-American Treaty of 1898. This results in a claim to territorial waters 285 nautical miles wide at one point.

34 Although Cambodia and Viet Nam did contest sovereignty over a number of islands, this dispute was resolved though the 1982 Agreement on Historic Waters (see below).

States can nonetheless be attributed to contrasting treatment of islands when unilateral claims were advanced.

With regard to the classification of insular features, Article 121(1) of the LOS Convention offers a reasonably clear and uncontroversial definition of an island. To qualify as an island under these terms, a feature must be "naturally formed," be an "area of land," be "surrounded by water" and, "above water at high tide."[35] Only when all four conditions are met can a feature enjoy the status of an island. The relevance of this is obvious: maritime jurisdictional zones are measured from land, which of course includes islands. A feature that does not qualify as an island cannot, therefore, be used as the basis of a claim to maritime jurisdiction.

One major challenge to the apparent simplicity of this formulation comes from the process of building on reefs, low-tide elevations and the like, and claiming that the results of this process is the emergence of an island. Numerous such structures exist among the Spratly Islands, a prime example being Mischief Reef. However, the requirement that a feature be "naturally formed" serves to exclude from consideration artificial islands or structures.[36] Such a view is confirmed by reference to the provisions of the LOS Convention dealing specifically with artificial islands and installations which unequivocally disqualify such features from generating maritime zones, save for safety zones.[37] It might also be pointed out that, given the extraordinary biological diversity of the South China Sea, building on reefs may well constitute a violation of the Convention on Biological Diversity, which requires parties to take steps to preserve and protect areas considered to be of significance.[38]

The second major issue involving insular features concerns the application of Article 121(3) of the LOS Convention. This much criticized provision reads as follows: "Rocks which cannot sustain human habitation

---

[35] The "above high tide" requirement is potentially problematic as no particular vertical datum specifying high or low water lines is included in the LOS Convention (see discussion related to normal baselines above).

[36] J. R. V. Prescott and C. H. Schofield, *The Maritime Political Boundaries of the World* (Leiden/Boston: Martinus Nijhoff Publishers, 2005), pp. 58–61.

[37] LOS Convention, *supra* note 2, Art. 60(8) states: "Artificial islands, installation and structures do not possess the status of islands. They have no territorial sea of their own, and their presence does not affect the delimitation of the territorial sea, the exclusive economic zone or the continental shelf."

[38] *The Convention on Biological Diversity*, Rio de Janerio, 22 May 1992, 31 *I.L.M.* 823.

or economic life of their own shall have no exclusive economic zone or continental shelf." The intent of Article 121(3) is clear – it is designed to restrain the use of small, uninhabited and remote insular features from generating disproportionate maritime claims. Instead, such features are restricted to generating a territorial sea of up to 12 nautical miles in breadth and potentially a contiguous zone out to 24 nautical miles. The term "rock" is not defined in the LOS Convention, and not used elsewhere in Article 121 (the Regime of Islands). The definition of the term, as well as what exactly is meant by "economic life" and "human habitation" has excited considerable scholarly debate and a variety of tests have been suggested.[39] It has, however, been suggested that the wording of Article 121(3) is deliberately ambiguous.[40] While this complex debate is beyond the scope of this contribution, some observations on the potential applicability of the regime of islands in the South China Sea are appropriate.

It has been suggested that, of the 150–180 insular features making up the Spratly Islands group, only 48 rise above high-tide to form consistently small, and in most cases tiny, islands or rocks.[41] Thus only these 48 features can potentially be subject to the regime of islands under the LOS Convention. It seems plausible to suspect that some, perhaps most, of these features should be most appropriately categorized as rocks rather

[39] See, for example, J. I. Charney, "Rocks that Cannot Sustain Human Habitation," *American Journal of International Law* 93(4) (1999): 863–878; A. O. Elferink, "Clarifying Article 121 (3) of the Law of the Sea Convention: The Limits Set by the Nature of International Legal Processes," *Boundary and Security Bulletin* 6(2) (1998): 58–68; B. Kwiatkowska and A. Soons, "Entitlement to Maritime Areas of Rocks which Cannot Sustain Human Habitation or Economic Life of their Own," *Netherlands Yearbook of International Law* 21 (1990): 139–181; J. M. Van Dyke, J. Morgan and J. Gurish, "The Exclusive Economic Zone of the Northwestern Hawaiian Islands: When do Uninhabited Islands Generate an EEZ?," *San Diego Law Review* 25(3) (1988): 425–494; and J. M. Van Dyke and R. A. Brooks, "Uninhabited Islands: Their Impact on the Ownership of the Oceans' Resources," *Ocean Development International Law Journal* 12 (1983): 265–284.

[40] See Prescott and Schofield, *supra* note 36, at pp. 57–91.

[41] Prescott and Hancox provide the figure of 48. See D. Hancox and J. R. V. Prescott, *A Geographical Description of the Spratly Islands and an Account of Hydrographic Surveys Amongst Those Islands*, Maritime Briefing 1(6) (Durham: International Boundaries Research Unit, University of Durham, 1995). However, some commentators offer lower figures. For example Dzurek offers the number 36. See Dzurek, *supra* note 24, at p. 1.

than islands proper. Furthermore, even if it can be maintained that some of these insular features can be classified as islands capable of supporting extended claims to maritime jurisdiction, there is little reason to suspect that they would be accorded full weight in a maritime boundary delimitation situation, especially one pitting small islands with short relevant coastlines versus long, mainland coasts (see below).

Competing Claims in the Gulf of Thailand

The Gulf of Thailand is also host to some interesting historic claims. In 1959 Thailand issued a decree claiming the Bight of Thailand, the northern extremity of the Gulf, as a "historic gulf". Thailand's claim has not resulted in any known international protests.[42] Furthermore, on 7 July 1982, Cambodia and Viet Nam signed the Agreement on Historic Waters of Viet Nam and Kampuchea.[43] A substantial zone of joint "historic waters" was defined. Unlike most claims to historic waters, which usually relate to bays, the claimed Cambodian and Vietnamese zone projects offshore from the two countries mainland coasts. The Agreement helped to facilitate the integration of the parties' straight baseline systems which meet at an undefined, floating "Point O" on the southwestern limit of the joint historic waters area. A particular virtue of this Agreement was that it served to clarify sovereignty over previously contested islands and thus resolved a contentious dispute.[44] Perhaps unsurprisingly, this remarkable claim to joint historic waters has provoked strong international protests.[45]

With regard to extended maritime jurisdictional claims, all four of the Gulf of Thailand littoral States advanced unilateral claims to continental

[42] The remarkable aspect of this historic claim is perhaps that it numbers among the select group of three claims to historic bays or waters (of 18) that the United States has *not* protested (Roach and Smith, *supra* note 21, at pp. 33–34).

[43] J. I. Charney and L. M. Alexander, eds, *International Maritime Boundaries*, Vol. III (Dordrecht: Martinus Nijhoff, 1998), pp. 2364–2365.

[44] The Poulo Wei group of islands was confirmed as Cambodian and the large island of Phu Quoc and the Tho Chu (Poulo Panjang) as Vietnamese.

[45] The United States protested on the grounds of the claim's recent nature and asserted that the United States and the international community in general had not acquiesced to the claim. See Roach and Smith, *supra* note 21, at pp. 39–40.

shelf in the 1970s.[46] As a consequence of the Gulf's relatively small size, allied to some creative and self-serving claims on the part of the coastal States, these continental shelf claims overlapped to a considerable degree. In large part these overlaps in maritime jurisdictional claims stemmed from contrasting treatment of islands. Claimants often accorded full effect to their own insular features, whilst ignoring or discounting those of their neighbours and vice versa. Indeed, multiple overlaps resulted in parts of the Gulf being subject to claim from three or even all four littoral States.[47]

Existing Agreements and Cooperative Efforts

Despite expansive national claims to maritime jurisdiction, the resulting broad areas of overlapping claims and the contentious sovereignty disputes over islands, the South China Sea and Gulf of Thailand are not wholly devoid of maritime boundary agreements. These delimitation agreements tend to be the exceptions rather than the norm and, perhaps unsurprisingly, have been spatially confined to margins.

Within the South China Sea proper, the earliest maritime boundaries are three boundary lines defined by the British colonial authorities in 1958 which delineated the three nautical mile territorial seas of Brunei and the neighbouring Malaysian territories of Sabah and Sarawak, as well as seabed rights out to the 100 fathom (182 metres) isobath in accordance with British claims at the time.[48] In 1969, Malaysia and Indonesia also defined a pair of boundaries relating to their respective continental shelf

---

[46] The first claimant was South Viet Nam in 1971 (a claim inherited by post-war reunited Viet Nam), followed by Cambodia in 1972, Thailand in 1973, and finally Malaysia in 1979.

[47] The extent of these overlaps was, at least implicitly, reduced considerably by the resolution of the Cambodia-Viet Nam dispute over islands through the parties' 1982 Joint Historic Waters agreement mentioned above. The continental shelf claims dating from the 1970s have not, however, subsequently been revised.

[48] While one of these lines is confined within Brunei Bay, the other two project out into the South China Sea. The westernmost of these two lines extends for approximately 60 nautical miles offshore and the eastern line extends for almost 35 nautical miles before the 100 fathom isobath is reached. See J. R. V. Prescott and C. H. Schofield, *The Undelimited Maritime Boundaries of the Asian Rim in the Pacific Ocean*, Maritime Briefing 3(1) (Durham: International Boundaries Research Unit, University of Durham, 2001), pp. 6–8.

rights between peninsula Malaysia and the Indonesia's Natuna Islands and also between Malaysia's Sarawak province and Indonesia's Kalimantan province.[49] More recently in 2000, China and Viet Nam concluded a maritime boundary agreement relating to the Gulf of Tonkin,[50] which includes the definition of a joint fishing zone straddling the boundary line. Additionally, as referred to above, Indonesia and Viet Nam have apparently reached agreement on the delimitation of their continental shelf boundary located to the north of the Natuna Islands group belonging to Indonesia and the southern coast of Viet Nam.

With regard to the Gulf of Thailand, as previously noted, the claims to continental shelf advanced by the littoral States resulted in broad areas of overlapping claims. These conflicting claims have proved difficult to resolve by the apportionment of rights through the definition of maritime boundaries. Indeed, the only maritime boundaries to be delimited to date have been a territorial sea border[51] and relatively short, partial section of continental shelf boundary[52] between Malaysia and Thailand concluded in 1979 and a delimitation concerning continental shelf and EEZ rights in the central Gulf reached between Thailand and Viet Nam in 1997.[53]

---

[49] See U.S. Department of State, *Continental Shelf Boundary: Indonesia-Malaysia*, Limits in the Seas, No. 1, (Washington D.C.: Bureau of Intelligence and Research, 21 January 1970).

[50] *Agreement between the People's Republic of China and the Socialist Republic of Viet Nam on the Delimitation of the Territorial Seas, the Exclusive Economic Zones and Continental Shelves in Beibu Bay/Bac Bo Gulf*, 25 December 2000 (entry into force: 30 June 2004). For background and analysis See G. Xue, *China and International Fishery Law and Policy* (The Hague: Martinus Nijhoff, 2005), pp. 204–228.

[51] *Treaty between the Kingdom of Thailand and [the Republic of] Malaysia Relating to the Delimitation of the Territorial Seas of the Two Countries* of 24 October 1979 (entry into force: 15 July 1982). See J. I. Charney and L. M. Alexander, eds, *International Maritime Boundaries*, Vol. I (Dordrecht: Martinus Nijhoff, 1993), pp. 1096–1098.

[52] *Memorandum of Understanding between Malaysia and the Kingdom of Thailand on the Delimitation of the Continental Shelf Boundary between the Two Countries in the Gulf of Thailand* of 24 October 1979 (entry into force: 30 June 2004). *Ibid.*, pp. 1105–1107.

[53] *Agreement between the Government of the Kingdom of Thailand and the Government of the Socialist Republic of Viet Nam on the Delimitation of the Maritime Boundaries between the Two Countries in the Gulf of Thailand* of 9 August 1997 (entry into force: 27 February 1998). See J. I. Charney and R. W. Smith, eds, *International Maritime Boundaries*, Vol. IV (Dordrecht: Martinus Nijhoff, 2002), pp. 2692–2694.

There has, however, been considerably greater progress in the conclusion of cooperative alternatives to the delimitation of boundary lines – maritime joint development zones. In addition to the above-mentioned joint "historic waters" Agreement by Cambodia and Viet Nam, the Gulf of Thailand littoral States have entered into two joint arrangements – between Malaysia and Thailand[54] and Malaysia and Viet Nam.[55] Furthermore, in 2001 Cambodia and Thailand signed a memorandum of understanding indicating that, while the two States would negotiate towards the delimitation of a maritime boundary to the north of the 11° North parallel of latitude, with regard to the remainder of their large area of overlapping claims discussions would simultaneously proceed with the objective of creating a maritime joint development zone.[56] At the time of writing, the negotiations between Cambodia and Thailand had yet to result in agreement.[57]

These joint agreements are consistent with Articles 74(3) and 83(3) of the LOS Convention, dealing with the delimitation of the exclusive economic zone and continental shelf, which provide in identical terms that:

> Pending agreement... the States concerned, in a spirit of understanding and cooperation, shall make every effort to enter into provisional arrangements of a practical nature and, during this transitional period, not to jeopardize

[54] *Memorandum of Understanding between the Kingdom of Thailand and [the Republic of] Malaysia on the Establishment of a Joint Authority for the Exploitation of the Resources of the Sea-Bed in a Defined Area of the Continental Shelf of the Two Countries in the Gulf of Thailand* of 21 February 1979 (entry into force: 24 October 1979). See Charney and Alexander, *supra* note 51, at pp. 1107–1111. A further agreement was required to put this memorandum of understanding into practice: *Agreement between the Government of Malaysia and the Government of the Kingdom of Thailand on the Constitution and Other Matters Relating to the Establishment of the Malaysia-Thailand Joint Authority*, concluded on 30 May 1990. See *ibid.*, pp. 1111–1123.

[55] *Memorandum of Understanding between Malaysia and the Socialist Republic of Viet Nam for the Exploration and Exploitation of Petroleum in a Defined Area of the Continental Shelf Involving the Two Countries*, signed on 5 June 1992 (entry into force: on 4 June 1993). See Charney and Alexander, *supra* note 43, at pp. 2335–2344.

[56] *Memorandum of Understanding regarding the Area of their Overlapping Claims to the Continental Shelf.* See Colson and Smith, *supra* note 32, at pp. 3743–3744.

[57] See, for example, C. H. Schofield, "Unlocking the Seabed Resources of the Gulf of Thailand," *Contemporary Southeast Asia* 29(2) (2007): 286–308, at p. 303. Disputes concerning the land boundary between the two countries, which flared up from mid-2008, appear likely to undermine efforts towards reaching agreement on maritime claims in the Gulf of Thailand.

> or hamper the reaching of the final agreement. Such arrangements shall be without prejudice to the final delimitation.

Joint development has proven to be a particularly attractive mechanism to deal with overlapping claims to maritime jurisdiction in the Gulf of Thailand because of the existence of substantial proven offshore oil and gas reserves in undisputed waters and, thus, the well-founded suspicion that hydrocarbon resources were and are present in the disputed areas of continental shelf. This knowledge, when coupled with uncertainty as to the precise location of the resources, raised concerns among the littoral States that were a boundary to be drawn, seabed resources might prove to be located on the "wrong side of the line." This made reaching a compromise boundary lines difficult to achieve, but simultaneously made joint solutions more attractive.[58]

Dispute Settlement, Preventative Diplomacy, and International Law

As previously noted, all the littoral States in the area of study, with the exception of Cambodia and Thailand are parties to the LOS Convention (and these two countries appear willing to embrace it). As such, all the States are committed to the protection of the marine environment and the optimal utilization and conservation of ocean resources. Indeed, they are legally obliged to find means of achieving these objectives. International law offers numerous options for the resolution of maritime disputes, or, failing that, accommodations between the littoral States in the interest of the maintenance of law and order at sea and associated matters. State practice in various parts of the world shows that, given the political will, difficulties can be surmounted either temporarily or permanently. Indeed, it can be assessed that there are no serious obstacles to such cooperation in the South China Sea and Gulf of Thailand.[59]

Nonetheless, sovereignty and jurisdictional issues still loom large in the South China Sea and Gulf of Thailand. Some reference has already been made to claims made by or on behalf of some of the littoral States. One of the questions most commonly asked is why these countries do not simply submit the question of sovereignty to the International Court of

---

58 *Ibid.*, pp. 289–290.

59 This point has been made repeatedly by participants at the various Workshops on Managing Potential Conflicts in the South China Sea.

Justice or some other arbitral body. The most direct answer is that since sovereignty is an absolute concept, for the disputed islands at least, there will be one winner and several losers. Furthermore, each of the claimant States is well aware, even if they would not care to say so openly, that their own claims, in international law terms at least, are far from flawless. Similarly, even though sovereign rights rather than sovereignty are at stake with regard to claims to continental shelf and EEZ claims, these claims are linked, or are perceived to be linked, to sovereignty over the South China Sea islands. Submitting the South China Sea island and ocean sovereignty disputes to international third party settlement therefore tends to be viewed as an unacceptable risk. Accordingly, it is highly improbable that judicial settlement will become an acceptable option in the South China Sea context, even though it has proved acceptable in the resolution of bilateral territorial disputes such as the International Court of Justice cases between Indonesia and Malaysia concerning Pulau Sipadan and Pulau Ligitan[60] and between Malaysia and Singapore over Pedra Branca/Pulau Batu Puteh.[61] Other options have been suggested, including the formation of an eminent persons group, which could engage in conciliation or mediation. However, it is unlikely that the claimants would agree to any measure that appeared to suggest that their claims were not absolute, final, and thus not open to discussion.

The prospects for a negotiated settlement of the ocean and island sovereignty issues appear equally discouraging. States are seldom keen to retreat from assertions of sovereignty especially, as is the case here, where they have been backed up with a great deal of high-charged rhetoric. Indeed, senior officials from China and Viet Nam are on record as having stated quite unequivocally that the sovereignty of their respective States is absolute and not subject to negotiation or discussion.

---

[60] In which it was determined that sovereignty over the islands in question belongs to Malaysia. See *Sovereignty over Pulau Ligitan and Pulau Sipadan (Indonesia/Malaysia)*, Judgment available at <www.icj-cij.org/icjwww/idecisions.htm>.

[61] On 23 May 2008 the International Court of Justice rendered its Judgment in this Case, finding that sovereignty over Pedra Branca/Pulau Batu Puteh rests with Singapore and sovereignty over Middle Rocks belongs to Malaysia. The Court did not rule directly on the question of ownership over South Ledge instead stating that "sovereignty over South Ledge…belongs to the State in the territorial waters of which it is located." See *Sovereignty over Pedra Branca/Pulau Batu Puteh, Middle Rocks and South Ledge (Malaysia/Singapore)*, <www.icj-cij.org/docket/index.php?p1=3&p2=1&code=&case=130&k=2b>.

In short, the prospects for settlement of the sovereignty disputes are bleak, to say the least. In these circumstances, it might be more profitable and certainly constructive to consider setting the sovereignty questions aside, while not requiring any country to abandon its claim, in favour of some form of joint accommodation. It is also possible that a cooler and more considered examination of the issues at stake will make this easier than it once seemed.

First, the oil factor. China, the Philippines and Viet Nam are engaged in joint exploration of the disputed parts of the South China Sea. Assuming the areas selected for exploration are thought to be the most highly prospective, rather than the opposite, it is perhaps significant that the good news expected by some of confirmation of the mineral wealth of the area has yet to arrive. Although no official pronouncement has been made, it is understood that results thus far have been disappointing.[62] This will come as no surprise to those who have always thought that the geology – what is known of it – is not especially promising. While a successful, joint energy exploration initiative has the potential to foster cooperation, this raises the question as to how a largely unsuccessful exploration campaign will affect the way in which the claimant States view sovereignty and jurisdictional issues in the South China Sea. It will certainly mean that much less is at stake than was hitherto thought. Having said this, States do not readily abandon claims to sovereignty over land, no matter how ostensibly valueless it may be. But the value of maritime jurisdiction in the South China Sea does not rest on oil and gas alone. Apart from living resources, there is also a wealth of potential in the abundance of the marine biodiversity of the area that is still largely untapped.

The second means of at least simplifying the sovereignty/jurisdictional question is to take a considered look at the potential impact of islands on what is accepted as the starting point for any apportion of rights over water and seabed: claims generated from mainland or archipelagic baselines. As noted above, regardless of competing sovereignty claims, only a few of the islands of the Spratlys group are potentially subject to the regime of islands under the LOS Convention. Fewer still can plausibly be characterized as fully-fledged islands rather than rocks within the meaning of Article 121(3).

[62] This information was supplied by a senior official of one of the States concerned on condition of confidentiality.

With regard to these features, the question then arises of the potential effect of any zones of maritime jurisdiction generated by these features on zones generated by mainland or archipelagic baselines. State practice and the jurisprudence of the International Court of Justice, as well as various ad hoc arbitral bodies have repeatedly accorded small islands with short relevant coasts reduced effect vis-à-vis long mainland coasts. Perhaps the most striking parallels are with the Channel Islands, a British possession located just off the coast of France in the Gulf of Normandy, and the French islands of St. Pierre and Miquelon, located in the northwest Atlantic just off the island of Newfoundland, Canada. In the former case, the Channel Islands were given an enclave of territorial sea, and in the latter, a very attenuated maritime zone.[63] Both of these sets of islands are very much bigger than any feature in the Spratly or Paracel chains. Both have viable populations measured in the thousands, which is not the case with regard to the Spratlys. Accordingly, it seems highly likely that the island features of the Spratlys area will be unlikely to generate anything much more than territorial seas. This is not inconsiderable, but it does suggest that the maritime zone generative capacity of the Spratly Islands is often overstated and is a far cry from oft-repeated and extravagant claims that the Spratlys "hold the key to the South China Sea."

It seems only sensible to urge a freezing of the current jurisdictional situation, together with a call for no more development on reef systems. Not only is this jurisdictionally meaningless, it is ecologically irresponsible.

The South China Sea Workshop Process

Few independent observers of the jurisdictional picture in the South China Sea have thought that resolution was possible in the short or medium term. This in and of itself does not matter, but the disputes have made it impossible to rationally and efficiently police these waters. Fishing is basically unregulated, and destructive methods are rife. Tankers discharge contaminated water with impunity. Badly needed marine scientific research more often than not cannot be undertaken. Additionally, the military occupation and associated fortification of many of the Spratlys results in

[63] See C. Carleton and C. H. Schofield, *Developments in the Technical Determination of Maritime Space: Delimitation, Dispute Resolution, Geographical Information Systems and the Role of the Technical Expert*, Maritime Briefing 3(4) (Durham: International Boundaries Research Unit, University of Durham, 2002), pp. 19–23.

unknown damage to the unique coral reef habitat that is vital to the sustainability of regional fisheries. It is also known that some marine charts have serious inaccuracies. All this compounds the problem and reinforces the contention that the status quo is an unacceptable state of affairs.

In 1989, Dr. Hasjim Djalal, in distinguished Indonesian diplomat and law of the sea expert, suggested to the second author that there might be a glimmer of hope if the countries of the region were encouraged to turn aside from the issues that divided them and focus on those that united them – their shared concern for the South China Sea and all it has to offer. They persuaded Canada to support a series of informal workshops, attended by officials, researchers, and naval officers, all in their personal capacities, to see if cooperative projects might be designed. It is important to note that the search was for opportunities for cooperation where there was a functional requirement for working together, not merely for the sake of cooperation. The projects could also be seen as confidence-building measures. Indeed, the Workshop series itself could be and was seen in that light.

Fifty meetings or so on, there are signs that a cooperative ethic is emerging with China, the Philippines and Viet Nam conducting joint petroleum exploration activities, naval exercises, marine scientific research expeditions, data-base compilations, and the like. Some of the work of the Technical Working Group of Marine Scientific Research found it way into the proposal documents for the United Nations Environment Programme/Global Environment Facility project on "Reversing Environmental Degradation Trends in the South China Sea and Gulf of Thailand".[64] Furthermore, the attention of the countries concerned, and others, has shifted from the rumoured, but unproven, petroleum resources to more serious matters: namely, that more than six hundred million people derive the greater part of their protein from the South China Sea.[65] If there is a serious degradation of the water or its resources, how are these people going to survive?

The Workshop process was always seen as a stepping stone to the first track – inter-governmental discussions. As noted above, this is starting to

---

[64] See UNEP, *UNEP/GEF Project "Reversing Environmental Degradation Trends in the South China Sea and Gulf of Thailand"* (UNEP/GEF SCS Project), available at <www.unepscs.org>.

[65] See Fortes, *supra* note 9.

happen, but there remains a role for informal diplomacy. The Workshops were always attended by very senior officials, in their private capacities, though that fiction was stretched a little thin sometimes. It is possible, though, that officials who travelled to a meeting with settled views on the primacy of the jurisdictional issue, or the undesirability or impossibility or even the lack of need to work with maritime neighbours, heard something to change their minds. Certainly, the quality and diversity of the information presented had an impact on participants. At an environmental meeting in Shanghai, for example, only the most intransigent individual would fail to be persuaded by senior officials of the United Nations Environment Programme when they said that failure to work together to prevent reef degradation and destruction would mean a bleak future for their grandchildren.

The SEAPOL Gulf of Thailand Project

The Gulf of Thailand Meeting of Experts was established under the auspices of the Southeast Asian Programme on Ocean Law, Policy and Management (SEAPOL), which owed its existence and continuation to the tireless efforts of Douglas Johnston. SEAPOL was established in 1997 to "examine policy recommendations based on scientific data."[66] The desirability and necessity for a sub-regional cooperative mechanism devoted to the Gulf of Thailand was recognized and support for this grew through meetings in 1998 with "growing consensus" emerging on the issue in 1999.[67]

At the end of the 2000 meeting, agreement was reached on a "summary of discussion" which recommended a twin-track process whereby the four Gulf of Thailand States "should consider entering into formal inter-state negotiations" to create a Gulf of Thailand cooperative mechanism (i.e.,

---

[66] F. Lai, "A Report on the SEAPOL Gulf of Thailand Project," Paper presented to the Third Meeting of the Regional Task Force on Legal Matters for the UNEP/GEF Project Reversing Environmental Degradation in the South China Sea and Gulf of Thailand, Annex 6, UNEP/GEF/SCS/RTF-L.3/3, available at <www.unepscs.org/SCS_Documents/Download/11_-_Regional_Task_Force_on_Legal_Matters/Reports_of_the_Meetings_of_the_Regional_Task_Force_on_Legal_Matters/3-Report-Third-Meeting-Regional-Task-Force-Legal-Matters.html>.

[67] *Ibid.* These meetings yielded four volumes of SEAPOL *Integrated Studies of the Gulf of Thailand.* See <www.seapol.net>.

track one). Simultaneously, a track two, non-governmental Gulf of Thailand Consultative Group comprising national coordinators and SEAPOL was to run in parallel in order to "maintain the momentum towards a formal consultative mechanism."[68]

Although the Foreign Ministry of Thailand circulated two diplomatic dispatches to their Gulf of Thailand partners including a draft treaty for negotiation, little subsequent progress has been reported. Key obstacles to progress were identified as being bureaucratic and legislative inertia coupled with changes in the officials involved at the track one level, which served to inhibit momentum. A further constraint was related to the differing approaches favoured among the parties, for example, between a general framework agreement versus a project specific approach. Furthermore, capacity issues, especially in financial terms, hampered participation in the project, especially regarding delegates from Cambodia and Viet Nam.

Concluding Thoughts – Charting the Way Forward?

The foregoing discussion demonstrates unequivocally that the conflicting and overlapping maritime jurisdictional claims that persist in the South China Sea and Gulf of Thailand have a number of negative consequences. With regard to seabed hydrocarbon resources, exploration activities are forestalled as a result of legal and fiscal uncertainty, preventing developments potentially of national economic significance to the (often developing) countries concerned. Similarly, with regard to important living resources, competing claims can lead to uncoordinated policies and potentially destructive and unsustainable competition between rival States. Such activities pose a significant threat to regional environmental and, ultimately, food security. Overlapping maritime claims also pose a clear threat to peace and security among the claimant States as there is a real risk of friction and even confrontation between the fishing fleets, and particularly security forces of each side, as these vessels each attempt to utilize or exert their enforcement rights in what they regard as rightfully "their" maritime space.

The logic for cooperation in the waters in question is therefore abundantly clear. Participants at the South China Sea Workshops repeatedly

[68] *Ibid.*

affirmed that "there are no serious obstacles to cooperation (in those waters)." Cooperation is either mandated or encouraged at international law and State practice has furnished many examples of how this can be achieved, and what it can accomplish. It is also worth emphasizing that working together can be "without prejudice" meaning that jurisdictional or other sensitive positions need not be compromised. Such cooperation has, however, been slow in eventuating in meaningful ways.

So what is missing? It is perhaps what might be termed a culture of cooperation, a way of thinking about the role of the State in ocean endeavours that is not discouraged by the prospect of having to adopt a consensual approach to a particular task. It is the role of researchers and commentators to foster such a culture and few have done more to make a contribution by way of scholarship and institution-building than Douglas Johnston. In this context it is worth highlighting the fact that laudable and practical initiatives such as the UNEP/GEF project on "Reversing Environmental Degradation Trends in the South China Sea and Gulf of Thailand" have been realized, at least in part, from Johnston's efforts, and, as noted above, the Workshop process.

It must also be said that Johnston showed the way with cooperation of another sort, namely, between the different disciplines whose expertise had to be brought to bear on a given issue. The result was that international lawyers had at least some notion of the context in which their work was situated. For their part, economists, resource managers, geographers and others could appreciate in their turn that they had to work within a legal framework which could, and certainly should, be responsive to their needs. It supported their work, rather than seeming to throw up continuous obstacles to it.

There are those who suggest that the track-two processes are largely irrelevant. However, the evidence pointing the other way is overwhelming. Initiatives such as SEAPOL, the South China Sea Workshops, and the Council for Security Cooperation in the Asia Pacific (CSCAP), could have made themselves irrelevant had they chosen different courses from the ones they followed and continue to following. Track-two fori in the Asia-Pacific region remain the principal means whereby new ideas are presented and considered, and then disseminated. This is not to say that complacency is justified: attention to the work of a network or process has to be earned. The initiatives whose work has been touched on in this essay may perhaps serve as part of the Johnston legacy. It remains to be seen whether this bequest is built on or wasted.

Regime-Building in East Asia: Recent Progress and Problems*

Mark J. Valencia**

Introduction

Much of East Asia is essentially maritime encompassing a maze of peninsulas, disputed islands, archipelagos, strategic straits and sea lanes. These features surround and are embedded in a series of semi-enclosed seas – the Sea of Japan (East Sea), the Yellow Sea, the East China Sea and the South China Sea. These seas harbour large marine ecosystems. According to Article 123 of the 1982 United Nations Convention on the Law of the Sea (LOS Convention)[1] their environment and living resources should be co-managed cooperatively by the littoral States. However, in general this is not the case.

For many years East Asian nations attempted to avoid escalating tension with their neighbours by either refraining from extending their maritime jurisdiction or foregoing provocative activities in maritime frontier areas. However, in the last two decades, all East Asian coastal countries have claimed 200 nautical mile (nm) exclusive economic zones (EEZs) and/or continental shelves. This has resulted in a sea change in the maritime arena in which disputes between unfriendly nations have seriously exacerbated relations between them. Indeed, given the geography and the plethora of maritime boundary and territorial disputes stimulated by this wave of

---

* Portions of this paper are derived and/or updated from Mark J. Valencia, "Maritime Confidence and Security Building in East Asia: Recent Progress and Problems," *Ocean Policy Studies*, The Nippon Foundation, Vol. 3, 2006, pp. 25–45.

** Maritime Policy Analyst, Kaneohe, Hawaii, USA.

[1] *United Nations Convention on the Law of the Sea*, Montego Bay, 10 December 1982, UN Doc. A/Conf.62/122 (1982); 21 *I.L.M.* 1261 (1982) [hereinafter LOS Convention].

extended maritime jurisdiction, it is not surprising that maritime issues have risen to the forefront of regional security concerns.[2]

Moreover, the increasing prominence of issues like environmental pollution and resource ownership, as well as illegal maritime activities, now including piracy and "terrorism," dictates broader responsibilities and changing priorities for military force structure, operations and training. Together with the requirements of self-reliance and force modernization, these concerns are reflected in the significant maritime dimension of current arms acquisition programmes in the region. This includes maritime surveillance and intelligence collection systems, multi-role fighter aircraft with maritime attack capabilities, modern surface combatants, submarines, anti-ship missiles, naval electronic warfare systems, and mine warfare capabilities. Because some of these new systems have offensive capabilities, they can be seen as provocative, and thus destabilizing, particularly by those countries that do not have similar capabilities and lack the means to acquire them. Moreover, possession of these new systems undoubtedly increases the risk of inadvertent escalation in time of conflict. In short, maritime East Asia has become an increasingly dangerous frontier where the building of trust and confidence is sorely needed.

The first step toward the peaceful settlement and avoidance of conflict is the creation of a sense of community.[3] The creation of such a community presupposes at least the mitigation and minimization of conflict, so that shared interests and common needs outweigh the factors that separate parties. A functional approach can help the growth of positive and constructive common work and of common habits and interests, decreasing the significance of artificial agreed boundaries and barriers – or the lack thereof – by overlaying them with a natural growth of common activities and administrative agencies. The challenge for Asia is to develop a variety of such arrangements that will demonstrate that a habit of dialogue and working together can build common, and eventually, cooperative security. Tactical learning, in which behaviour towards cooperation is changed, must be replaced by complex learning in which values and beliefs about reaching goals through cooperation are changed. In this context, cooperation in the maritime sphere can be a means of building confidence, reducing

---

[2] M. J. Valencia, *A Maritime Regime for North-East Asia* (Hong Kong: Oxford University Press, 1996), pp. 7–12.

[3] *Ibid.*, pp. 10–11.

tension and eliminating points of conflict, and it can have spillover effects in other fields. But confrontation and conflict in the maritime sphere can have the opposite effect. In assessing the current situation, this contribution provides an update on the significant progress and the problems in building multilateral maritime regimes in East Asia.

Pre-existing Maritime Cooperation

There is a pre-existing foundation in East Asia upon which robust maritime regimes and confidence can be built.[4] The fundamental regime governing management of the ocean environment and its resources is the 1982 LOS Convention. Among East Asian coastal States, only Cambodia, North Korea and Thailand have not ratified this "constitution for the oceans." The United States, an influential State actor in the region, is also a non-party. Nevertheless, the 1982 LOS Convention provides the conceptual framework and specific mandates for regime-building. Although these apply globally, this contribution is concerned with indigenous interpretations and development of multilateral maritime regimes particular to the natural and political environment of East Asia.

In Southeast Asia, permanent committees of the Association of Southeast Asian Nations (ASEAN) of marine importance include: Fisheries, Meteorology, Science and Technology, and Shipping. There is also a 1975 ASEAN Agreement for the Facilitation of Search for Ships in Distress and Rescue of Survivors of Ship Accidents.[5] In the area of environmental protection, there is an ASEAN Sub-regional Environment Programme and an ASEAN Committee on Petroleum, which has developed contingency plans for combating oil spills. The United Nations Environment Programme for East Asian Seas has enhanced marine environmental awareness and capabilities.

There are also several marine-relevant international organizations operating in the region, such as the Indo-Pacific Fisheries Commission, the Southeast Asia Fisheries Development Centre, the International

[4] M. J. Valencia, "Regional maritime regime-building: prospects in Northeast and Southeast Asia," *Ocean Development and International Law* 31 (2000): 223–247.

[5] *Agreement for the Facilitation of Search of Ships in Distress and Rescue of Survivors of Ship Accidents*, Kuala Lumpur, 15 May 1975, <http://www.aseansec.org/1401.htm>.

Center for Living Aquatic Resources Management, the Committee for Coordination of Joint Prospecting for Mineral Resources in Asian Offshore Areas, and the Working Group for the Western Pacific of the Intergovernmental Oceanographic Commission. These organizations are not indigenously derived or funded and include among their members both extra-ASEAN and Southeast Asian States. Nevertheless, they may serve as models, platforms, or stimuli for indigenously initiated marine regional arrangements.

Several specialized United Nations agencies whose terms of reference include marine problems also have offices in the region, including the UNESCO Regional Office for Science and Technology for Southeast Asia in Jakarta, divisions of the Economic and Social Commission for Asia and the Pacific (ESCAP) concerned with transportation and natural resources, and the United Nations Environment Programme (UNEP) Regional Office in Bangkok. The latter has recently made great strides in formulating a plan for cooperative scientific research in the South China Sea, although its implementation is fraught with political difficulties. The many U.N.-sponsored or supported national projects and bilateral assistance programmes in the marine sphere also have helped to stimulate and support national marine awareness and build capacity in the region. The now defunct non-governmental Southeast Asian Programme in Ocean Law and Policy (SEAPOL) helped build capacity and understanding regarding law of the sea matters in Southeast Asia.

In Southeast Asia, there are already several multilateral regimes with maritime relevance. These include the 1995 ASEAN Treaty on the Southeast Asia Nuclear-Weapon-Free Zone (The Bangkok Treaty),[6] the 1976 ASEAN Treaty of Amity and Cooperation, and ASEAN's 1971 declared goal of a Zone of Peace, Freedom and Neutrality. Much to the chagrin of the United States and other extraregional maritime powers, the 1995 Bangkok Treaty defines the Southeast Asia Nuclear-Weapon-Free Zone as including the signatories' continental shelves and EEZs.[7] It also commits States parties to refrain from transporting nuclear weapons inside or

6 *Treaty on the Southeast Asia Nuclear-Weapon-Free Zone*, Bangkok, 15 December 1995, <www.nuclearfiles.org/menu/library/treaties/nuclear-free-zones/trty_bangkok-south-east-asia-nuclear-free_1995-12-15.htm>, 12 March 2008.

7 *Treaty on the Southeast Asia Nuclear-Weapon Free Zone*, Bangkok, 15 December 1995, <www.aseansec.org/2082.htm>, Art. 1(a) and 2.

outside the Zone and to not allowing other States to have control over such weapons in their territory.[8] There are loopholes that preserve freedom of the high seas, innocent passage, archipelagic sea lane passage and transit passage, and allow States parties to make exceptions regarding visits and transits to and through their territory, waters and airspace by foreign ships and aircraft. Nevertheless, the United States remains wary and will likely not accede to the Treaty. Nevertheless, ASEAN has recently agreed on a plan to strengthen the implementation of the 1995 Bangkok Treaty.[9]

More fundamental to a specific management regime for the South China Sea are the 1992 ASEAN Declaration on the South China Sea,[10] and the repeated Indonesian-led South China Sea Workshop Statements reaffirming agreement by the claimants to resolve any territorial or maritime disputes through peaceful means.[11] The Council for Security Cooperation in the Asia-Pacific (CSCAP), a track-two confidence-building exercise, has a Working Group on Maritime Cooperation. This Working Group's efforts, particularly its Guidelines for Regional Maritime Cooperation,[12] could provide the foundation for a process of comprehensive maritime security regime-building.

Despite these beginnings, nation building still has primacy in Southeast Asia. With the exception of Thailand, all the nations in and around the region have achieved independence or have experienced society-transforming movements within the past 50 years. Many are still struggling with the basic problems of nationhood, thus inserting a nationalistic fervour into regional and maritime affairs. Within ASEAN itself, relations are cordial but competitive and perhaps unstable in the long term. The ASEAN States produce many of the same raw materials and the resulting direct competition for credit, investment markets, and development assistance

---

[8] *Ibid.*, Arts. 2 and 3(1).

[9] "SE Asia to adopt new plan on nuclear-free zone," *Reuters*, 28 July 2007.

[10] *ASEAN Declaration on the South China Sea*, Manila, 22 July 1992, <www.aseansec.org/1196.htm>, 12 March 2008.

[11] See compilation of meeting statements from the Informal Working Group on Managing Potential Conflict in the South China Sea, circulated at the 10th Workshop, available from Centre for Asian Legal Studies, University Of British Columbia, <faculty.law.ubc.ca/scs/>.

[12] Council for Security Cooperation in the Asia-Pacific, Working Group on Maritime Cooperation, *No. 4: Guidelines for Regional Maritime Cooperation* (Kuala Lumpur: CSCAP, December 1997).

may increase with advancing economic development. Southeast Asian nations only now are beginning to perceive clearly their own national marine interests and how these differ from those of neighbouring States or outside maritime powers. At this juncture, commonalities are neglected and differences tend to be emphasized. The maintenance of newly-acquired national sovereignty over ocean resources and jurisdiction over ocean space has underlying national security connotations, and any proposed coordination or cooperation might be perceived in some quarters as impinging on national security.

Moreover, extension of maritime jurisdiction has imposed inequities and imbalances in marine endowments upon an already economically and politically competitive milieu. The marine area and attendant resource base of Singapore, Cambodia, Laos, and Brunei are negligible compared with the great gains of Indonesia, the Philippines and Vietnam. Thailand perceives that it has lost access to important fishery resources that made its large fleet profitable. Ironically, the windfall increases in the area and resources for Vietnam and Indonesia raise the possibility of competition and conflict between these new maritime neighbours. This redistribution of area and potential wealth, as well as the concomitant potential for conflict, could even ultimately endanger ASEAN's progress.

In Northeast Asia, regional cooperation is in a much earlier stage of development and there are several serious obstacles to maritime regime-building. In political terms, there are four countries with six governments, with little history or experience in multilateral cooperation. In this context, a primary obstacle to truly regional cooperation is the difficulty of involving both China and Taiwan in a multilateral marine policy regime covering areas claimed by both, particularly given the increasing tension in their relationship.

Another difficulty for Northeast Asia is the isolation and pugnacious, non-participatory stance of North Korea. Since North Korea borders and claims continental shelf, "security zones," and EEZs in the Japan (East) and Yellow (West) Seas, its eventual participation in functional marine policy regimes is important. And Russia's Far East maritime policy is not yet clear, other than to fiercely enforce its resource ownership, nor is it certain how stable and steadfast its policy is or will be. Although the participation of the major powers, China and Japan, is critical to a successful regime in Northeast Asia, both may be reluctant to participate unless they can dominate.

Other than boundary and island sovereignty issues which stimulate nationalism, there is a rather low marine "awareness" in Northeast Asia.[13] Despite growing interest and obvious need, Northeast Asian States continue to ignore most of the opportunities currently available in the marine sphere. They have seemed largely incapable of resolving the growing multiple use conflicts in their own EEZs, let alone those conflicts that are transnational in character. Furthermore, the ocean as a whole continues to play a role in the national and regional development process far below its potential for most of the coastal States of Northeast Asia.

Maritime issues are generally only a ripple in the great ebb and flow of economic and political relations in Northeast Asia. But many national frontiers are now maritime in nature and nationalism can elevate these issues into symbols of national pride, integrity and legitimacy. Indeed, some maritime issues may be so crucially situated in time or substance vis-à-vis the balance of much greater issues that they could act like a rogue wave or surge that significantly disturbs political relations in Northeast Asia. Disputes over islands or boundaries in areas of high petroleum potential are the epitome of this problem. Considering the tenuous or even hostile relations between most of the States in the region and the likelihood of petroleum in disputed continental shelf areas, such maritime issues could become the "tail that wags the dog" of international relations in Northeast Asia.

Multilateral Maritime Regime-Building

The North-West Pacific Region Action Plan

The North-West Pacific Region Action Plan (NOWPAP) was adopted as part of UNEP's Regional Seas Programme in 1994 by its four Member States: China, Japan, the Republic of Korea and Russia. NOWPAP's objectives are to forge concrete cooperation in marine environmental protection. However, it continues to face many problems, including lack of agreement regarding the specific geographic definition of the region for cooperation, the priorities for projects, the means of their implementation, and the

[13] Valencia, *supra* note 2, at pp. 213–214.

all-important allocation of costs. Moreover, centrally located North Korea is not a member. Thus NOWPAP's progress has been severely retarded. Indeed, despite its international architecture and activities, it is doubtful that a true transnational regime[14] has been established.

The 11th NOWPAP Intergovernmental Meeting (IGM) was held in Moscow on 20–21 December 2006.[15] The IGM approved the report of Executive Director of UNEP on the progress in the implementation of NOWPAP activities. The Meeting also discussed the reports on the activities of the four Regional Activity Centres (RACs) of NOWPAP; the revised Terms of Reference of the NOWPAP Regional Coordinating Unit (RCU); and the Focal Points for and Meetings of the Special Monitoring and Coastal Environmental Assessment Regional Activity Center (CEARAC), Marine Environmental Emergency Preparedness and Response Activity Center (MERRAC), and the Pollution Monitoring Regional Activity Centre (POMRAC). To further enhance the efficiency and effectiveness of NOWPAP activities, the IGM decided to carry out an evaluation of the performance of the RACs in 2007. The Twelfth NOWPAP Intergovernmental Meeting was held in China in 2007.[16]

In terms of regime-building, NOWPAP's most significant accomplishment has been the establishment of a minimum level of preparedness for and cooperative approach to oil spills through the adoption of the 2003 NOWPAP Regional Oil Spill Contingency Plan and the 2004 Memo-

---

[14] Formally, regimes are sets of implicit or explicit principles (beliefs of fact, causation, and rectitude), norms (standards of behaviour defined in terms of rights and obligations), rules (specific prescriptions or proscriptions for action), and decision-making procedures (prevailing practices for making and implementing collective choice) around which actor expectations converge. S. D. Krasner, "Structural causes and regime consequences: regimes as intervening variables," in S. D. Krasner, ed., *International Regimes* (Ithaca: Cornell University Press, 1983), pp. 1–21. A more practical definition, and that used herein, is multilateral agreements among States which aim to regulate national actions within an issue area. O. R. Young, *International Cooperation: Building Regimes for International Resources and the Environment* (Ithaca: Cornell University Press, 1989), pp. 196–198. Mere cooperation can exist without a regime.

[15] "11th NOWPAP Intergovernmental Meeting held in Moscow, Russia on 20–21 December 2006," *PEMSEA E-Updates* 8(1) (2007), <www.pemsea.org/pdf-documents/e-updates-archives/Vol-VIII-Issue-01-2007.pdf>, 12 March 2008.

[16] The Final Report of the 12th NOWPAP Intergovernmental Meeting is available at <www.nowpap.org/data/IGM12%20report.pdf>.

randum of Understanding (MOU) on Regional Cooperation Regarding Preparedness and Response to Oil Spill, and related activities.[17]

A similar arrangement has been agreed to for Southeast Asia. In May 1993, six ASEAN countries and Japan signed an MOU on ASEAN Oil Spill Preparedness and Response to improve marine oil spill combat capabilities, as well as an Action Plan for its implementation.[18] Many meetings have been held, the focus has been expanded to include hazardous and noxious substances, and the membership has also expanded. But capacities and regional cooperation have not greatly improved and effective implementation of the action plan remains in doubt. A similar agreement was reached between Cambodia, Thailand and Vietnam for the Gulf of Thailand, but it is only in its infancy.[19]

Cooperation in the Malacca and Singapore Straits

The Strait of Malacca Safe Navigation Scheme between Malaysia, Indonesia and Singapore, and the concomitant revolving fund established by Japan and these nations to cover costs of preventing and cleaning up oil spills from tankers, are the most concrete examples of indigenous Southeast Asian maritime regime-building.[20]

Although policy statements abound regarding anti-piracy cooperation among the Straits States and between them and user States, the most visually operationalized regime is the July 2004 agreement on trilateral coordinated Malacca Strait Sea Patrols[21] and the September 2005 Eyes in

---

[17] See, for example, the Project on Oil Spill Preparedness and Response in the ASEAN Seas Area (ASEAN-OSPAR Project), <www.nmc.com.sg/asean-ospar.html>.

[18] A. Syah Maparessa, "ASEAN-Oil Spill Preparedness and Response (ASEAN-OSPAR) Challenges and Opportunities Towards Operational and Effective Regional Cooperation in Oil Spill Preparedness and Response," Paper presented at The East Asian Seas Congress, 12–16 December 2006, Haikou City, China.

[19] P. Prasetwong, "Regional Agreements in Preparedness and Response: Gulf of Thailand," Paper presented at The East Asian Seas Congress, 12–16 December 2006, Haikou City, China.

[20] Valencia, *supra* note 2, at p. 238.

[21] MINDEF Singapore, "Launch of Trilateral Coordinated Patrols–MALSINDO Malacca Straits Coordinated Patrol," 20 July 2004, <www.mindef.gov.sg/imindef/news_and_events/nr/2004/jul/20jul04_nr.html>.

the Sky agreement.[22] The effectiveness of these trilateral endeavours has been questioned regarding the fulfilment of respective duties and the loose link between air surveillance and surface action. However, these inefficiencies are being addressed. There is also an unpublicized regime of limited hot pursuit, at least between the Malaysian and Singapore navies.

ReCAAP

At Japan's initiative, the Regional Cooperation Agreement on Combating Piracy and Armed Robbery Against Ships in Asia (ReCAAP) was ratified by 14 countries and launched in November 2006.[23] The anti-piracy initiative includes major maritime nations such as China, India, South Korea and Japan, as well as seven ASEAN States. Its coordinating centre is in Singapore. Although its objective is to ensure the safety and security of key shipping lanes like the Malacca Strait, neither Indonesia nor Malaysia are members. Moreover its focus, indeed, its mandate, is only information sharing.[24] As a result, there are doubts as to its effectiveness in reducing the incidence of piracy in the region. In terms of regime-building, there is an expectation to provide relevant information to the centre. Whether sufficient critical information will be forthcoming from all parties is questionable.

East Asia's Contribution to the Implementation of the Regime of "Provisional Arrangements of a Practical Nature" In Lieu of Boundary Delimitation

Articles 74(3) and 83(3) of the LOS Convention, on the delimitation of the exclusive economic zone and continental shelf respectively, state that pending agreement, the States concerned are to make every effort to enter into "provisional arrangements of a practical nature" and, dur-

---

[22] MINDEF Singapore, "Launch of Eyes In the Sky (EIS) Initiative," 13 September 2005, <www.mindef.gov.sg/imindef/news_andevents-/nr/2005/sep/13sep05_nr.html>.

[23] "Factsheet on the Regional Cooperation Agreement on Combating Piracy and Armed Robbery Against Ships in Asia (ReCAAP)," <app.mot.gov.sg/data/ReCaap%20 factsheet%20_nov06_%20%5BFINAL%5Das%20of%20281106.pdf>.

[24] DPA, "Anti-piracy Centre Opens in Singapore," *Daily India.com*, 29 November 2006.

ing this transitional period, not take action to jeopardize or hamper the reaching of a final agreement. Given the maze of overlapping boundary claims and disputed islands in East Asia and the pre-existing political tensions, it is not surprising that a regime is emerging that governs State behaviour in such situations. The components are many: declarations of conduct in such situations; no drilling in disputed areas; prior notification of "research" in disputed areas; and joint development of living and non-living resources in areas of overlapping claims. This regime is not formally agreed; it is in most instances bilateral rather than multilateral; and it is sometimes violated. Nevertheless, it can be argued that the frequency of use of these provisional arrangements, especially by major powers like China and Japan, makes similar approaches expected State behaviour in similar circumstances.

The 2002 ASEAN/China Declaration on the Conduct of Parties in the South China Sea[25]

In November 1999, the Philippines proposed a specific code of conduct covering the entire South China Sea, which included a commitment to no further occupations of the features there. But the ASEAN claimants could not agree among themselves regarding the geographic coverage of the proposed code. Vietnam wanted it to include the Paracels, which have been occupied by China since it took them by force from South Vietnam in 1974. This definition of the proposed code's coverage was of course unacceptable to China. Malaysia wanted the code restricted to the area of overlapping claims. In addition, China wanted a general declaration of principles rather than specific prohibitions. Malaysia, in a bid to break the impasse and to get China to accept the proposal, suggested at the ASEAN annual meeting in Brunei in August 2002 that the "code" be replaced by a political "declaration" and that the coverage be non-specific as well.

On 4 November 2002, ASEAN and China signed a ten point Declaration on the Conduct of Parties in the South China Sea.[26] It was hoped

[25] For a thorough treatment of the South China Sea issues, see M. J. Valencia, J. Van Dyke and N. Ludwig, *Sharing the Resources of the South China Sea* (New York: Kluwer Law International, 1997).

[26] *Declaration on the Conduct of Parties in the South China Sea*, Phnom Penh, 4 November 2002, <www.aseansec.org/13163.htm> (hereinafter ASEAN-China Declaration).

that this would reduce tension and improve the general political climate in the South China Sea. Two bilateral codes of conduct (Vietnam/Philippines; China/Philippines) already existed, as did numerous platitudes in statements issued by the track-two Indonesian-led South China Sea Workshops. However, violations of these undertakings by most parties had been frequent.

Important points reiterated in the 2002 Declaration included commitments to the U.N. Charter, the 1982 LOS Convention, the 1976 ASEAN Treaty of Amity and Cooperation in Southeast Asia, and the 1954 Five Principles of Peaceful Co-existence.[27] In particular, the parties again undertook to resolve jurisdictional disputes by peaceful means without resorting to the threat or use of force and in accordance with international law including the 1982 LOS Convention. To this end they also repledged to exercise self restraint in the conduct of activities that would complicate or escalate disputes and affect peace and stability in this area.

The new points of the Declaration included a statement that the parties will refrain from "action of inhabiting [sic] on the presently uninhabited islands, reefs, shoals, cays and other features."[28] Actually, this clause was watered down from a ban on erecting new structures such as those China has built in spurts on Mischief Reef from 1995 onwards.

The 2002 Declaration also commits the parties, albeit voluntarily, to hold dialogues and exchange views and information between their defence and military officials, and to prior notification to the other parties of impending joint/combined military exercises, including those with States from outside the region. And it reaffirms the parties' commitment to freedom of navigation in and over-flight above the South China Sea, a matter of great interest to maritime powers. The Declaration also lists topical areas that, pending a comprehensive and durable settlement, may be the subject of cooperative endeavours, for example, marine environmental protection; marine scientific research; safety of navigation; search and rescue; and combating transnational crime, including trafficking in illicit drugs, piracy and armed robbery at sea, and arms smuggling. The Declaration establishes a detailed regime, but one whose effectiveness has yet to be evaluated.

---

[27] "Five Principles of Peaceful Co-Existence," Wikpedia (May 2007), <en.wikipedia.org/wiki/Five_Principles_of_Peaceful_Coexistence>.

[28] ASEAN-China Declaration, *supra* note 26.

The downside to the 2002 Declaration is that:

1. It is a political declaration, not a legally binding detailed code of conduct.
2. It makes no mention of the geographic scope or even of the "Spratlys" or "Paracels."
3. It repeats worn and ineffective platitudes from previous bilateral codes of conduct and the South China Sea Workshops.
4. Most important, the parties themselves publicly acknowledged that a binding code of conduct is important for peace and stability in the region and agreed to continue to work towards achieving one.[29]

In the months preceding and following the Declaration the Philippines moved civilians onto Pag-asa in the Spratlys;[30] Vietnam rescued Chinese fishermen whose boats sank in the Spratlys but then delayed releasing them;[31] Vietnamese soldiers fired at two Philippine bombers flying reconnaissance at about 1000 feet over Vietnamese-occupied but Philippine-claimed Pugad;[32] Vietnamese forces were alleged to have occupied Parola after Philippine forces abandoned it;[33] China continued to beef up its military presence in the area;[34] and the destruction of coral reefs in the area increased.[35] Vietnam has undertaken construction of a gas pipeline

[29] "ASEAN and China Sign Landmark Pact on Spratlys Now," *Channel News Asia*, 4 November 2002; " 'Code of Conduct' for Disputed Islands," *BBC News*, <news.bbc.co.uk/1/hi/world/asia-pacific/2399425.stm>, 12 March 2008; "ASEAN, China Sign Document on Territorial Dispute," *Kyodo News Service*, Tokyo, 4 November 2002; V. Torsricharoen, "ASEAN, China to Ink Declaration over Territorial Row Mon," *Kyodo News Service*, Tokyo, 30 October 2002.

[30] L. Baguioro, "Philippines Move Civilians to Spratlys," *The Straits Times*, Singapore, 10 October 2002.

[31] C. Arthurs, "Vietnam Detain Chinese Fishing Group," *BBC News Asia-Pacific*, London, 9 September 2002.

[32] "Philippines Official Plays Down Vietnamese Shooting Incident," *Deutsche Presse-Agentur*, Berlin, 5 September 2002.

[33] "Philippine Defense Chief Denies Vietnam Occupying Disputed Island in Spratlys," *BBC Monitoring Asia-Pacific Political*, London, 11 September 2002.

[34] *Ibid.*

[35] PEMSEA, "Scientists Alarmed on Destruction of Spratlys Environment," *Newsletter of PEMSEA*, 7 July 2002.

that would remove gas from an area claimed by China,[36] and Malaysia has occupied three additional features. China, the Philippines and Vietnam are contemplating joint exploration of part of the disputed area despite reservations by other claimants.[37] On 9 July 2007, Chinese forces killed at least one Vietnamese fisherman in a clash in the Spratly area.[38] Thus the regime is "leaking." Indeed ASEAN and China are still struggling to transform the 2002 Declaration into a legally-binding code and have agreed to intensify efforts to do so.[39]

Prior Notification

"Illegal" scientific research in another country's EEZ is not usually considered a security issue. But when the countries in question are rivals for power and leadership in Northeast Asia, such activities can stir national sentiments and become the progenitor of potential conflict. Chinese marine scientific research and navy ships have been frequenting the area of the East China Sea claimed by both China and Japan since 1998, causing great concern and consternation in Japan. Indeed, Japan claims that activities carried out by Chinese ships in Japan's claimed EEZ over the past five years include collection of data for military purposes as well as exploration of natural resources – both in violation of the 1982 LOS Convention.[40] China argued that it was conducting research in its claimed EEZ and on its claimed continental shelf which is allowed by the 1982 Convention.

---

[36] *Reuters*, Beijing, 10 April 2007; A. Symon, "China, Vietnam spar over gas," *Asia Times*, 30 April 2007.

[37] J. Pangco Panares, "Code of Conduct on Spratlys Sought," *Manila Standard Today*, 27 October 2006.

[38] "Vietnam-China Tussle Over Spratlys," *China Knowledge*, 19 July 2007, <www.chinaknowledge.com/news/news-print.aspx?ID=9291&cet=general>; "Chinese Navy Fires on Vietnam Fishing Boats in Oil-Rich Area," *Strait News*, 19 July 2007, <app.mfa.gov.sg/pr/read_content.asp?View,7744,>; "ASEAN Foreign Ministers Begin Annual Meeting," *Kyodo*, 29 July 2007.

[39] "China-ASEAN Agree to Discuss Territorial Dispute," *The Star Online*, 26 October 2006; B. Wain, "ASEAN Faces Difficulty in Beijing Negotiation – Divisions Within Group Block Consensus on Code for the South China Sea," *Asian Street Journal*, Hong Kong, 1 August 2002.

[40] "China Said Conducting Drills in Japan's EEZ," *The Daily Yomiuri*, Tokyo, 26 July 2001.

On 13 February 2001, Japan and China agreed on a mutual prior notification system.[41] The Agreement cleverly avoids specifying any line beyond which advance notification is required. It simply says that China is to give Japan at least two months notice when its research ships plan to enter waters "near Japan and in which Japan takes interest" and that similarly, Japan is to inform China before its vessels enter waters "near" China. The notification must include the name of the organization conducting the research, the name and type of vessels involved, the responsible individual, the details of the research, such as its purpose and equipment to be used, the planned length of the survey, and the areas to be surveyed. Japan has claimed on several occasions that China continues to violate the Agreement.[42]

Voluntary prior notification of joint military exercises in the disputed area is also a part of the 2002 ASEAN-China Declaration.[43] Moreover, such courtesy or expectation has been proposed and is being discussed between Japan and South Korea regarding their overlapping claims in the Sea of Japan (East Sea)[44] and for the naval activities of Indonesia and Malaysia in their area of overlapping claims in the Sulawesi Sea.[45]

Joint Development

Although the concept of joint development of mutually claimed resources originated in the Middle East,[46] Asia leads the world in its implementation, innovation and variety. Regarding hydrocarbon deposits in disputed areas in Southeast Asia, joint development arrangements have been

---

41 "Japan, China Agreement on Maritime Notice System Detailed," *BBC Monitoring Asia Pacific-Political*, London, 13 February 2001.

42 Ministry of Defense, *Japan Security Outlook* 2002, 2003, 2004, Tokyo; M. Mochizuki and M. O'Hanlon, "Put a Lid on Raising Sino-Japanese Tension," *The Japan Times*, 20 April 2005.

43 ASEAN-China Declaration, *supra* note 26.

44 "Aso, Song agree to hold EEZ talks this month," *The Japan Times*, 4 June 2007.

45 Personal communication, Maritime Institute of Malaysia, June 2007.

46 F. Fesharaki, "Joint Development of Offshore Petroleum Resources: The Persian Gulf Experience," in M. J. Valencia, ed., *The South China Sea: Hydrocarbon Potential and Possibilities of Joint Development* (New York: Pergamon Press, 1981), pp. 1325–1334.

implemented between Malaysia and Thailand,[47] Malaysia and Vietnam,[48] and Indonesia and Australia[49] (upon independence East Timor replaced Indonesia in the arrangement). There is also an agreement between China, the Philippines and Vietnam to "undertake research" in the eastern part of the disputed South China Sea although this is now the subject of serious controversy in the Philippines and may be abrogated.[50] However, if the research indicates there are economic deposits of hydrocarbons in the area, exploration may follow.

On 18 June 2001, Cambodia and Thailand signed an MOU regarding the Area of their Overlapping Maritime Claims to the Continental Shelf.[51] The MOU recorded the intent of the two governments to agree on a treaty for joint development of the hydrocarbon resources in the "Joint Development Area," and a maritime border for the "Area to be Delimited."

In Northeast Asia, joint development between South Korea and Japan has been ongoing for over 30 years.[52] China and North Korea have agreed

---

47 *Memorandum of Understanding between Malaysia and Thailand*, 24 October 1979, reprinted in P. Polahan, "Thailand-Malaysia Memorandum of Understanding," in Valencia, *ibid.*, p. 1355; see M. J. Valencia, *South-East Asian Seas, Oil Under Troubled Waters* (Oxford: Oxford University Press, 1985), p. 62.

48 *Memorandum of Understanding between Malaysia and Vietnam*, May 5, 1992, copy on file with author.

49 *Treaty Between Australia and the Republic of Indonesia on the Zone of Cooperation in an Area Between the Indonesian Province of East Timor and Northern Australia*, Timor Sea, 11 December 1989, [1991] *Australian Treaty Series* No. 9 [hereinafter The Timor Gap Treaty].

50 *A Tripartite Agreement for Joint Marine Scientific Research in Certain Areas in the South China Sea*, <www.newsbreak.com.ph/index.php?option=com_remository&Itemd=88889273$func=startdown&id=219>; Miriam Grace A. Go, "Malacanang Knows More Than Spratlys Deal," <abs-cbnNews.com/Newsbreak>, 10 March 2008.

51 The Duong Tara, "Country Report for Cambodia," 7th ASCOPE Conference, 5–7 November 2001, Kuala Lumpur.

52 *Agreement Concerning Joint Development of the Continental Shelf between Japan and Korea*, 30 January 1978, reprinted in C.-H. Park, "Joint Development Between Korea and Japan in the East China Sea," in Valencia, *supra* note 46, at p. 1335; M. Miyoshi, "The Japan-South Korea Agreement on Joint Development of the Continental Shelf," in M. J. Valencia, ed., *Geology and Hydrocarbon Potential the South China Sea and Possibilities of Joint Development* (New York: Pergamon Press, 1985), p. 533.

on joint maritime oil development in the northern Yellow Sea[53] and China and Japan have agreed in principle to joint development in the East China Sea.[54] South Korea and China continue to discuss a similar provisional arrangement for their overlapping boundary claims in the Yellow Sea.

The typical structure of a joint development arrangement includes: (1) a treaty that identifies the shared area and establishes the legal basis for cooperation; (2) the establishment of a bilateral joint commission usually composed of equal members from the two parties to govern the venture; and (3) the granting of exploitation rights to one or more private enterprises by either the contracting States directly or by the joint commission.[55] The governing joint commission can be strong, as in the Thai-Malaysia and Australia-Indonesia arrangements, "with full juridical personality and powers to license, stipulate terms and exemptions and enter into development contracts with foreign operators." It can also be weak, as in the South Korea-Japan and Malaysia-Vietnam Agreements, operating solely as a consultative body to coordinate cooperative activities or as a liaison between national oil companies. The Japan-Republic of Korea Joint Commission, for instance, has the power only to recommend action to the two contracting parties. The process of selecting and then monitoring the licensees or concessionaires is perhaps the most important element of such agreements. The original Timor Gap Treaty between Australia and Indonesia was the most robust and detailed in the administrative structure it created, perhaps because the two countries had a history of mutual suspicion toward each other and needed clear procedural rules to avoid difficulties.

Several significant specific issues need to be resolved for a joint development authority to be established: (1) the boundaries of the area to be governed by any authority that is established; (2) the status of the claims to this area during the joint development period; (3) the resources to be managed, e.g., oil and gas, fisheries, environment, maritime space; (4) the mechanism of management and its governing structure; (5) the

---

[53] Y. Sang-mein, "Bohai Oil and North Korea," <english.hai.co.kr/popups/print.hani?ksn=208111>, 9 May 2007.

[54] "Special Report: Premier Wen Visits ROK, Japan," *Xinhua*, 11 April 2007, <news.xinhuanet.com/english/2007-04/12/content_5964748.htm>; "No Breakthrough with Beijing in Gas Rights Talk," *The Japan Times*, 25 May 2007.

[55] Valencia, et al., *supra* note 25, at pp. 184–186.

law that would apply in the area governed by the authority; (6) the fiscal and production arrangements; and (7) the length of the agreement. These are formidable – but not insurmountable – issues. It is important to emphasize that it is not necessary for the contracting parties to resolve their conflicting maritime boundary disputes in order to enter into a joint development venture. Moreover, the legal regime does not have to be uniform. For example, the South Korea-Japan and Australia-Indonesia joint development arrangements divide their areas into subzones where different legal regimes govern. Several of the joint development arrangements now functioning are between countries with substantial suspicion or enmity between them. Nevertheless, the countries agreed to depoliticize the dispute to explore for resources and provide benefits for the people of both countries.

Bilateral Fisheries Agreements as "Provisional Arrangements"

China/Vietnam Fisheries Agreement

China and Vietnam signed agreements on fisheries management and boundary delineation in the Gulf of Tonkin that took effect in 2004.[56] The Sino-Vietnamese Fisheries Agreement was the first in East Asia to establish a cooperative fisheries management programme within a demarcated zone, the Common Fisheries Zone. The Joint Fishery Committee established by the Agreement is quite robust and may make rules and regulations governing the Common Fishery Zone to enforce its allocations and limits. It also contains a dispute settlement mechanism.

China/Japan Fisheries Agreement

The China-Japan Fisheries Agreement of November 11, 1997, entered into force in June 2000. Faced with overlapping EEZ (and continental shelf) claims in the East China Sea and burgeoning fisheries disputes, China and Japan agreed to establish three different zones where different fisheries regimes apply:

[56] D. Rosenberg, "China's bilateral fisheries agreements with Japan, South Korea and Vietnam," ZNet, 1 July 2005, <www.zmag.org/content/print_article.cfm?itemID=8205.&sectionID=1>, 12 March 2008.

- Exclusive fishing zones in their EEZs up to 52 nm from their respective baselines in the area between 27° N and 30° 40' N;
- Joint regulation in the area beyond 52 nm from each State's baselines and between 27° N and 30° 40' N; and
- Exclusion of the application of the fisheries agreement to the area south of 27° N.

Each country manages its fisheries with 52 nm of its baselines; beyond 52 nm and between 27° N and 30° 40' N, boats of the two countries may fish without prior approval of the other's government.[57] The area south of 27° N, including the area around the disputed Diaoyu/Senkaku islets, remains unregulated high seas.

The 1997 Agreement mandates the establishment of a China-Japan Joint Fisheries Commission that makes recommendations on: matters relating to catch quotas, and other terms and conditions of fishing operations for the nationals and fishing vessels of each signatory State in the other's EEZ; matters regarding the maintenance of the "fisheries order"; and matters relating to fisheries cooperation between the two States.

Although the 1997 bilateral Agreement is a step in the right direction, there remain several fundamental problems with potential negative implications for both fisheries and international relations in the East China Sea. For example, the EEZ and continental shelf boundaries have not been agreed. Moreover, the relevance of Taiwan's and South Korea's claims and roles remains an open question. There is no effective dispute settlement mechanism built into the Agreement. Specific quotas and conditions of operations must be negotiated every year and may be held hostage to the quality of political relations between the parties. Nevertheless, this Agreement is a remarkable step towards the building of at least a bilateral regime governing maritime activities in areas of overlapping or unclear jurisdiction.

Japan-South Korea Fisheries Agreement

Also faced with overlapping EEZ and territorial claims and increasingly frequent and bitter fisheries incidents in the Sea of Japan (East Sea), Japan and South Korea established a joint fishing area there and in a

[57] H. Seiguyama, "Japan, China Conclude Bilateral Fisheries Treaty," *The Daily Yomiuri*, Tokyo, 28 February 2000.

small area south of Cheju Island.[58] They also agreed on the establishment of a South Korea-Japan Joint Fisheries Commission to implement their cooperation for the conservation and management of the living resources in their joint fishing areas. The Commission will make recommendations to the States on matters relating to the conditions of fishing operations, the maintenance of the "fisheries order," the condition of the marine living resources, cooperation between the two States in the fisheries field, the conservation and management of the marine living resources in the joint fishing area, and other matters relating to the implementation of the Agreement. The signatory States are supposed to respect the recommendations of the Commission in their determination of the terms and conditions for allowing the fishing vessels of other States in their EEZ.

The Agreement was very unpopular in both countries, particularly in South Korea where it became a hot political issue. The situation is also complicated by the inclusion of the small area south of Cheju, which overlaps China's claimed EEZ. Indeed, the Commission will be exercising prescriptive jurisdiction for the conservation and rational management of living resources in an area covered by the Japan-China Agreement. Presumably South Korea and Japan agreed to establish the small joint fishing area there to preserve their legal position with regard to their respective claims with China.

South Korea-China Fisheries Agreement

On 11 November 1998, after five years and 19 rounds of negotiations, South Korea and China reached an agreement on fisheries in the Yellow Sea. The Agreement established three different zones between the two countries: EEZs, joint, and transitional. In the EEZ, the coastal State exercises its sovereign rights over resources. The transitional areas are zones of about 20 nautical miles in width on both sides of a joint fishing area where the nationals and fishing vessels of the two States are allowed to fish. The zones extend northward to 29° 45' latitude. These zones were incorporated into the countries' respective EEZs after four years of joint management. The two countries also agreed to gradually reduce their harvesting so as to maintain balanced fishing in the transitional areas; to take measures for the conservation and management of the living resources in accordance with the decisions of the joint fisheries commission; to

[58] Rosenburg, *supra* note 56.

jointly conduct surveillance to ensure that the conditions for fishing are observed; and to exchange a list of fishing vessels for effectively carrying out these obligations.

Problems remained. South Korea felt it could not wait four years until the transitional area was incorporated into each EEZ because of the rapidly increasing number of Chinese fishing vessels engaged in indiscriminate over-exploitation of the living resources in the transitional area. Thus, South Korea tried to incorporate the area into its EEZ ahead of the four-year time schedule through negotiations. Transgressions continue to be reported by both sides. Most important, the Agreement is provisional in that it can be finalized only when the boundary delimitation is settled, and it is incomplete in that it will need many more negotiations on the terms and conditions for fishing and catch quotas in each other's EEZ.

Commonalities and Limitations

The above agreements address three key issues. First, they reaffirm each country's exclusive rights over fishery resources and fishing activities in its own EEZ. Second, they establish general principles for reciprocal fishing access in each other's EEZ. Third, the agreements create a cooperative management regime for their shared fishery resources. Thus the core aspect of these fisheries agreements is their establishment of joint fishing areas as a provisional arrangement for dealing with overlapping prescriptive and enforcement jurisdictions of the coastal States in their overlapping EEZs.

For Northeast Asia, the agreements are the result of the inability of the States to agree on delimitation of their EEZ boundaries. In fact, if they could have clearly delimited their boundaries, these contrived and complicated arrangements for joint fishing areas might not have been necessary.

From a resource management perspective, the main limitation of these agreements is that they focus on managing fishing activity in designated areas that only comprise part of the fishery ecosystem.[59] Unregulated waters still exist for unrestricted exploitation of fish stocks. For example, the Sino-Japanese Agreement provides a "Current Fishing Pattern Zone" around the disputed Sinkaku/Diaoyu Island thereby avoiding the territorial dispute over the ownership of the island. However, many fish stocks

---

[59] *Ibid.*

migrate seasonally from the adjacent management zone in the East China Sea to these unrestricted fishing waters. Thus, in Northeast Asia, the continuity of the ecosystem of the Yellow and East China Seas and the migration of the fish stocks through all three littoral countries' claimed zones dictate that a trilateral unified management system will eventually be necessary.[60]

Each agreement establishes a Joint Fishery Committee (JFC) comprised of representatives from each country appointed by each government, as well as several commissioners.[61] Although each JFC has somewhat different scope and authority, they all have several common functions, such as research on the status of fisheries, consultation with fishing industry interests, and the making of recommendations to fisheries management authorities on access to fishing zones. They may make recommendations on fishing quotas, types of species to be caught, and other conditions for fishing.

Each JFC establishes an area for joint fisheries management in the common seas between the coastal countries. In the Tonkin Gulf, China and Vietnam have designated a "Common Fishery Zone" and "Waters in Transitional Arrangements." In the Yellow Sea, China and South Korea have designated "Provisional Waters" and the "Waters in Transitional Arrangements." The "Waters in Transitional Arrangements," located on either side of the joint resource management area, provide some flexibility for each country in restricting fishing in formerly open waters. After four years, each country has to phase out its fishing activities in the Transitional Zone of the other country and gradually conform to the coastal State's EEZ jurisdiction. The different types of management zones represent an effort to preserve some traditional fishing communities and to mitigate the impact of fishing restrictions necessary to achieve sustainable yields.

The JFCs have the authority to decide on conservation and management measures, including the allocation of fishing quotas and the maintenance of fishing order. They all have the responsibility to ensure that fisheries are not endangered through over-exploitation. Recommendations and

[60] J. Choi, "A New Sub-regional Fisheries Cooperation System in Northeast Asia," Paper presented at SEAPOL Inter-Regional Conference on Ocean Governance and Sustainable Development Millennium, Bangkok, Thailand, 21–23 March 2001.

[61] Rosenberg, *supra* note 56.

decisions are made by consensus. The JFCs determine annually how many fishing vessels of each country to permit in these joint resource management areas. This is accomplished using a "quantity control approach" that quantifies the total allowable catch (TAC) of several target species, the status of each resource, and the extent of traditional fishing and modern fishing methods and management. Based on these considerations, the JFCs then derive the allowable number of vessels.

Fishing vessels of one country need to apply for a license to fish in the other country's EEZ. These vessels have to comply with the terms of the joint fisheries agreement as well as the domestic laws and regulations of that country. Any violation is subject to legal procedures of the country controlling the EEZ where the fishing takes place. In the case of seizure or detention, the fishing vessels and crew must be promptly released upon posting a bond or other form of security.

The biggest difference among the agreements is that the Sino-Vietnamese Agreement for the Tonkin Gulf fisheries also incorporates a permanent maritime boundary delimitation. By contrast, there is no permanent maritime boundary agreement between China and Japan or between South Korea and China. The boundaries used in the latter two agreements are provisional, pending final delimitation of the overlapping EEZ claims.

In the Sino-Japanese and Sino-South Korean Agreements, enforcement in the joint resource management area is to be carried out by the flag State of each fishing boat. In the Sino-Vietnamese Agreement, enforcement is carried out by each coastal State within its EEZ. Each country has started to take the painful steps necessary to shrink fishing grounds, cut back fishing fleets, and recycle redundant labour in order to conserve and manage the vital resources.

The contracting parties carry out periodical joint patrols to prevent illegal fishing. They also conduct monitoring, surveillance and control of fishing vessels including boarding and inspection. However, no workable enforcement mechanism has been established, except for the enforcement arrangement in the China-South Korea Transitional Zones. There are no contact points for the exchange of information about violations, nor is there a joint programme to provide information to fishermen about the laws and regulations. Many fishermen find it difficult to accept that they cannot fish in waters where they have been doing so for years. Hence, monitoring and enforcement efforts need to be strengthened to improve the effectiveness of the agreements.

Another limitation is that the JFCs have little transparency or accountability. They make their decisions in private with no public participation or dispute settlement mechanisms for redress of grievances. The JFCs generally do not publish their deliberations, the data upon which their decisions are based, or the results of scientific findings. Hence, it is difficult to fully understand the rationale for the regulations.

Some important issues remain unresolved. For example, South Korea does not recognize the Sino-Japanese fisheries management regime. South Korea contends that its own EEZ includes part of the northern East China Sea and that it was not consulted in the negotiation of the Sino-Japanese Agreement. Some migratory fish stocks are therefore targeted by South Korean fishermen in these contested waters.

The major significance of all these agreements is that they were signed in accordance with the 1982 LOS Convention and based on two main objectives: the peaceful settlement of fishery disputes, and the establishment of a system for sustainable fishing for the communities bordering the China Seas. They are all the result of political compromise among the countries with very different levels of economic development, domestic political systems, and foreign policy concerns. Despite their limitations, the agreements establish an expectation that States will agree to share mutually claimed fishery resources pending delimitation of boundaries. Indeed, even North and South Korea have agreed in principle to a similar mechanism for their disputed boundary in the Yellow (West) Sea to avoid more accidental naval clashes there. However, the details are still being negotiating.[62]

Analysis[63]

The absence of robust multilateral maritime regimes in East Asia in any issue area reflects political calculations by the States regarding the rewards and risks and losses and benefits of maintaining the status quo versus developing regimes acceptable and beneficial to all sides involved. Asian

---

[62] L. Chi-dong, "Two Koreas Agree on Safe Passage for Trains Cross-border Railway Tests," *Yonhap News*, 9 May 2007; The Associated Press, "Koreas Hold Military Talks Amid Sea Border Dispute," *International Herald Tribune*, 9 July 2007, <www.ith.com/articles/ap/2007/07/10/asia/AS-GEN-Koreas-Military.php>.

[63] Valencia, *supra* note 2, at p. 19.

countries are simply not yet sufficiently aware of the seriousness of the need for multilateral maritime regimes that focus on the management of fisheries resources and maritime environmental protection, let alone security. Taken together, this situation argues strongly for an ad hoc, issue-specific, evolutionary process for multilateral maritime regime-building.

In Northeast Asia, each country must weigh the pros and cons of regime participation. For China, regime participation could lead to a fairer international legal environment, technology transfer from Japan, South Korea and Taiwan, and confidence-building in itself as a member of the international community. But China would have to limit its flexibility in the marine sphere and commit scarce resources to fulfil its regime responsibilities. Furthermore, Beijing generally prefers bilateral relationships which it can dominate.

For Taiwan, regime participation would expand its channels for discussions with China and enable maritime issues to be addressed. It would also enhance its status vis-à-vis China. On the other hand, Taiwan might have to share its technological know-how and possibly sensitive resources and other data with the other participants, including China. Taiwan would probably have to pay more than its share for the implementation of the regime. If China and Taiwan are to be included in a regime, it must be non-governmental – and much of the East China Sea must initially be excluded.

For Japan, participation in a maritime regime is favoured by its economic and technological dominance, its knowledge and experience, and its web of bilateral maritime agreements. Benefits include the protection of regional fisheries resources and the environment; elimination of the transaction costs of annual bilateral fisheries negotiations; and enhancement of its status in the region. But a prominent role for Japan in a multilateral regime is inhibited by its preference for bilateral relationships which it can dominate; the Kuriles dispute with Russia; the memories in East Asia of Japan's wartime behaviour; Japan's priority on immediate national economic gain; and its bureaucratic conservativeness and generally reactive posture regarding international affairs. However, Japan has begun to exert leadership in international affairs and perhaps the recent creation of a cabinet-level Department of Ocean Affairs will provide the impetus for Japan's regional leadership in maritime affairs in particular.[64]

---

[64] "Basic Sea Law Set to Unify Ocean Policies," *The Yomiuri Shimbun*, 21 July 2007, <www.yomiuri.co.jp/dy/national/20070721TDY04004.htm>.

North Korea could use participation in a regime as a "coming out" into the international community; to feel out potential regional partners; to increase its financial, technical, and knowledge capacity; and to gain a cleaner environment. But, for North Korea, opening of its society to foreigners and their cultures and practices could undermine domestic control. Moreover, scarce resources would have to be diverted to fulfill its obligations to the regime.

For South Korea, participation in a multilateral maritime regime could help avoid costly fisheries disputes; enhance conservation and management of fishery resources; enhance coordination with neighbouring countries regarding the management of transnational fish stocks; possibly improve its access to neighbours' stocks; increase the efficiency of its fishing effort; improve its international stature; and expand its points of diplomatic contact with North Korea.

South Korea recognizes that environmental degradation, depletion of fishery resources, maritime anarchy, and political conflict are not in its own, or any other nation's, interest. Indeed, South Korea supports a multilateral marine environmental protection regime and might even be willing to exercise leadership thereof. Although South Korea is the only State to border all three Seas, it puts more emphasis on the Yellow Sea because it is more polluted and because Korea's western coast is more developed than its eastern shore. Moreover, a regime for the Yellow Sea could employ China as a go-between to help broaden its contacts with North Korea.

Russia has much to gain and little to lose politically and economically by participating in a regional maritime regime, especially for the Sea of Japan (East Sea). It could gain technological and financial assistance as well as diminished poaching of marine resources by foreigners. But fisheries cooperation as well as cooperation in environmental protection is hampered by bureaucratic confusion, ineffectiveness, lack of infrastructure, economic malaise, and a relative lack of interest in maritime affairs, particularly in Northeast Asian seas. Nevertheless, Russia does have considerable experience in fisheries arrangements and international cooperation in marine environmental protection, a legislative base upon which to build, and considerable fisheries and oceanographic expertise, and it is interested in fulfilling its proper role as a member of the international community. The rise of a domestic environmental movement and pressure from aid organizations could promote its participation in a regional marine environmental protection regime. However, Russia's participation in regional maritime regimes will likely be determined by the progress of Russian market reform, political relations between Russia and Northeast Asia, particularly Japan, and the impact of incipient non-governmental organizations.

To move the process forward in Northeast Asia, the entities concerned might agree to begin discussing objectives and principles for multilateral maritime regimes in Northeast Asia. Indeed, there appears to be a confluence of incentives for such a dialogue. Russian foreign ministry officials would like to enhance Russia's role in multilateral dialogue in the region, specifically including maritime cooperation. South Korea's Ministry of Maritime Affairs and Fisheries would presumably be supportive of such a dialogue. Even China appears to be willing to consider limited multilateral approaches on a sea-by-sea basis. Although Japan may still be reluctant to lead such an effort, it would probably participate if it is initiated by another State or a U.N. agency.

In Southeast Asia, regional cooperation on marine issues is clearly more advanced than in Northeast Asia. But even there it is incipient or superficial at best. It may be premature to hope for extensive cooperation in a region composed of increasingly nationalistic developing States. Perhaps necessity will be the mother of cooperation. Although political constraints limit the possibilities for cooperation with regard to marine issues, political relationships can and have improved over time. The underlying economic potential of the marine resources of the region remains a part of its geography. With improving political conditions, an increasing share of this potential might become available to be tapped. Clear analysis of these possibilities may help motivate and to accelerate the improvement of political relationships throughout Southeast Asia.

Conclusion

The last few years have witnessed both significant positive and negative developments in regime-building in Asia. With the extension of jurisdiction, several bilateral agreements on fisheries in Northeast Asia have been renegotiated. They set the stage and form the basis for a more comprehensive regional agreement harmonizing the bilateral agreements and expanding the regime to include all the region's nations. Although progress has been slow regarding a regional marine environmental protection regime, a multilateral dialogue could speed up the process.

There seem to be several common factors encouraging the positive developments. The fisheries and prior notification agreements in Northeast Asia are clearly conflict avoidance mechanisms. They were adopted as a result of a past practice of maritime conflict avoidance, the introduction of the EEZ regime, conflicting claims, an increasing frequency and intensity of incidents, and, above all, domestic political pressure for action in both Japan and South Korea. In sum, leaders considered the political relationships between China, Japan and South Korea too important to be undermined by the rising nationalism accompanying these ocean disputes. These decision-makers concluded that it was in these countries' common interest to compromise and reach at least interim solutions. Drawing on regime theory,[65] the major factors contributing to regime formation in these examples are national leadership, clearly defined benefits, the necessity to avoid conflict, and, to some extent, public awareness and concern.

Just as commonalities of interest account for progress, fundamental diversity and even conflicts of interests explain the obstacles to maritime regime formation. Whereas it was in the common interest of ASEAN and China to reach a political agreement on the South China Sea, China and Vietnam's refusal to compromise on the specific inclusion of the Paracels in the geographic scope of the 2002 Declaration, and China's refusal to accept specific prohibitions, prevented agreement on a robust code of conduct. Moreover, it led to a thinly papered-over split within ASEAN on this issue.

The increased acquisition of maritime power and the festering disputes over islands and maritime space continue as an integral part of the security mosaic in East Asia. According to regime theory, the obstacles to maritime regime formation there include diverse fundamental national interests and high politics. It is clear that where relations are poor and colonial era or Cold War disputes linger, e.g., North-South Korea, China-Taiwan and the Northern Territories/Southern Kuriles, regime-building will be difficult at best.

There are two trends regarding the building of maritime regimes, one positive and one negative. The positive trend is the construction of a

[65] M. J. Valencia, ed., *Maritime Regime-building Lessons Learned and Their Relevance for Northeast Asia* (Boston: Martinus Nijhoff Publishers, 2001), pp. 160–162.

diverse web of bilateral understandings and exchanges which can expand and have a spillover effect on relations in general and maritime regime creation in particular. Given this network of arrangements, a multilateral agreement on a "code of conduct" for both Southeast Asian and Northeast Asian seas (or for the latter at least a declaration of intent) would be a natural next step. Initially, multilateral arrangements should address common maritime problems like search and rescue, environmental protection, drug trafficking, and smuggling. Further out to sea, in time and space, an international naval or "self-defense" force might ensure ocean peacekeeping, including safety of navigation. This joint force could focus on areas outside national jurisdiction and emphasize protection of fisheries, air-sea rescue and environmental monitoring. The harder issues of comprehensive fisheries management and common security would be next.

However, there are hidden constraints even in these positive trends.[66] There is no one comprehensive issue, institution or initiative and while the multiple parallel dialogues overlap in participants, objectives, and focus, some critical issues are neglected. Nevertheless this structure of multiple communities, communities within communities, seems to work, albeit inefficiently, because there are linkages between them.[67]

This positive trend, however, is overlain on a larger security dilemma. There is a pre-existing pattern of suspicion and distrust based on fundamental political and ideological differences which extend like sensitive tentacles into the maritime sphere. Thus, the extension of jurisdiction and the nationalism that accompanies it, combined with enhanced armament and technology, have ultimately made the maritime frontier in East Asia more challenging for maritime regime-building and more dangerous for political relations.

---

66 *Key Issues in Asia-Pacific Security* (Honolulu: East-West Center, 2001), p. 22, <www.eastwestcenter.org/fileadmin/stored/pdfs/SeniorPolicySeminar2001.pdf>, 12 March 2008.

67 *Ibid.*

Notes on the Historic Waters Regime and the Bay of Fundy

Ted L. McDorman*

Introduction

Historic waters are an orphaned offshore international legal regime. All the other offshore international legal regimes, e.g., internal waters, territorial sea, 200-nautical mile zones, are grounded in multilateral treaties, in particular, the 1982 United Nations Convention on the Law of the Sea (LOS Convention).[1] The historic waters regime, however, was left out of the LOS Convention, as well as the 1958 Convention on the Territorial Sea and Contiguous Zone.[2] Both multilateral treaties note that "historic bays" (aligned with historic waters)[3] exist, albeit in a round-about manner, where it is stated that the juridical bay provisions in the conventions do not apply to historic bays.[4] Neither convention provides any indication of when historic waters exist or the consequence of waters being historic.[5] As stated by the International Court of Justice in the 1982 *Libya-Tunisia Case*: "It is clearly the case that... the notion of historic rights or

* Professor, Faculty of Law, University of Victoria, Canada.

1 *United Nations Convention on the Law of the Sea*, Montego Bay, 10 December 1982, entered into force 16 November 1994, 1833 *U.N.T.S.* 397 [hereinafter LOS Convention].

2 *Convention on the Territorial Sea and Contiguous Zone*, Geneva, 29 April 1958, entered into force 22 September 1964, 516 *U.N.T.S.* 205.

3 See *Case Concerning the Land, Island and Maritime Frontier Dispute (El Salvador/Honduras: Nicaragua Intervening)*, [1992] *International Court of Justice Reports* 351, at para. 383 [hereinafter the *Gulf of Fonseca Case*].

4 LOS Convention, *supra* note 1, at Art. 10(6) and Convention on the Territorial Sea and the Contiguous Zone, *supra* note 2, at Art. 7(6). See generally, C. R. Symmons, *Historic Waters in the Law of the Sea: A Modern Re-appraisal* (Leiden: Martinus Nijhoff, 2008), at pp. 21–22.

5 *Case Concerning the Continental Shelf (Tunisia/Libya)*, [1982] *International Court of Justice Reports* 18, at para. 100 [hereinafter *Libya/Tunisia Case*].

waters...are governed by...customary international law."[6] Symmons, in his recent study, points out that the international legal rules regarding historic waters are "vague and difficult to apply" precisely because of the reliance that must be had on customary international law.[7] The 1962 U.N. Secretariat Study "Juridical Regime of Historic Waters, Including Historic Bays" notes that:

> [T]he concept of "historic waters" has its root in the historic fact that States through the ages claimed and maintained sovereignty over maritime areas which they considered vital to them without paying much attention to divergent and changing opinions about what general international law might prescribe....[8]

As international law of the sea evolved with 3-nautical mile territorial seas and then 12-nautical mile territorial sea, the historic waters concept was seen as being "necessary in order to maintain a State's title to some areas of water which might escape the codification formula..."[9]

Regarding what rights attach to historic waters, the International Court in the 1951 *Anglo-Norwegian Fisheries Case* stated:

> By "historic waters" are usually meant waters which are treated as internal waters but which would not have that character were it not for the existence of an historic title.[10]

The Court in the 1982 *Tunisia-Libya Continental Shelf Case* commented, however, that:

> international law...does not provide for a single "regime" for "historic waters" or "historic bays," but only for a particular regime for each of the concrete, recognized cases of "historic waters" or "historic bays."[11]

---

[6] *Ibid.*

[7] Symmons, *supra* note 4, at p. 286 and see pp. 285–287.

[8] "Juridical Regime of Historic Waters, Including Historic Bays," Study prepared by the U.N. Secretariat, U.N. Doc. A/CN.4/143, 9 March 1962, reprinted in *Yearbook of the International Law Commission*, Vol. II, 1962 (New York, 1964), at p. 7, para. 38.

[9] D. P. O'Connell, *The International Law of the Sea*, Vol. I (edited by I. A. Shearer) (Oxford: Clarendon Press, 1982), at p. 425, referring to the U.N. Secretariat Study, *supra* note 8.

[10] *Fisheries Case (United Kingdom v. Norway)*, [1951] *International Court of Justice Reports* 116, at p. 130.

[11] *Libya-Tunisia Case, supra* note 5, at para. 100. This was cited with approval in the *Gulf of Fonseca Case, supra* note 3, at para. 384.

In the 1992 *Gulf of Fonseca Case* before the Court, El Salvador, Honduras and Nicaragua agreed that the Gulf of Fonseca, on which all three States had coasts, was an historic bay and the Court found that a right of innocent passage existed noting that "rights of innocent passage are not inconsistent with a regime of historic waters..."[12] The 1962 U.N. Secretariat Study commented that "the scope of the historic title emerging from the continued exercise of sovereignty should not be wider in scope than the scope of the sovereignty actually exercised."[13] Thus, the legal effect of historic waters, for example on navigation, can depend on what was and is being asserted by the claiming State. Nevertheless, historic waters are usually associated with historic internal waters, with the claiming State asserting authority over foreign vessel navigation.

That the historic waters regime can be moulded to specific circumstances is part of its lure for some coastal States. The same flexibility makes the regime suspect by other States.

It is not the intention of this contribution to unravel the many legal intricacies of the historic waters regime. Rather the goal is to raise questions about the regime in the context of the Bay of Fundy, the water with the mighty tides pinched between Nova Scotia and New Brunswick. Canada and the United States have differing views of the international legal status of the Bay of Fundy. In 1973, Canada indicated that it claimed the Bay of Fundy as being "internal waters of Canada, on an historical basis."[14] The U.S. view on the status of the Bay of Fundy was restated in the early 1980s:

> Canada long has maintained an inchoate claim that the Bay of Fundy constitutes Canadian "historic" or "internal" waters. The United States always has reserved its position in this respect.[15]

---

[12] *Gulf of Fonseca Case, supra* note 3, at para. 393.

[13] U.N. Secretariat Study, *supra* note 8, at p. 23, para. 164.

[14] Canada, Department of External Affairs, Bureau of Legal Affairs, letter dated 17 December 1973, reprinted in (1974), 12 *Canadian Yearbook of International Law* 277, at p. 279. It is not to be understood that this 1973 letter is the first indication by Canada of a claim of the Bay of Fundy as historic internal waters. The 1973 letter, however, is a clear articulation of the claim.

[15] United States, "Reply of the United States," in *Pleadings, Oral Arguments, Documents, in the Case Concerning Delimitation of the Maritime Boundary in the Gulf of Maine (Canada/United States)*, Vol. 5, at p. 484, footnote 4.

The Bay of Fundy is one of several ocean areas that Canada asserts as being historic internal waters, with the United States not accepting the various claims. The most prominent of the bilateral historic waters disputes concerns the waters of Canada's Arctic archipelago, which includes the Northwest Passage.[16] The Bay of Fundy is somewhat unique, however, in that there has been a long history of interaction between Canada/Great Britain and the United States respecting the Bay.

Locating the Bay of Fundy

Canada has not identified the location of the Bay of Fundy for purposes of historic internal waters. The question is – where is the line at the mouth of the Bay of Fundy? La Forest notes the history of a possible line from Brier Island, Nova Scotia to the mouth of the St. Croix River in New Brunswick dating back to 1621.[17] Such a line is muddied by the uncertainty that existed until 1817 respecting the ownership of islands in the waters adjacent to the St. Croix River in Passamaquoddy Bay.[18] In the early 1900s, the maritime boundary between Canada and the United States within Passamaquoddy Bay and beyond the Bay into the Grand Manan Channel, between Maine and Grand Manan Island, was settled.[19] In 1963 La Forest wrote that "Canada should...solidify its

---

[16] See generally, D. Pharand, "The Arctic Waters and The Northwest Passage: A Final Revisit," *Ocean Development and International Law* 38 (2007): 3–69.

[17] G. V. La Forest, "Canadian Inland Waters of the Atlantic Provinces and the Bay of Fundy Incident," *Canadian Yearbook of International Law* 1 (1963): 149–171, at p. 164.

[18] See "Decision of the Commissioners under the Fourth Article of the Treaty of Ghent," 24 November 1817, reprinted in International Boundary Commission, *Joint Report Upon the Survey and Demarcation of the Boundary between the United States and Canada from the Source of the St. Croix River to the Atlantic Ocean* (Washington: U.S. Printing Office, 1934), Appendix II, "Treaties and Conventions Pertaining to the Boundary Previous to the Treaty of 1908," at p. 164.

[19] See *Treaty between Great Britain and the United States respecting the Demarcation of the International Boundary with Canada*, Washington, 11 April 1908, entered into force 4 June 1908, Article I, reprinted in C. Parry, ed., *The Consolidated Treaty Series*, Vol. 206 (Dobbs Ferry, N.Y.: Oceana Publications, 1969), at pp. 377–391; *United States-Great Britain Treaty Fixing the Boundary Line in Passamaquoddy Bay*, Washington, 21 May 1910, entered into force 20 August 1910, Article I, *U.S.T.S.* 551, reprinted in C. I. Bevans, *Treaties and Other International Agreements of the United States of America,*

claim [to the Bay of Fundy] by defining the exact location of the line closing the bay…"[20]

In 1971 Canada constructed a line across the entry of the Bay of Fundy in order to delineate Fishing Zone 2, "Bay of Fundy."[21] This line, referred to as a "fisheries closing line," extends from Whipple Point on Brier Island (Nova Scotia) to Gannet Rock, and then to Yellow Ledge, Machias Seal Island and North Rock, and along Grand Manan Island (New Brunswick) to the Canada-U.S. boundary in the Grand Manan Channel. The main and longest segment of the line, from Whipple Point to Gannet Rock, is 23.27-nautical miles and then the line moves to the east to Yellow Ledge and Machias Seal Island. Gannet Rock is barren but has a lighthouse that was manned from 1831 until 1996. The approximate distance between Grand Manan Island (New Brunswick) and Whipple Point (on Brier Island, Nova Scotia) is 32-nautical miles.

There are two important geographic features of the 1971 Bay of Fundy fisheries closing line: (i) it has as a terminus the end point of the Canada-U.S. maritime boundary in the Grand Manan Channel; and (ii) it utilizes as points Machias Seal Island and North Rock, both of these islets are subject to a sovereignty dispute between Canada and the United States.[22]

It was clear in 1971 that the fisheries closing lines[23] were not of the same legal character as a straight baseline in that the lines were not to

---

1776–1949, Vol. 12 (U.S. Department of State, 1974), at p. 341; and *United States-Canada Treaty regarding the Demarcation of the Boundary between the United States and Canada*, Washington, 24 February 1925, entered into force 17 July 1925, 43 *L.N.T.S.* 239, Art. III.

[20] La Forest, *supra* note 17, at p. 165.

[21] Canada, Fishing Zones of Canada (Zone 1, 2 and 3) Order, *Canada Gazette Part II*, Vol. 105, SOR/71–81, 25 February 1971, at pp. 363–364, came into effect on 10 March 1971, currently found in *Consolidated Regulations of Canada* (1978), Vol. XVIII, chapter 1547.

[22] See D. H. Gray, "Canada's Unresolved Maritime Boundaries," *Geomatica* 48(2) (1994): 131–144, at pp. 138–140.

[23] In addition to the Bay of Fundy, fisheries closing lines were used across the Cabot Strait and Strait of Belle Isle regarding the Gulf of St. Lawrence and on the west coast respecting Hecate Strait and Queen Charlotte Sound. See Fishing Zones of Canada (Zone 1, 2 and 3) Order, *supra* note 21.

affect the limits of Canada's internal waters and territorial seas.[24] The direct effect of the 1971 fisheries closing lines was to exclude foreign fishers from the enclosed waters. In 1971 the *Canada Shipping Act* was amended such that vessel-source pollution laws and regulations applied to foreign vessels within the newly created fishing zones.[25]

Respecting the relationship of the Bay of Fundy fisheries closing line to Canada's claim of the Bay of Fundy as historic waters, Johnson writes:

> External Affairs Minister Sharp claimed…that the closing lines were tied to historic Canadian rights…[26]

Perhaps more pertinently, Herman writes:

> [T]he waters of the Bay of Fundy that are considered to be part of Canada's internal waters [based on historic reasons] could be said to be all those waters bounded by a notional straight line drawn across the mouth of the Bay from Brier Island to Machias Seal Island.… [T]hese notional lines would presumably follow the so-called fishery closing lines promulgated in 1971…[27]

The fisheries closing line has the effect of including as part of Fishing Zone 2 Passamaquoddy Bay, which is framed by the coasts of New Brunswick and Maine at the mouth of the St. Croix River. As already noted, there is a maritime boundary between Canada and the United States within Passamaquoddy Bay.[28] The principal navigation route into the Maine coastal area of Passamaquoddy Bay, which includes the port of Eastport, is through Head Harbour Passage, which is Canadian waters, and the adjacent waters of the Bay of Fundy landward of the 1971 fisheries closing line.

---

[24] See L. H. J. Legault, "Maritime Claims," in R. St. J. Macdonald, G. L. Morris and D. M. Johnston, eds, *Canadian Perspectives on International Law and Organization* (Toronto: University of Toronto Press, 1974), at p. 384 and L. L. Herman, "Proof of Offshore Territorial Claims in Canada," *Dalhousie Law Journal* 7 (1982): 3–38, at p. 13.

[25] Canada, Amendment of the Canada Shipping Act, *Statutes of Canada 1970–71–72*, chapter 27, section 2. See Legault, *ibid.*, at p. 384.

[26] B. Johnson, "Canadian Foreign Policy and Fisheries," in B. Johnson and M. W. Zacher, eds, *Canadian Foreign Policy and the Law of the Sea* (Vancouver: University of British Columbia Press, 1977), at p. 68.

[27] Herman, *supra* note 24, at p. 7.

[28] See *supra* note 19.

Canada-U.S. Relations Regarding the Bay of Fundy

La Forest details the legislation and action of the governments of Great Britain, Canada, Nova Scotia, and New Brunswick respecting the Bay of Fundy and other Atlantic coastal bays up to 1900[29] and concludes that:

> [U]p to the turn of the century, Great Britain and her colonies had claimed and enforced exclusive jurisdiction over all bays in the Atlantic Provinces,...and had also passed specific statutes exercising jurisdiction over several of the more important bays. The United States, however, had constantly objected to this claim...[30]

La Forest cites British sources that "many seizures" of U.S. fishing vessels in the Bay of Fundy had taken place up to 1845.[31] Various treaties and other fisheries arrangements with the United States are explained by La Forest as being respectful of U.K. rights by being accommodating to the United States.[32]

The legal status of the Bay of Fundy was commented upon in the arbitral award concerning *The Washington* in 1853.[33] *The Washington* was a U.S. vessel seized 10-nautical miles from shore for alleged illegal fishing in the Bay of Fundy. The controversy was over the meaning of the word "bays" in the 1818 Anglo-American Convention.[34] The view of Great Britain was that the Bay of Fundy was "rightfully claimed by Great Britain as a bay;"[35] a position rejected by the United States. The decision fell to the Umpire Joshua Bates, who wrote:

---

[29] La Forest, *supra* note 17, at pp. 150–158.

[30] *Ibid.*, p. 159.

[31] *Ibid.*, p. 153.

[32] *Ibid.*, pp. 153–154.

[33] "*The Washington*" reprinted in J. B. Moore, *History and Digest of International Arbitrations to which the United States has been a Party*, Vol. IV (Washington, 1898), at p. 4342. The version of *The Washington Award* consulted for this contribution is an excerpt from the *Award* in J. B. Moore, *A Digest of International Law*, Vol. I (Washington, 1906), at pp. 785–787.

[34] *Convention of Commerce between Great Britain and the United States*, London, 20 October 1818, entered into force 30 January 1819, reprinted in C. Parry, ed., *The Consolidated Treaty Series*, Vol. 69 (Dobbs Ferry, N.Y.: Oceana Publications, 1969), at pp. 294–297.

[35] See O'Connell, *supra* note 9, at p. 357, footnote 99.

> It was urged on behalf of the British Government that by coasts, bays, etc., is understood an imaginary line, drawn along the coast from headland to headland, and that the jurisdiction of Her Majesty extends three marine miles outside the line; thus closing all the bays on the coast or shore, and that great body of water called the Bay of Fundy against Americans and others, making the latter a British bay.[36]

Ultimately, Bates did not accept this position and sided with the United States that the Bay of Fundy was not a bay within the meaning of the 1818 Treaty and, thus, not a British bay. Supporting this view, Bates made several points:

- the "headlands doctrine" proposed by the United Kingdom was new and had been limited in an 1839 French-British Treaty to only to those bays having mouths less than ten miles;
- the Bay of Fundy has several bays within it and was a "great body of water" with the proper analogy being with the Bay of Biscay and the Bay of Bengal "over which no nation can have the right to assume the sovereignty;"[37]
- "One of headlands of the Bay of Fundy is in the United States;"[38] and
- "ships bound to Passamaquoddy must sail through a large space of it [the Bay of Fundy]."[39]

The question of bays in the 1818 Treaty was part of the 1910 Great Britain – United States *North Atlantic Fisheries Arbitration*.[40] The Tribunal examined the meaning to be given bays in the 1818 Treaty indicating that:

> The interpretation must take into account all the individual characteristics which for any one of the different bays are to be appreciated, the relation of its width to the length of penetration inland, the possibility and necessity of its being defended by the State in whose territory it is indented; the

[36] "*The Washington*," *supra* note 33, at p. 786.

[37] *Ibid.*

[38] *Ibid.*

[39] *Ibid.* Note generally regarding *The Washington Award*: M. P. Strohl, *The International Law of Bays* (The Hague: Martinus Nijhoff, 1963), at pp. 283–284 and 380–383.

[40] *Award of the Tribunal of Arbitration in the Question Relating to the North Atlantic Coast Fisheries*, 7 September 1910, United Nations, Vol. XI, *Reports of International Arbitral Awards* 173.

> special value which it has for the industry of the inhabitants of its shores; the distance which it is secluded from the highways of nations on the open sea and other circumstances not possible to enumerate in general.[41]

The Tribunal declined to apply a mechanical approach (a mileage limit on the mouth of a bay) to the definition of bays[42] and as a result the Bay of Chaleur, Miramichi Bay and a number of other bays were seen as being subject to British control.[43] The dissenting Judge Luis Drago elaborated on the concept of historic bays and that the above-named bays fit into this class.[44] By agreement the Bay of Fundy was not considered by the Tribunal.[45] While the exclusion of the Bay of Fundy from the 1910 *North Atlantic Fisheries Arbitration* can be read as meaning that its status had been resolved, it appears that the reason for exclusion was one of pragmatism rather than one with legal significance.[46]

La Forest argues that the reasoning in *The Washington Award* regarding the Bay of Fundy "cannot…be accepted because it is clearly inconsistent with that of the court in the *North Atlantic Coast Fisheries Arbitration*."[47] He takes the view that the understanding of a bay within the 1818 Treaty set out in the 1910 *Arbitration* "is clearly applicable" to the Bay of Fundy and "that the decision in *The Washington* case was wrong, for the Bay of Fundy is obviously a bay" that fits the wording for a bay set out in the *Arbitration*.[48] La Forest argues that the Bay of Fundy is comparable to the Bay of Chaleur and Miramichi Bay, which were seen by the 1910 Arbitral Tribunal as being bays under British sovereignty.[49] La Forest also argues that the reasoning of *The Washington Award* respecting the headlands of the Bay of the Fundy not being in the same country "is one of fact" and

---

41 *Ibid.*, p. 199.

42 *Ibid.*, p. 196.

43 *Ibid.*, pp. 199–200. Note: Strohl, *supra* note 39, at pp. 284–286.

44 *Ibid.*, p. 206.

45 *Ibid.*, p. 200.

46 "[T]he British claimed that it was strictly for policy reasons that they had not pressed their arguments" vis-à-vis the Bay of Fundy. E. C. Foley, "Nova Scotia's Case for Coastal and Offshore Resources," *Ottawa Law Review* 13 (1981): 281–308, at p. 306 with sources cited therein at footnote 155.

47 La Forest, *supra* note 17, at p. 162.

48 *Ibid.*

49 *Ibid.*, pp. 163 and 161.

"is simply wrong."[50] He notes that the headlands of the Bay of Fundy are Brier Island and Grand Manan Island.[51] O'Connell supports this view referring to an United Kingdom Law Officer's Opinion of 1844 and noting that: "The true closing line of the Bay links Manan Island (sic) in New Brunswick with Whipple Point in Nova Scotia...."[52] Moreover, O'Connell notes that, as the Canada-United States ocean boundary is the median line between Grand Manan Island and Maine, "this takes the United States coastline out of the Bay altogether."[53]

La Forest also asserts "that Great Britain never accepted the correctness of *The Washington* decision."[54] O'Connell indicates the opposite noting that Great Britain "had given effect to the Award in the case of the *Washington*, and American fisherman had been permitted to treat the waters of the bay as high seas...."[55] Jessup notes, somewhat tentatively, that: "It seems that the British Government...acquiesced in Mr. Bate's decision."[56] Following *The Washington Award*, several bilateral fisheries treaties between Great Britain and the United States provided for U.S. fishing rights in waters adjacent to Canada.[57] It is unclear whether the United States felt compelled to enter into such arrangements, the British felt it was granting a special right to the United States or they were a pragmatic response to the reality of fishing activity.

---

50 *Ibid.*, p. 163.

51 *Ibid.*, p. 164. La Forest does not refer to Machias Seal Island, sovereignty over which is contested by Canada and the United States, as a headland.

52 O'Connell, *supra* note 9, at p. 357.

53 *Ibid.*, p. 357, footnote 98.

54 La Forest, *supra* note 17, at pp. 164–165.

55 O'Connell, *supra* note 9, at p. 362 and see also p. 357.

56 P. C. Jessup, *The Law of Territorial Waters and Maritime Jurisdiction* (New York: Jennings and Co., 1927, reprint by New York: Kraus Reprint, 1970), at p. 411.

57 See *Treaty regarding the North Atlantic Fisheries, Commercial Reciprocity, etc., between Great Britain and the United States*, Washington, 5 June 1854, entered into force 9 September 1854, Articles I and II, reprinted in C. Parry, ed., *The Consolidated Treaty Series*, Vol. 112 (Dobbs Ferry, N.Y.: Oceana Publications, 1969), at pp. 31–37 and *Treaty between Great Britain and the United States for the Amicable Settlement of All Causes of Difference between the Two Countries*, Washington, 8 May 1871, entered into force 17 June 1871, Articles XVIII and XIX, reprinted in C. Parry, ed., *The Consolidated Treaty Series*, Vol. 143 (Dobbs Ferry, N.Y.: Oceana Publications, 1969), at pp. 145–162. See generally, Canada, "Historical Introduction," in *Gulf of Maine, Pleadings, supra* note 15, Vol. 1, Annex to Memorial of the Canada, at pp. 188–190.

In a 1957 memorandum prepared by the U.S. Department of State, the Bay of Fundy was listed as one "of a number of bodies of water considered (or claimed) to be historic bays."[58] Specifically regarding the Bay of Fundy, the memorandum reiterates the findings of the 1853 *The Washington Award*.[59]

La Forest sets out the Canadian legislation up to the 1960s that applied to fishing activities within the Bay of Fundy and other east coast Canadian bays and the arrangements entered into with the United States.[60] Of particular note are the amendments made to the 1927 *Customs Act* in 1936[61] that contained a definition of "Canadian waters." Maps approved by orders-in-council were prepared that depicted Canadian waters.[62] The maps did not show the Bay of Fundy as being "Canadian waters;" rather what was shown were bays within the Bay of Fundy as being "Canadian waters."[63] La Forest argues that these maps, "though amounting to a claim of jurisdiction by the Canadian government, cannot be construed as a statement of Canada's full rights to Canadian waters."[64] In another work, the legislation, maps and the work in their preparation were referred to as "the first intensive and officially recognized determination of Canada's jurisdiction in the offshore."[65]

As the result of Russian vessels fishing in the Bay of Fundy in 1962, the Prime Minister of Canada indicated that the Bay of Fundy had always been considered part of Canada.[66] Legault writes that the Prime Minister "made clear... that this bay [Bay of Fundy] constituted Canadian internal

---

58 Memorandum prepared by Frank Boas, Office of the Legal Advisor, U.S. Department of State, August 1957, reproduced in M. M. Whiteman, ed., *Digest of International Law*, Vol. 4 (Washington, 1965), at p. 233.

59 *Ibid.*, p. 236.

60 La Forest, *supra* note 17, at pp. 165–169.

61 Canada, *Act to Amend the Customs Act*, S.C. 1936, c. 30, section 2, amending the *Customs Act*, R.S.C. 1927, c. 42.

62 See B. Calderbank, A. M. MacLeod, T. L. McDorman and D. H. Gray, *Canada's Offshore: Jurisdiction, Rights, and Management* (Victoria: Trafford Publishing, 3rd edition, 2006), at pp. 103–105 for a discussion of the maps and the work of the Interdepartmental Committee on Territorial Waters. Note also: La Forest, *supra* note 17, at p. 167.

63 La Forest, *supra* note 17, at p. 167.

64 *Ibid.*, p. 168.

65 Calderbank, et al., *supra* note 62, at p. 105.

66 Canada, John Diefenbaker, Prime Minister, House of Commons, *Debates*, 15 November 1962, at p. 1650.

waters."[67] La Forest notes that it was reported that the Russians agreed to respect Canada's claim.[68] There is no indication of whether the United States was aware of or, if aware, reacted to the purported Canadian statement respecting the Bay of Fundy.

As noted above, in 1971 Canada established a fisheries closing line across the mouth of the Bay of Fundy. The U.S. State Department issued a statement specifically on the fisheries closing lines which referred to the Canadian measure as a "unilateral act...totally without foundation in international law."[69]

In 1977 both Canada and the United States adopted 200-nautical mile exclusive national fishing zones.[70] Thus, the long-running focus on Canadian assertions of fishing (and other) authority in the Bay of Fundy and U.S. fishing activities in the Bay as demonstrating or not demonstrating the existence of historic waters changed.

In the 1970s, there was a proposal considered within the United States to construct an oil refinery near Eastport, Maine, that would have necessitated oil tanker traffic through the Canadian waters of the Bay of Fundy and Passamaquoddy Bay.[71] Canada informed the United States that the

[67] Legault, *supra* note 24, at p. 383.

[68] La Forest, *supra* note 17, at p. 150.

[69] United States, "Statement on Canadian Fisheries Closing Lines Announcement," 18 December 1970, reprinted in S. H. Lay, R. Churchill and M. Nordquist, eds., *New Directions in the Law of the Sea*, Vol. I (Dobbs Ferry, N.Y.: Oceana Publications, 1973), at p. 74, at paragraph 2.

[70] United States, *Fishery Conservation and Management Act* (renamed the *Magnuson Fishery Conservation and Management Act*), 90 *U.S. Stat.* 331, *Public Law* 94–265, codified at 16 *U.S.C.* 1801–1882.

Canada, Fishing Zones of Canada (Zones 4 and 5) Order, *Canada Gazette Part II*, Vol. 111, SOR/77–62, 1 January 1977, at pp. 115–120 replaced by Fishing Zones of Canada (Zones 4 and 5) Order, *Consolidated Regulations of Canada* (1978), Vol. XVIII, chapter 1548, at pp. 13741–13746 and Fishing Zones of Canada (Zone 6) Order, *Canada Gazette Part II*, Vol. 111, SOR/77–173, 24 February 1977, at pp. 652–655, replaced by Fishing Zones of Canada (Zone 6) Order, *Consolidated Regulations of Canada* (1978), chapter 1549, at pp. 13747–13750. Zones 4 and 5 are on the east and west coasts and zone 6 is the Arctic.

[71] See generally, J. E. Carroll, *Environmental Diplomacy: An Examination and a Prospective of Canadian-U.S. Transboundary Environmental Relations* (Ann Arbor: University of Michigan Press, 1983), at pp. 62–72; A. L. Springer, *The International Law of Pollution: Protecting the Global Environment in a World of Sovereign States* (Westport, Connecticut: Quorum Books, 1983) at pp. 185–200; and J. Spiller and J. Roanowicz,

proposed tanker traffic created an "unacceptable risk in the transport of a large volume of pollutants" through Head Harbour Passage[72] and indicated that Canada was "prepared to regulate, if necessary, to prevent it."[73] The United States responded claiming that a right of innocent passage existed through Head Harbour Passage.[74] In 1982, Canada adopted a regulation which restricted the amount of oil that could be carried by an oil tanker through Head Harbour Passage.[75] The regulation was crafted to deal only with the proposed tanker traffic and not that of the oil needs of the Eastport population.[76] The United States took the view that the regulation exceeded Canada's international jurisdictional authority.[77] In 1983, the refinery proponent opted not to proceed with the proposed Eastport refinery. More recently, the United States has expressed the view that Head Harbour Passage is a strait used for international navigation through which a right of innocent passage exists.[78]

"Effectiveness of Canadian and U.S. Environmental Assessment Processes in Evaluating Transboundary Impacts," in A. Reiser, J. Spiller and D. VanderZwaag, eds., *Environmental Decisionmaking in a Transboundary Region* (New York: Springer-Verlag, 1986), at pp. 116–131.

72 Canada, Canadian Embassy, "Note No. 206 to the United States," 7 June 1973, at paragraph 7, copy on file with the author.

73 Canada, Department of External Affairs, "Letter to Edmund Muskie," 7 November 1979, at paragraph 2, copy on file with the author.

74 United States, "Aide-Memoire," 12 March 1975, as referenced in United States, Department of State, *Digest of United States Practice in International Law, 1975* (Washington, 1976), at p. 432.

75 Canada, Oil Carriage Limitation Regulations, *Canada Gazette Part II*, Vol. 116, SOR/82–244, 11 February 1982, at pp. 892–893. The Regulation was revoked in May 1987 on the grounds that it was *ultra vires* the enabling legislation. Note was made that there was no longer need for the Regulation since the plans to build the facility in Eastport had been abandoned. See Canada, Oil Carriage Limitation Regulations, revocation, *Canada Gazette Part II*, Vol. 121, SOR/87–268, 7 May 1987, at pp. 2077–2078.

76 Carroll, *supra* note 71, at pp. 70–71.

77 See *ibid.*, at p. 71 and "U.S. Irked by Limits on Shipping Passage," *Montreal Gazette*, 20 February 1982.

78 United States, "Commentary – The 1982 United Nations Convention on the Law of the Sea and the Agreement on Implementation of Part XI" attached to "Message from the President of the United States transmitting United Nations Convention on the Law of the Sea, with Annexes, done at Montego Bay, December 10, 1982 (the Convention) and the Agreement Relating to the Implementation of Part XI of the United Nations Convention on the Law of the Sea of 10 December 1982, adopted at New York, July

In the 1984 Canada-United States *Gulf of Maine Case*[79] before a chamber of the International Court of Justice, the Bay of Fundy was discussed in the context of the effect the length of the coasts within the Bay may have on the maritime boundary in the middle of the Gulf of Maine. As already noted, in submissions in the *Case*, the United States made it clear that it "reserved its position" respecting the Canadian claim that the Bay of Fundy was historic internal waters.[80] Judge Schwebel (the U.S. Judge) in his separate opinion, quoted from the argument of Canadian counsel that: "Canada maintains for historical reasons its right to treat the waters of the Bay of Fundy as internal waters."[81] The legal status of the Bay of Fundy, however, was not commented upon by the majority, in the Separate Opinion of Judge Schwebel or the Dissenting Opinion of Judge Gros.

The relevance of the Bay of Fundy was in determining the length of coastline of Canada that was to be compared with the length of coastline of the United States. Canada took the view that the coasts of New Brunswick and Nova Scotia within the Bay of Fundy should be included as part of the relevant coastline of Canada. The United States opposed this, taking the view that a straight line from the terminus of the Canada-United States maritime boundary in the Grand Manan Channel to Cape Sable (off the coast of Shelburne County, Nova Scotia) was the relevant coastline of Canada regarding the Bay of Fundy.[82] The Court held that the relevant coastline of Canada included large areas of the coasts of both Nova Scotia and New Brunswick within the Bay of Fundy,[83] without referencing any closing line across the mouth of the Bay of Fundy. Schwebel, in his Separate Opinion, would have drawn a line from Saint John, New Brunswick, to Brier Island, Nova Scotia, as being the line across the entry of the Bay of Fundy for the purposes of coastal length

28, 1994 (the Agreement) and signed by the United States, subject to ratification, on July 29, 1994," Senate, Treaty Document 39, 103d Congress, 2d Session IV (1994), at p. 19, reprinted in (1995), 34 *I.L.M.* 1393–1447.

[79] *Case Concerning Delimitation of the Maritime Boundary in the Gulf of Maine (Canada/United States)*, [1984] *I.C.J. Reports* 246 [hereinafter *Gulf of Maine Case*].

[80] United States, "Reply of the United States," *supra* note 15, at p. 484, footnote 4.

[81] *Gulf of Maine Case, supra* note 79, at p. 354.

[82] United States, "Reply of the United States," *supra* note 15, at p. 484, para. 265.

[83] *Gulf of Maine Case, supra* note 79, at p. 336, para. 221.

calculation.[84] Despite extensive argument regarding the Bay of Fundy, the *Gulf of Maine Case* has little to say on the legal status of the Bay, except that the two States disagree.

Of note is a listing made by Roach and Smith, of the U.S. State Department, in 1996 of claims made to historic bays that have been protested by the United States where the Bay of Fundy is not mentioned.[85] Roach and Smith do, however, note the disagreement over the legal status of Head Harbour Passage within Passamaquoddy Bay.[86]

In December 2006, proposals were presented to the United States Federal Energy Regulatory Agency (FERC) to construct liquefied natural gas (LNG) receiving terminals, pipelines, and associated facilities near Eastport, Maine. These proposals are premised on the use of tankers to transport LNG through Head Harbour Passage and the adjacent areas of the Bay of Fundy. In February 2007, the Canadian Ambassador to the United States informed the chair of FERC of Canada's view that the potential passage of LNG tankers through "environmentally-sensitive and navigationally-challenging marine and coastal areas of the sovereign Canadian waters of Head Harbour Passage" posed an unacceptable risk and that Canada was "prepared to use domestic legal means to... prevent such passage from occurring."[87]

The Bay of Fundy as a Juridical Bay

Article 10 of the LOS Convention provides for the construction of baselines across the mouths of certain bays with the result being that the waters

---

[84] *Ibid.*, p. 355.

[85] J. A. Roach and R. W. Smith, *United States Responses to Excessive Maritime Claims* (The Hague: Martinus Nijhoff, 2nd ed., 1996), at pp. 33–34. Note can also be made of the listing "Claimed and Potential Historic Bays," in L. M. Alexander, *Navigational Restrictions within the New LOS Context: Geographical Implications for the United States* (Rhode Island: Offshore Consultants, 1986), at p. 89, where again the Bay of Fundy is not noted.

[86] Roach and Smith, *ibid.*, p. 290.

[87] Canada, Canadian Embassy, "Letter from Ambassador Michael Wilson to Joseph T. Kelliher," 14 February 2007, reprinted in *Saint John Telegraph Journal*, 15 February 2007, at p. A8.

landward of the baselines are internal waters.[88] The maximum length of the baselines "between the low-water marks of the natural entrance points of a bay" is to be 24-nautical miles.[89] Where, because of islands, a bay has more than one mouth, it is the total length of the lines across the mouths that is to be measured.[90] These provisions are the same as those in Article 7 of the 1958 Territorial Sea and Contiguous Zone Convention.

Canada was not a party to the 1958 Territorial Sea and Contiguous Zone Convention when the fisheries closing line was adopted for the Bay of Fundy in 1971. Canada has made no public comment regarding the international legal status of the Bay of Fundy since its 1973 statement on historic waters. Thus, Canada has not publicly relied on Article 10 of the 1982 LOS Convention, to which Canada became a party in 2003, respecting the Bay of Fundy and appears to continue to assert that the Bay of Fundy is historic internal waters.

As noted above, the length of Canada's fisheries line across the mouth of the Bay of Fundy from Nova Scotia to Grand Manan Island exceeds the 24-nautical mile limit applicable to juridical bays. However, other baselines, if constructed by Canada, may lead to all or part of the Bay of Fundy being considered a juridical bay.

Another wrinkle on the possibility of the Bay of Fundy, as defined by the fisheries closing line, being a juridical bay pursuant to the 1982 LOS Convention is that Article 10 "only applies to bays the coasts of which belong to a single State." While there is little question that Nova Scotia and New Brunswick form the primary coasts for the Bay of Fundy, the United States presence within Passamaquoddy Bay, enclosed by the Canadian fisheries line, may matter.

The Bay of Fundy as Shared Waters

As noted above, in the 1853 *The Washington Arbitral Award*, the Umpire supported his view that the Bay of Fundy was not a bay within the meaning of the 1818 Anglo-American Fisheries Convention noting that: "One of the headlands of the Bay of Fundy is in the United States, and ships

[88] LOS Convention, *supra* note 1, Art. 10(4)

[89] *Ibid.*

[90] R. R. Churchill and A. V. Lowe, *The Law of the Sea* (Manchester: Manchester University Press, 3rd edition, 1999), at p. 42.

bound to Passamaquoddy must sail through a large space of it."[91] La Forest has argued that the comments in *The Washington Award* respecting the headlands of the Bay of the Fundy not being in the same country "is one of fact" and "is simply wrong."[92] He notes that the headlands of the Bay of Fundy are Brier Island, Nova Scotia, and Grand Manan Island, New Brunswick.[93]

While the principal headlands at the mouth of the Bay of Fundy are both Canadian (Grand Manan Island (or possible Machias Seal Island)[94] and Brier Island), there is a secondary entrance to the Bay of Fundy through Grand Manan Channel and this is clearly a shared waterway between Maine and New Brunswick, albeit for only a short distance until the maritime boundary is reached. A line fully enclosing the Bay of Fundy would need to go from Nova Scotia to Grand Manan Island (or Machias Seal Island) and from Grand Manan Island (or Machias Seal Island) to the Canada-United States maritime boundary. The 1971 Canadian fisheries closing line extended to the Grand Manan Channel and thus closed off the secondary entrance to the Bay of Fundy. Arguably, as a matter of geography, the Bay of Fundy may be shared since the closing line (if not the headlands) directly engages the United States.[95]

Even if only the principal headlands (or the "low-water marks of the natural entrance points")[96] were considered to be sufficient to close the Bay of Fundy as historic waters, Passamaquoddy Bay may be a shared bay. The current Canadian definition of the Bay of Fundy, albeit primarily for fishing purposes, is that found in the 1971 regulations and the Bay of Fundy includes Passamaquoddy Bay. Bouchez refers to two categories of "bays enclosed by the territory of more than one State" as being: (i) where two States each have a headland; and (ii) where the entry of a bay is within the authority of one State, but the other State borders an inner

---

91 "*The Washington*," *supra* note 33, at p. 786.

92 La Forest, *supra* note 17, at p. 163.

93 *Ibid.*, p. 164. La Forest does not refer to Machias Seal Island, sovereignty over which is contested by Canada and the United States, as a headland.

94 This was suggested by Herman, *supra* note 24, at p. 7.

95 This situation is unchanged even if Machias Seal Island (should it be determined to be Canadian as opposed to American) is used as the headland rather than Grand Manan Island as proposed by Herman, *supra* note 24, at p. 7.

96 LOS Convention, *supra* note 1, Art. 10(5) and Convention on the Territorial Sea and the Contiguous Zone, *supra* note 2, Art. 7(5).

part of the bay.[97] Based on the second category, it can be argued that it is the claimed waters as a whole that needs to be looked at as being shared and not just the headlands or "natural entrance points."

Beyond the geography that may define the waters as shared, there may be a functional consideration respecting whether a bay or water area is shared. In the 1853 *The Washington Award*, the Umpire noted that "ships bound to Passamaquoddy must sail through it [the Bay of Fundy]."[98] This is the normal routeing for vessels to Eastport, Maine, south of Grand Manan Island and through Head Harbour Passage. There is no other way for commercial and government vessels to access Eastport, Maine, and the U.S. coastal area within Passamaquoddy Bay.

In the context of historic waters considerations, the 1962 U.N. Secretariat Study[99] makes no specific reference to the role of functional considerations in determining whether bays or waters are shared. Blum, however, places considerable weight on functionality.

> [I]t becomes evident that one of the major considerations which permit a given bay to be turned into an historic bay is the fact that by its incorporation into the national domain of the littoral State no harm is done, or is likely to be done, to another State.... It is, however, abundantly clear that these considerations do not hold good in those cases where the shores of a bay are possessed by more than one littoral State. In such cases it can hardly be maintained that the bay "is not and cannot be made a roadway from one nation to another". This consideration, which seems to be implicit in all those cases where multinational bays are involved, was given explicit recognition [in *The Washington Award*].[100]

Assessing an Historic Waters Claim

The 1962 U.N. Secretariat Study outlines three criteria for assessing historic waters:

> In determining whether or not a title to "historic waters" exists, there are three factors which have to be taken into consideration, namely,

---

[97] L. J. Bouchez, *The Regime of Bays in International Law* (Leyden: A. Y. Sythoff, 1964), at p. 116.

[98] "*The Washington,*" *supra* note 33, at p. 786.

[99] U.N. Secretariat Study, *supra* note 8.

[100] Y. Z. Blum, *Historic Titles in International Law* (The Hague: Martinus Nijhoff, 1965), at p. 270. See also, Bouchez, *supra* note 97, at pp. 174–175 and 184.

(i) the authority exercised over the area by the State claiming it as "historic waters";
(ii) the continuity of such exercise of authority;
(iii) the attitude of foreign States.

First, effective exercise of sovereignty over the area by the claiming State is a necessary requirement for title to the area as "historic waters" of that State. Secondly, such exercise of sovereignty must have continued during a considerable time so as to have developed into a usage. Thirdly, the attitude of foreign States to the activities of the claiming State in the area must have been such that it can be characterized as an attitude of general toleration.

The burden of proof of title to "historic waters" is on the State claiming such title, in the sense that, if the State is unable to prove to the satisfaction of whoever has to decide the matter that the requirements necessary for the title have been fulfilled, its claim to the title will be disallowed.[101]

The U.N. Secretariat Study indicates that "there seems to be fairly general agreement" on the three criteria[102] and that in the 1951 *Anglo-Norwegian Fisheries Case* the two States "agreed that the onus of proof was on the State claiming a historic title."[103] While there has been no international adjudication that has pronounced definitively on the applicability of the above considerations for evaluating a claim of historic internal waters, Churchill and Lowe assert that the three criteria "were implicitly accepted by the International Court" in the *Gulf of Fonseca Case*.[104] The factors have been used in numerous U.S. domestic cases on historic waters claims raised by states.[105]

While the factors for consideration in an historic waters claim can be easily stated, the detailed application in any given situation is fraught with uncertainties.[106] As regards the Bay of Fundy, specific comment needs to be made on the third criteria "the attitude of foreign states."[107]

---

[101] U.N. Secretariat Study, *supra* note 8, at p. 25, paras. 185, 186 and 188.

[102] *Ibid.*, at p. 13, para. 80.

[103] *Ibid.*, at p. 21, para. 152.

[104] Churchill and Lowe, *supra* note 90, at p. 44.

[105] See, for example, *United States v. Louisiana*, (1968), 394 *U.S.* 11; *United States v. Alaska*, (1975), 422 *U.S.* 184; *United States v. Louisiana (Mississippi/Alabama Boundary)*, (1985), 470 *U.S.* 93; and *Alaska v. United States*, (2005), 545 *U.S.* 75.

[106] See generally, Symmons, *supra* note 4, at pp. 111–116.

[107] See generally, Symmons, *supra* note 4, at pp. 213–245.

Not surprisingly, the attitude of foreign States is of particular importance where a historic waters claim exists respecting shared waters. The 1962 U.N. Secretariat Study notes:

> the persistent opposition by one or more of the riparian States to the exercise of sovereignty over the bay by one or more of the other riparian States must naturally be of great if not decisive importance in evaluating whether or not the requirement of toleration had been fulfilled.[108]

At a later stage the Study notes: "it would seem reasonable, in the case of a State (or States) claiming historic title to waters bordered by two or more States, to accord special importance to the attitude of the other riparian State (or States)."[109] Academic writers take the view that there is a strong presumption against a shared waterway being historic waters.[110] In the *Gulf of Fonseca Case*, the International Court accepted the claim by the three States (El Salvador, Honduras and Nicaragua) that the shared waters were historic waters.[111] This suggests the existence of a threshold of explicit assent by the bordering States that shared waters are historic waters.

The more usual situation in assessing foreign State attitudes to a historic waters claim is where there has been a claim to historic waters adjacent to a single State. There has been debate regarding whether it is necessary for States with a direct interest in the claimed historic waters to consent or indicate recognition of the claim or whether it is sufficient for these same States to remain silent and therefore acquiesce to claim.[112] As the 1962 U.N. Secretariat Study notes, "there is substantial agreement that

---

[108] U.N. Secretariat Study, *supra* note 8, at p. 25, paras. 185, 186 and 188.

[109] *Ibid.*, p. 25, para. 186.

[110] See Churchill and Lowe, *supra* note 90, at p. 45 and Blum, *supra* note 100, at p. 270.
The U.S. Department of State Geographer has commented that "at a minimum, the states bordering the bay must all agree that the bay is an 'historic bay'" and noted that Vietnam's claim that the Gulf of Tonkin was historic waters was questionable since China, which borders the Gulf, does not claim the area as historic waters. United States, Department of State, Bureau of Intelligence and Research, *Limits in the Seas: Straight Baselines – Vietnam* (No. 99), (Washington, 12 December 1983), at pp. 9–10.

[111] *Gulf of Fonseca Case, supra* note 3, at para. 394.

[112] See D. Pharand, *Canada's Arctic Waters in International Law* (Cambridge: Cambridge University Press, 1988), at p. 99, and see also, U.N. Secretariat Study, *supra* note 8, at p. 16, paras. 106–110.

inaction on the part of foreign States is sufficient to permit an historic title to a maritime area to arise by effective and continued exercise of sovereignty over it by the coastal State during a considerable period of time..."[113] As a corollary of this, it is accepted that "effective" protests on the part of interested States of an historic waters claim may negate acquiescence and defeat the claim.[114] Less clear is the number of such protests that are necessary and whether a single protest can undermine an historic waters claim.[115] O'Connell writes that "virtually total toleration of those nations whose interests are clearly affected" by an historic waters claim is required.[116] The 1962 U.N. Secretariat Study is less categorical commenting that "how wide-spread the opposition must be to preclude the emergence of historic title... is a matter of appreciation in light of the circumstances in each case."[117]

Conclusion

The historic waters regime is an orphan regime of the law of the sea. Both its content and the manner of successfully making a claim are within the realm of customary international and State practice. Although texts of draft provisions on historic waters were discussed during the negotiation of the 1982 LOS Convention, according to the U.S. delegation: "There was strong opposition to a general doctrine of historic waters..."[118] No wording was adopted.

---

[113] U.N. Secretariat Study, *supra* note 8, at p. 17, para. 112.

[114] Pharand, *supra* note 112, at p. 105 and O'Connell, *supra* note 9, at p. 39.

In the historic waters context, both Bouchez, *supra* note 97, at pp. 272–273 and Pharand, *supra* note 112, at pp. 100–102, place significant emphasis on the necessity of "effective protest" for which not just a paper protest but "more forceful steps" or "conduct in accordance with the protest" is necessary. The U.N. Secretariat Study, *supra* note 8, at p. 17, paras. 113–115 does not go this far.

[115] See Symmons, *supra* note 4, at pp. 236–241.

[116] O'Connell, *supra* note 9, at p. 39.

[117] U.N. Secretariat Study, *supra* note 8, at p. 18, para. 130 and see pp. 17–18, paras. 116–120.

[118] United States, "Delegation Report of Third Session, 1975," in *Reports of the United States Delegation to the Third United Nations Conference on the Law of the Sea* (Honolulu: Law of the Sea Institute, 1983), at p. 94. More generally on the historic waters discussions during the negotiation of the LOS Convention, see L. F. E. Goldie, "Historic Bays

The rich history of Canadian-U.S. interactions respecting the Bay of Fundy, and more specifically in recent years, Head Harbour Passage within Passamaquoddy Bay, provides an opportunity for analysis and differing views on the international legal status of the Bay as historic waters. The sharing of Passamaquoddy Bay and the navigational necessity and long-time practice of commercial and U.S. government vessels accessing the port of Eastport by traversing Canadian waters adds an important complexity to the juridical bay and the historic waters regime analysis.

The dispute settlement regime of the LOS Convention explicitly recognizes the orphan status of historic waters by providing State parties with the opportunity to exclude from compulsory settlement "disputes... involving historic bays or titles."[119] When Canada ratified the LOS Convention in 2003, it exercised its option to use this provision as a shield against potential contrary claims.[120] It appears that the United States, when it becomes a party to the Convention, will also take advantage of the shield as regards historic waters.[121] Thus, even if the United States becomes a party to the LOS Convention, the international legal dispute regarding the Bay of Fundy is unlikely to be resolved through adjudication pursuant to the LOS Convention.

---

in International Law – An Impressionistic Overview," *Syracuse Journal of International Law and Commerce* 11 (1984): 211–273, at pp. 265–267 and S. Rosenne, "Historic Waters in the Third United Nations Conference on the Law of the Sea," in T. D. Gill and W. P. Heere, eds., *Reflections on Principles and Practice of International Law* (The Hague: Martinus Nijhoff, 2000), at pp. 195–197. Note also: F. Francioni, "The Status of the Gulf of Sirte in International Law," *Syracuse Journal of International Law and Commerce* 11 (1984): 311–326, at p. 321 who notes that the proposals were "quickly discarded."

119 LOS Convention, *supra* note 1, Art. 298(1)(a)(i). See generally, Rosenne, *supra* note 118, at pp. 197–200.

120 Canada, "Declaration Made Upon Ratification," 7 November 2003, reprinted in United Nations, *Law of the Sea Bulletin*, No. 53 (New York, 2004), at p. 15.

121 In December 2007, the U.S. Senate Foreign Relations Committee voted to recommend U.S. accession to the LOS Convention and reported this to the full Senate for consideration. Attached to the Advice and Consent Resolution before the Senate was a declaration (section 2) outlining the exemptions that were proposed to be attached by the United States to its instrument of accession. United States, Senate, Executive Report 110–9, Part VIII – Text of Resolution of Advice and Consent to Ratification, at pp. 19–24 (2007).

The Legacy of the Dalhousie University Ocean Studies Programme (DOSP) in the Caribbean

Carlyle L. Mitchell*

Introduction

This essay addresses the contribution of the Dalhousie University Ocean Studies Programme (DOSP)[1] to fisheries and ocean management in the Caribbean. DOSP was unique in its focus on ocean management and on fostering a multidisciplinary approach. Douglas Johnston, Edgar Gold, and others championed this approach in anticipation of the entry into force of the United Nations Convention of the Law of the Sea (LOS Convention) and its impact on ocean management problems of developing countries, in particular, the management of their exclusive economic zones (EEZs). The small islands in the Eastern Caribbean, now Member States of the Organisation of Eastern Caribbean States (OECS), presented a special case, and DOSP conducted seminal work with these islands in the early 1980s. This work impacted heavily on future developments, not only in the Eastern Caribbean but in the entire Caribbean Sea. This essay traces the DOSP influence and, in some cases, lack of it, on initiatives by the Caribbean Community (CARICOM) and the Association of Caribbean

---

* Adjunct Professor, Telfer School of Management, University of Ottawa, Canada.

[1] DOSP was terminated in 1988 when it was combined with the Canadian Maritime Transportation Centre (CMTC) to become the International Institute for Transportation and Ocean Policy Studies (IITOPS). After a few years, IITOPS changed into the Oceans Institute of Canada (OIC). This merged with Elizabeth Mann Borgese's Dalhousie University–based International Ocean Institute (IOC) and became the International Ocean Institute of Canada (IOIC). Much of the traditional DOSP/ITOPS/OIC work has reverted back to the Dalhousie Law School under its Marine and Environmental Law Institute established in 2002/03. IOI has provided annual training courses in ocean management to students from many countries, including the Caribbean. To this day, ex-DOSP personnel continue to be involved with developments in the marine area in the Caribbean, mainly under the auspices of CIDA programmes and projects.

States (ACS) and to current initiatives to establish a management regime for this sea.

DOSP's interest in the Caribbean Sea sprang from the problems of maritime boundary delimitation that would emanate from the Third United Nations Conference on the Law of the Sea (UNCLOS III). Because of the many island States in the Caribbean Sea and their proximity to each other, it was recognized that delimitation would be both difficult and costly, and that these States might have difficulties managing their EEZs. The LOS Convention would provide sovereign rights over extended marine space for these islands, but would also impose more responsibilities on States. With funding from the Canadian International Development Agency (CIDA), DOSP initiated studies for some of the States in the Eastern Caribbean, namely all members of the Organization of Eastern Caribbean States (OECS), dealing with delimitation and ocean management issues. These constituted seminal work for the entire Commonwealth Caribbean Region.

The DOSP studies in the 1980s left a legacy that has implications for recent developments for the Caribbean Sea, such as the international recognition in 2006 of the Caribbean Sea as a special area for sustainable development and the greater economic integration of Caribbean and Central American States. These developments have resulted in initiatives to establish a fisheries and oceans management regime for the Caribbean Sea and have highlighted the importance of maritime boundary delimitation in the Caribbean. These developments can be traced back to the DOSP studies and to the participation of many ex-DOSP personnel (DOSPIES) in their strategy and formulation. This study therefore examines the DOSP legacy in the Caribbean and covers the following aspects: 1) the DOSP seminal studies in the Eastern Caribbean; 2) DOSP's influence on CIDA's involvement with regional initiatives in the Caribbean; and 3) the implications of DOSP initiatives for the current management regime of the Caribbean Sea.

Seminal DOSP Studies in the Eastern Caribbean

During 1980/81, DOSP carried out its first study in the Windward Islands on a sub-contractual basis to the Foundation Reshaping the International Order (RIO) of the Netherlands, but funded by CIDA. In 1981, a draft of the study, *The Integration of Marine Space in the National Development Strategies of Small Island States: The Case of the Caribbean States of Grenada and St. Lucia,*[2] was circulated to governments in the Windward and Leeward islands and then presented at a regional seminar in Kingstown under the auspices of the West Indies Associated States Council of Ministers, the forerunner of the OECS. The proceedings from this seminar resulted in the report entitled, *Problems of Development and Ocean Management in the Eastern Caribbean: Economic, Legal, Environmental, and Planning Aspects.*[3]

Other studies and one seminar in the Leeward Islands followed during the next three years. A Leeward draft study, *Development and Ocean Management in the Eastern Caribbean: the Case of the Leeward Islands*, was submitted to delegates at a regional seminar held in Basterre, St. Kitts, in June 1983 and was published after the seminar in 1984.[4] The following year, DOSP published the *Management and Utilization of the Marine Resources of the British Virgin Islands.*[5] In effect, the research and studies covered all the present Member States of the OECS, since the BVI was not a Member State in 1985.

The DOSP studies were unique in that they were 1) multidisciplinary in nature, combining legal, environmental and economic disciplines, 2) concentrated on a new field, i.e., ocean management, 3) provided an economic model of ocean sector development (the Dalhousie model), and 4) identified the requirements and strategy for delimitation and EEZ management. The main findings of the studies and the two OECS regional seminars are:

---

2 Dalhousie Ocean Studies Programme (DOSP), *The Integration of Marine Space in the National Development Strategies of Small Island States: The Case of the Caribbean States of Grenada and St. Lucia* (Halifax: DOSP, Dalhousie University, 1981).

3 E. Gold, N. G. Letalik and C. L. Mitchell, *Problems of Development and Ocean Management in the Eastern Caribbean: Economic, Legal, Environmental and Planning Aspects* (Halifax: Dalhousie Ocean Studies Programme, Dalhousie University, 1981).

4 Dalhousie Ocean Studies Programme (DOSP), *Development and Ocean Management in the Eastern Caribbean: the Case of the Leeward Islands* (Halifax: DOSP, Dalhousie University, 1984).

5 Dalhousie Ocean Studies Programme, *Management and Utilization of the Marine Resources of the British Virgin Islands* (Halifax: DOSP, Dalhousie University, 1985).

1) The studies provided EEZ boundaries for the OECS States, based on emerging principles, criteria and methods for delimitation, namely the method of equidistance from baselines, even though there have been few formal boundary delimitation agreements in the OECS region since 1983. The studies identified problems stemming from delimitation of islands that were so close to each other that 200 nautical mile EEZ claims were impossible, except for the eastern maritime zones in the Atlantic Ocean. Some of the States had potential north-south boundaries that were only a few nautical miles apart, e.g., Grenada, and St. Vincent and the Grenadines. However, DOSP suggested that internal boundaries between OECS States should not be problematic. It warned that the greatest threat was external, such as that posed by Venezuela. Venezuela's claim to Aves Island (Bird Rock in the English-speaking Caribbean) could reduce the EEZs of OECS Member States in the Caribbean Sea significantly if it were given the status of an island. This would entitle Venezuela to a 200 nautical mile EEZ instead of considering it a rock, which would entitle it to a 12 nautical mile territorial sea only.[6]
2) The studies highlighted the economic importance of the ocean sector in the OECS economies.[7] The DOSP model of the ocean sector included fishing, shipping, and tourism industries. The inclusion of tourism as an ocean industry and the concern for the protection of the marine environment were significant in that most States did not view the ocean as an important economic sector. Coastal areas such as swamps were being filled in for developmental purposes, and the fishing industries were neglected or poorly managed by governments. The DOSP-model findings indicated that the ocean sector constituted a major sector in OECS Member State economies and that their EEZs could provide a new frontier for development.[8]

---

[6] *United Nations Convention on the Law of the Sea*, Montego Bay, 10 December 1982, UN Doc. A/Conf.62/122 (1982); 21 *I.L.M.* 1261 (1982), Art. 121, para. 3.

[7] See, for example, C. L. Mitchell, "Ocean Development and Marine Resources in the Eastern Caribbean," in E. Gold., ed., *A New Law of the Sea for the Caribbean: An Examination of Marine Law and Policy Issues in the Lesser Antilles* (New York: Springer-Verlag, 1988), pp. 54–82.

[8] See, for example, E. Gold and C. L. Mitchell, "The New Law of the Sea in the Eastern Caribbean: Towards a New Ocean Development Action Plan," in Gold, *ibid.*, pp. 265–276.

3) The studies identified the need for fisheries resource management, particularly since countries would exercise sovereign rights to fisheries resources within their EEZs. It was pointed out that this would permit some OECS Member States to develop an industrial fishery along with their traditional artisanal fisheries. It was therefore essential that scientific resource assessments be made. The presence of highly migratory fish stocks in the region meant that these assessments had to be made on a regional basis, and the basic requirements for regional management were outlined.
4) The seminars associated with the studies provided exposure, training and experience to both OECS officials and Canadian researchers. OECS fisheries officers, foreign affairs officials, ports officials, and lawyers met and discussed national and regional ocean issues for the first time in an integrated forum. The recommendations that emanated from these seminars included the following:

    a) the establishment of a commission for regional cooperation in management, exploration and exploitation of resources, surveillance and enforcement in offshore areas;
    b) the need for OECS Member States to support the research and developmental activities of international organizations in the fisheries area and, where necessary, for Member States to join international organizations that promoted fisheries development as soon as possible; and
    c) the adoption of internationally-accepted standards against pollution for the territorial sea and the EEZ.

The short-term impact of the DOSP studies was that they provided the OECS Secretariat with a focus and strategy for delimitation and for ocean and fisheries development. At the end of DOSP's activities in the Eastern Caribbean in 1985, awareness had been created at the political and governmental levels, e.g., the OECS, of the implications of the LOS Convention for small island States. This led to an interest in taking some action to bring about some of the changes necessary to enable the OECS Member States to improve their fisheries and related ocean management capacities. Some States also began to prepare for delimitation negotiations, e.g., Antigua and Barbuda with St. Kitts and Nevis, or to undertake delimitation negotiations, e.g., St. Lucia with France on its northern boundary with Martinique, and Dominica with France, i.e., with the European Economic Community. Between 1985 and 1987, the OECS Secretariat, with support from DOSP, and in collaboration with the Commonwealth Secretariat, organized four workshops on "Maritime Boundary Delimitations and Negotiation in the OECS." In 1987, the Secretariat established the OECS Fisheries Desk in St. Vincent and the Grenadines, the first regional indigenous organization for fisheries management and development in the CARICOM region. This unit commenced work on fisheries statistics, coordinated surveillance of OECS States in the Windward and Leeward islands, provided small project assistance in fisheries, and brought fisheries issues to the attention of the OECS Authority, i.e., to the Heads of State.

CIDA's Involvement and Regional Initiatives

Before the 1980s, CIDA's bilateral programmes in the Caribbean were predominantly associated with fisheries development. Since the 1980s, when Caribbean States became increasingly concerned about the implications of UNCLOS III, CIDA's involvement widened to oceans and was influenced by DOSP. This involvement was exemplified by an evolution in CIDA's strategy in this region, which was characterized by 1) a decreasing emphasis on fisheries development and an increasing emphasis on fisheries management and on wider environmental and oceans management, and 2) moving away from bilateral agreements and programmes toward regional programmes. CIDA's approach also evolved to emphasize a) medium- to long-term project duration, b) considerable interaction between CIDA and the recipients at the planning and implementation stages, and c) indigenous institutional strengthening and capacity-building.[9] Most of these elements were characteristics of the DOSP approach.

The Government of Canada had also established the International Centre for Ocean Development (ICOD), an initiative of then Prime Minister Pierre Trudeau, to assist the developing countries with implementing the provisions of UNCLOS III. One of ICOD's contributions was providing assistance in establishing the OECS Fisheries Unit mentioned earlier. Both CIDA and ICOD operated in the OECS region in the 1980s and into the

[9] Canadian International Development Agency (CIDA), *Strategy for Ocean Management and Development* (Hull: Policy Branch, CIDA, 1998).

early 1990s, until ICOD was terminated in 1992. Canadian government support to the OECS led to a request from CARICOM to CIDA for a project that would provide a resource assessment of the fisheries resources within the jurisdiction of CARICOM Member States. CIDA responded by supporting the Caribbean Fisheries Resource Assessment and Management Programme (CFRAMP).[10] This programme played a crucial role in establishing indigenous regional capability for fisheries management.

CFRAMP was developed in close collaboration and cooperation with CARICOM Member States. It initially covered the period 1991–1998 and had a total budget of $20.9 million, of which CIDA contributed $17.6 million (84 per cent) and the CARICOM States $3.3 million (16 per cent).[11] The project's goal was to promote the management and conservation of CARICOM countries' fisheries resources for the purpose of providing the CARICOM region with the basic information and institutional capacity necessary to manage and develop those resources. Specifically, the objectives of CFRAMP were a) to strengthen fisheries management structures and improve management capabilities and technical expertise through training and advisory assistance, b) to provide information on fishing resource abundance for management purposes, and c) to identify and establish a suitable regional fisheries management advisory mechanism.

CFRAMP encountered a number of implementation problems, and some programmes fell behind schedule. Consequently, a mid-programme review[12] recommended that CFRAMP be extended because it was attaining its major objectives: it increased management capacity and capability for fisheries by means of institutional support and training of key personnel; it provided resource information and management plans for fisheries; and it contributed to fisheries community development. However, since one of its major objectives was to establish a regional fisheries management

---

[10] C. L. Mitchell and H. Charles, *A Strategy for Co-operation in Sustainable Oceans Management and Development, Commonwealth Caribbean* (London: Commonwealth Secretariat, October 1996).

[11] Caribbean Regional Fisheries Mechanism (CRFM), *The CARICOM Fisheries Resource Assessment and Management Program (CFRAMP)* (Belize City: CRFM, 2006).

[12] C. L. Mitchell, *The Monitoring and Evaluation of the CARICOM Fisheries Resource and Management Program (CFRAMP)* (Hull: Policy Branch, Canadian International Development Agency, 1997).

advisory mechanism, CFRAMP commissioned a study[13] in 1996 that made a positive recommendation for the establishment of a regional fisheries mechanism (RFM), which would provide indigenous institutional capacity for fisheries management. In order to attain this objective, CARICOM commissioned a further study in 1997 to analyze the rationale for an RFM and to make recommendations for its establishment. The study made recommendations for the establishment of a CARICOM RFM and indicated how the extension of the CFRAMP programme could facilitate its implementation.[14] CIDA did extend CFRAMP, and in 2003 the Caribbean Regional Fisheries Mechanism (CRFM) was established by CARICOM. However, contrary to the expectation that CRFM would operate as a fisheries management unit, it instead has been functioning as a coordinating and advisory unit for the region.

Prior to the study on the RFM, CIDA co-financed with the Commonwealth Secretariat a study entitled, *A Strategy for Cooperation in Sustainable Oceans Development for CARICOM*, which was presented at the CIDA Oceans Initiatives Workshop at the Summit of the Sea in St. John's, Newfoundland, in 1997.[15] This study provided a strategy for CARICOM to follow in meeting the requirements of Agenda 21 for sustainable development and the Small Island Developing States (SIDS) Programme of Action. One of its major recommendations was that a CARICOM Regional Centre for Oceans Management (RCOM) be established. It was pointed out that the RCOM did not conflict with the RFM in that the RFM would be an integral part of the wider oceans mechanism. This study attained endorsement and approval at the level of the Standing Committee of Ministers, and recommendations based on it were presented to the CARICOM Council. The Council, however, had centred attention on having the Caribbean Sea declared a special area for sustainable development. At the tenth inter-sessional meeting of the CARICOM Council in 1999, the Council endorsed a proposal to have the Caribbean Sea internationally recognized as a special area for sustainable

[13] R. Mahon, B. Scotland and M. Scipio, *Concept Paper for a Regional Fisheries Advisory Mechanism*, (Belize City: CARICOM Fisheries Resource Assessment and Management Program, 1996).

[14] R. Rainford and C. L. Mitchell, *A Regional Fisheries Advisory Mechanism for CARICOM* (Georgetown: CARICOM Secretariat, January 1998).

[15] Mitchell and Charles, *supra* note 10.

development.[16] The submission of this proposal to the United Nations was encouraged and supported by the ACS, a supra-regional organization founded in 1994 with the aim of promoting consultation, cooperation and concerted action among all the countries of the Caribbean. Its objectives were to strengthen the regional cooperation and integration processes, preserve the environmental integrity of the Caribbean Sea, and promote the sustainable development of the Greater Caribbean.[17]

On 15 February 2000, the United Nations General Assembly approved Resolution 54/225, "Promoting an Integrated Management Approach to the Caribbean Sea Area in the Context of Sustainable Development." Subsequent resolutions supported this resolution and the efforts of Caribbean countries to implement a sustainable fisheries management programme by strengthening the CRFM.[18] Finally, in December 2006, the General Assembly adopted a resolution entitled "Towards the Sustainable Development of the Caribbean Sea for Present and Future Generations" (A/C.2/61/L.30) that secured the recognition of the Caribbean Sea as a special area in the context of sustainable development by the international community.[19] The resolution saluted the work and initiatives taken by the ACS, including the establishment of a Caribbean Sea Commission and the adoption of a definition of the concept of the Caribbean Sea as a special zone.[20]

Prior to, during, and after CFRAMP, there were CIDA projects or programmes[21] dealing with the ocean environment and with coastal zone

---

16 Communique issued at the Conclusion of the Tenth Inter-Sessional Meeting of the Conference of Heads of Government of the Caribbean Community, 4–5 March 1999, Paramaribo, Suriname, <www.caricom.org/jsp/communications/communiques/10inthgc_1999_communique.jsp>.

17 The ACS consists of 25 Member States and three Associate Members. The main organs of the Association are the Ministerial Council and a Secretariat in Port of Spain Trinidad and Tobago. See <www.acs-aec.org>.

18 United Nations General Assembly Resolutions 55/203 (20 December 2000), 57/261 (27 February 2003), and 59/230 (16 February 2005).

19 W. R. Denis, "New UN Resolution on the Caribbean Sea," The Greater Caribbean this Week, 17 January 2007, <www.acs-aec.org/column/greatCarib01172007_en.htm>.

20 See Association of Caribbean States (ACS), "ACS Intensifies Its Efforts on Caribbean Sea Initiative," *News Release* NR/023/2006, 24 July 2006, <www.acs-aec.org/PressCenter/NewsReleases/2006/nr0232006_en.htm>.

21 Two other major projects besides CFRAMP were the OECS Environmental Capacity Development Project (ENCAPD) and the Jamaica Environmental Action Plan

management that have had strong institutional strengthening components and have encouraged partnership arrangements between Canadian universities and some of the universities in the Caribbean, particularly the University of the West Indies (UWI). In 1984, UWI established the Centre for Resource Management and Environmental Studies (CERMES), which continues to provide research and training in marine environmental studies and in marine resource management. More indirectly, CIDA supported the Oceans Institute of Canada (OIC) in a programme on sustainable oceans covering legal and policy issues, not only in the Caribbean but in Asia and the South Pacific as well, and also funded the International Ocean Institute's (IOI) training programmes in oceans management conducted at Dalhousie University, which have been attended by many Caribbean participants. CIDA's investments in the ocean sector in the Caribbean amounted to about CDN $40 million over the period 1991 to 2004.[22] By identifying problem areas and priorities that enabled the OECS and CARICOM to mount successful applications for Canadian assistance in the marine area, DOSP can claim some credit for this level of investment.

Current Developments: Towards a Management Regime for the Caribbean Sea

Recent developments in the Caribbean have brought to the fore the significance and relevance of DOSP's advice on fisheries and ocean management. The most important of these developments are the United Nations recognition of the Caribbean Sea as a special area for sustainable development, the CARICOM fisheries policy and regime, and a renewed interest in maritime boundary delimitation. These developments were, to some extent, spurred by greater integration in the Caribbean region brought about by a) the establishment of the ACS, b) the widening of

(ENACT). These projects were designed to improve the capacity for environmental management, in which coastal and marine resource management were significant features. S. Free, "CIDA's Caribbean Programs for Coasts, Oceans and the Environment," Paper presented at International Workshop on New Directions in Management Capacity-Building for Coastal and Ocean Management in the Wider Caribbean, Havana, Cuba, July 1998.

22 Free, *ibid.*

CARICOM's membership by the inclusion of Suriname, Haiti, and the Dominican Republic, c) the pursuit of the CARICOM Single Market and Economy (CSME), and d) linkages between the ACS and CARICOM with the Central American Integration System (SICA) and the Central American Economic Integration System (SIECA). The integration of Caribbean and Central American States augurs well for initiatives to manage the Caribbean Sea.

The UN Recognition of the Caribbean Sea as a Special Area for Sustainable Development

The UN's recognition of the Caribbean Sea as a special area for sustainable development implies that a management regime for this sea is essential. Since the Caribbean Sea is affected by global issues such as global warming and sea-level rise, trans-shipment of radioactive and hazardous wastes, and the over-exploitation of its fisheries resources, it is obvious that an ocean management regime is required. The ACS has taken the lead in this regard by establishing the Caribbean Sea Commission for protecting and safeguarding the Caribbean Sea. The Commission consists of a multidisciplinary body that seeks to ensure greater harmonization of the planning and implementation of the numerous activities that directly and indirectly impact on the Caribbean.

CARICOM's Regional Fisheries Policy and Regime Initiative

The CARICOM Heads, at the fourteenth inter-sessional meeting in February 2003, endorsed proposals on the imperative of elaborating a Common Fisheries Regime (CFR).[23] In March 2003, the Caribbean Regional Fisheries Mechanism (CRFM) was officially launched, and it was decided that the CRFM would take responsibility for the CFR and a working group was established for this purpose. CARICOM Heads, at their fifteenth inter-sessional meeting in March 2004,[24] charged the CARICOM

[23] Caribbean Regional Fisheries Mechanism, "Towards a Common Fisheries Policy and Regime for CARICOM," <www.caricom-fisheries.com/website_content/publications/documents/Towards_a_Common_Fisheries_Policy_and_Regime.pdf>.

[24] Communique issued at the Conclusion of the Fifteenth Inter-Sessional Meeting of the Conference of the Heads of Government of the Caribbean Community, 25–26 March 2004, Basterre, St. Kitts and Nevis, <www.caricom.org/jsp/communications/communiques/15inthgc_2004_communique.jsp>.

Secretariat with producing proposals on the CFR and on maritime delimitation. A subsequent study commissioned by CARICOM outlined the requirements for regional fisheries management for the Caribbean Sea based on the premise that if CARICOM established a management regime, other Caribbean States would have to comply with management measures.[25] The study found that, although the CFRM was functioning as a coordinating and advisory unit, it could be strengthened to become the regional management unit for the Caribbean Sea. Recommendations were made on the CFR and maritime delimitation. The working group has had meetings since 2004 on the common fisheries regime. It has produced a draft agreement covering areas such as a common fisheries zone, implementation arrangements including the powers of implementing agencies, conservation and management of fisheries resources, data collection and research, registration of fishing vessels, and the marketing and trading of fisheries resources.[26] It therefore seems likely that CARICOM will establish a regional fisheries organization for the Caribbean Sea, one that can be expanded in the future to fulfil the role of a wider ocean management organization for this sea.

Maritime Boundary Delimitation

Since DOSP's initial research on preparing the OECS and CARICOM region for maritime boundary delimitation, boundary delimitation has proceeded slowly and intermittently and has been problematic, leading to disputes that have resulted in costly international tribunal arbitration. It has been reported that less than one quarter of the potential boundaries have been settled. The potential maritime boundaries claims yet to be settled can be conveniently grouped into four categories: 1) CARICOM Member States and metropolitan powers; 2) CARICOM Member States and other Caribbean States; 3) intra-CARICOM (other than OECS Members); and 4) intra-OECS Members.[27] It is of interest to note that the majority of maritime boundaries delimited thus far by CARICOM countries fall within the first and second categories. Further, no delimitation agreement has been concluded between an OECS and a non-OECS

---

[25] C. Dundas and C. L. Mitchell, *A Common Fisheries Regime for the Caribbean Sea* (Georgetown: CARICOM Secretariat, 2004).

[26] "New Fishing Rules Coming," *Barbados Advocate*, 10 August 2007.

[27] Dundas and Mitchell, *supra* note 25.

CARICOM Member State, or between two OECS Member States. It is incongruous that the OECS, despite the background work done by DOSP, CIDA, COMSEC and other agencies, and the agreement among Member States that delimitation would not be an issue, has made so little progress to date.

Boundary disputes between Trinidad and Tobago and Barbados and between Guyana and Suriname have recently been resolved by international adjudication under the LOS Convention.[28] Barbados disputed the 1990 Trinidad and Tobago-Venezuela maritime boundary treaty, which entered into force in 1991.[29] After five rounds of maritime boundary delimitation talks and four fisheries sessions, which began in 2000, they failed to reach an agreement. In 2004, Barbados took its offshore boundary dispute to the Permanent Court of Arbitration (PCA) which ruled in Barbados' favour in 2006.[30] Guyana and Suriname followed the same route when boundary commissions from both countries failed to resolve an offshore maritime boundary dispute. The PCA settled this dispute in 2007.[31] The problem of Aves Island with regard to maritime boundary delimitation with OECS Member States articulated by DOSP remains a major obstacle to delimitation of maritime boundaries between these States and Venezuela. The OECS is endeavouring to negotiate with Venezuela on a regional basis, but, so far, this has not been accepted by Venezuela. It is

---

28 Ad hoc arbitration under Annex VII of the LOS Convention, *supra* note 6. The LOS Convention, in part XV, sets out rules for the resolution of disputes between States arising out of the interpretation or application of the Convention. States can choose between the following four means for settling disputes: the International Tribunal for the Law of the Sea (ITLOS) in Hamburg, Germany; the International Court of Justice in the Hague, The Netherlands; ad hoc arbitration (in accordance with Annex VII of the LOS Convention); and a "special arbitral tribunal" constituted for certain categories of disputes (established under Annex VII of the LOS Convention).

29 *Treaty between the Republic of Trinidad and Tobago and the Republic of Venezuela on the Delimitation of Marine and Submarine Areas*, 18 April 1990, reprinted in the *Law of the Sea Maritime Boundary Agreements (1985–1991)* (New York: Office for Ocean Affairs and the Law of the Sea, United Nations, 1992), pp. 25–29.

30 *Barbados and the Republic of Trinidad and Tobago: Award of the Arbitral Tribunal*, The Hague, Permanent Court of Arbitration, 11 April 2006, <www.pca-cpa.org/upload/files/Final%20Award.pdf>.

31 *Guyana and Suriname: Award of the Arbitral Tribunal*, The Hague, Permanent Court of Arbitration, 17 September 2007, <www.pca-cpa.org/upload/files/Guyana-Suriname%20Award.pdf>.

likely that this issue will also have to be settled by international arbitration. The current situation and initiatives in the Caribbean indicate that maritime boundary delimitation will pose a major challenge to the future management and development of the Caribbean Sea.

Conclusion

The DOSP studies during the 1980s have had an impact on fisheries and oceans management and development initiatives that has endured to the present day. Commencing in the Eastern Caribbean, the impact spread to CARICOM and then to the entire Caribbean Sea. DOSP highlighted the necessity of regional management, which led to the establishment of the OECS Regional Fisheries Unit (since amalgamated into the OECS Resource Management and Environmental Unit) and the CFRM. Although the CFRM has always functioned as a fisheries advisory unit, and not as a fisheries management unit, it can be strengthened to become the regional fisheries management unit for CARICOM and the Caribbean Sea.

Recent developments in the Caribbean have brought to the fore the significance and relevance of DOSP's advice on fisheries and ocean management. That advice continues to be as relevant today as it was during the 1980s, mainly because the issues faced today are virtually the same as those faced twenty years ago. DOSP's efforts are exemplified by the successes experienced by the OECS and CARICOM, which include: 1) institutional strengthening of regional and national fisheries management units; 2) increasing attention paid to environmental concerns; 3) more trained personnel in environmental, fisheries, and oceans management; 4) greater regional cooperation by means of regional seminars and workshops; and 5) improved public information on marine issues. These successes can be considered a fitting tribute to Douglas Johnston, Edgar Gold and the "DOSPIES."

Part III

The Publications of Douglas Milllar Johnston

Prepared by Christian L. Wiktor* and Ted L. McDorman**

Introduction

Douglas M. Johnston was a creative and insightful internationally-renowned scholar. Writing to convey innovative concepts, ideas and to persuade a reader of new approaches to topics was his passion. Words, their choice and placement, were his craft. Where conventional words failed to convey precise meaning or sufficient nuance, new words or borrowed words (i.e., words imported from different disciplines) were required.

Douglas wrote with a pen in long-hand. He often wrote in libraries undisturbed by students and other activities around him. He invariably prepared only a single draft, with minimal cross-outs – had this typed – and on perusal of the typed-version perhaps corrected a comma or two beyond typist misspellings. In some cases footnotes were prepared simultaneously with the text; more often, a text was finished and then the footnotes were discovered and added. His work on footnotes was painstaking, sparing no effort in glossing each point as appropriate.

Before putting pen to paper, there was research, reading and mulling. Mulling involved contemplation, but also lots of conversation with a large number of colleagues from a wide variety of disciplines, depending on the topic. Much of the mulling process involved reflection upon and engagement with the interdisciplinary approach and policy orientation of Douglas' most influential teacher, Professor Myres S. McDougal at Yale University.

From this, as demonstrated below, came an impressive record of writing and scholarly inquiry of a large breadth of subject matters.

* Professor Emeritus, Dalhousie Law School, Halifax, Canada, and colleague and close friend of Douglas Johnston for 35 years.

** Professor, Faculty of Law, University of Victoria, Canada.

The Publications of Douglas M. Johnston

This bibliography includes 143 titles of essays, collections of essays, proceeding papers, periodical articles, and books. Johnston started his publishing career in 1960 with an article on "The International Law of Fisheries" published in two installments in 1960 and 1963 in the *Current Law and Social Problems*. This was an outline of his doctoral dissertation (J.S.D.) submitted to Yale Law School in 1962, and published by the prestigious Yale University Press in 1965 under the title: *The International Law of Fisheries: A Framework for Policy-Oriented Inquiries*. This work, reprinted in 1987 containing an extensive new introductory essay, became the best known publication on the subject for generations. It is interesting to note that international fisheries was the topic of his M.C.L thesis presented at the University of Montreal in 1958 under the title: *A Juridical Approach to the Problem of the World Fisheries*.

During the forty-six years of his publishing career (1960–2006), the most prolific years were when Johnston held a permanent appointment as a member of university law schools. While Douglas published 17 titles during twelve years (an average of 1.4 titles per year) when he was affiliated with six varied institutions of higher education (1960–1972), he published 66 titles during fifteen years (an average of 4.4 titles per year) when he was professor of law at Dalhousie University Law School (1972–1987), and 61 titles during eighteen years (an average of 3.4 titles per years) when he was professor of law (and later professor emeritus) at the University of Victoria Law School (1988–2006).

Of the 143 titles, Johnston published 94 titles alone as author (88 titles) or editor (6 titles), which represent 65.7 per cent of all his publications. He published the remaining 49 titles with co-authors (26 titles) and co-editors (23 titles). These joint publications demonstrate the ability of Johnston to cooperate with a large number of people with diverse backgrounds. The list of individuals with whom Johnston cooperated in the publications of the 49 titles includes 36 names which in descending order includes Ronald St. John Macdonald (10 titles), Gerald L. Morris of the University of Toronto Law School (7 titles), Edgar Gold of Dalhousie Law School (5 titles), Ankana Sirivivatnanon of SEAPOL (4 titles), Phillip Saunders of Dalhousie Law School (3 titles), A. Paul Pross of Dalhousie University (3 titles), and the others with two or one titles each.

The closest, most productive, and long lasting was the cooperation of Johnston with Ronald St. John Macdonald, based on a special relationship which developed during a period of over forty-five years. This produced ten major books and articles. Among their outstanding publications, (the first two collections of essays were co-edited also with Gerald L. Morris), the following works should be mentioned: *Canadian Perspectives on International Law and Organization* (1974); *The International Law and Policy of Human Welfare* (1978); *The Structure and Process of International Law: Essays in Legal Philosophy* (1983); and *Towards World Constitutionalism: Issues in the Legal Ordering of the World Community* (2005). On the special relationship between Johnston and Macdonald, see *Tribute to Ronald St. John Macdonald,* by C. L. Wiktor in the *Canadian Yearbook of International Law* (Volume 44, pp. 485–487; 2006).

By type of publications – 34 titles were books, 31 periodical articles, 67 essays and papers, and 11 reports and proceedings. A small selection of Johnston's major publications demonstrates the depth and the wide range of topics of his scholarly work. They includes such highly praised books of long-lasting value as: *The International Law of the Fisheries: A Framework for Policy-Oriented Inquiries* (1965); *Canadian Perspectives on International Law and Organization* (co-editor; 1974); *The International Law and Policy of Human Welfare* (co-editor; 1978); *The Structure and Process of International Law: Essays in Legal Philosophy* (co-editor; 1986); *The Theory and History of Ocean Boundary-Making* (1988); *Consent and Commitment in the World Community: The Classification and Analysis of International Instruments* (1997); *Towards World Constitutionalism: Issues in the Legal Ordering of the World Community* (co-editor, 2005), and his last work *The Historical Foundations of World Order: The Tower and the Arena* (2008). This last work, a monumental and pioneer work on the history of international law from the antiquity to the present times, was finalized posthumously by his son, Keith Johnston, in December 2007 and published by Martinus Nijhoff in the Netherlands.

In addition, Johnston's enormous contribution to international law in general, and the law of the sea and international environmental law in particular, is further substantiated by the articles, essays and papers enumerated in this list of his publications.

The Subjects of Publications

Almost all the 143 titles of Douglas M. Johnston's publications deal with public international law, including law of the sea and international environmental law (only one title concerns domestic insurance law). The publications deal with the more specific topics as follows:

A. *International Law (29 Titles)*

- Diplomacy and Foreign Relations
  - China (2)
  - East Asia (1)
- History (1)
  - Scotland (1)
- Human Rights (2)
- International Organizations (2)
- Legal Profession (1)
- Regional Perspectives
  - Asia Pacific (3)
  - Canada (3)
  - North Pacific (1)
- Theoretical Aspects (7)
- Treaties (2)
  - China (3)

B. *Law of the Sea (94 Titles)*

- General (8)
  - Arctic Ocean (2)
  - Canada (3)
  - Japan (1)
  - Southeast Asia (3)
- Law of the Sea Conventions (5)
  - Economic Zones (1)
    - North America (1)
  - Deep Ocean Mining (1)
  - Fisheries (13)
    - Northeast Asia (1)
    - Northwest Atlantic (3)

- Marine Environment (4)
  - Marine Pollution (5)
    - Arctic (2)
    - Asia Pacific (1)
    - East Asia (1)
- Marine Resources (1)
    - North Pacific (1)
  - Offshore Industry (1)
    - British Columbia (2)
    - Canada (1)
- Maritime Boundaries (9)
- Maritime Transit (1)
  - Southeast Asia (2)
  - Northwest Passage (2)
- Ocean Management
  - Coastal Zone Management (5)
    - Atlantic Canada (1)
    - Canada (2)
    - Southeast Asia (8)
- Research and Technology (3)

C. *International Environmental Law (20 Titles)*

- General (9)
  - Arctic Ocean (1)
  - Canada (1)
  - South China Sea (1)
  - Southeast Asia (2)
- Pollution (1)
  - Acid Precipitation
    - North America (1)
  - Singapore (4)

Chronological Listing

This bibliography of publications by Douglas M. Johnston is arranged chronologically by the date of publication, and subdivided by title. The bibliography is divided into six parts representing the six periods of Johnston's academic career.

I. Lecturer and Assistant Professor of Law, Faculty of Law, University of Western Ontario, London, Ontario (1959–1964)

1960

"*The International Law of Fisheries*: A Policy-Oriented Inquiry in Outline," 1 *Current Law and Social Problems* 19–67 (1960).
This is the first of two articles constituting an outline of Johnston's doctoral thesis (J.S.D.) submitted to Yale Law School in 1962.

1963

"The International Law of Fisheries: A Policy-Oriented Inquiry in Outline (Concluded)," 3 *Current Law and Social Problems* 146–237 (1963).

1964

"Legalism and Realism in Insurance Law: Two Recent Canadian Decisions," 3 *Western Law Review* 196–209 (1964).

II. Teaching, Studying and Research Work (1964–1969): Louisiana State University Law Center, Baton Rouge (1964–1965); Yale Law School, New Haven, Conn. (1965–1966); Harvard Law School, Cambridge, Mass. (1966–1969); New School for Social Research, New York (1968–1969)

1965

The International Law of the Fisheries: A Framework for Policy-Oriented Inquiries (New Haven: Yale University Press, 1965), xxiv, 554 p.
Based on the author's thesis (J.S.D.), presented at Yale Law School in 1962 under a slightly different title: *The International Law of Fisheries: A Policy Oriented Inquiry* (typewritten, 249, 176 leaves). The publication was reprinted in 1987 by New Haven Press and Martinus Nijhoff in a series entitled: "The New Haven Studies in International Law and World Public Order." See also Johnston's thesis (M.C.L.) presented at the McGill University in Montreal in 1958 under title: *A Juridical Approach to the Problems of the World Fisheries* (typewritten, 312 leaves).

1967

"Law, Technology and the Sea," 55 *California Law Review* 449–472 (1967).

"New Uses of International Law in the North Pacific," 43 *Washington Law Review* 77–114 (1967).

"Treaty Analysis and Communist China: Preliminary Observations," 61 *American Society of International Law, Proceedings* 126–134 (1967). Address delivered at a panel during the sixty-first annual meeting of the American Society of International Law in Washington, D.C., on 28 April 1967.

1968

Agreements of the People's Republic of China, 1949–1968: A Calendar (co-editor with Hungdah Chiu) (Cambridge, Mass.: Harvard University Press, 1968), xvii, 286 p. (Harvard Studies in East Asian Law, 3).

III. Professor of Political Science, Department of Political Economy, University of Toronto (1969–1972)

1969

"The Legal Theory of Fishery Organization," in L. M. Alexander, ed., *The Law of the Sea: International Rules and Organization for the Sea* (Proceedings of the Third Annual Conference of the Law of the Sea Institute) (Kingston, R.I.: University of Rhode Island, 1969), pp. 431–435.

1970

"Canada's Arctic Marine Environment: Problems of Legal Protection," 29 *Behind the Headlines* 1–7 (Nos. 5–6, 1970).

1971

"The Arctic Marine Environment: A Managerial Perspective," in L. M. Alexander, ed., *The Law of the Sea: The United Nations and Ocean Management* (Proceedings of the Fifth Annual Conference of the Law of the Sea Institute) (Kingston, R.I.: University of Rhode Island, 1971), pp. 312–318.

"The Canadian Initiative to Establish a Maritime Zone for Environmental Protection: Its Significance for Multilateral Development of International Law" (co-author with Ronald St. J. Macdonald and Gerald L. Morris), 21 *University of Toronto Law Journal* 247–251 (1971). Statement presented at the annual conference of the American Society of International Law meeting in New York on 24 April 1970.

A Century of Struggle: Canadian Essays on Revolutionary China (co-editor with James M. Gibson) (Toronto: Canadian Institute of International Affairs, 1971), xii, 177 p.
Contains eighteen essays by various authors. Also published in 1975 (xii, 191 p.).

"Chinese Diplomacy," in J. M. Gibson and D. M. Johnston, eds., *A Century of Struggle: Canadian Essays on Revolutionary China* (Toronto: Canadian Institute of International Affairs, 1971), pp. 145–154.

"Marginal Diplomacy in East Asia," 27 *International Journal* 469–506 (1971).

IV. Professor of Law, Dalhousie Law School, Halifax (1972–1987); Co-Founder and Principal Investigator, Dalhousie Ocean Studies Programme (1979–1985); Co-Founder and Planning and Programme Co-Ordinator, Southeast Asian Programme in Ocean Law, Policy and Management (SEAPOL), Thailand (1982–2006).

1972

"Marine Pollution Control: Law, Science and Politics," 28 *International Journal* 69–102 (1972–1973).

"Recent Canadian Marine Legislation: An Historical Perspective," in L. M. Alexander and G. R. S. Hawkins, eds., *Canadian-U.S. Maritime Problems* (Kingston, R.I.: Law of the Sea Institute, University of Rhode Island, 1972), pp. 63–67.
Paper presented at a workshop held in Toronto, 15–17 June 1971.

1973

The Economic Zone in the Law of the Sea: Survey, Analysis and Appraisal of Current Trends (co-author with Edgar Gold) (Kingston, R.I.: Law of the Sea Institute, University of Rhode Island, 1973), iii, 53 p. (Law of the Sea Institute, Occasional Paper, No. 17).

"International Law and Society in the Year 2000" (co-author with Ronald St. J. Macdonald and Gerald L. Morris), 51 *Canadian Bar Review* 316–332 (1973).

1974

"Canadian Approaches to International Law" (co-author with Ronald St. J. Macdonald and Gerald L. Morris), in R. St. J. Macdonald, G. L. Morris and D. M. Johnston, eds., *Canadian Perspectives on International Law and Organization* (Toronto: University of Toronto Press, 1974), pp. 940–954.

Canadian Perspectives on International Law and Organization (co-editor with Ronald St. J. Macdonald and Gerald L. Morris) (Toronto: University of Toronto Press, 1974), xx, 972 p.
Contains thirty-eight papers in English and French by various authors.

"Chinese Treaty Behaviour: Experiments in Analysis," in R. L. Dial, ed., *Advancing and Contending Approaches to the Study of Chinese Foreign Policy* (Halifax: Centre for Policy Studies, Dalhousie University, 1974), pp. 385–396.

"International Environmental Law: Recent Developments and Canadian Contributions," in R. St. J. Macdonald, G. L. Morris and D. M. Johnston, eds., *Canadian Perspectives on International Law and Organization* (Toronto: University of Toronto Press, 1974), pp. 555–611.

The International Law of Pollution (co-editor with James Barros) (New York: Free Press, 1974), xvii, 476 p.
Also published in Spanish in Argentina under title: *Contaminación y derecho internacional* (Buenos Aires, Marymar, 1977, 559 p.).

"The Regional Consequences of a Global Fisheries Convention," in G. Pontecorvo, ed., *Fisheries Conflicts in the North Atlantic: Problems of Management and Jurisdiction* (Cambridge, Mass.: Ballinger, 1974), pp. 35–51.
Paper presented at a workshop conducted by the Law of the Sea Institute held in Hamilton, Bermuda, 14–17 January 1974.

1975

Coastal Zone: Framework for Management in Atlantic Canada (co-editor with A. Paul Pross and Ian McDougall) (Halifax: Institute of Public Affairs, Dalhousie University, 1975), xiii, 201, 48 p.

This report was commissioned by Inland Waters Directorate, Environmental Management Service, Department of the Environment, Ottawa, Canada; originally issued on 22 November 1974.

"The Coastal Zone Management Challenge" (co-author with A. Paul Pross), in D. M. Johnston, A. P. Pross and I. McDougall, eds., *Coastal Zone: Framework for Management in Atlantic Canada* (Halifax: Institute of Public Affairs, Dalhousie University, 1975), pp. 1–14.

"Development, Environment and Marine Resources in the North Pacific," in H. Bull, ed., *Asia and the Western Pacific: Towards a New International Order* (Sydney: Thomas Nelson, and the Australian Institute of International Affairs, 1975), pp. 232–261.
Paper presented at the international conference organized by the Australian Institute of International Affairs held at the Australian National University in Canberra, 14–17 April 1973.

"The Economic Zone in North America: Scenarios and Options," 3 *Ocean Development and International Law* 53–68 (1975).

"The Existing Statutory Framework in Canada and the Atlantic Provinces," in D. M. Johnston, A. P. Pross and I. McDougall, eds., *Coastal Zone: Framework for Management in Atlantic Canada* (Halifax: Institute of Public Affairs, Dalhousie University, 1975), pp. 29–53.

"The International Legal Environment," in D. M. Johnston, A. P. Pross and I. McDougall, eds., *Coastal Zone: Framework for Management in Atlantic Canada* (Halifax: Institute of Public Affairs, Dalhousie University, 1975), pp. 15–27.

"The New Equity in the Law of the Sea," 31 *International Journal* 79–99 (1975–1976).

"The New Lawyer in a Transnational World" (co-author with Ronald St. J. Macdonald and Gerald L. Morris), 25 *University of Toronto Law Journal* 343–357 (1975).

"Options and Recommendations" (co-author with A. Paul Pross and Ian McDougall), in D. M. Johnston, A. P. Pross and I. McDougall, eds., *Coastal Zone: Framework for Management in Atlantic Canada* (Halifax: Institute of Public Affairs, Dalhousie University, 1975), pp. 149–163.

"The Options for LOS III: Appraisals and Proposal," in F. T. Christy, Jr., and others, eds., *Law of the Sea: Caracas and Beyond* (Proceedings of the Ninth Annual Conference of the Law of the Sea Institute) (Cambridge, Mass.: Ballinger, 1975), pp. 357–372.
Paper presented at the annual conference held in Miami, 6–9 January 1975.

"Some Treaty Law Aspects of a Future International Fishing Convention," in H. Gary Knight, ed., *The Future of International Fisheries Management* (St. Paul, Minn.: West, 1975), pp. 103–157.

1976

"Equity and Efficiency in Marine Law and Policy," in D. M. Johnston, ed., *Marine Policy and the Coastal Community* (New York: St. Martin's Press; London: Croom Helm, 1976), pp. 297–327.

"Facts and Value in the Prevention and Control of Marine Pollution," in W. M. Reisman and B. H. Weston, eds., *Toward World Order and Human Dignity: Essays in Honor of Myres S. McDougal* (New York: Free Press, 1976), pp. 534–561.

Marine Policy and the Coastal Community: The Impact of the Law of the Sea (editor) (New York: St. Martin's Press; London: Croom Helm, 1976), 338 p.
Contains thirteen essays on the marine economy, ocean politics and the coastal community by various authors. Also published in Portuguese in Brazil under title: *A política marítima e a comunidade litorânea: o impacto do direito marítimo* (São Paulo, Editora Cultrix, 1980, 265 p.).

A New Regime for the Oceans, A Report of the Trilateral Task Force on the Oceans, Paris, December 1975 (co-rapporteur with Shigeru Oda, Michael Hardy, Ann L. Hollick, and Johan Jørgen Holst) (New York, N.Y.: Trilateral Commission, 1976), viii, 54 p. (The Triangle Papers, 9).

1977

"Coastal Zone Management in Canada: Purposes and Prospects," 20 *Canadian Public Administration* 140–151 (1977).

"Conditions for Effective Fisheries Management in the Northwest Atlantic" (co-author with Giulio Pontecorvo and Maurice Wilkinson), in Lee G. Anderson, ed., *Economic Impacts of Extended Fisheries Jurisdiction* (Ann Arbor, Mich.: Ann Arbor Science Publishers, 1977), pp. 51–103.

Paper presented at a symposium held at the University of Delaware in Newark, Delaware, in April 1977.

"The Law of the Sea: Recent Trends in the Literature," 71 *American Journal of International Law* 539–555 (1977).
This is a review article of nine books on the law of the sea.

"Legal and Diplomatic Developments in the Northwest Atlantic Fisheries," 4 *Dalhousie Law Journal* 37–61 (1977).

1978

"The Foundations of Justice in International Law," in R. St. J. Macdonald, D. M. Johnston and G. L. Morris, eds., *The International Law and Policy of Human Welfare* (Alphen aan den Rijn: Sijthoff & Noordhoff, 1978), pp. 111–146.

The Future of the Offshore: Legal Developments and Canadian Business (co-editor with Donald J. Patton and Clare Beckton) (Halifax: Centre for International Business Studies, Dalhousie University, 1978), viii, 243 p.
Proceedings of a conference sponsored by the Centre for International Business Studies and the Public Services Committee, Faculty of Law, Dalhousie University.

The International Law and Policy of Human Welfare (co-editor with Ronald St. J. Macdonald and Gerald L. Morris) (Alphen aan den Rijn: Sijthoff & Noordhoff, 1978), xvii, 690 p.
Contains twenty-five essays by various authors.

"The International Law of Human Welfare: Concept, Experience and Priorities" (co-author with R. St. J. Macdonald and G. L. Morris), in R. St. J. Macdonald, D. M. Johnston and G. L. Morris, eds., *The International Law and Policy of Human Welfare* (Alphen aan den Rijn: Sijthoff & Noordhoff, 1978), pp. 3–79.

Regionalization of the Law of the Sea (editor) (Proceedings of the Eleventh Annual Conference of the Law of the Sea Institute) (Cambridge, Mass.: Ballinger, 1978), ix, 346 p.
The annual conference was held at the University of Hawaii in Honolulu, 14–17 November 1977.

"The Scottish Tradition in International Law," 16 *Canadian Yearbook of International Law* 3–45 (1978).

1979

"Deep Ocean Mining: Interim Arrangements and Alternative Outcomes," in S. Allen and J. P. Craven, eds., *Alternatives in Deepsea Mining* (Proceedings of a Workshop of the Law of the Sea Institute) (Honolulu: Law of the Sea Institute, 1979), pp. 57–74.
Paper presented at the workshop held at the Aspen Institute Hawaii, 11–14 December 1978.

1980

"Impact of the New Law of the Sea on Japanese-Canadian Relations," in K. A. J. Hay, ed., *Canadian Perspectives on Economic Relations with Japan* (Montreal: Institute for Research on Public Policy, 1980), pp. 95–127.
Paper presented at a conference held at York University in Toronto in May 1979. See also comment on the paper by B. Applebaum (pp. 128–132).

1981

"Acid Precipitation in North America: The Case for Transboundary Cooperation" (co-author with Peter Finkle), 14 *Vanderbilt Journal of Transnational Law* 787–843 (1981).
Reprinted, Calgary, Canadian Institute of Resources, University of Calgary, 1983, 75 p.

"The Conduct of Oceanographic Research in the '80s," in *Pacem in Maribus X, Proceedings* (Msida, Malta: International Ocean Institute, 1981), pp. 37–44.
Comments presented at the tenth annual meeting of Pacem in Maribus held in Vienna, 27–31 October 1980.

The Environmental Law of the Sea (editor) (Gland, Switzerland: International Union for Conservation of Nature and Natural Resources, 1981), 419 p. (IUCN Environmental Policy and Law Paper, No. 18).

"The Environmental Law of the Sea: Conclusions and Recommendations" (co-author with Cyrille de Klemm, in collaboration with Francoise Burhenne-Guilmin and Daniel B. Navid), in D. M. Johnston, ed., *The Environmental Law of the Sea* (Gland, Switzerland: International Union for Conservation of Nature and Natural Resources, 1981), pp. 387–419.

"The Environmental Law of the Sea: Historical Development," in D. M. Johnston, ed., *The Environmental Law of the Sea* (Gland, Switzerland: International Union for Conservation of Nature and Natural Resources, 1981), pp. 17–70.

The Marine Environment and the Caracas Convention on the Law of the Sea (A Study of the Third United Nations Conference on the Law of the Sea and Other Related Marine Environmental Activities) (co-author with Ted L. McDorman, Norman G. Letalik, Hal Mills and Edgar Gold) (Halifax: Dalhousie Ocean Studies Programme, 1981), ix, 127 p.

"Regional Approaches to the Protection and Conservation of the Marine Environment" (co-author with Lawrence M. G. Onomoto), in D. M. Johnston, ed., *The Environmental Law of the Sea* (Gland, Switzerland: International Union for Conservation of Nature and Natural Resources, 1981), pp. 285–385.

1982

Arctic Ocean Issues in the 1980's (editor) (Proceedings a Workshop co-sponsored by Law of the Sea Institute and Dalhousie Ocean Studies Programme) (Honolulu: Law of the Sea Institute, 1982), iii, 60 p.
The workshop was held at Mackinac Island, Michigan, 10–12 June 1981.

"Emerging Legislative Trends in Southeast Asia" (co-author with Norman G. Letalik), in M. J. Valencia, E. Gold, Chia Lin Sien and N. G. Letalik, eds., *Shipping, Energy and Environment: Southeast Asian a Perspectives for the Eighties* (Halifax: Dalhousie Ocean Studies Programme, 1982), pp. 231–250.

Environmental Management in the South China Sea: Legal and Institutional Developments (Honolulu: East-West Center, 1982), vii, 114 p. (East-West Environment and Policy Institute, Research Paper No. 10).

"Extended Jurisdiction: The Impact of UNCLOS III on Coastal State Practice" (co-author with Edgar Gold), in T. A. Clingan, Jr., ed., *Law of the Sea: State Practice in Zones of Special Jurisdiction* (Proceedings of the Thirteenth Annual Conference of the Law of the Sea Institute) (Honolulu: Law of the Sea Institute, 1982), pp. 3–56.
This paper was presented at the annual conference held in Mexico City, 15–18 October 1979. Originally issued in Halifax in 1979 by the Dalhousie Ocean Studies Programme: "New Directions in Ocean Law, Policy and Management" (typewritten, 66 leaves).

"International Environmental Law: A Canadian Perspective on Recent Developments," in P. Z. R. Finkle and A. R. Lucas, eds., *Environmental Law in the 1980s: A New Beginning* (Calgary: Canadian Institute of Resources Law, University of Calgary, 1982), pp. 207–220.
Originally issued in Halifax by the International Institute for Transportation and Ocean Policy Studies (Dalhousie University, Marine Research Report Series No. 4; typewritten, 24 leaves).

"Ship-Generated Pollution: The Creator of Regulated Navigation" (co-author with Edgar Gold), in T. A. Clingan, Jr., ed., *Law of the Sea: State Practice in Zones of Special Jurisdiction* (Proceedings of the Thirteenth Annual Conference of the Law of the Sea Institute) (Honolulu: Law of the Sea Institute, 1982), pp. 156–197.
This paper was presented at the annual conference held in Mexico City, 15–18 October 1979. See also "Introduction" by Johnston (pp. 153–155). Originally issued in Halifax in 1979 by the Dalhousie Ocean Studies Programme: "New Directions in Ocean Law, Policy and Management" (typewritten, 59 leaves).

"UNCLOS III: Environmental Issues in Southeast Asia," in M. J. Valencia, E. Gold, Chia Lin Sien and N. G. Letalik, eds., *Shipping, Energy and Environment: Southeast Asian a Perspectives for the Eighties* (Halifax: Dalhousie Ocean Studies Programme, 1982), pp. 209–214.

1983

Dalhousie Ocean Studies Programme: Programme Development Prospects and Opportunities, 1984–1989. Halifax: Dalhousie Ocean Studies Programme, October 1983. 32 leaves.

"The Heritage of Political Thought in International Law," in R. St. J. Macdonald and D. M. Johnston, eds., *The Structure and Process of International Law: Essays in Legal Philosophy, Doctrine and Theory* (The Hague: Martinus Nijhoff, 1983; 1986), pp. 179–225.

"International Legal Theory: New Frontiers of the Discipline" (co-author with Ronald St. J. Macdonald), in R. St. J. Macdonald and D. M. Johnston, eds., *The Structure and Process of International Law: Essays in Legal Philosophy, Doctrine and Theory* (The Hague: Martinus Nijhoff, 1983; 1986), pp. 1–14.

International Symposium on the New Law of the Sea in Southeast Asia: Developmental Effects and Regional Approaches (co-editor with Edgar Gold and Phiphat Tangsubkul), (Halifax: Dalhousie Ocean Studies Programme, 1983), ix, 239 p.

Maritime Boundary Delimitation: An Annotated Bibliography (co-editor with Ted L. McDorman and Kenneth P. Beauchamp) (Lexington, Mass.: Lexington Books, 1983), xiii, 207 p.
Recipient of the Joseph L. Andrews Bibliographical Award presented by the American Association of Law Libraries in 1983.

"Maritime Boundary Delimitation and UNCLOS III," in D. M. Johnston, E. Gold and P. Tangsubkul, eds., *The New Law of the Sea in Southeast Asia: Development Effects and Regional Approaches* (Halifax: Dalhousie Ocean Studies Programme, 1983), pp. 139–145.

The Structure and Process of International Law: Essays in Legal Philosophy, Doctrine and Theory (co-editor with R. St. J. Macdonald) (The Hague: Martinus Nijhoff, 1983; second printing, 1986), vii, 1234 p. (Development in International Law, Vol. 6).
Contains thirty-six essays by leading international scholars.

1984

The Law of the Sea and Ocean Industry: New Opportunities and Restraints (co-editor with Norman G. Letalik) (Proceedings of the Sixteenth Annual Conference of the Law of the Sea Institute) (Honolulu: Law of the Sea Institute, 1984), xxiii, 605 p.
The annual conference was held in Halifax, 21–24 June 1982.

1985

Canada and the New International Law of the Sea (Toronto: University of Toronto Press, 1985), xviii, 124 p. (Canada. Royal Commission on the Economic Union and Development Prospects for Canada. Collected Research Studies, Vol. 54).
This is a printed version of a study commissioned by the Royal Commission on the Economic Union and Development Prospects for Canada, dated at Halifax, September 1984 (typewritten, 167 leaves, with separate footnotes). Isssued also in French under title: *Le Canada et le nouveau droit international de la mer* (xviii, 142 p.) (Commission royale sur l'union économique et les perspectives de développement du Canada).

"The Common Heritage and the Law of the Sea," in *International Regulation and Deregulation: Emerging Trends in the Role of International Institutions* (Proceedings of the Fourteenth Annual Conference of the Canadian Council on International Law) (Ottawa: Canadian Council on International Law, 1985), pp. 237–242.
Paper presented at the annual conference in Ottawa, 19 October 1985.

"Conservation and Management of the Marine Environment Responsibilities and Required Initiatives in Accordance with the 1982 U.N. Convention on the Law of the Sea," in R. B. Krueger and S. A. Riesenfeld, eds., *The Developing Order of the Oceans* (Proceedings of the Eighteen Annual Conference of the Law of the Sea Institute) (Honolulu: Law of the Sea Institute, 1985), pp. 133–179.
This paper, with notes, presented at the annual conference held in San Francisco, 24–27 October 1984, consists mainly of a chart, prepared by the Dalhousie Ocean Studies Programme.

"Systemic Environment Damage: The Challenge to International Law and Organization," 12 *Syracuse Journal on International Law and Commerce* 255–282 (1985).

1986

An Overview on the New Law of the Sea in Southeast Asia: Problems and Prospects of Implementation (SEAPOL Monograph Series No. 1) (Bangkok: Southeast Asian Project in Ocean Law, Policy and Management (SEAPOL) [1986?], 63 p.

1987

"The International Law of Fisheries: A Post-Classical Review and Assessment," in D. M. Johnston, *The International Law of Fisheries: A Framework for Policy Oriented Inquiries* (New Haven: New Haven Press, and Dordrecht: Martinus Nijhoff, 1987), pp. xxv–lxxx.

"Pollution Control in the Semi-Enclosed Seas of East Asia" (co-author with Edgar Gold), in C.-H. Park and J. K. Park, eds., *The Law of the Sea: Problems from the East Asian Perspective* (Honolulu: Law of the Sea Institute, 1987), pp. 96–105.

"Problems of Maritime Delimitation," in C.-H. Park and J. K. Park, eds., *The Law of the Sea: Problems from the East Asian Perspective* (Honolulu: Law of the Sea Institute, 1987), pp. 83–88.

V. Professor of Law and Chair in Asia-Pacific Legal Relations, Faculty of Law, University of Victoria (1987–1995)

1988

"The Designing of a Transit Management System," in C. Lamson and D. L. VanderZwaag, eds., *Transit Management in the Northwest Passage: Problems and Prospects* (Cambridge: Cambridge University Press, 1988), pp. 279–308.

"Functionalism in the Theory of International Law," 26 *Canadian Yearbook of International Law* 3–59 (1988).

"Marine Pollution Agreements: Successes and Problems," in J. E. Carroll, ed., *International Environmental Diplomacy: The Management and Resolution of Transfrontier Environmental Problems* (Cambridge, England: Cambridge University Press, 1988), pp. 199–206.

"Ocean Boundary Issues and Developments in Regional Perspective" (co-author with Phillip M. Saunders), in D. M. Johnston and P. M. Saunders, eds., *Ocean Boundary Making: Regional Issues and Developments* (London: Croom Helm, 1988), pp. 313–349.

"[Ocean Boundary Making:] Introduction" (co-author with Phillip M. Saunders), in D. M. Johnston and P. M. Saunders, eds., *Ocean Boundary Making: Regional Issues and Developments* (London: Croom Helm, 1988), pp. 1–15.

Ocean Boundary Making: Regional Issues and Developments (co-editor with Phillip M. Saunders) (London: Croom Helm, 1988), 356 p., maps. "This collection of studies is one of a series of three works produced under the Ocean Boundary Project of Dalhousie Ocean Studies Programme (DOSP)." It "consists chiefly of seven regional studies of ocean boundary-making issues and developments."

"The Theoretical Bases of International Law: A Canadian Perspective," in *Canada and Europe: An Evolving Relationship* (Proceedings of the Seventeenth Annual Conference of the Canadian Council on Inter-

national Law) (Ottawa: Canadian Council on International Law, 1988), pp. 46–63.
Paper presented at the annual conference in Ottawa, 21 October 1988.

The Theory and History of Ocean Boundary-Making (Kingston and Montreal: McGill-Queen's University Press, 1988), xiii, 445 p.

1990

"The Driftnetting Problem in the Pacific Ocean: Legal Considerations and Diplomatic Options," 21 *Ocean Development and International Law* 5–39 (1990).

"Science Policy and Ocean Management," in D. L. VanderZwaag and C. Lamson, eds., *The Challenge of Arctic Shipping: Science, Environmental Assessment, and Human Values* (Montreal and Kingston: McGill-Queen's University Press, 1990), pp. 211–218.

"Strains in the Theory of International Law," in *Canada, Japan and International Law* (Proceedings of the Nineteenth Annual Conference of the Canadian Council on International Law) (Ottawa: Canadian Council on International Law, 1990), pp. 196–208.
Paper presented at the annual conference in Ottawa, October 19, 1990

1991

"Is Coastal State Fishery Management Successful or Not?" 22 *Ocean Development and International Law* 199–208 (1991), and in T. Kuribayashi and E. L. Miles, eds., *The Law of the Sea in the 1990s: A Framework for Further International Cooperation* (Proceedings of the Twenty-fourth Annual Conference of the Law of the Sea Institute) (Honolulu: Law of the Sea Institute, 1992), pp. 113–126.
Paper presented at the annual conference held in Tokyo, July 24–27, 1990.

Pacific Ocean Boundary Problems: Status and Solutions (co-author with Mark J. Valencia) (Dordrecht: Martinus Nijhoff, 1991), vii, 214 p. (Publications on Ocean Development, Vol. 6).

1992

"Theory, Consent, and the Law of Treaties: A Cross-Disciplinary Perspective," 12 *Australian Year Book of International Law* 109–124 (1992).

1993

"An Evaluation of the Regional Seas Approach: Introduction," in A. Cooper and Edgar Gold, eds., *The Marine Environment and Sustainable Development: Law, Policy, and Science* (Proceedings of the Twenty-Fifth Annual Conference of the Law of the Sea Institute) (Honolulu: Law of the Sea Institute, 1993), pp. 505–509.
The annual conference was held in Malmö, Sweden, August 6–9, 1991.

"International Fishery Disputes: A Challenge to Modern Diplomacy," in K. I. Matics and T. L. McDorman, eds., *SEAPOL International Workshop on Challenges to Fishery Policy and Diplomacy in South-East Asia* (Bangkok: Southeast Asian Programme in Ocean Law, Policy and Management (SEAPOL), 1993), pp. 79–82.

"Vulnerable Coastal and Marine Areas: A Framework for the Planning of Environmental Security in the Ocean," 24 *Ocean Development and International Law* 63–79 (1993).

1994

"Maritime Boundary Disputes," in P. T. Haydon and A. L. Griffiths, eds., *Maritime Security and Conflict Resolution at Sea in the Post-Cold War Era* (Halifax: Center for Foreign Policy Studies, Dalhousie University, 1994), pp. 27–31.

"Regional Fishery Arrangements: Options for Northeast Asia," in R. St. J. Macdonald, ed., *Essays in Honour of Wang Tieya* (Dordrecht: Martinus Nijhoff, 1994), pp. 407–423.

1995

"Canada and the Law of the Sea Convention," in F. W. Crickard and G. L. Witol, eds., *The Niobe Papers: Oceans Policy in the 1990s: A Pacific Perspective* (Nepean, Ont.: Naval Officers' Association of Canada, 1995), Vol. 6, pp. 1–10.
Paper presented at the annual conference of the Naval Officers' Association of Canada held in Vancouver on 10 June 1994.

"Conservation and Management of the Marine Environment: Required Initiatives and Responsibilities under the 1982 United Nations Convention on the Law of the Sea" (co-author with Phillip M. Saunders and Peter Payayo), in *The Law of the Sea: Priorities and Responsibilities in Implementing the Convention* (Gland, Switzerland: International Union for Conservation of Nature and Natural Resources, 1995), pp. 121–155.

"Fisheries" (co-author with Mark J. Valencia), in M. J. Valencia, ed., *The Russian Far East in Transition: Opportunities for Regional Economic Cooperation* (Boulder: Westview Press, 1995), pp. 147–182.

"Stresses and Mind-sets in Fishery Management," 18 *Dalhousie Law Journal* 154–167 (1995).

"UNCED: The Coastal and Ocean Challenge," in K. L. Koh, R. C. Beckman and Chia Lin Sien, eds., *Sustainable Development of Coastal and Ocean Areas in Southeast Asia: Post-Rio Perspectives* (Singapore: National University of Singapore, 1995), pp. 1–52.

VI. Post Retirement (1995–2006): Professor Emeritus, Faculty of Law, University of Victoria (1995–2006); Adjunct Professor of Law, Dalhousie Law School, Halifax (1995–2006); Visiting Professor, National University of Singapore (1995–1999)

1996

Land-Based Marine Pollution Problems in the Asia-Pacific Region: Status and Legal Developments (co-editor with Twesukdi Piyakarnchana and Vitit Muntarborn) (Bangkok: Institute of Asian Studies, Chulalongkorn University, 1996), 351 p.

"UNCLOS III and UNCED: A Collision of Mind Sets?" in L. K. Kriwoken, M. Haward, D. VanderZwaag and B. Davis, eds., *Oceans Law and Policy in the Post-UNCED Era: Australian and Canadian Perspectives* (The Hague: Kluwer Law International, 1996), pp. 11–24.

1997

Consent and Commitment in the World Community: The Classification and Analysis of International Instruments (Irvington-on-Hudson, N.Y.: Transnational Publishers, 1997), xxxviii, 346 p.

1998

Asia-Pacific Legal Development (co-editor with Gerry Ferguson) (Vancouver: UBC Press, 1998), vi, 611 p.
Includes eleven essays by eighteen contributors.

"Conclusion: Reflections on the Convergence and Divergence of Legal Systems" (co-author with Gerry Ferguson), in D. M. Johnston and G. Ferguson, eds., *Asia-Pacific Legal Development* (Vancouver: UBC Press, 1998), pp. 548–594.

"Environmental Law as 'Sacred Text': Western Values and Southeast Asian Prospects," in D. M. Johnston and G. Ferguson, eds., *Asia-Pacific Legal Development* (Vancouver: UBC Press, 1998), pp. 405–465.

"Introduction: The Evolution of the Project," in D. M. Johnston and G. Ferguson, eds., *Asia-Pacific Legal Development* (Vancouver: UBC Press, 1998), pp. 3–14.

SEAPOL Integrated Studies of the Gulf of Thailand: Selected Papers (editor) (Bangkok: Southeast Asian Programme in Ocean Law, Policy and Management (SEAPOL), 1998), Vol. 1, 234 p.
Includes four papers presented at the First Meeting of Experts of the SEAPOL Gulf of Thailand Project held in Hua Hin, Thailand, 20–21 October 1997.

"Towards Management for the Gulf of Thailand: Charting the Course of Cooperation" (co-author with David L. VanderZwaag), in D. M. Johnston, ed., *SEAPOL Integrated Studies of the Gulf of Thailand* (Bangkok: Southeast Asian Programme in Ocean Law, Policy and Management (SEAPOL), 1998), Vol. 1, pp. 69–135.

1999

"The Concept of System Compliance," in D. M. Johnston and A. Sirivivatnanon, eds., *System Compliance in Ocean Management: Summary Proceedings and Selected Papers* (Bangkok: Southeast Asian Programme in Ocean Law, Policy and Management (SEAPOL), 1999), pp. 102–115.

"Fishery Diplomacy and Science and the Judicial Function," 10 *Yearbook of International Environmental Law* 33–39 (1999).

SEAPOL Integrated Studies of the Gulf of Thailand: Selected Papers (editor) (Bangkok: Southeast Asian Programme in Ocean Law, Policy and Management (SEAPOL), 1999), Vol. 2, 317 p.

System Compliance in Ocean Management: Summary Proceedings and Selected Papers (co-editor with Ankana Sirivivatnanon), (Bangkok: Southeast Asian Programme in Ocean Law, Policy and Management (SEAPOL), 1999), 260 p.
Contains selected papers and summary proceedings of the International Conference on "Ocean Governance and System Compliance in the Asia-Pacific Context," held at Rayong in eastern Thailand, 9–11 December 1997.

2000

"Innocent Passage: Treaty Text and Subsequent State Practice," in D. M. Johnston and A. Sirivivatnanon, eds., *Maritime Transit and Port State Control: Trends in System Compli-ances* (Bangkok: Southeast Asian Programme in Ocean Law, Policy and Management (SEAPOL), 2000), pp. 11–24.

Maritime Transit and Port State Control: Trends in System Compliance (co-editor with Ankana Sirivivatnanon), (Bangkok: Southeast Asian Programme in Ocean Law, Policy and Management (SEAPOL), 2000), 149 p.
Includes papers and proceedings of the International Conference on "Maritime Transit Issues Revisited" held in Manila, Philippines, 17–18 November 1999 (Part I), followed by a report based on records of the International Symposium on "Port State Control: Pasts Successes, Future Challenges," held at Hong Kong, 6–7 December 1999 (Part II).

"The Ocean and International Environmental Law: Swimming, Sinking, and Treading Water at the Millennium" (co-author with David L. VanderZwaag), 43 *Ocean & Coastal Management* 141–161 (2000).

SEAPOL Integrated Studies of the Gulf of Thailand (co-editor with Ankana Sirivivatnanon), (Bangkok: Southeast Asian Programme in Ocean Law, Policy and Management (SEAPOL), 2000), Vol. 3, 113 p.
Contains the report of the Third SEAPOL Meeting of Experts on the Gulf of Thailand, held at Cha-Am in southern Thailand, 12–14 July 1999.

2001

B.C. Offshore Hydrocarbon Development: Issues and Prospects (co-editor with Erin N. Hildebrand) (Victoria: Maritime Awards Society of Canada; Ocean Centre of Excellence, British Columbia Institute of Technology, March 2001), 66 p.
"Background Report Prepared by the Maritime Awards Society of Canada."

"Fires and Haze: Singapore's Costs and Responsibilities" (co-author with Euston Quah), in L. Low and D. M. Johnston, eds., *Singapore Inc: Public Policy Options in the Third Millennium* (Singapore: Asia Pacific Press, 2001), pp. 157–177.

"Forest Fires and Environmental Haze in Southeast Asia: Using the 'Stakeholder' Approach to Assign Costs and Responsibilities" (co-author with Euston Quah), 63 *Journal of Environmental Management* 181–191 (2001).

"Public Policy Challenges and Opportunities: An Editorial Introduction," in L. Low and D. M. Johnston, eds., *Singapore Inc: Public Policy Options in the Third Millennium* (Singapore: Asia Pacific Press, 2001), pp. 1–15.

Singapore Inc: Public Policy Options in the Third Millennium (co-editor with Linda Low) (Singapore: Asia Pacific Press, 2001), xiv, 423 p.

"Southeast Asia: Lessons Learned," in M. J. Valencia, ed., *Maritime Regime Building: Lessons Learned and Their Relevance for Northeast Asia* (The Hague: Martinus Nijhoff, 2001), pp. 73–86.

2002

B.C. Offshore Development Issues: The 2002 Dunsmuir Symposium Report (Victoria: Maritime Awards Society of Canada, 2002), 23, [6] p.
Report on the MASC Symposium "Current Offshore Policy and Related Issues," dated August 2002, held at the Dunsmuir Lodge (University of Victoria), outside Sidney, B.C., 11 May 2002. Appendix I (4 pages) contains a summary of the two-volume *Report of the Scientific Review Panel, British Columbia Offshore Hydrocarbon Development* (available on-line), submitted to the B.C. Minister of Energy and Mines on 15 January 2002.

"Marine Regionalism in Comparative Perspective," in D. M. Johnston and A. Sirivivatnanon, eds., *Ocean Governance and Sustainable Development in the Pacific Region: Selected Papers, Commentaries and Comments* (Bangkok: Southeast Asian Programme in Ocean Law, Policy and Management (SEAPOL), 2002), pp. 53–63.

"The Northwest Passage Revisited," 33 *Ocean Development and International Law* 145–164 (2002).

Ocean Governance and Sustainable Development in the Pacific Region: Selected Papers, Commentaries and Comments (co-editor with Ankana Sirivivatnanon) (Bangkok: Southeast Asian Programme in Ocean Law, Policy and Management (SEAPOL), 2002), 584 p.
Proceedings of the SEAPOL Inter-Regional Conference on "Ocean Governance and Sustainable Development in the East and Southeast Asian Seas: Challenges in the New Millennium," held at Bangkok, Thailand, 21–23 March 2001.

"The Role of Land-locked Countries in Subregional Coastal Management," in M. I. Glassner, ed., *Resource Management and Transit to and from the Sea* (Bangkok: Southeast Asian Programme in Ocean Law, Policy and Management (SEAPOL), 2002), pp. 54–65.

2003

"The Future of the Arctic Ocean: Competing Domains of International Public Policy," 17 *Ocean Yearbook* 596–624 (2003).

Revisiting the Law of the Sea (Victoria: Maritime Awards Society of Canada, May 2003), 40 p.
Report based on discussions of a Meeting of Experts organized by the Maritime Awards Society of Canada held at the Dunsmuir Lodge (University of Victoria) outside Sidney, B.C., 21–22 March 2003.

"RUNCLOS: The Case for and against Revision of the Law of the Sea Convention," 109 *Chuo Law Review* 135–159 (2003).

Straddling Stocks in the Northwest Atlantic: Conservation Concerns and Options (co-author with Aldo Chircop and Hugh R. Williamson) (Halifax: Dalhousie Law School, Marine and Environmental Programme, April 2003), 91 leaves.
Report to the Fisheries Council of Canada, prepared in cooperation with the International Oceans Institute of Canada.

2005

International Ocean Law: Materials and Commentaries (co-editor with Ted L. McDorman, Alexander J. Bolla, and John Duff) (Durham, N.C.: Carolina Academic Press, 2005), xiv, 430 p.

Towards World Constitutionalism: Issues in the Legal Ordering of the World Community (co-editor with Ronald St. J. Macdonald) (Leiden: Martinus Nijhoff, 2005), xviii, 968 p.

"World Constitutionalism in the Theory of International Law," in R. St. J. Macdonald and D. M. Johnston, eds., *Towards World Constitutionalism: Issues in the Legal Ordering of the World Community* (Leiden: Martinus Nijhoff, 2005), pp. 3–29.

2006

"The Challenge of International Ocean Governance: Institutional, Ethical and Conceptual Dilemmas," in D. R. Rothwell and D. L. VanderZwaag, eds., *Towards Principled Oceans Governance: Australian and Canadian Approaches and Challenges* (London: Routledge, 2006), pp. 349–399.

2008

The Historical Foundations of World Order: The Tower and the Arena (Leiden, Boston: Martinus Nijhoff, 2008), xxi, 875 p.
Contains a list of publications of Douglas M. Johnston (1931–2006), pp. 837–844.

Book Reviews By Douglas M. Johnston (2001–2005)

Beaulac, Stephane, *The Power of Language in the Making of International Law: The World Sovereignty in Bodin and Vattel and the Myth of Westphalia* (Leiden; Boston: Martinus Nijhoff, 2004; xiv, 200 p.), in 33 *International Journal of Legal Information* 140–141 (2005).

Carty, Anthony, and Smith A. Richard, *Sir Gerald Fitzmaurice and the World Crisis: A Legal Adviser in the Foreign Office, 1932–1945* (The Hague; Boston: Kluwer Law International, 2000; xi, 690), in 3 *Journal of the History of International Law* 255–260 (2001).

Eyffinger, Arthur, *The 1899 Hague Peace Conference: The Parliament of Man, the Federation of the World* (The Hague: Kluwer Law International, 1999; xiv, 480 p.), in 29 *International Journal of Legal Information* 144–147 (2001).

Koskenniemi, Martti, *The Gentle Civilizer: The Rise and Fall of International Law, 1870–1960* (Cambridge, England; New York: Cambridge University Press, 2001; xiv, 569 p.), in 31 *International Journal of Legal Information* 539–541 (2003).

Index